AF352786

Advances in Theoretical and Computational Energy Optimization Processes

Advances in Theoretical and Computational Energy Optimization Processes

Volume 1

Editors

Ferdinando Salata
Iacopo Golasi

MDPI • Basel • Beijing • Wuhan • Barcelona • Belgrade • Manchester • Tokyo • Cluj • Tianjin

Editors
Ferdinando Salata
University of Rome "Sapienza"
Italy

Iacopo Golasi
Sapienza University of Rome
Italy

Editorial Office
MDPI
St. Alban-Anlage 66
4052 Basel, Switzerland

This is a reprint of articles from the Special Issue published online in the open access journal *Processes* (ISSN 2227-9717) (available at: https://www.mdpi.com/journal/processes/special_issues/energt_optimization).

For citation purposes, cite each article independently as indicated on the article page online and as indicated below:

LastName, A.A.; LastName, B.B.; LastName, C.C. Article Title. *Journal Name* **Year**, *Article Number*, Page Range.

Volume 1
ISBN 978-3-03936-638-5 (Hbk)
ISBN 978-3-03936-639-2 (PDF)

Volume 1-2
ISBN 978-3-03936-684-2 (Hbk)
ISBN 978-3-03936-685-9 (PDF)

Contents

About the Editors

Ferdinando Salata was born in Rome in 1977, and is currently a researcher at the Department of Astronautical Engineering, Electrical and Energy (DIAEE) for the Scientific Sector ING-IND/11, at the University of Rome "Sapienza" (Italy) (according to the Italian university). He earned his degree in mechanical engineering, specialising in energy, at the University of Rome "Sapienza" in 2003, and went on to complete his PhD in "Technical Physic" in 2007 at the same university. For his thesis, he studied the use of ultraviolet radiation, coupled with HEPA filtration, for the disinfection of biological airborne contaminants in air conditioning systems. After completing his PhD, Ferdinando accepted a research grant from the Department where he completed his thesis. During this period, Ferdinando was named "Expert in the field" for the Scientific Sector ING-IND/11, at the Faculties of Civil and Industrial Engineering of the University of Rome "Sapienza". In 2017, he completed his national academic qualification (required for becoming an associate professor) at the "Ministry of University and Research". That same year, he joined DIAEE as an assistant professor. He is a member of the National Association of Italian Technical Physics. In recent years, he has studied CHP systems; the energy demand optimization of buildings; the energy and reliability optimization of conditioning and lighting systems; natural ventilation in buildings; desalination through absorption machines; urban microclimate and outdoor thermal comfort; thermal conductivity in soils. He has the role of docent (at Faculty of Architecture of University of Rome "Sapienza") for the course "Environmental Applied Physics" and teaching assistant (at Faculty of Civil and Industrial Engineering of University of Rome "Sapienza") for the courses: "Applied Physics" for Electrical Engineering, "Environmental Applied Physics" for Building Engineering—Architecture. He taught a course on "Energy certification of buildings" on behalf of the Lazio Region and on behalf of the Kyoto Club Italia. He is, and has been, the tutor or co-tutor of several MSc thesis dissertations at his faculty. He is a member of the Academic Board of the Ph.D. in Energy and Environment at DIAEE and a member of the International Advisory Board for the *Thermal Science Journal, Journal of Daylighting, Atmosphere Journal, Sustainable Cities and Society Journal.*

Iacopo Golasi was born in Rome in 1987 and achieved, in March 2014, his MSc, with honors in mechanical engineering, from the University of "Roma Tre". He completed his doctorate in "Energy and Environment" in the Department of Astronautics, Electrical and Energetics Engineering (DIAEE) at the University of Roma "Sapienza". At present, his interests lie in the development of new empirical outdoor comfort indices able to predict the thermal perception of people in the analysis of the influence of materials' thermo-physical properties on outdoor thermal comfort and buildings' energy demands, in the evaluation of natural ventilation inside buildings (with particular focus on innovative solutions as the solar chimney), in the buildings' energy efficiency assessment, through building a dynamic simulation analysis, also considering the inter-building effect and in the analysis of lighting installations with LED-type light sources (performing an energy, economic and maintenance comparison with traditional installations). He provides consulting services to private and public companies, such as Aeroporti di Roma S.p.A., Lamaro Appalti S.p.A. and ANCI (National Association of Italian Municipalities). He has been the co-tutor of various thesis dissertations, covering different topics related to the subjects of Applied Physics, Thermotechnics, Lighting and Thermodynamics. He is the co-author of more than 40 papers published in international peer review journals or presented at conferences.

Editorial

Advances in Theoretical and Computational Energy Optimization Processes

Ferdinando Salata * and Iacopo Golasi *

Department of Astronautics, Electrical and Energetics Engineering, University of Rome "Sapienza", 00184 Rome, Italy
* Correspondence: ferdinando.salata@uniroma1.it (F.S.); iacopo.golasi@uniroma1.it (I.G.)

Received: 1 June 2020; Accepted: 1 June 2020; Published: 4 June 2020

Industry, construction and transport are the three sectors that traditionally lead to the highest energy requirements. This is why, over the past few years, all the involved stakeholders have widely expressed the necessity to introduce a new approach to the analysis and management of those energy processes characterizing the aforementioned sectors. The objective is to guide production and energy processes to an approach aimed at energy savings and a decrease in environmental impact. Indeed, all of the ecosystems are stressed by obsolete production schemes deriving from an unsustainable paradigm of constant growth and related to the hypothesis of an environment able to absorb and accept all of the anthropogenic changes.

Leading the production processes of industry, construction and transport to a revision of their energy requirements is necessary and the research activity is called to carry out its natural innovative function.

The industrial sector is in full transition and transformation towards its version 4.0 and is therefore called to review its management and the supply costs of energy and raw materials to limit its environmental impact. Research activity must support best practices in energy management and encourage a reduction in greenhouse gas emissions. The construction sector should apply future retrofit solutions, able to increase energy efficiency and taking into account environment and climate change at the same time. The transport sector is moving towards new mobility with respect to the past, thanks to the transition from fossil fuels to the electrification and the use of artificial intelligence, thus increasing the level of automation. In the context of great attention towards the sustainable and respectful future of the planet, this study and the diffusion of the results provided by the scientific community concerning the most recent signs of progress in energy optimization are expected to play a key role.

With the aim of proposing the next generation of energy processes and leading to positive implications for the environment, climate and sustainability, this Special Issue, "Advances in Theoretical and Computational Energy Optimization Processes" has aimed to collect sophisticated contributions on all these aspects, highlighting current state-of-the-art research with respect to the results of the main research groups. Studies on energy processes, production methods and innovative mechanisms related to research based on computational optimization methods are part of this scientific collection. This Special Issue has also aimed to encourage a debate on future scenarios in each of those sectors currently characterized by significant energy requirements.

In this Special Issue, numerous articles have found a home and they have been proposed by researchers from countries belonging to geographical areas over the world. In particular, the affiliation of the authors sees nations represented according to the following percentages: China, 69%; Pakistan, 9%; Malaysia, 5%; Mexico, 3%; USA, 3%; Spain, 2%; Iran, Taiwan and Vietnam, 1.5%; Norway and UK, 1%; Chile, Denmark, Ghana, Oman and South Africa, 0.5%. The topics covered range across all energy sectors. Starting from the production of energy up to its final consumption, the authors discuss and propose ideas and opportunities to optimize processes, methods, equipment and machinery to minimize energy needs from non-renewable sources and the environmental pollution derived from it.

The authors, in presenting their scientific works, have shown that it is possible to intervene in multiple sectors to try and optimize numerous energy processes. This is proof that the scientific community is active in producing ideas that will allow a transition to a low-carbon future and apply new theories and models based on innovative algorithms.

In particular, in the field of studies concerning energy optimization in the civil construction sector, the following publications have found space in this collection of scientific works:

- Smart Community Energy Cost Optimization Taking User Comfort Level and Renewable Energy Consumption Rate into Consideration (Keywords: smart communities; user comfort levels; renewable energy consumption rate) [1].
- Cogeneration Process Technical Viability for an Apartment Building: Case Study in Mexico (Keywords: cogeneration; technical viability; apartment building) [2].
- Efficient Energy Management in Offices Using Bio-Inspired Energy Optimization Algorithms (Keywords: appliance scheduling techniques; bacterial foraging algorithm (BFA); energy management; system; energy optimization algorithms; grasshopper optimization algorithm (GOA); smart grid) [3].
- Multi-Objective Optimal Scheduling Method for a Grid-Connected Redundant Residential Microgrid (Keywords: redundant residential microgrid (RR-microgrid); optimal scheduling; virtual energy storage system (VESS); non-dominant sorting genetic algorithm II (NSGA-II); analytic hierarchy process (AHP)) [4].
- Optimized Energy Management Strategies for Campus Hybrid PV–Diesel Systems During Utility Load Shedding Events (Keywords: hybrid PV–diesel generator systems; digital resource management; energy management) [5].
- A Modular Framework for Optimal Load Scheduling Under Price-Based Demand Response Scheme in Smart Grid (Keywords: smart grid; demand response; load scheduling; home energy management; enhanced differential evolution; hybrid gray wolf-modified enhanced differential evolutionary algorithm) [6].

The application of innovative methodologies to encourage more efficient transport has led to the production of the following scientific works:

- Hybrid Energy Feature Extraction Approach for Ship-Radiated Noise Based on CEEMDAN Combined with Energy Difference and Energy Entropy Complete Ensemble Empirical Mode Decomposition with Adaptive Noise (CEEMDAN) (Keywords: energy difference (ED); energy entropy (EE); hybrid energy feature extraction; ship-radiated noise (S-RN)) [7].
- Energy-Efficient Train Driving Strategy with Considering the Steep Downhill Segment (Keywords: rail transit; train control; energy-efficient driving strategy; steep downhill segment; local optimization) [8].
- Numerical Investigation of SCR Mixer Design Optimization for Improved Performance (Keywords: selective catalyst reduction system; emission control; marine diesel engine; urea; ammonia) [9].
- A Rotor-Sync Signal-Based Control System of a Doubly-Fed Induction Generator in the Shaft Generation of a Ship (Keywords: shaft generator; DFIG; shipboard; power; control) [10].
- Intelligent Energy Management for Plug-In Hybrid Electric Bus with Limited State Space (Keywords: plug-in hybrid electric bus; energy management; Q-learning; limited state space; Hardware-in-Loop (HIL) simulation) [11].

The research sector that operates in the production, transport and dispatching of energy has been enriched by the considerations contained in the following articles:

- Integrated Delphi-AHP and Fuzzy TOPSIS Approach Toward Ranking and Selection of Renewable Energy Resources in Pakistan (Keywords: Delphi; analytical hierarchy process; fuzzy technique

- for order of preference by similarity to ideal solution techniques; renewable energy (RE) resources; sustainable energy planning) [12].
- Flexible Responsive Load Economic Model for Industrial Demands (Keywords: demand-side management; economic demand response model; consumer utility function; electricity market restructuring) [13].
- Implementation of Maximum Power Point Tracking Based on Variable Speed Forecasting for Wind Energy Systems (Keywords: maximum power tracking (MPT); wind speed forecasting; wind energy system (WES); state feedback controller) [14].
- Numerical Investigation of Influence of Reservoir Heterogeneity on Electricity Generation Performance of Enhanced Geothermal System (Keywords: reservoir heterogeneity; enhanced geothermal system; electricity generation; performance; influence) [15].
- Modeling of Future Electricity Generation and Emissions Assessment for Pakistan (Keywords: electricity demand; emissions; LEAP model; fossil fuels; renewable energy) [16].
- Power Transmission Congestion Management Based on Quasi-Dynamic Thermal Rating (Keywords: transmission line; meteorological parameter; quasi-dynamic thermal rating (QDR); transmission congestion) [17].
- Control Strategy of Electric Heating Loads for Reducing Power Shortage in Power Grid (Keywords: power shortage; electric heating load; electric water heater; demand response; virtual energy storage (VES), virtual state of charge (VSOC)) [18].
- Productivity Models of Infill Complex Structural Wells in Mixed Well Patterns (Keywords: complex structural well; mixed well pattern; productivity evaluation; semi-analytical model; well location optimization) [19].
- Influence and Optimization of Geometrical Parameters on Coast-Down Characteristics of Nuclear Reactor Coolant Pumps (Keywords: reactor coolant pump; coast-down characteristics; geometrical parameters; multiple linear regression; transition process) [20].
- Model for Optimizing Location Selection for Biomass Energy Power Plants (Keywords: biomass energy; site selection; optimization; MCDM; FMCDM; FAHP; TOPSIS) [21].
- Wind Energy Generation Assessment at Specific Sites in a Peninsula in Malaysia Based on Reliability Indices (Keywords: reliability indices; wind farms; sequential Monte Carlo simulation; Malaysia) [22].
- Temporal Feature Selection for Multi-Step Ahead Reheater Temperature Prediction (Keywords: reheat steam temperature; temporal feature selection; delay order prediction; deep neural network; genetic algorithm) [23].
- Investigating the Dynamic Impact of CO_2 Emissions and Economic Growth on Renewable Energy Production: Evidence from FMOLS and DOLS Tests (Keywords: carbon emissions; economic growth; energy; renewable energy; fully modified ordinary least square (FMOLS); dynamic panel cointegration model) [24].
- Mathematical Modeling and Simulation on the Stimulation Interactions in Coalbed Methane Thermal Recovery (Keywords: coalbed methane thermal recovery; thermal stimulation interaction; heat–gas–coal model; modeling and simulation) [25].
- Economic Dispatch of Multi-Energy System Considering Load Replaceability (Keywords: multi-energy system; economic dispatch; load replaceability; multi-energy conversion) [26].
- Using Pso Algorithm to Compensate Power Loss Due to the Aeroelastic Effect of the Wind Turbine Blade (Keywords: Aeroelastic effect; Optimization model; Power loss; Pre-twist angle; Pre-twisting method) [27].
- Energy Model for Long-Term Scenarios in Power Sector under Energy Transition Laws (Keywords: electricity model; power plants prospective; Mexican prospective) [28].

- Multi-Agent Consensus Algorithm-Based Optimal Power Dispatch for Islanded Multi-Microgrids (Keywords: islanded multi-microgrids; real-time power dispatch; multi-agent; consensus algorithm) [29].
- Simulation-Based Design and Economic Evaluation of a Novel Internally Circulating Fluidized Bed Reactor for Power Production with Integrated CO_2 Capture (Keywords: chemical looping combustion; power production; carbon capture; internally circulating reactor; reactor design; fluidization; techno-economics; computational fluid dynamics; filtered two-fluid model; coarse-grid simulations) [30].
- A Time-Sequence Simulation Method for Power Units' Monthly Energy-Trade Scheduling with Multiple Energy Sources (Keywords: monthly energy-trade scheduling; time-sequence simulation method; feasibility; fairness; consumption of renewable energy) [31].
- Reliability Evaluation Method Considering Demand Response (DR) of Household Electrical Equipment in Distribution Networks (Keywords: demand response; household electrical equipment; real-time electricity price; incentive mechanism; capacity constraint; reliability evaluation) [32].
- A Dispatching Optimization Model for Park Power Supply Systems Considering Power-To-Gas and Peak Regulation Compensation (Keywords: park power supply system; power-to-gas; peak regulation compensation; ancillary service; wind/photovoltaic generation consumption) [33].
- Salp Swarm Optimization Algorithm-Based Controller for Dynamic Response and Power Quality Enhancement of an Islanded Microgrid (Keywords: microgrid; optimization; voltage and frequency regulation; dynamic response enhancement; salp swarm optimization algorithm; power quality) [34].
- Off-Grid Solar PV Power Generation System in Sindh, Pakistan: A Techno-Economic Feasibility Analysis (Keywords: off-grid Solar PV power generation; remote rural regions; economic feasibility; CO_2 mitigation; Pakistan) [35].

Research in the industrial production sector and its associated processes have found space in the following works:

- A Novel Robust Method for Solving CMB Receptor Model Based on Enhanced Sampling Monte Carlo Simulation (Keywords: CMB receptor model; effective variance weighted least squares algorithm; enhanced sampling Monte Carlo simulation) [36].
- Mold Level Predictor of Continuous Casting Using Hybrid EMD-SVR-GA Algorithm (Keywords: empirical mode decomposition; support vector regression; genetic algorithm; mold level; continuous cast) [37].
- Artificial Neural Networks Approach for a Multi-Objective Cavitation Optimization Design in a Double-Suction Centrifugal Pump (Keywords: multi-objective optimization; artificial neural network; NPSHr prediction; cavitation optimization; CFD) [38].
- Determination of Acidity of Waste Cooking Oils by Near-Infrared Spectroscopy (Keywords: free acidity; NIRS; partial least squares; waste cooking oil) [39].
- Extended State Observer-Based Predictive Speed Control for Permanent Magnet Linear Synchronous Motor (Keywords: permanent magnet linear synchronous motor (PMLSM); extended state observer (ESO); predictive function control (PFC); composite control; robustness) [40].
- Direct Speed Control of Pmsm Based on Terminal Sliding Mode and Finite Time Observer (Keywords: non-singular terminal sliding mode control (NTSMC); finite-time observer (FTO); mismatched/matched disturbance/uncertainties; permanent magnet synchronous motor (PMSM)) [41].
- Numerical and Experimental Study of a Vortex Structure and Energy Loss in a Novel Self-Priming Pump (Keywords: vortex structure; energy loss; entropy production; self-priming pump) [42].

Processes **2020**, *8*, 669

These studies, and the researchers who participated in them are the cornerstone for human well-being of tomorrow, supporting economic development and reducing fuel poverty. The future will be sustainable because it can only be such. The sustainable future is closer than we can imagine.

I must congratulate the authors and reviewers for their outstanding work, and thank the editors, assistants, and all the staff of MDPI for the quality of their work.

Conflicts of Interest: The authors declare no conflicts of interest.

References

1. Shi, K.; Li, D.; Gong, T.; Dong, M.; Gong, F.; Sun, Y. Smart community energy cost optimization taking user comfort level and renewable energy consumption rate into consideration. *Processes* **2019**, *7*, 63. [CrossRef]
2. Valdés, H.; Leon, G. Cogeneration process technical viability for an apartment building: Case study in Mexico. *Processes* **2019**, *7*, 93. [CrossRef]
3. Ullah, I.; Khitab, Z.; Khan, M.N.; Hussain, S. An efficient energy management in office using bio-inspired energy optimization algorithms. *Processes* **2019**, *7*, 142. [CrossRef]
4. Liu, W.; Liu, C.; Lin, Y.; Bai, K.; Ma, L.; Chen, W. Multi-objective optimal scheduling method for a grid-connected redundant residential microgrid. *Processes* **2019**, *7*, 296. [CrossRef]
5. Maritz, J. Optimized energy management strategies for campus hybrid PV-diesel systems during utility load shedding events. *Processes* **2019**, *7*, 430. [CrossRef]
6. Hafeez, G.; Islam, N.; Ali, A.; Ahmad, S.; Usman, M.; Saleem Alimgeer, K. A Modular Framework for Optimal Load Scheduling under Price-Based Demand Response Scheme in Smart Grid. *Processes* **2019**, *7*, 499. [CrossRef]
7. Li, Y.; Chen, X.; Yu, J. A hybrid energy feature extraction approach for ship-radiated noise based on CEEMDAN combined with energy difference and energy entropy. *Processes* **2019**, *7*, 69. [CrossRef]
8. Liu, W.; Tang, T.; Su, S.; Yin, J.; Cao, Y.; Wang, C. Energy-efficient train driving strategy with considering the steep downhill segment. *Processes* **2019**, *7*, 77. [CrossRef]
9. Mehdi, G.; Zhou, S.; Zhu, Y.; Shah, A.H.; Chand, K. Numerical investigation of SCR mixer design optimization for improved performance. *Processes* **2019**, *7*, 168. [CrossRef]
10. Nguyen, T.-T. A rotor-sync signal-based control system of a doubly-fed induction generator in the shaft generation of a ship. *Processes* **2019**, *7*, 188. [CrossRef]
11. Guo, H.; Du, S.; Zhao, F.; Cui, Q.; Ren, W. Intelligent energy management for plug-in hybrid electric bus with limited state space. *Processes* **2019**, *7*, 672. [CrossRef]
12. Solangi, Y.A.; Tan, Q.; Mirjat, N.H.; Valasai, G.D.; Khan, M.W.A.; Ikram, M. An integrated Delphi-AHP and fuzzy TOPSIS approach toward ranking and selection of renewable energy resources in Pakistan. *Processes* **2019**, *7*, 118. [CrossRef]
13. Sharifi, R.; Anvari-Moghaddam, A.; Fathi, S.H.; Vahidinasab, V. A flexible responsive load economic model for industrial demands. *Processes* **2019**, *7*, 147. [CrossRef]
14. Zhang, Y.; Zhang, L.; Liu, Y. Implementation of maximum power point tracking based on variable speed forecasting for wind energy systems. *Processes* **2019**, *7*, 158. [CrossRef]
15. Zeng, Y.; Tang, L.; Wu, N.; Song, J.; Zhao, Z. Numerical investigation of influence of reservoir heterogeneity on electricity generation performance of enhanced geothermal system. *Processes* **2019**, *7*, 202. [CrossRef]
16. Mengal, A.; Mirjat, N.H.; Walasai, G.D.; Khatri, S.A.; Harijan, K.; Uqaili, M.A. Modeling of future electricity generation and emissions assessment for Pakistan. *Processes* **2019**, *7*, 212. [CrossRef]
17. Wang, Y.; Sun, Z.; Yan, Z.; Liang, L.; Song, F.; Niu, Z. Power transmission congestion management based on quasi-dynamic thermal rating. *Processes* **2019**, *7*, 244. [CrossRef]
18. Xue, S.; Che, Y.; He, W.; Zhao, Y.; Zhang, R. Control strategy of electric heating loads for reducing power shortage in power grid. *Processes* **2019**, *7*, 273. [CrossRef]
19. Sun, L.; Li, B.; Li, Y. Productivity models of infill complex structural wells in mixed well patterns. *Processes* **2019**, *7*, 324. [CrossRef]
20. Zhao, Y.; Si, X.; Wang, X.; Zhu, R.; Fu, Q.; Zhong, H. The influence and optimization of geometrical parameters on coast-down characteristics of nuclear reactor coolant pumps. *Processes* **2019**, *7*, 327. [CrossRef]

21. Wang, C.-N.; Tsai, T.-T.; Huang, Y.-F. A model for optimizing location selection for biomass energy power plants. *Processes* **2019**, *7*, 353. [CrossRef]

22. Ali Kadhem, A.; Abdul Wahab, N.I.; Abdalla, A. Wind energy generation assessment at specific sites in a Peninsula in Malaysia based on reliability indices. *Processes* **2019**, *7*, 399. [CrossRef]

23. Gui, N.; Lou, J.; Qiu, Z.; Gui, W. Temporal Feature Selection for Multi-Step Ahead Reheater Temperature Prediction. *Processes* **2019**, *7*, 473. [CrossRef]

24. Khan, M.W.A.; Panigrahi, S.K.; Almuniri, K.S.N.; Soomro, M.I.; Mirjat, N.H.; Alqaydi, E.S. Investigating the Dynamic Impact of CO2 Emissions and Economic Growth on Renewable Energy Production: Evidence from FMOLS and DOLS Tests. *Processes* **2019**, *7*, 496. [CrossRef]

25. Teng, T.; Wang, Y.; He, X.; Chen, P. Mathematical Modeling and Simulation on the Stimulation Interactions in Coalbed Methane Thermal Recovery. *Processes* **2019**, *7*, 526. [CrossRef]

26. Zheng, T.; Dai, Z.; Yao, J.; Yang, Y.; Cao, J. Economic dispatch of multi-energy system considering load replaceability. *Processes* **2019**, *7*, 570. [CrossRef]

27. Zhao, Y.; Liao, C.; Qin, Z.; Yang, K. Using PSO algorithm to compensate power loss due to the aeroelastic effect of thewind turbine blade. *Processes* **2019**, *7*, 633. [CrossRef]

28. Hernández-Luna, G.; Romero, R.J.; Rodríguez-Martínez, A.; Ponce-Ortega, J.M.; CerezoRomán, J.; Toledo Vázquez, G.D. Energy model for long-term scenarios in power sector under energy transition laws. *Processes* **2019**, *7*, 674. [CrossRef]

29. Zhai, X.; Wang, N. Multi-agent consensus algorithm-based optimal power dispatch for islanded multi-microgrids. *Processes* **2019**, *7*, 679.

30. Cloete, J.H.; Khan, M.N.; Cloete, S.; Amini, S. Simulation-based design and economic evaluation of a novel internally circulating fluidized bed reactor for power production with integrated CO_2 capture. *Processes* **2019**, *7*, 723. [CrossRef]

31. Sun, L.; Zhang, Q.; Zhang, N.; Song, Z.; Liu, X.; Li, W. A Time-Sequence Simulation Method for Power Unit's Monthly Energy-Trade Scheduling with Multiple Energy Sources. *Processes* **2019**, *7*, 771. [CrossRef]

32. Chen, H.; Tang, J.; Sun, L.; Zhou, J.; Wang, X.; Mao, Y. Reliability evaluation method considering demand response (DR) of household electrical equipment in distribution networks. *Processes* **2019**, *7*, 799. [CrossRef]

33. Qin, Y.; Lin, H.; Tan, Z.; Yan, Q.; Li, L.; Yang, S.; De, G.; Ju, L. A Dispatching Optimization Model for Park Power Supply Systems Considering Power-to-Gas and Peak Regulation Compensation. *Processes* **2019**, *7*, 813. [CrossRef]

34. Jumani, T.A.; Mustafa, M.; Anjum, W.; Ayub, S. Salp swarm optimization algorithm-based controller for dynamic response and power quality enhancement of an islanded microgrid. *Processes* **2019**, *7*, 840. [CrossRef]

35. Xu, L.; Wang, Y.; Solangi, Y.; Zameer, H.; Shah, S. Off-Grid Solar PV Power Generation System in Sindh, Pakistan: A Techno-Economic Feasibility Analysis. *Processes* **2019**, *7*, 308. [CrossRef]

36. Hou, W.; Yang, Y.; Wang, Z.; Hou, M.; Wu, Q.; Xie, X. A novel robust method for solving CMB receptor model based on enhanced sampling Monte Carlo simulation. *Processes* **2019**, *7*, 169. [CrossRef]

37. Lei, Z.; Su, W. Mold level predict of continuous casting using hybrid EMD-SVR-GA algorithm. *Processes* **2019**, *7*, 177. [CrossRef]

38. Wang, W.; Osman, M.; Pei, J.; Gan, X.; Yin, T. Artificial neural networks approach for a multi-objective cavitation optimization design in a double-suction centrifugal pump. *Processes* **2019**, *7*, 246. [CrossRef]

39. García Martín, J.F.; López Barrera, M.C.; Torres García, M.; Zhang, Q.-A.; Álvarez Mateos, P. Determination of the acidity of waste cooking oils by near infrared spectroscopy. *Processes* **2019**, *7*, 304. [CrossRef]

40. Wang, Y.; Yu, H.; Che, Z.; Wang, Y.; Zeng, C. Extended state observer-based predictive speed control for permanent magnet linear synchronous motor. *Processes* **2019**, *7*, 618. [CrossRef]

41. Wang, Y.; Yu, H.; Che, Z.; Wang, Y.; Liu, Y. The direct speed control of PMSM based on terminal sliding mode and finite time observer. *Processes* **2019**, *7*, 624. [CrossRef]

42. Chang, H.; Agarwal, R.K.; Li, W.; Zhou, L.; Shi, W. Numerical and experimental study of a vortex structure and energy loss in a novel self-priming pump. *Processes* **2019**, *7*, 701. [CrossRef]

Article

Salp Swarm Optimization Algorithm-Based Controller for Dynamic Response and Power Quality Enhancement of an Islanded Microgrid

Touqeer Ahmed Jumani [1,2,*], Mohd. Wazir Mustafa [1], Madihah Md. Rasid [1], Waqas Anjum [1,3] and Sara Ayub [1,4]

[1]   School of Electrical Engineering, Universiti Teknologi Malaysia, Johor Bahru 81310, Malaysia;
      wazir@utm.my (M.W.M.); madihahmdrasid@utm.my (M.M.R.); Waqas.anjum@iub.edu.pk (W.A.);
      sara.ayub@buitms.edu.pk (S.A.)
[2]   Department of Electrical Engineering, Mehran University of Engineering and Technology SZAB Campus
      Khairpur Mirs, Khairpur 66020, Pakistan
[3]   Department of Electronic Engineering, The Islamia University of Bahawalpur, Bahawalpur 63100, Pakistan
[4]   Department of Electrical Engineering, Balochistan University of Information Technology, Engineering and
      Management Sciences, Quetta 87300, Pakistan
*   Correspondence: atouqeer2@graduate.utm.my; Tel.: +60-187-908-744

Received: 8 October 2019; Accepted: 7 November 2019; Published: 10 November 2019

Abstract: The islanded mode of the microgrid (MG) operation faces more power quality challenges as compared to grid-tied mode. Unlike the grid-tied MG operation, where the voltage magnitude and frequency of the power system are regulated by the utility grid, islanded mode does not share any connection with the utility grid. Hence, a proper control architecture of islanded MG is essential to control the voltage and frequency, including the power quality and optimal transient response during different operating conditions. Therefore, this study proposes an intelligent and robust controller for islanded MG, which can accomplish the above-mentioned tasks with the optimal transient response and power quality. The proposed controller utilizes the droop control in addition to the back to back proportional plus integral (PI) regulator-based voltage and current controllers in order to accomplish the mentioned control objectives efficiently. Furthermore, the intelligence of the one of the most modern soft computational optimization algorithms called salp swarm optimization algorithm (SSA) is utilized to select the best combination of the PI gains (k_p and k_i) and dc side capacitance (C), which in turn ensures optimal transient response during the distributed generator (DG) insertion and load change conditions. Finally, to evaluate the effectiveness of the proposed control approach, its outcomes are compared with that of the previous approaches used in recent literature on basis of transient response measures, quality of solution and power quality. The results prove the superiority of the proposed control scheme over that of the particle swarm optimization (PSO) and grasshopper optimization algorithm (GOA) based MG controllers for the same operating conditions and system configuration.

Keywords: microgrid; optimization; voltage and frequency regulation; dynamic response enhancement; salp swarm optimization algorithm; power quality

---

1. Introduction

Due to the recent developments in power electronics and artificial intelligence technology, the flexible deployment and performance of distributed generators (DGs) have been improved significantly. This is necessary in order to cope up with the increasing power demand along with improvement in power quality and reliability of the power system. Furthermore, these DGs combine together with a group of loads to form a small entity of an electrical network called Microgrid (MG) [1]. MG can be interconnected with the utility grid through a static switch at point of common coupling (PCC) or operated isolated. In the grid-tied mode, the power can be imported or exported between the MG and main grid as needed while in the islanded mode the MG operates to support the load at its own without any external support from the utility grid. Furthermore, in grid-tied mode, the voltage magnitude and frequency are dictated by the giant power system, while power flow between DGs and the main grid is the major control concern. Contrary to that, one of the very basic tasks of any control architecture in islanded mode is to regulate the voltage magnitude and frequency at their rated values. A general structure of the MG is shown in Figure 1.

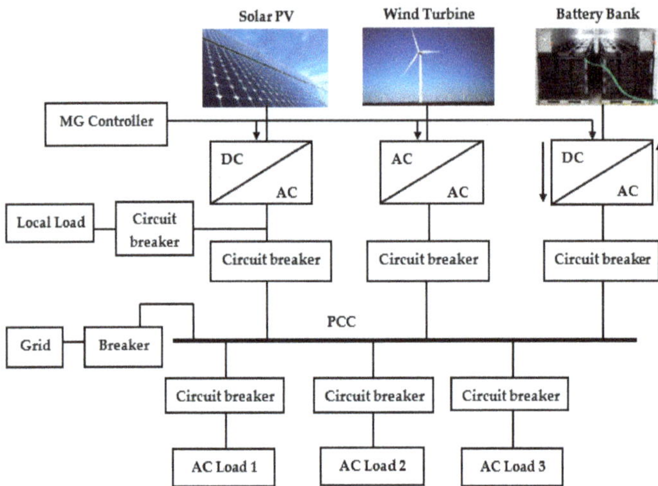

Figure 1. A basic structure of Microgrid [2].

Figure 1 portrays a very general structure of an MG where the DGs (wind and solar in this case), battery banks and loads are connected to the point of common coupling through three-phase circuit breakers. To inject a controlled amount of power at rated voltage and frequency, generally, a controlled non-linear power electronic device is utilized. These devices are non-linear in nature and hence need to be properly controlled in order to achieve a higher power quality along with the proper accomplishment of control objectives such as voltage, frequency and power flow regulation. In grid-connected mode, the sole objective of the MG controller is to ensure proper sharing of power among DGs and between MG and utility grid while in islanded mode the prime control objective is to regulate voltage and frequency at their rated values along with controlled power-sharing among DGs and power quality. Since MG faces more challenges in the islanded mode as compared to that of the grid-connected mode, this research is purely dedicated to the islanded mode of MG operation.

The study aims to improve the transient response of an islanded MG under different operating conditions. The mentioned task is accomplished by using the intelligence of the salp swarm optimization algorithm (SSA) to obtain the optimal combination of Proportional plus Integral (PI) controller parameters and dc-link side capacitance value provides the least settling time and overshoot during DG injection and load switching conditions. SSA is one of the most intelligent optimization algorithms introduced by Mirjalili et al. [3]. As compared to Genetic Algorithm (GA) and Particle Swarm Optimization (PSO), SSA is a more evolved algorithm for solving the different optimization problems [3]. It has been applied and found better than its competitor algorithms in solving many engineering problems like parameter identification of Photo-Voltaic (PV) modules [4], optimal allocation and capacity of DGs [5], extracting optimal parameters of fuel cells [6], training ANN for pattern recognition [7] and load-frequency control [8]. The mentioned studies have duly verified the effectiveness of SSA in solving the studied optimization problems better than conventional optimization techniques. This research work utilizes the intelligence of SSA for obtaining the optimal combination of PI gains and dc-link capacitance value to improve the dynamic response of studied islanded MG and regulating voltage and frequency under the DG insertion and load change conditions. The controller utilizes the droop control in the control structure as a power-sharing controller along with the voltage and current control loops. To authenticate the effectiveness of the proposed control technique, its performance is compared with that of the PSO and GOA based controllers for the identical operating conditions.

The subsequent sections are described as follows; Section 2 discusses the modern MG control architectures and a brief literature review of the available techniques used to tackle the presented problem. Section 3 describes the considered islanded MG along with its complete mathematical model and control architecture. In Section 4 the proposed methodology for obtaining an optimal transient response of studied MG system and basics of SSA along with the justification for considered fitness function has been provided. Finally, in Section 5, the obtained results are presented.

2. Modern Microgrid Control Architectures

Recently, several control architectures have been developed for the islanded MGs using proportional plus integral (PI) regulator-based power, voltage and current control loops. The generalized control architecture of the MG based on the above-mentioned control loops is shown in Figure 2.

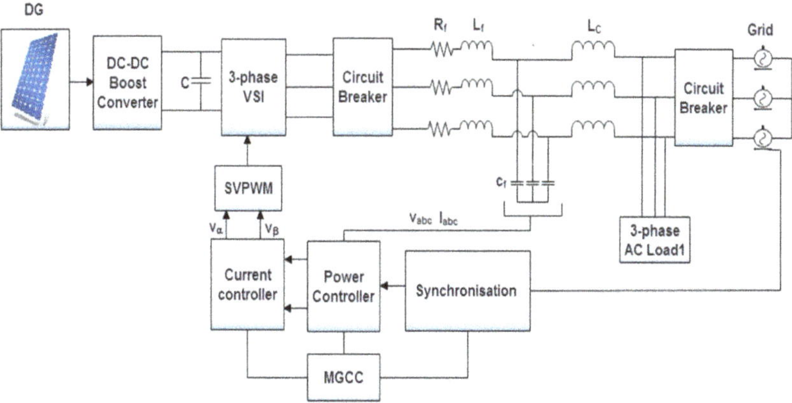

Figure 2. A generalized control structure for MGs.

For the grid-tied mode, there is no requirement of the voltage controller as it is dictated by the main grid while the power controller may be replaced by a voltage controller in case of the autonomous mode. Due to the robust structure and simple control action, PI controllers are found to be the most widely used regulators in these control loops. The main drawback of using PI regulators in modern MG control architectures is that their performance is dictated by their proportional and integral gains (k_p and k_i). Hence, selection of these gains decides the overall performance of the controller and consequently, the studied system. Over the last years, these gains were selected by using the "trial and error" or Ziegler-Nicolas (Z-N) method. These conventional methods of PI tuning suffer from many disadvantages, such as excessive time consumption, uncertainty in gain selection and complex calculations which restricted PI controller application in the latest MG control architectures up to a large extent. Most recently, with the development in the area of artificial intelligence (AI) and its applications in optimization field, several AI methods such as fuzzy logic (FL), genetic algorithm (GA) and particle swarm optimization (PSO) were explored to optimize PI controller parameters for dynamic response enhancement of islanded ac MGs. The results presented in the mentioned research papers clearly show the importance of the AI techniques in obtaining optimal PI parameters, which led to the enhanced transient response of the studied MG systems.

Many researchers around the world have worked on solving the mentioned optimization problem using different AI techniques. Al-Saedi et al. [9] developed an MG controller for controlling voltage and frequency of an autonomous MG. The PSO was utilized to optimize the system parameters and hence dynamic response. The developed controller successfully regulated the voltage magnitude and managed to keep it within standard limits ($\pm$5% of the rated value). However, the frequency level was not up to the standard ($\pm$1% of the rated value) during transient conditions. An attempt was made to solve the mentioned problem in reference [10] where the PSO based controller was developed for regulating voltage and frequency of an islanded MG. The designed controller successfully regulated the frequency and kept in well within the standard limits despite the huge load and source variations. However, the study did not consider the voltage profile, which is one of the very important parameters in the islanded mode of MG. Authors in the references [11,12] developed an AI-based controller for the voltage regulation of an islanded MG using Pareto based BB-BC and hybrid big-bang big-crunch (BB-BC) algorithms, respectively. In both case studies, the frequency declined to 59.7 Hz from its rated value (60 Hz) during DG insertion and load change conditions. Most recently, Qazi et al. [13] used the whale optimization algorithm (WOA) for regulating frequency and voltage independently by formulating two different fitness functions in an islanded MG. However, it may be noted that the voltage and frequency are two interrelated parameters [12], and therefore, it is not practically possible to independently optimize these two inter-dependent parameters. Most recently, in November 2018, the authors in reference [14] explored the grasshopper optimization algorithm (GOA) in order to regulate voltage and frequency along with the optimal transient response of an islanded MG. The authors achieved a better dynamic response as compared to previous quoted literature; however, as compared to WOA and PSO, no significant improvement was observed in overshoot and settling time during load change conditions. To address the stated limitation, for the very first time as per the best of authors' knowledge, an additional parameter that is dc side capacitance has been adopted as optimization variable along with four gains of two PI regulators in islanded MG controls. In almost all previous MG control architectures the dc side capacitance value has been adopted based on "trial and error" method whose disadvantages are already discussed earlier.

3. Proposed Islanded MG Architecture

In this section, a detailed insight of the islanded MG model along with the proposed SSA based voltage-frequency (v-f) controller is provided. A comprehensive block diagram of the power and control circuit of the studied islanded infinite bus MG system is shown in Figure 3.

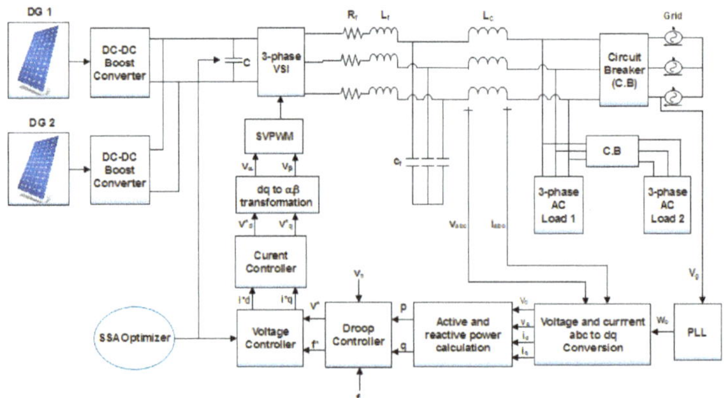

Figure 3. Studied islanded MG with power and control architecture.

The power circuit contains two solar photovoltaic (PV) modules, a three-phase voltage source inverter (VSI), RLC filter, coupling inductor and two RL loads. Since the studied MG is operated in an islanded mode, the utility grid disconnected from the rest of the network using a three-phase circuit breaker. In order to control the voltage and frequency of the studied MG system, the voltage and current signals are converted from *abc* to *dq* reference frame by using well-known Park's transformation as given in Equations (1) and (2), respectively:

$$
\begin{bmatrix} v_d \\ v_q \\ v_0 \end{bmatrix} = \sqrt{\frac{2}{3}} \begin{bmatrix} \cos\theta & \cos\left(\theta - \frac{2\pi}{3}\right) & \cos\left(\theta + \frac{2\pi}{3}\right) \\ -\sin\theta & -\sin\left(\theta - \frac{2\pi}{3}\right) & -\sin\left(\theta + \frac{2\pi}{3}\right) \\ \frac{1}{\sqrt{2}} & \frac{1}{\sqrt{2}} & \frac{1}{\sqrt{2}} \end{bmatrix} \begin{bmatrix} v_a \\ v_b \\ v_c \end{bmatrix}
\tag{1}
$$

$$
\begin{bmatrix} i_d \\ i_q \\ i_0 \end{bmatrix} = \sqrt{\frac{2}{3}} \begin{bmatrix} \cos\theta & \cos\left(\theta - \frac{2\pi}{3}\right) & \cos\left(\theta + \frac{2\pi}{3}\right) \\ -\sin\theta & -\sin\left(\theta - \frac{2\pi}{3}\right) & -\sin\left(\theta + \frac{2\pi}{3}\right) \\ \frac{1}{\sqrt{2}} & \frac{1}{\sqrt{2}} & \frac{1}{\sqrt{2}} \end{bmatrix} \begin{bmatrix} i_a \\ i_b \\ i_c \end{bmatrix}
\tag{2}
$$

The basic aim of this transformation is to achieve an equivalent two-phase orthogonal stationary reference frame from three-phase rotating signals. This conversion is necessary because the PI controllers do not work properly on sinusoidal reference signals. A more detailed version of the proposed control architecture is depicted in Figure 4.

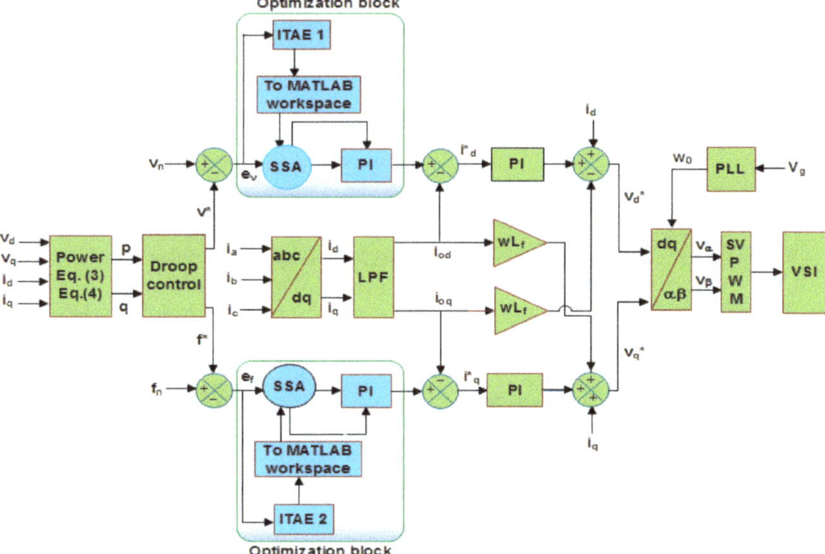

Figure 4. Detailed diagram of the salp swarm optimization algorithm based controller for islanded MG.

After obtaining the voltage and current signals in *dq* reference frame, the active and reactive power is measured using Equations (3) and (4), respectively:

$$
p = v_d i_d + v_d i_q
\tag{3}
$$

$$
q = v_d i_q - v_q i_q
\tag{4}
$$

In order to remove the harmonic contents from the fundamental active and reactive power curves, a low pass filter (LPC) with the following transfer function models is used:

$$\frac{P}{p} = \frac{w_c}{s + w_c} \tag{5}$$

$$\frac{Q}{q} = \frac{w_c}{s + w_c} \tag{6}$$

It may be noted that in conventional synchronous generator-based power systems, the voltage and frequency of the system are regulated by an auto-voltage regulator (AVR) and governor system respectively. However, these methods do not work for DGs such as solar PV, where the output power from panels is dc and fluctuating in nature due to unpredictable weather conditions and continuous load switching. Hence, in this case, a droop controller is utilized to compensate for the voltage and frequency sags and swells during any source and load change conditions. Droop control can instantly compensate for the voltage and frequency drop without any communication link with the central MG controller. It does so by continuously measuring the power output and accordingly changing the reference signals (v^* and f^*) for the voltage controller. The dynamics for the droop control are given in Equations (7) and (8):

$$f^* = f_n - k_w P \tag{7}$$

$$v^* = v_n - k_v Q \tag{8}$$

The values for the v_n and f_n are taken as 220 V and 50 Hz, respectively. The two output signals of the droop controller, i.e., which v^* and f^*, are then compared with their respective nominal values using a comparator and errors (e_v and e_f) are fed to two PI controllers whose gains are optimized by SSA optimizer. The sole aim of optimizing PI gains is to achieve an optimal dynamic response of the studied MG system with minimum overshoot and settling time in voltage and frequency curves. In order to precisely track the current reference signals (i_d^* and i_q^*), two more PI controllers are used in the current controller block of the proposed controller. Since the gains for the PI controller used in voltage controller block were optimized using SSA, optimizing gains for PI regulators in the current controller block will increase the complexity and duration of the overall optimization process and may result in un-optimal results at the end due to the excessive number of optimization variables. The transfer function equations for the current controller as derived from Figure 4 are as in Equations (9) and (10):

$$v_d^* = k_{pv} + \frac{k_{iv}}{s}\left(i_{d_{ref}} - i_{od}\right) - w * L_f * i_{oq} + v_d \tag{9}$$

$$v_q^* = k_{pf} + \frac{k_{if}}{s}\left(i_{d_{ref}} - i_{od}\right) + w * L_f * i_{od} + v_q \tag{10}$$

Finally, the space vector pulse width modulation (SVPWM) technique is used to generate the firing pulses for VSI so that a regulated quantity of power with high power quality can be injected to feed the connected load.

4. Proposed Methodology

This section is further divided into two subsections. The first section discusses the mathematical formulation of SSA for optimizing MG control parameters, while the fitness function to be minimized by the SSA is presented in the second section.

4.1. Salp Swarm Algorithm and Its Implementation

In this section, the motivation behind the proposed optimization algorithm along with its complete mathematical model is discussed in detail. Salps are the family member of the Salpidae group. They possess the translucent barrel-shaped figure and move similar to that of jellyfish. To model their movement mathematically, their initial positions are initialized randomly as portrayed in Equation (11) [3]:

$$K_1^{1:n} = rand\ (\dots) * (ub_j - lb_j) + lb_j, \quad \forall j \in no.\ of\ variables \tag{11}$$

where $K_1^{1:n}$ shows the initial positions of the salps, ub_j and lb_j represents the upper and lower limit, respectively, and *rand* (...) is the command for generating random numbers between 0–1.

Afterward, two groups are formed: leader and followers. All the salps except those at the rear are labeled as followers while the front ones are called leaders. The leader salps guide the whole swarm while the followers follow them continuously in order to search for a food source labeled as M in this case. Similar to other metaheuristic optimization techniques, the location of each salp is defined in a defined search space having n number of dimensions that are equal to the number of controlled parameters in an optimization problem. Furthermore, the location of the entire number of salps is kept in a matrix K, which is defined in Equation (12) [3]:

$$K_j^1 = \begin{cases} M_i + c_1\big((ub_j - lb_j)c_2 + lb_j\big) c_3 \geq 0 \\ M_i - c_1\big((ub_j - lb_j)c_2 + lb_j\big) c_3 < 0 \end{cases} \tag{12}$$

where the symbol K_j^1 symbolizes the position of leader salp in the jth dimension, M_i shows the location of the food source in the jth dimension and c_1, c_2 and c_3 denote the random numbers. It is significant to note that, unlike conventional optimization methods such as GA and PSO, the SSA effectively manages to avoid trapping into local minimum due to its adaptive optimization mechanism. SSA updates the position of follower salps with respect to each other and allows them to move gradually towards the leading salp, which prevents the algorithm from stagnating into the local optima. Thus, the algorithm produces an optimal or near-optimal solution precisely during an optimization process [3]. Furthermore, SSA has better exploration versus exploitation balancing capability, which is the fundamental requirement for reaching the best available solution for an optimization problem. As can be seen from Equation (12), the leader salp upgrades its location with reference to the food source only. The coefficient c_1 in Equation (12) is one of the very important parameters in SSA since it helps in balancing the exploitation and exploration characteristics of SSA and is defined in Equation (13):

$$c_1 = 2e^{-\left(\frac{4l}{L}\right)^2} \tag{13}$$

where l represents the current number of iteration, while L denotes the number of maximum iterations. The symbols c_2 and c_3 used in Equation (12) represents the random numbers between 0 and 1. In order to upgrade the location of the follower salps, a similar equation to that of Newton's law of motion is used:

$$K_j^i = \frac{1}{2}\,at^2 + v_0\,t \tag{14}$$

where $i \geq 2$ and K_j^i is the positions of ith follower salp in jth dimension, t denotes the time and v_0 is the symbol used for the velocity at the start of the optimization process, which is generally taken as 0. As the time in the optimization procedure can be replaced by the iterations and the variance between two successive iterations cannot be in a fractional number, hence by assuming $v_0 = 0$, the Equation (14) can be re-written as given underneath:

$$K_j^i = \frac{K_j^i + K_j^{i-1}}{2} \tag{15}$$

where $i \geq 2$ and K_j^i denotes the position of ith follower salp in jth dimension. It is important to note that the salp chain possesses the potential of moving towards the continuously changing global optimum (food source) in order to find a better solution by exploring and exploiting the defined search space. Some of the very important features of SSA which make it different from the conventional optimization algorithms are listed as follows [3];

1. The algorithm keeps the best-obtained solution after each iteration and assigns it to the global optimum (food source) variable. Hence, it can never be wiped out even if the whole population deteriorates.
2. SSA updates the position of the leading salp with respect to the food source only, which is the best solution obtained thus far; therefore, the leader salp always explores and exploits the space around it for a better solution.
3. SSA updates the position of follower salps with respect to each other in order to let them move towards the leading salp gradually.
4. Gradual movements of follower salps prevent the SSA from being easily stagnating into local optima.
5. Parameter c_1 is decreased adaptively over the course of iterations, which helps the algorithm to explore the search space at starting and exploits it at the ending phase.
6. SSA has only one main controlling parameter (c_1), which reduces the complexity and makes it easy to implement.

The above-mentioned merits of SSA make it potentially able to solve the optimization problems better than conventional optimization methods and hence became the motivation for the current research work. In addition, the adaptive mechanism of SSA allows this algorithm to find an accurate estimation of the best solution by continuously avoiding the trapping into local solutions.

Further details of SSA along with its pseudo-code and benchmarking against the conventional optimization methods can be found in reference [3]. It may be noted that for tuning the characteristics of the SSA for solving a specific problem, two parameters may be adjusted, namely the number of searching salps and the total iterations' number. The complete methodology of obtaining optimal PI parameters and dc-link capacitance for enhancing the dynamic response of the grid-tied MG system using SSA is depicted in the flow-chart shown in Figure 5.

Similar to other metaheuristic optimization methods, SSA spread a specified number of random search agents within searching space in the first stage. Based on the operating mechanism of SSA as given in Figure 5, these search agents are permitted to change their position in the restricted search space to minimize the stated Fitness Function (FF) subjected to a defined set of constraints. It is important to note that the mathematical model for simulating salp chains cannot be directly employed to solve optimization problems. In other words, there is a need to tweak the model according to the studied system configuration in order to make it applicable to solve the specific optimization problem. The ultimate goal of any optimization algorithm is to determine the global optimum which is generally carried out by minimization or maximization process of a defined FF. In SSA, the mentioned objective is accomplished by making the leading salp to move towards the food source and the follower salps follow the leading salp as per the governing equations of the algorithm. In the context of the optimization process, the food source is treated as the global optimum; therefore, the salp chain automatically moves towards it. It is assumed that the best solution obtained so-far is the global optimum and the food source to be chased by the salp chain until the termination criterion is reached. In the current case study, the completion of the maximum number of iterations is set as the termination criteria of the optimization process. Hence, once the pre-decided maximum number of iterations is reached, the obtained optimized PI gains and capacitance value are collected from the MATLAB workspace and are exported to the PI controller and dc side capacitor block in MG Simulink model. After that, the MG Simulink model is executed for 0.5 s to obtain the optimized power, voltage and current curves.

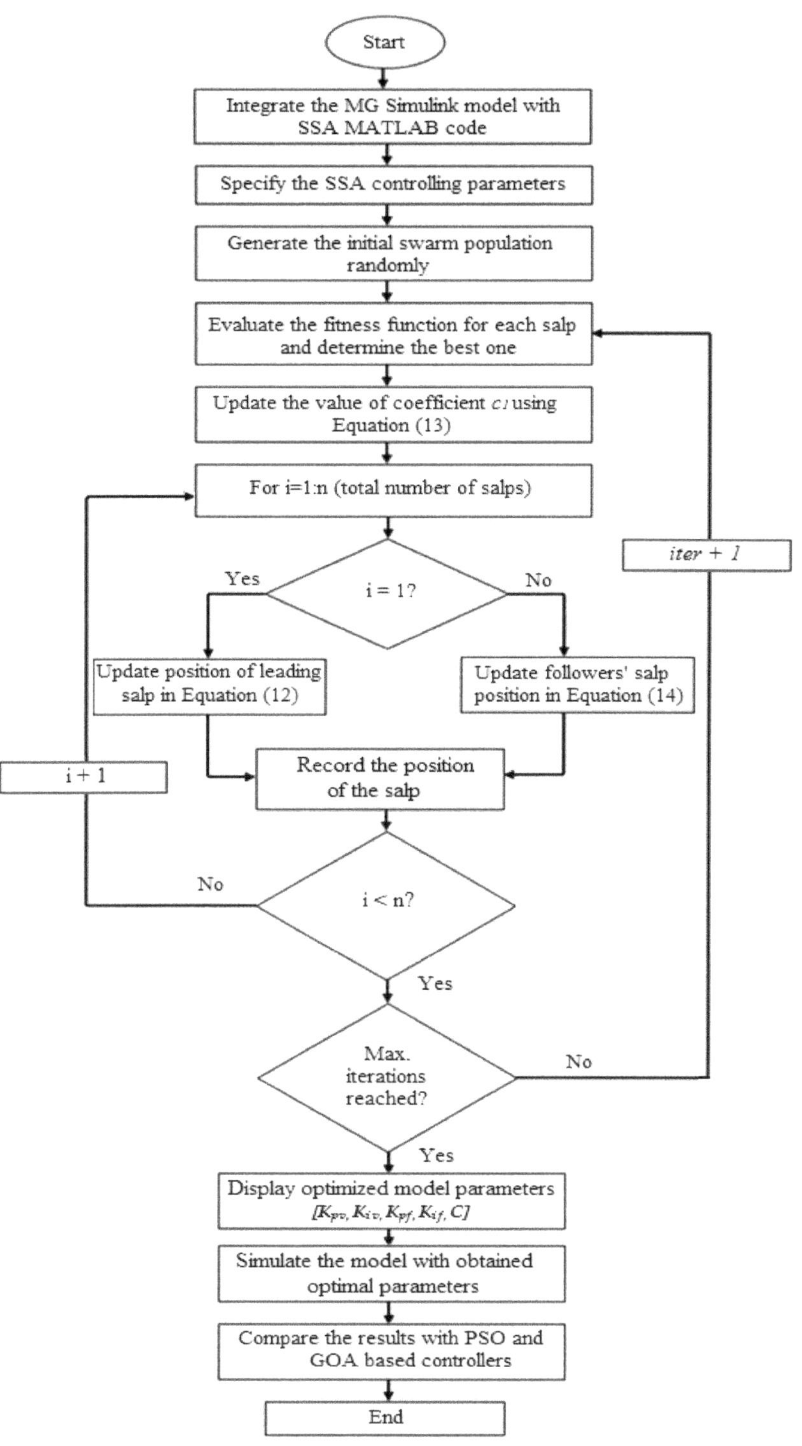

Figure 5. Proposed SSA based optimal plus integral (PI) parameter selection in studied islanded MG.

4.2. Fitness Function Formulation

The operating point of the power system fluctuates continuously and abruptly due to continuous and sudden load variations. Furthermore, in grid-connected MG power systems, the sudden insertion and desertion of DGs into the power system may push it into an unstable region of operation due to large power fluctuations and overshoots. One of the solutions to avoid such a situation is to use optimized controller parameters that can provide an optimum transient response of the system. Keeping in view the mentioned issues in the power system, the two gains of each PI controller (K_{pv}, K_{iv}, K_{pf}, K_{if}) placed in the first control loop of the studied MG controller along with the dc link capacitance (C) are optimized by SSA. In fact, for all metaheuristic and evolutionary optimization methods, the presence of the FF is a compulsory requirement that evaluates the fitness of each and every search agent and selects the best one for further comparison during the next iteration. In the context of the current study, where the optimal tuning of PI controller is required, four FF criterions that are generally considered in the literature are Integral Square Error (ISE), Integral Absolute Error (IAE), Integral Time Square Error (ITSE) and Integral Time Absolute Error (ITAE). However, ITAE is the most widely used FF criterion than its compilators due to the easy implementation, realistic error indexing and better outcomes [15,16]. The ITSE and ISE used to square the error which produces large perturbation in results even for a very small change in error signal, and hence generates impractical results. In addition, due to the continuous-time multiplication with the absolute value of error, the ITAE produces more realistic error indexing as compared to IAE. Hence looking at the prominent features of the ITAE criterion, it is adopted as the FF to be minimized in this study. Mathematically, the ITAE is expressed as provided in Equations (16) and (17):

$$ITAE1 = \int_0^\infty t|e_v|\,dt \tag{16}$$

$$ITAE2 = \int_0^\infty t|e_f|\,dt \tag{17}$$

where t refers to the total time of simulation, $|e_v|$ and $|e_f|$ are the absolute values errors in active and reactive power measurement, respectively, which are figured out by arithmetically deducting the measured power values from the setpoint power values. The overall FF is formed by mathematically adding two measured ITAE values as given in Equation (18):

$$FF = Min\left\{ \int_0^\infty t* |e_v|dt + \int_0^\infty t* |e_f|dt \right\} \tag{18}$$

It is important to understand that the magnitude of the FF in Equation (18) needs to be minimized by SSA in order to acquire the optimal PI parameter values that provide an optimum transient response of the MG system. The islanded MG modeling along with its controller and formulation of FF is carried out in MATLAB/SIMULINK version 2018a, while the coding of SSA along with its parameters like the total number of salps, iterations and optimization variables are depicted in the MATLAB editor window. Furthermore, the Simulink model is made to run for the pre-set maximum number of iterations using the "sim ( )" command from the editor window. Finally, once the simulation is finished, the optimized PI gains are placed into the corresponding PI block in MATLAB/SIMULINK model to achieve the optimal dynamic behavior of the designed MG power system.

5. Results and Discussion

The evaluation of the dynamic response for the developed islanded MG system is studied by using MATLAB/SIMULINK version 2018a. In order to make a justified evaluation of PSO, GOA and SSA the identical system parameters are used for both simulations and are depicted in Table 1.

Table 1. Studied MG system parameters.

Parameter	Symbol	Value
Solar PV rating	P_s	150 kW
Filter capacitance	C_f	2.5 mF
Filter inductance	L_f	95 mH
Switching frequency	f_{sw}	10 kHz
Sampling frequency	f_s	500 kHz
Load 1	P_1, Q_1	50 kW, 30 kVAR
Load 2	P_2, Q_2	40 kW, 20 kVAR
Load 3	P_3, Q_3	40 kW, 20 kVAR

The optimized parameters obtained at the end of the optimization process are provided in Table 2.

Table 2. The optimized parameters obtained at the end of the simulation.

Optimization	K_{pv}	K_{iv}	K_{pf}	K_{if}	C (*mF*)
PSO	0.2571093	25.6392019	0.9374905	9.3847852	23.817
GOA	0.9441557	12.8365850	26.768654	1.2474575	17.458
SSA	1.5485963	0.87302975	2.1385992	15.583932	19.954

The considered MG is purely operated in islanded mode, and hence, no grid-connection is made throughout its operation. Furthermore, the voltage magnitude and frequency curves are measured at the load connection point. In addition, since the operating conditions, the number of iterations and simulation time is pre-defined for studied islanded MG model, the optimized variables obtained through PSO, GOA and SSA at the end of the simulation ensure optimal dynamic response of the studied MG system for its complete operation. The obtained optimized parameters are inserted into the islanded MG SIMULINK model and the developed model is then evaluated for the three cases which are thoroughly discussed in the subsequent subsection.

5.1. Voltage and Frequency Regulation during DG Insertion and Load Change

As discussed earlier, the regulation of voltage and frequency is one of the major control concerns in any islanded MG due to the absence of support from the main grid. Achieving a stable rated voltage and frequency after the DG insertion and abrupt load changes with minimum overshoot and settling time needs fine-tuning of the system parameters. Hence, three different metaheuristic techniques, i.e., PSO, GOA and SSA, were utilized separately to obtain the optimal set of PI parameters and dc-link capacitance in order to minimize the overshoot and settling time after a disturbance in the MG system. Once the simulation started from the MATLAB editor, the SSA initiated its searching process for the most suitable PI gains and capacitance combination that delivers the least value of FF in order to achieve optimal dynamic response with least possible overshoot and settling time. It simulates the developed SIMULINK MG model for the pre-set number of iterations. Finally, at the end of the simulation, optimal values of four PI gains along with the capacitance value were obtained that provided minimum error integrating the FF value, which in turn guarantees the optimal dynamic behavior of the developed MG model. At exactly 0.05 s, the solar PV modules were switched on through a three-phase CB. Once the DGs were injected into the power system, they caused overshoots in the system voltage, as depicted in Figure 6.

It may be noted from Figure 6 that the system voltage observed an overshoot during the DG insertion at 0.05 s of simulation. The magnitude of the overshoot depended on the DG rating as well as the selection of controller parameters. In order to have a fair comparison between the three optimal parameter selection methods, i.e., PSO, GOA and SSA, the DG rating and other system parameters were taken identical for all three cases. At 0.25 s of the simulation, a load of 40 kW, 20 kVAR was added to the system, and as a result of that, the system voltage observed a dip in voltage. Similarly, at 0.55 s, the load-3 was disconnected from the system. The corresponding response shows a swell in voltage at the mentioned simulation time. A zoomed version of Figure 6 during DG injection, load injection and load detachment is shown in Figure 7a–c, respectively.

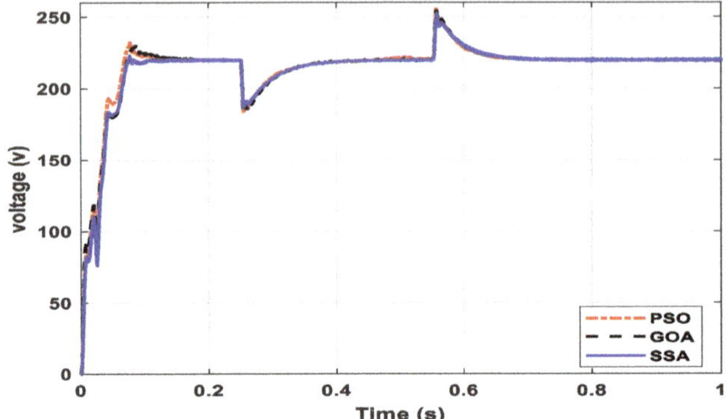

Figure 6. The voltage response of the system during DG insertion and load change.

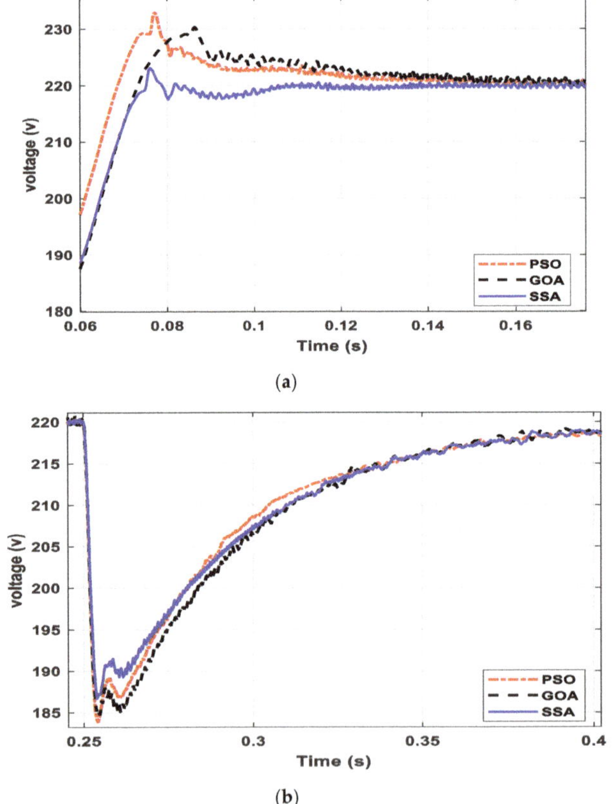

(a)

(b)

Figure 7. *Cont.*

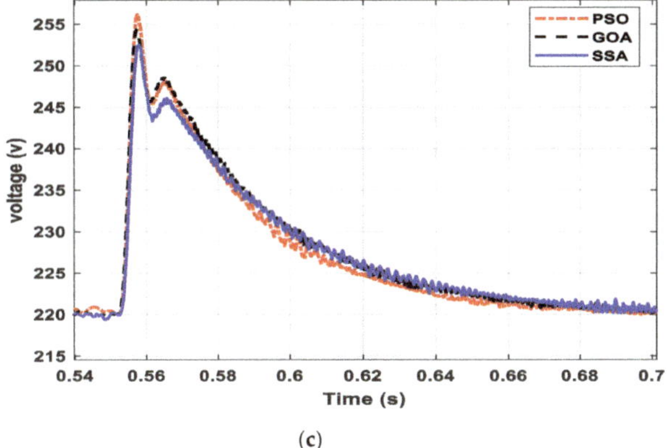

(c)

Figure 7. Voltage profile at (**a**) DG insertion (**b**) abrupt Load increment (**c**) abrupt load decrement.

It is obvious from Figure 7a–c that the optimal parameters obtained by the SSA optimization method provide better results compared to PSO and GOA in terms of overshoot and settling time for all three studied conditions. Another parameter that needs to be regulated during the islanded mode of MG operation is the frequency of the system. The frequency response of the system for PSO, GOA and SSA based MG system is shown in Figure 8.

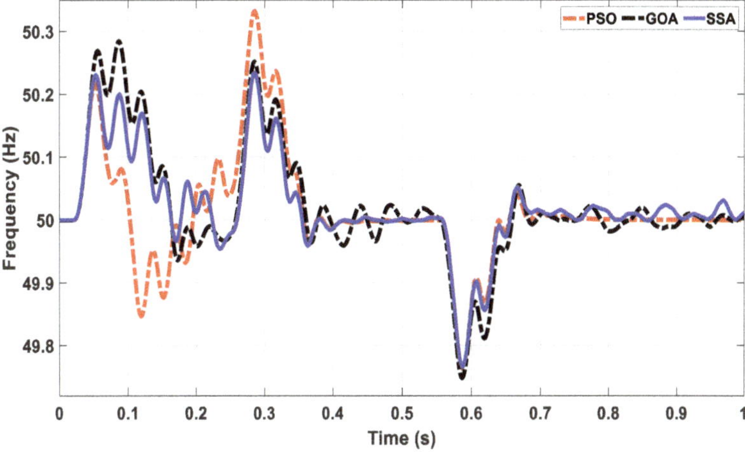

Figure 8. The frequency response of the MG system for the studied optimization methods.

Figure 8 shows the frequency response of the MG system for all three optimization methods. The result shows that all the metaheuristic methods provide a stable system frequency response within the allowable deviation range (±1%). However, the SSA based MG provides a better dynamic response as compared to its compotators. Table 3 shows a comparative analysis of all the studied methods for the voltage and frequency regulation of the studied islanded MG.

Table 3. Dynamic response evaluation of the proposed controller for voltage and frequency regulation.

	Studied Condition	Method	Maximum Overshoot/Undershoot (%)	Peak Time (ms)	Settling Time (ms)
Voltage	MG insertion	PSO	5.86	27.2	37.7
		GOA	4.68	36.3	64.5
		SSA	1.45	26.2	26.36
	Load injection	PSO	16.45	4.00	94.21
		GOA	16.00	4.70	94.20
		SSA	15.04	3.90	94.19
	Load detachment	PSO	16.41	7.70	73.50
		GOA	15.59	7.50	78.50
		SSA	14.77	7.80	77.40
Frequency	MG injection	PSO	0.44	2.05	-
		GOA	0.54	5.58	-
		SSA	0.46	2.30	-
	Load injection	PSO	0.66	35.2	-
		GOA	0.50	34.8	-
		SSA	0.46	35.0	-
	Load detachment	PSO	0.50	36.4	-
		GOA	0.48	36.7	-
		SSA	0.46	36.8	-

It may be noted from Table 3 that the SSA based controller provided better results for the most important dynamic response indicators as compared to PSO and GOA based controllers for the same operating conditions. Furthermore, it provided the most stable operation of the studied MG system and maintained the voltage within ±5% and frequency within ±1% of their nominal values, and hence, satisfied the IEEE standards. It is important to note that the settling time for the frequency is not provided. This is due to the reason that the frequency curve did not cross the ±2% of the rated value, and hence, settling time calculation is not applicable in this case.

5.2. Performance Evaluation of Studied Optimization Algorithms

In this section, the results obtained from the performance evaluation of the studied optimization algorithms are presented. Three different optimization algorithms, namely PSO, GOA and SSA, were tested to minimize the stated fitness function under identical operating conditions and system parameters. Furthermore, in order to carry out a fair comparison among the mentioned algorithms, all algorithms were tested for an identical number of iterations, i.e., 50 iterations, and number of search agents, i.e., 50 number of search agents. Since all the metaheuristic algorithms were initiated by spreading the search agents randomly in the bounded search area, the current study was tested for 20 simulation runs for each algorithm and the best (minimum) fitness function values were adopted for comparison. The convergence curve for the tested algorithms is shown in Figure 9.

It can be seen from Figure 9 that the SSA achieves the least value of fitness function (0.5840618) in the 17th iteration, whereas for the PSO and GOA the least magnitude obtained was recorded as 0.9211586 and 0.8748774 in the 21st and 25th iteration, respectively. Hence, the SSA converges faster and provides a higher quality of solution as compared to its competitors.

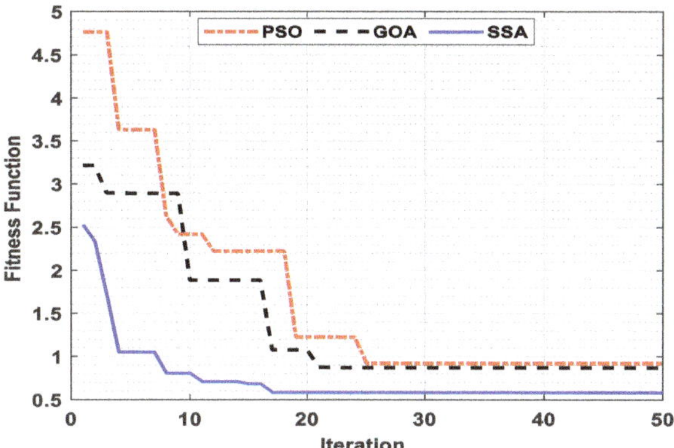

Figure 9. Convergence profile for PSO, GOA and SSA.

5.3. Power Quality Analysis

Due to the presence of non-linear power electronic devices like VSI and absence of the utility grid, maintaining the power quality and hence sinusoidality of the supply voltage and current is a very challenging task. A proper control architecture with optimal parameters can ensure pure sinusoidal voltage and current waveforms along with the high-power quality in such cases. The three-phase output current waveform for the studied system during all studied conditions is shown in Figure 10a, while its zoomed version from 0.1 to 0.22 s of simulation run is given in Figure 10b.

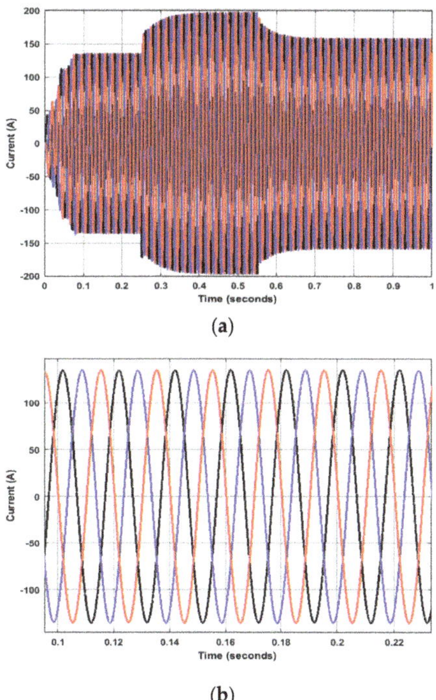

(a)

(b)

Figure 10. Three-phase sinusoidal current waveform (**a**) complete operation (**b**) zoomed version from 0.1–0.22 s.

It can be seen from Figure 10 that the solar PV dc output current is inverted into almost pure sinusoidal waveform by the proposed controller with the least possible distortion. Furthermore, to analyze the harmonic contents present in the obtained current waveform, the Fast Fourier Transform (FFT) analysis has been carried out for all three operating conditions of the system and the results are depicted in Figure 11.

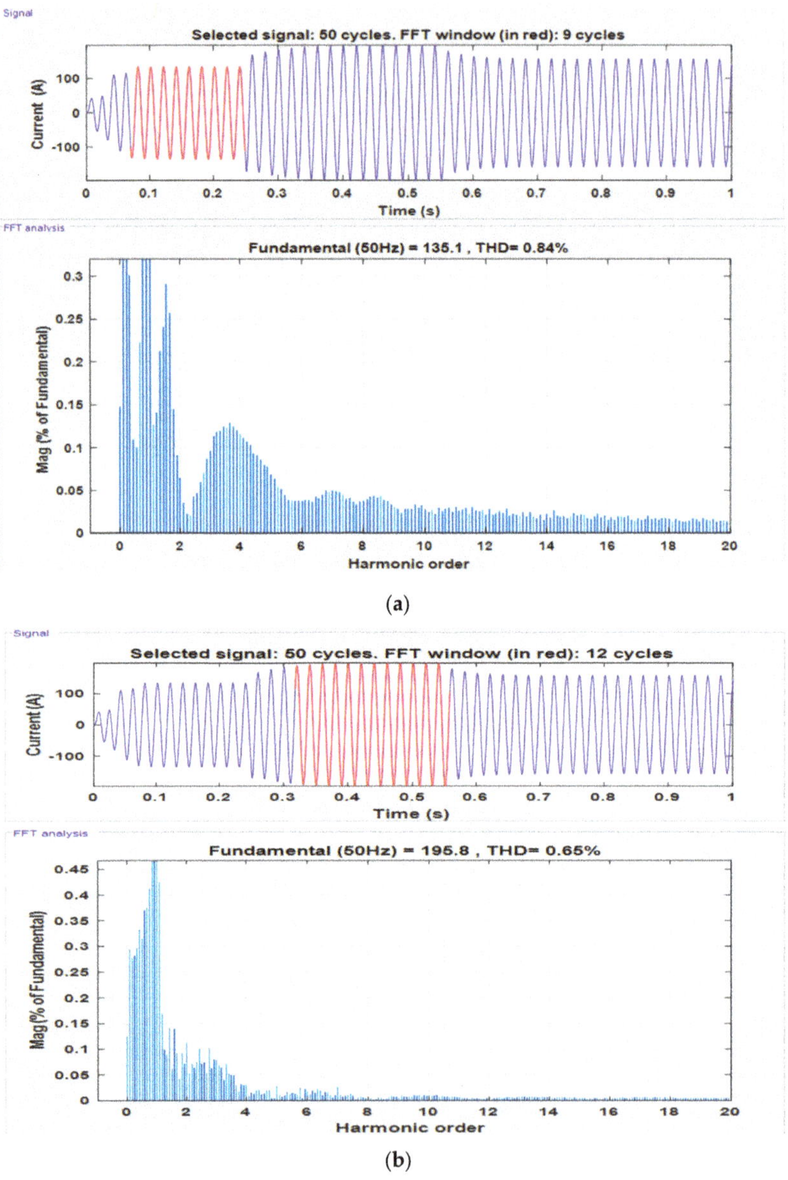

(a)

(b)

Figure 11. *Cont.*

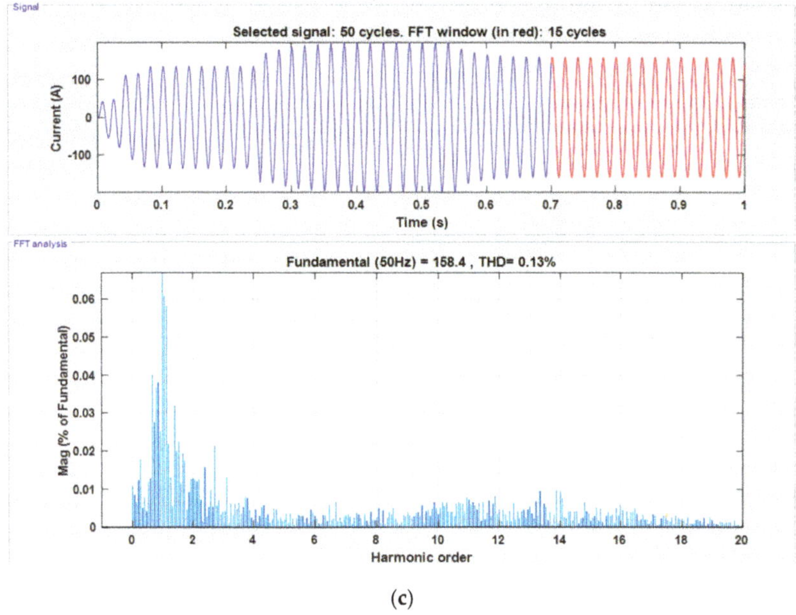

(**c**)

Figure 11. FFT analysis of studied MG system during (**a**) DG injection, (**b**) load injection and (**c**) load detachment.

It is clear from the FFT analysis of the studied power system that the SSA based controller duly satisfies the power quality standards set by IEEE 1547–2003 [17]. The results from Figure 11 are explained in tabular form in Table 4.

Table 4. FFT analysis understudied operating conditions.

Operating Condition	Percentage Harmonics (%)
MG injection	0.84
Load injection	0.65
Load detachment	0.13

It is obvious from Figure 11 and Table 4 that the proposed controller provides a high quality of power supply under all studied operating conditions and provides pure sinusoidal waveform, which shows the effectiveness of the proposed controller in maintaining optimal transient response with high power quality.

6. Conclusions

In this paper, an optimal controller for islanded MG has been successfully developed using the intelligence of SSA. The developed controller has successfully maintained the voltage and frequency at their nominal values with minimum possible overshoot and settling time during MG injection and load change conditions. The convergence behavior of the studied optimization algorithms proved that the SSA offers a higher quality of solution and faster optimization capabilities as compared to its counterpart optimization algorithms. Furthermore, the results from the power quality analysis show that the developed controller achieves high power quality and maintained almost perfect sinusoidal waveform of the voltage and current. The outcomes of the proposed controller were compared with that of the PSO and GOA based controller for the same operating conditions and system configuration.

The results show that the SSA based parameter selection provides an optimal dynamic response during all studied conditions and has superior performance as compared to its competitors.

Author Contributions: Each author has contributed significantly. The first and the corresponding author has initiated the idea and carried out the simulations along with preparation of the initial draft. The second author (M.W.M.) supervised the whole project and made significant corrections in the initial version of the article. M.M.R. has reviewed the article and helped in validation process. W.A. and S.A. helped in software and writing the revised version of the article respectively.

Funding: No funding was opted for this project.

Conflicts of Interest: The authors declare no conflict of interest.

List of Symbols

Symbol	Name
C_f	Low pass filter capacitance
$K_1^{1:n}$	initial positions of the salps
K_j^1	Position of leader salp
L_f	Low pass filter inductance
M_i	Location of the food source in the jth dimension
R_f	Low pass filter resistance
V_g	Grid voltage
e_f	Frequency error
e_v	Voltage error
f^*	Reactive frequency
f_n	Nominal frequency
i_{abc}	Three-phase current
i_d^*	Direct reference current
i_q^*	Quadrature reference current
k_v	Droop constant for voltage
k_w	Droop constant for frequency
lb_j	Lower bound of search boundary
ub_j	Upper bound of search boundary
v^*	Reference voltage
v_{abc}	Three-phase voltage
v_d^*	Direct reference voltage
v_n	Nominal or rated voltage
v_q^*	Quadrature reference voltage
v_α, v_β	Reference voltage in $\alpha\beta$ frame
a	Acceleration of the leading salp
i	Salp number
C	dc-link capacitance
c_1, c_2, c_3	Random numbers
K_{pf}, K_{if}	Gains for the lower arm PI controller
K_{pv}, K_{iv}	Gains for the upper arm PI controller
l	Number of iterations
L	Number of maximum iterations
M	Food source position
θ	Reference angel
p	Active power
q	Reactive power
$rand$	Random number
t	Total simulation time
ω	Angular frequency
ω_c	Filter cut-off frequency
v_0	Initial velocity of leading salp

References

1. Hatziargyriou, N. *Microgrids: Architectures and Control*; John Wiley & Sons: Chichester, UK, 2014.
2. Jumani, T.A.; Mustafa, M.W.; Rasid, M.M.; Mirjat, N.H.; Baloch, M.H.; Salisu, S. Optimal Power Flow Controller for Grid-Connected Microgrids using Grasshopper Optimization Algorithm. *Electronics* **2019**, *8*, 111. [CrossRef]
3. Mirjalili, S.; Gandomi, A.H.; Mirjalili, S.Z.; Saremi, S.; Faris, H.; Mirjalili, S.M. Salp Swarm Algorithm: A bio-inspired optimizer for engineering design problems. *Adv. Eng. Softw.* **2017**, *114*, 163–191. [CrossRef]
4. Abbassi, R.; Abbassi, A.; Heidari, A.A.; Mirjalili, S. An efficient salp swarm-inspired algorithm for parameters identification of photovoltaic cell models. *Energy Convers. Manag.* **2019**, *179*, 362–372. [CrossRef]
5. Tolba, M.; Rezk, H.; Diab, A.; Al-Dhaifallah, M. A novel robust methodology based Salp swarm algorithm for allocation and capacity of renewable distributed generators on distribution grids. *Energies* **2018**, *11*, 2556. [CrossRef]
6. El-Fergany, A.A. Extracting optimal parameters of PEM fuel cells using Salp Swarm Optimizer. *Renew. Energy* **2018**, *119*, 641–648. [CrossRef]
7. Abusnaina, A.A.; Ahmad, S.; Jarrar, R.; Mafarja, M. Training neural networks using salp swarm algorithm for pattern classification. In Proceedings of the 2nd International Conference on Future Networks and Distributed Systems, Amman, Jordan, 26–27 June 2018; p. 17.
8. Kumari, S.; Shankar, G. A Novel Application of Salp Swarm Algorithm in Load Frequency Control of Multi-Area Power System. In Proceedings of the 2018 IEEE International Conference on Power Electronics, Drives and Energy Systems (PEDES), Chennai, India, 18–21 December 2018; pp. 1–5.
9. Al-Saedi, W.; Lachowicz, S.W.; Habibi, D.; Bass, O. Voltage and frequency regulation based DG unit in an autonomous microgrid operation using Particle Swarm Optimization. *Int. J. Electr. Power Energy Syst.* **2013**, *53*, 742–751. [CrossRef]
10. Vinayagam, A.; Alqumsan, A.A.; Swarna, K.; Khoo, S.Y.; Stojcevski, A. Intelligent control strategy in the islanded network of a solar PV microgrid. *Electr. Power Syst. Res.* **2018**, *155*, 93–103. [CrossRef]
11. Moarref, A.E.; Sedighizadeh, M.; Esmaili, M. Multi-objective voltage and frequency regulation in autonomous microgrids using Pareto-based Big Bang-Big Crunch algorithm. *Control Eng. Pract.* **2016**, *55*, 56–68. [CrossRef]
12. Sedighizadeh, M.; Esmaili, M.; Eisapour-Moarref, A. Voltage and frequency regulation in autonomous microgrids using Hybrid Big Bang-Big Crunch algorithm. *Appl. Soft Comput.* **2017**, *52*, 176–189. [CrossRef]
13. Qazi, S.H.; Mustafa, M.W.; Sultana, U.; Mirjat, N.H.; Soomro, S.A.; Rasheed, N. Regulation of Voltage and Frequency in Solid Oxide Fuel Cell-Based Autonomous Microgrids Using the Whales Optimisation Algorithm. *Energies* **2018**, *11*, 1318. [CrossRef]
14. Jumani, T.A.; Mustafa, M.W.; Rasid, M.M.; Mirjat, N.H.; Leghari, Z.H.; Saeed, M.S. Optimal Voltage and Frequency Control of an Islanded Microgrid using Grasshopper Optimization Algorithm. *Energies* **2018**, *11*, 3191. [CrossRef]
15. Killingsworth, N.; Krstic, M. Auto-tuning of PID controllers via extremum seeking. In Proceedings of the American Control Conference, Portland, OR, USA, 8–10 June 2005; pp. 2251–2256.
16. Seborg, D.E.; Edger, T.F.; Mellichamp, D.A. *Process Dynamics and Control*, 2nd ed.; John Wiley & Sons: Chichester, UK, 2004.
17. Association, I.S. *IEEE 1547 Standard for Interconnecting Distributed Resources with Electric Power Systems*; IEEE Standards Association: Piscataway, NJ, USA, 2003.

 *processes*

Article

A Dispatching Optimization Model for Park Power Supply Systems Considering Power-to-Gas and Peak Regulation Compensation

Yunfu Qin [1], Hongyu Lin [1,2,*], Zhongfu Tan [1,2], Qingyou Yan [1], Li Li [3], Shenbo Yang [1,2], Gejirifu De [1,2] and Liwei Ju [1,2]

[1] School of Economics and Management, North China Electric Power University, Beijing 102206, China; qinyunfu815@163.com (Y.Q.); tzhf@ncepu.edu.cn (Z.T.); yanqingyou@ncepu.edu.cn (Q.Y.); ysbo@ncepu.edu.cn (S.Y.); dove@ncepu.edu.cn (G.D.); hdlw_ju@ncepu.edu.cn (L.J.)

[2] Beijing Key Laboratory of New Energy and Low-Carbon Development, North China Electric Power University, Beijing 102206, China

[3] State Grid Jibei Electric Power Economic Research Institute, Beijing 100000, China; jbjyyll@126.com

* Correspondence: hone@ncepu.edu.cn

Received: 6 October 2019; Accepted: 28 October 2019; Published: 4 November 2019

Abstract: To ensure the stability of park power supply systems and to promote the consumption of wind/photovoltaic generation, this paper proposes a dispatching optimization model for the park power supply system with power-to-gas (P2G) and peak regulation via gas-fired generators. Firstly, the structure of a park power system with P2G was built. Secondly, a dispatching optimization model for the park power supply system was constructed with a peak regulation compensation mechanism. Finally, the effectiveness of the model was verified by a case study. The case results show that with the integration of P2G and the marketized peak regulation compensation mechanism, preferential power energy storage followed by gas storage had the best effect on the park power supply system, which minimized the clean energy curtailment to 11.18% and the total cost by approximately \$120.190 and maximized the net profit by approximately \$152.005.

Keywords: park power supply system; power-to-gas; peak regulation compensation; ancillary service; wind/photovoltaic generation consumption

---

1. Introduction

Due to the fluctuation and randomness of wind/photovoltaic power generation, the issue of energy curtailment in China is still serious [1]. According to data from the National Energy Administration of China, the amount of wind power curtailment reached 10.5 billion kW·h in the first half of the year 2019 [2]. Wind/photovoltaic power curtailment is an obstacle to the development of clean energy in China and may cause greater economic losses. Therefore, research on the consumption of wind/photovoltaic power generation is still of great practical significance.

Power-to-gas (P2G) technology can, to a great extent, reduce wind/photovoltaic curtailment. Unconsumed electricity can be converted into natural gas via P2G and then into electricity in reverse through gas-fired generators (GFG), which plays a positive role in the consumption of clean energy generation [3]. Therefore, P2G can change the coupling modes of different energies [4]. In the previous studies on P2G, Gholizadeh et al. [5] evaluated and enhanced the security of P2G in networked energy hubs. Yang et al. [6] and Ye et al. [7] proposed optimal dispatching models for the multi-energy systems, which provided a new optimization method for the flexible operation of integrated energy systems. P2G technology can promote the level of renewable energy consumption in a hybrid system, as demonstrated by Marco et al. [8] and Hassan et al. [9]. Fischer al. [10] looked into the use of on-site storage and a model predictive controller and came up with possible solutions for P2G in a smart city context. All of the studies highlighted that P2G is an effective tool to promote the coupling of electricity and gas as well as clean energy consumption. Miguel et al. [11] analyzed the prospect of P2G in Portugal, while Garcia et al. [12] expanded the scope of research and extended it to Europe. Although the P2G is promising, it is still in the early stage of development [13]; thus, it is necessary to continue to study in this field.

Gas-fired power generation is the key for P2G to participate in the power-gas-power cycle. GFGs are quick to start/stop, have an easily adjustable output, and operate flexibly, which are suitable for peak regulation. Ghasemi et al. [14] looked into the importance of the gas-fired power plant location in a power-gas system with respect to peak regulation. Zhao et al. [15] studied the economy of peak regulation only via GFG, while Zhong et al. [16] analyzed the economy of peak regulation via GFG and electric vehicles under the battery-to-gas mode.

Peak regulation compensation is a method that is used to encourage the GFG to participate in peak regulation [17]. The importance of peak regulation compensation and its optimization have been discussed in the studies of Li et al. [18], Yang et al. [19], and Na et al. [20]. However, the above studies did not take P2G into consideration in their models.

In terms of the peak regulation of GFG, fuzzy clustering [21], mixed integer programming [22], and cooperative game models [16] were adopted to study its technological economy. With regard to P2G, Gil et al. [23] used the two-stage optimal power flow method to evaluate the impact of P2G technology on power and natural gas networks; the operational risks and costs of electrical systems due to wind power uncertainty were analyzed with the Conditional Value at Risk (CVaR) theory by Xu et al. [24], while Tan et al. [25] improved the CVaR theory to study the integrated energy system along with robust optimization. Weiler et al. [26] used three dimensional (3D) urban modeling tools in heat pump and co-generation systems.

Although previous research referred to integrated energy systems with P2G for energy coupling and power curtailment reduction, the role that the P2G plays in peak regulation to promote wind/photovoltaic consumption is absent, since peak regulation is so important in the current situation where China is trying to construct an electricity spot market and the production of clean energy is being encouraged to replace conventional fossil energy for sustainable development. Also, the economy of peak regulation via GFGs with the compensation mechanism was studied without the consideration of P2G's effect in prior literature. Hence, the research questions of this paper focus on three aspects. First, is the combination of P2G and peak regulation via GFG with compensation better for wind/photovoltaic consumption? Second, does this combination help with making more profit for industrial parks? Third, does the use of the energy storage device and P2G lead to different results? To answer the research questions, this study contributes in the following way, compared with the previously published papers:

First, more attention is paid to the optimization effect of the P2G on the park power supply system. In that case, whether the P2G, as an incentive, is conducive to clean energy utilization and the reduction of GFG's costs is found out. Second, the marketized peak regulation compensation mechanism, as another incentive, is introduced to better encourage GFG to cooperate with clean energy units for more clean energy consumption. Finally, with the integration of P2G and peak regulation compensation, scenarios, including different use orders of the energy storage device and the gas storage tank with the P2G, are designed in order to analyze which order is better for making the maximum profit for the system. The analysis result could serve as a reference for the practical application of these two devices.

2. Dispatching Optimization Model for the Park Power Supply System

To mitigate the clean energy curtailment without profit loss, a dispatching optimization model for the park power supply system needs to be constructed. In this section, the structure of the park power supply system is described, followed by the output model construction of different types of equipment. Finally, objective functions are established with constraints.

2.1. Structure of the Park Power Supply System with the P2G

The power supply system includes a wind turbine (WT), photovoltaic unit (PVU), energy storage device (ESD), P2G, gas storage tank, and GFG (see Figure 1). Priority is given to wind/photovoltaic power supply in the park, and the surplus power is input into the P2G or stored in the ESD. P2G is set to convert the surplus power into CH_4, and CH_4 is only used for power generation. The surplus CH_4 will be stored in the gas storage tank. Also, the WT, PVU, P2G, and gas storage tank (P2G and gas storage tank are defined as PGST in this paper) only participate in the internal transactions, and the GFG participates in the ancillary service market. When the wind/photovoltaic power generation cannot meet the power load of users in the park, the GFG provides the peak regulation service.

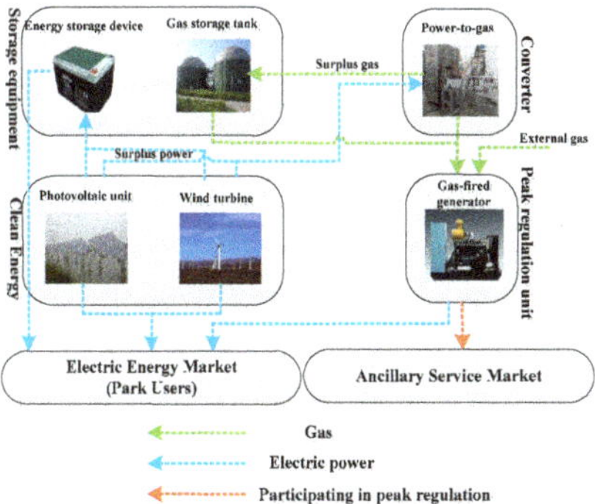

Figure 1. Park power supply system structure.

2.2. Output Models of Equipment

The available equipment in the constructed park power supply system includes clean energy units (WT and PVU), GFG, PGST, and ESD. Before building the dispatching optimization model, the output models of different kinds of equipment are given below.

The WT output is affected by the wind speed. When wind speed is within the acceptable range for the unit, the power output of the unit increases as the wind speed increases. However, if it exceeds the acceptable range, that is, the wind speed is too slow or too fast, the WT will not start in order to avoid damage to the body. The specific output model is detailed in [27]. The PVU output model generally coincides with the β distribution, which is detailed in [28].

The generation efficiency of GFG is greatly affected by its output, which decreases as the output decreases. The third-order efficiency model is adopted in this paper, which can better analyze the impact of output fluctuations on the system. The specific model [29] is

$$g_{i,g}(t) = F_{GT}(t) \cdot \eta_G(t) \tag{1}$$

where $F_{GT}(t)$ is the natural gas consumption of the GFG at time t (kW·h); $g_{i,g}(t)$ is the output of the GFG at time t (kW·h); and $\eta_G(t)$ is the generation efficiency of the GFG at time t (%).

P2G is one of the key conditions for realizing bidirectional energy coupling in the power-gas system. The reaction processes are detailed in [30]. The comprehensive energy conversion efficiency of power-to-CH_4 is about 45–60%. The power-to-CH_4 model and the state of gas storage [25] are illustrated as

$$Q_{P2G}(t) = E_{P2G}(t)\varphi_{P2G}/HCV \tag{2}$$

$$S_g(t) = S_g(T_0) + \sum_{t=1}^{T} (Q_{P2G}(t) - Q_o(t)) \tag{3}$$

where $Q_{P2G}(t)$ is the CH$_4$ production of the P2G at time t (m^3); $E_{P2G}(t)$ is the power consumption of P2G at time t (kW·h); φ_{P2G} is the P2G conversion efficiency (%); HCV is the high calorific value of natural gas (39 MJ/m^3) [31]; $S_g(t)$ is the gas volume in the gas storage tank at time t (m^3); $S_g(T_0)$ is the initial gas volume (m^3); and $Q_o(t)$ is the gas volume from the gas storage tank to the GFG at time t (m^3).

2.3. Peak Regulation Compensation Mechanism

Due to China's complicated electricity market, which is still developing, the peak regulation service (belonging to the ancillary service market) is an effective way to promote the consumption of clean energy generation. According to the "Market Regulation Measures for Power Peak Regulation Service" in a certain area of China, the peak regulation service is divided into basic (unpaid) and paid peak regulation based on a load rate of 52% (i.e., a peak regulation rate of 48%). The basic peak regulation service is provided when the peak regulation rate of a unit is less than 48%; the paid peak regulation service is provided when the peak regulation rate is more than 48% or a unit is started up or shut down for peak regulation.

The compensation mechanism involves units submitting the day-ahead quotations for peak regulation, and then the peak regulation service is provided according to the order from low to high. The current clearing price is taken as a settlement price, that is, the maximum quotation for the peak regulation compensation occurring in the unit in the statistical period of the same day. The peak regulation service follows the principle of "basic peak regulation taking precedence over paid peak regulation, and low-priced peak regulation taking precedence over high-priced peak regulation, on the premise of the safe operation of the power grid". The peak regulation compensation model is described as

$$P_a(t) = \begin{cases} P_1, & 45\% < R_{\text{load}} \le 52\% \\ P_2, & 40\% < R_{\text{load}} \le 45\% \\ P_3, & R_{\text{load}} \le 40\% \end{cases} \tag{4}$$

$$R_{\text{load}} = \frac{g(t)}{C(t)} \tag{5}$$

where $P_a(t)$ is the compensation price at time t ($/kW·h); P_1, P_2, and P_3 are the current clearing prices under different peak regulation rates ($/kW·h); R_{load} is the load rate of unit w at time t (%); $C(t)$ is the capacity of unit w (kW); and $g(t)$ is the available output of unit w at time t (kW·h).

2.4. Objective Functions

As illustrated above, to reduce clean energy wastage by GFG actively participating in peak regulation and gain more profit at the same time, the dispatching optimization model (including the objective functions and constraints) based on the system net profit, the wind/photovoltaic curtailment rate, and the GFG cost is constructed.

The economic benefit is the most important focus of the park power supply system; therefore, the net profits are considered. Firstly, the net profit model of WT and PVU is given by

$$Z_{wpv} = \sum_{j=1}^{J} \sum_{m=1}^{M} \sum_{t=1}^{24} [g_{j,w}(t) \times (p_{park}(t) - c_w) + g_{m,pv}(t) \times (p_{park}(t) - c_{pv})] \tag{6}$$

where Z_{wpv} is the net profit of the WT and PVU ($); $g_{j,w}(t)$ and $g_{m,pv}(t)$ are the outputs of the WT and the PVU at time t, respectively (kW·h); $p_{park}(t)$ is the internal power price in the park ($/kW·h); and c_w

and c_{pv} are the unit costs of the WT and the PVU (\$). The unit cost covers unit power generation, operation, and maintenance costs.

The GFG costs mainly cover power generation, operation, and maintenance costs, and the income is from selling electricity and obtaining peak regulation compensation, so the net profit model is given by

$$Z_g = \sum_{i=1}^{I} \sum_{t=1}^{24} \left[g_{i,g}(t) \times (p_{park}(t) - c_g) + G_{i,g} \times P_a(t) \right] \tag{7}$$

$$c_g = F_{GT}(t) \times \rho_g + c_g^* \times g_{i,g}(t) \tag{8}$$

where Z_g is the net profit of the GFG (\$); $P_a(t)$ is the peak regulation compensation price of the GFG at time t (\$/kW·h); $G_{i,g}$ is the compensable electricity at time t (kW·h); c_g is the power generation cost of the GFG (\$); ρ is the natural gas price (\$/m^3); and c_g^* is the unit operation and maintenance cost (\$).

Therefore, the net profit of the park is given by

$$f_1 = \max R = \max\left(Z_{wpv} + Z_g - c_{PGST} - c_{ESD}\right). \tag{9}$$

Therein,

$$c_{PGST} = g_{PGST}(t) \times c_{PGST}^* \tag{10}$$

$$c_{ESD} = g_{ESD}(t) \times c_{ESD}^* \tag{11}$$

where R is the system's net profit (\$); c_{PGST} and c_{ESD} stand for the costs of the PGST and the ESD respectively, both of which only refer to the operation and maintenance costs (\$); c_{PGST}^* and c_{ESD}^* are the unit operation and maintenance costs of the PGST and the ESD (\$); and $g_{ESD}(t)$ and $g_{PGST}(t)$ are the storage capacity (kW).

Following the concern on net profits in the park, the clean energy consumption is also what this paper mainly focuses on; therefore, the objective of clean energy curtailment rate is given by

$$f_2 = \min I = \min\left(\frac{U_c}{U_t} \times 100\%\right) \tag{12}$$

where I is the clean energy curtailment rate (%); U_c is the actual clean energy curtailment (kW·h); and U_t is the total clean energy power generation (kW·h).

To ensure more electricity from clean energy units can be consumed, the GFG acts as a bridge between clean energy units and P2G; therefore, its cost is one of this paper's focuses. The objective of the minimum GFG costs is analyzed to motivate more willing cooperation with clean energy units. The corresponding objective is given by

$$f_3 = \min c_g. \tag{13}$$

2.5. Constraints

The constraints used in this paper include the supply and load balance, the running constraints of different types of equipment, and the reserve capacity of the system.

The constraint of the system power balance is stated as

$$\sum_{i=1}^{I} g_{i,g} + \sum_{j=1}^{J} g_{j,w} + \sum_{m=1}^{M} g_{m,pv} = D \tag{14}$$

where $g_{i,g}$, $g_{j,w}$, and $g_{m,pv}$ are the available outputs of GFG, WT, and PVU respectively (kW·h); and D is the load demand (kW·h).

The running constraints of GFG, WT, and PVU are detailed in [32].

The constraint of the system's reserve capacity is

$$\sum_{g=1}^{G} [g^{\max}(t)(1-\theta)](1-l) \geq D(t) + R(t) \tag{15}$$

where $D(t)$ is the load demand at time t (kW·h); $R(t)$ is the system reserve demand at time t (kW); l is the line loss rate (%); θ is the unit self-use rate (%); and $g^{\max}(t)$ is the maximum output at time t (kW·h).

Since the objective functions with the above constraints are mixed integer programming problems, the General Algebraic Modeling System (GAMS) software is a good solution tool. The mixed-integer programming models and the GAMS are detailed in [33].

3. Case Study

3.1. Scenario Settings

To verify the effectiveness of the proposed model, four scenarios based on different types of storage equipment and the different use priorities were set up (see Table 1).

The different types of storage equipment were as follows:

- In Scenario 1 (S1), the park power supply system employed an energy storage device (ESD) to store the unconsumed wind/photovoltaic power generated to be sold in the next period;
- In Scenario 2 (S2), the park power supply system employed a P2G and a gas storage tank (PGST) to convert the unconsumed wind/photovoltaic power generated into natural gas to be stored in the gas storage tank as the fuel for power generation.

The different storage equipment use priorities were as follows:

- In Scenario 3 (S3), the park power supply system employed a PGST and an ESD. The unconsumed wind/photovoltaic power generated was preferentially converted into natural gas via the P2G and stored in the gas storage tank as the fuel for power generation, and then the surplus was stored in the ESD.
- In Scenario 4 (S4), the park power supply system employed a PGST and an ESD. The unconsumed wind/photovoltaic power generated was preferentially stored in the ESD, and then the surplus was converted into natural gas via the P2G and stored in the gas storage tank as the fuel for power generation.

Table 1. Four scenarios based on different types of storage equipment and their use priorities.

Scenario	Energy Storage Device (ESD)	P2G and Gas Storage Tank (PGST)	Use Priority
S1	√	/	/
S2	/	√	/
S3	√	√	PGST > ESD
S4	√	√	ESD > PGST

3.2. Basic Data

A park power supply system used in an area of China was selected for the case study. In this park, clean energy units, gas-fired generators (GFGs), an ESD, and a PGST were employed. The clean energy units contained a wind turbine (WT) and a photovoltaic unit (PVU), and the PGST included a P2G and a gas storage tank. Besides, there were four GFGs in the park (the specific parameters are shown in Tables 2 and 3). The capacity of the ESD was 1000 kW, the maximum charging/discharging power was 200 kW, the charging/discharging loss was 0.4, and the initial energy storage was set to be 0; the unit cost of operation and maintenance was 0.071 $/kW·h.

Table 2. Equipment parameters of the wind turbine (WT), the photovoltaic unit (PVU), and the P2G and gas storage tank (PGST).

Equipment	Capacity	Unit Cost ($/kW·h)	Current Limitation (km³/h)	Conversion Efficiency	Unit Operation and Maintenance Cost ($/kW·h)
Wind turbine (WT)	10 (MW)	0.049	/	/	/
Photovoltaic unit (PVU)	500 (kW)	0.085	/	/	/
P2G	400 (kW)	/	/	60%	0.078
Gas store tank	50 (km³)	/	20~50	/	

Table 3. Equipment parameters of the gas-fired generators (GFGs)

GFG	*a*	*b*	*c*	*d*
Capacity(kW)	1000	500	500	500
Generation efficiency	60%	60%	60%	60%
Unit operation and maintenance cost ($/kW·h)	0.098	0.098	0.098	0.098

The power load demand and the output of clean energy units in this park are illustrated in Figure 2. Given the uncertainty of wind and photovoltaic power generation, the output data of the clean energy units on a typical day were selected for this research. GFGs satisfied the load demand that WT and PVU could not cope with.

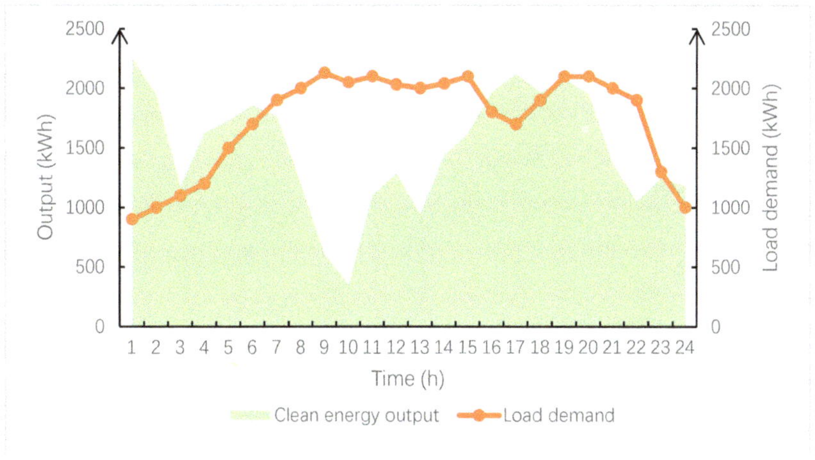

Figure 2. Power load demand of the park and output of clean energy units.

Based on the load demand in the park, a day was divided into three periods: 8:00–15:00 and 19:00–22:00 were the peak periods, 23:00–4:00 was the valley period, and 5:00–7:00 and 16:00–18:00 were the flat periods. For the given periods, the power price data are presented in Table 4, according to the actual prices of a province in China. In addition to power prices, the natural gas price was 0.488 $/m³ (for convenient calculation, it was converted into the unit calorific value price, 0.049 $/kW·h).

Table 4. Price data.

Category	Item	Price ($/kW·h)
Park power price	Peak period	0.170
	Valley period	0.113
	Flat period	0.141
External power price	Peak period	0.204
	Valley period	0.157
	Flat period	0.172

Besides selling electricity, GFGs participate in peak regulation to gain profit, which increases clean energy consumption. In the light of peak regulation compensation mechanism, GFGs propose their quotations the day before operation. The quotation standards under different peak regulation rates are shown in Table 5.

Table 5. Quotation intervals for peak regulation.

Peak Regulation Rate (PRR)	Lower Limit ($/kW)	Upper Limit ($/kW·h)
48% < PRR ≤ 55%	0.042	0.071
55% < PRR ≤ 60%	0.071	0.113
PRR > 60%	0.113	0.141

According to the load demand and peak regulation rate of each unit, the day-ahead peak regulation quotations of all GFGs during each period were generated randomly, and these are displayed in Table 6.

Table 6. gas-fired generators' (GFGs') day-ahead quotations for peak regulation (Unit: $/kW·h).

Times \ Unit	GFG *a*	GFG *b*	GFG *c*	GFG *d*
1	0.061	0.056	0.062	0.069
2	0.066	0.055	0.063	0.069
3	0.065	0.054	0.060	0.058
4	0.060	0.062	0.059	0.066
5	0.082	0.094	0.085	0.097
6	0.108	0.107	0.078	0.113
7	0.091	0.078	0.109	0.110
8	0.139	0.126	0.135	0.132
9	0.113	0.131	0.118	0.121
10	0.125	0.113	0.114	0.139
11	0.138	0.140	0.137	0.124
12	0.131	0.123	0.140	0.117
13	0.121	0.132	0.123	0.129
14	0.123	0.124	0.115	0.122
15	0.130	0.117	0.135	0.132
16	0.089	0.103	0.107	0.092
17	0.106	0.105	0.085	0.098
18	0.096	0.082	0.097	0.084
19	0.124	0.130	0.129	0.139
20	0.126	0.119	0.130	0.132
21	0.123	0.136	0.137	0.130
22	0.135	0.117	0.138	0.122
23	0.062	0.055	0.066	0.058
24	0.054	0.063	0.066	0.065

3.3. Results Analysis

3.3.1. Wind/Photovoltaic Curtailment

WT, PVU, GFG, and ESD were the units providing power. The power output of PGST was in the form of gas-fired power generation. Figure 3 shows the output of different types of equipment in each scenario, and Table 7 shows the clean energy curtailment rates of different scenarios. The results indicate that the available outputs of the WT and PVU were basically the same in different scenarios, but the wind/photovoltaic curtailment differed accordingly. The curtailment rate in the S1 was the highest, reaching 16.70%, whereas when the PGST was employed, the curtailment rate decreased (12.37% in S2), as did the GFG output. Moreover, in the scenario where both the PGST and ESD were employed, because of the different priorities of using the two devices, different results occurred for the wind/photovoltaic curtailment and the GFG output. The curtailment rate in S3 decreased to 11.59% compared with that in S2, but that in S4 decreased more—11.18%. Thus, the method of 'energy storage preferentially, then gas storage' has the best effect on reducing wind/photovoltaic curtailment.

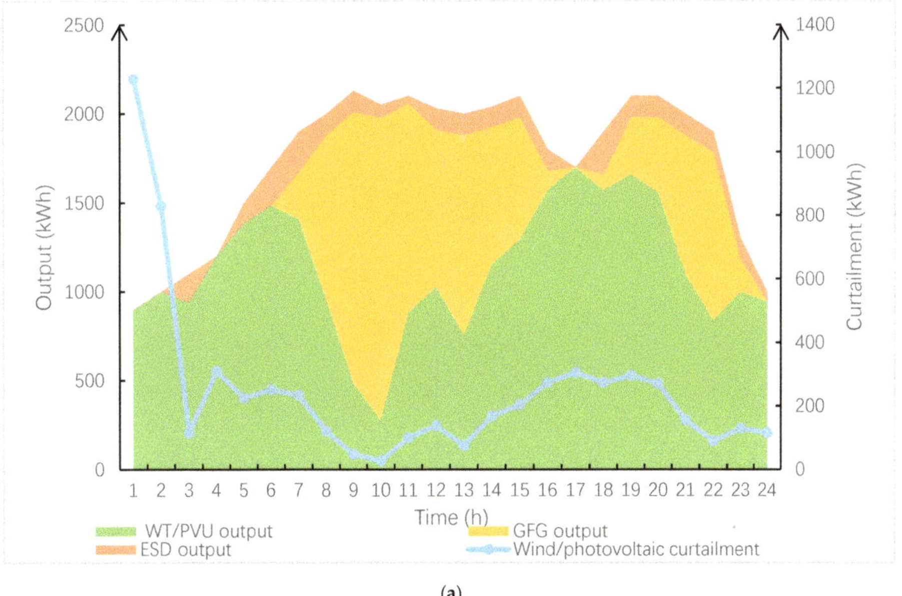

(**a**)

Figure 3. *Cont.*

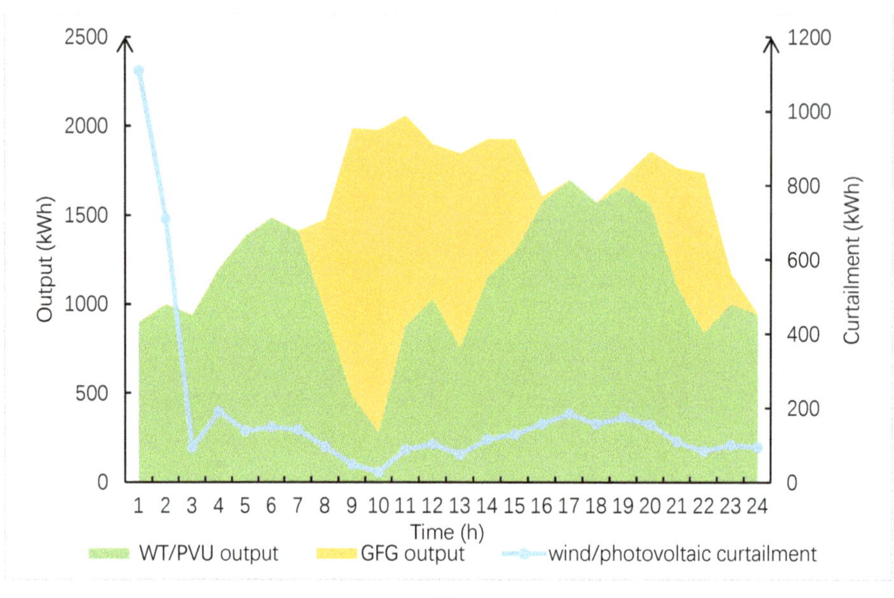

(**b**)

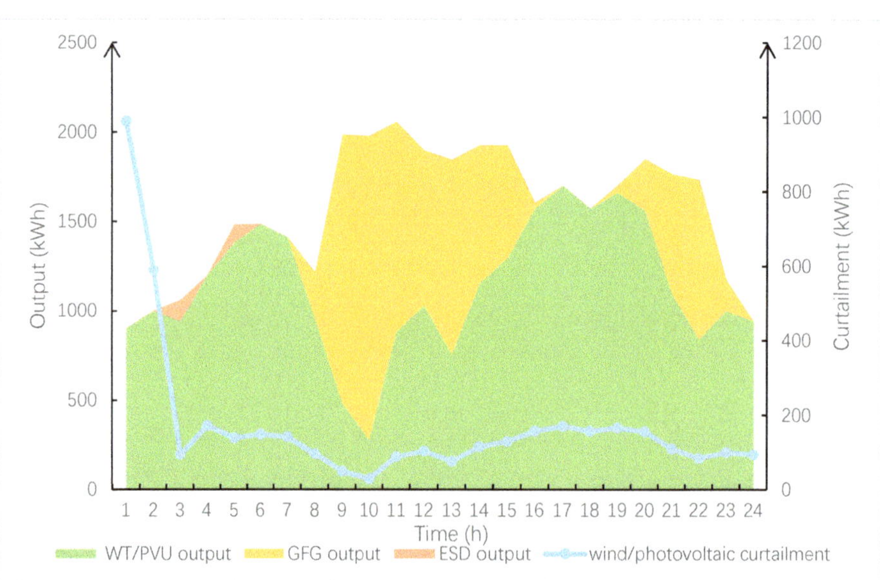

(**c**)

Figure 3. *Cont.*

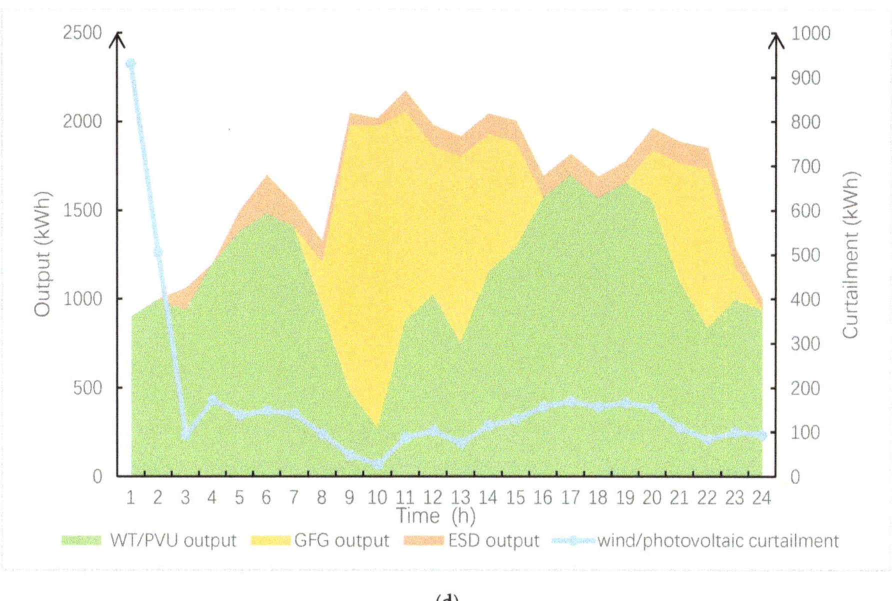

(d)

Figure 3. Unit output and wind/photovoltaic curtailment in different scenarios. (**a**) Unit output and wind/photovoltaic curtailment in S1. (**b**) Unit output and wind/photovoltaic curtailment in S2. (**c**) Unit output and wind/photovoltaic curtailment in S3. (**d**) Unit output and wind/photovoltaic curtailment in S4.

Table 7. Clean energy curtailment rates in different scenarios.

Scenario	1	2	3	4
Clean energy curtailment rate	16.70%	12.37%	11.59%	11.18%

3.3.2. GFG Costs

For different scenarios, the GFG costs during different periods are shown in Table 8.

Table 8. GFG costs during different periods (Unit: $).

	Scenario	S1	S2	S3	S4
	Valley period	26.442	25.593	25.593	25.593
	Flat period	66.458	4.666	4.666	0
Period	Peak period	1651.693	1494.174	1452.744	1426.302
	Total	1744.593	1524.433	1483.003	1451.754

Table 8 indicates that, with the participation of the PGST, the GFG costs decreased significantly, especially in flat and peak periods. Compared with S1, the GFG costs in S2 decreased by $220.160, because the GFG used gas from the P2G instead of purchasing it from the outside; thus, its generation cost decreased. The GFG costs in S3 and S4 were both lower than that of S2 by $41.430 and $72.680, respectively, which indicates that with the combination of the ESD and the GPST, the costs were further reduced. Additionally, compared with S3, the costs in S4 decreased by $31.249, so the different priorities of using the PGST and the ESD affect the GFG costs, and the method of 'energy storage preferentially, then gas storage' has a better effort.

3.3.3. System Economic Benefits

In view of the above four scenarios, taking the transactions in the electric energy market and ancillary service market into account, the results of system economic benefits in different scenarios are shown in Table 9.

Table 9. System total costs and net profit in different scenarios (Unit: $).

Scenario	S1	S2	S3	S4
Total costs	3728.577	3662.543	3641.050	3608.387
Net profit	2203.436	2319.526	2336.069	2355.441

As presented in Table 9, the scenario with PGST (namely, S2) earned $116.090 more and cost $66.034 less than S1 where the PGST was not considered. This demonstrates that the P2G can effectively reduce the total cost and increase the net profit. Under the joint operation of the PGST and the ESD, the method of 'energy storage preferentially, then gas storage' (namely, S4) minimized the total cost of the system and maximized its net profit.

4. Conclusions

To promote clean energy consumption to mitigate resource wastage and encourage gas-fired generators to collaborate with clean energy units, the construction of a dispatching optimization model with P2G and the peak regulation compensation mechanism for GFG was carried out in this paper. The model tends to pursue the maximum net profit of the park power supply system and the minimum clean energy curtailment rate and GFG costs. After conducting a case study, it was found out that systems with P2G can effectively promote clean energy consumption in comparison with ones involving no P2G. Additionally, with the joint operation of the PGST and the ESD, the method of 'energy storage preferentially, then gas storage' can maximize benefits.

Due to the high installation capacity of clean energy units but insufficient peak regulation sources and immature market mechanism development in China, marketized compensation is still an important means to encourage conventional generators to participate in peak regulation for clean energy consumption, and P2G not only increases the consumption but also reduces GFG costs, being a novel means in the power industry. As P2G technology develops (lower investment and higher conversion efficiency), it will be more practical and popular in the future.

In future research, the effect of P2G on CO_2 emission reduction (environmentally) will be taken into account and how much it can gain for the park power supply system in terms of green certificate trading and carbon trading (economically) will be investigated. Also, more marketized peak regulation compensation means and more types of peak regulation units based on local resource advantages will be discussed, respectively.

Author Contributions: Supervision, Z.T.; Writing—Original draft preparation, Y.Q. and H.L.; Writing—Review and editing, Q.Y. and L.J.; Formal analysis, L.L. and S.Y.; Software, G.D.

Funding: This work was partially supported by the Project funded by China Postdoctoral Science Foundation (2019M650024), the National Nature Science Foundation of China (Grant Nos. 71904049,71874053, 71573084), the Beijing Social Science Fund(18GLC058) and the 2018 Key Projects of Philosophy and Social Science Research, Ministry of Education, China (18JZD032).

Conflicts of Interest: The authors declare no conflict of interest.

References

1. Pu, L.; Wang, X.; Tan, Z.; Wu, J.; Long, C.; Kong, W. Feasible electricity price calculation and environmental benefits analysis of the regional nighttime wind power utilization in electric heating in Beijing. *J. Clean. Prod.* **2019**, *212*, 1434–1445. [CrossRef]
2. Renewable Energy Grid Operation in the First Half of 2019. Available online: http://www.nea.gov.cn/2019-07/25/c_138257185.htm (accessed on 25 July 2019).
3. Stephen, C.; Pierluigi, M. Integrated modeling and assessment of the operational impact of power-to-Gas (P2G) on electrical and gas transmission networks. *IEEE Trans. Sustain. Energy* **2015**, *6*, 1234–1244.
4. Ju, L.; Zhao, R.; Tan, Q.; Lu, Y.; Tan, Q.; Wang, W. A multi-objective robust scheduling model and solution algorithm for a novel virtual power plant connected with power-to-gas and gas storage tank considering uncertainty and demand response. *Appl. Energy* **2019**, *250*, 1336–1355. [CrossRef]
5. Gholizadeh, N.; Vahid-Pakdel, M.J.; Mohammadi-ivatloo, B. Enhancement of demand supply's security using power to gas technology in networked energy hubs. *Int. J. Electr. Power Energy Syst.* **2019**, *109*, 83–94. [CrossRef]
6. Yu, J.; Ma, M.; Guo, L.; Zhang, S. Reliability evaluation of integrated electrical and natural-gas system with power-to-gas. *Proc. CSEE* **2018**, *38*, 708–715.
7. Yang, D.; Xi, Y.; Cai, G. Day-Ahead Dispatch Model of Electro-Thermal Integrated Energy System with Power to Gas Function. *Appl. Sci.* **2017**, *7*, 1326. [CrossRef]
8. Marco, B.; Gabriele, F. Optimising energy flows and synergies between energy networks. *Energy* **2019**, *173*, 400–412.
9. Hassan, A.; Patel, M.K.; Parra, D. An assessment of the impacts of renewable and conventional electricity supply on the cost and value of power-to-gas. *Int. J. Hydrogen Energy* **2019**, *44*, 9577–9593. [CrossRef]
10. Fischer, D.; Kaufmann, F.; Oliver, S.L.; Christoper, V. Power-to-gas in a smart city context—Influence of network restrictions and possible solutions using on-site storage and model predictive controls. *Int. J. Hydrogen Energy* **2018**, *43*, 9483–9494. [CrossRef]
11. Miguel, C.V.; Mendes, A.; Madeira, L.M. An Overview of the Portuguese Energy Sector and Perspectives for Power-to-Gas Implementation. *Energies* **2018**, *11*, 3259. [CrossRef]
12. Garcia, D.A.; Barbanera, F.; Cumo, F.; di Matteo, U.; Nastasi, B. Expert Opinion Analysis on Renewable Hydrogen Storage Systems Potential in Europe. *Energies* **2016**, *9*, 963. [CrossRef]
13. Liu, W.; Wen, F.; Xue, Y.; Zhao, J.; Dong, Z.; Zheng, Y. Cost characteristics and economic analysis of power-to-gas technology. *Autom. Electr. Power Syst.* **2016**, *40*, 1–11.
14. Masoud, G.; Rajabi, M.H.; Amin, H. Importance of gas-fired power plants location in integrated operation of power and natural gas systems: Peak load condition analysis. In Proceedings of the 26th Iranian Conference on Electrical Engineering, Mashhad, Iran, 8–10 May 2018; pp. 1227–1232.
15. Zhao, C.; Zhang, H.; Li, F.; Yuan, J. Economic research on gas CHP and peaking generation in Beijing. *Int. Pet. Econ.* **2017**, *25*, 99–105.
16. Zhong, K.; Wang, D.; Yang, G.; Fu, J. Peak shaving analysis of electric vehicle and gas turbine under B2G mode. *Electrotech. Appl.* **2018**, *37*, 47–52.
17. Shen, W.; Zhen, W.; Zhou, D.; Gao, Y. Wind power peak Regulation pricing model under wind and fire alternative trading mechanism—A case study of wind Power Integration, Gansu Province. In Proceedings of the 2018 China International Conference on Electricity Distribution, Tianjin, China, 17–19 September 2018.
18. Li, Y.; Li, H. Improvement of compensation mechanism on peak-regulating auxiliary service for thermal power unit with energy saving dispatch. *Heilongjiang Electr. Power* **2014**, *26*, 194–197.
19. Yang, Z.; Li, K.; Wang, N.; Jin, Z.; Song, S.; Guo, X. A Model of Considering the Economic Analysis and Environmental Protection for Thermal Power Compensation on Peak Regulation. *J. Eng. Thermophys.* **2018**, *39*, 2124–2130.
20. Na, C.; Yuan, J.; Zhu, Y.; Xue, L. Economic Decision-Making for Coal Power Flexibility Retrofitting and Compensation in China. *Sustainability* **2018**, *10*, 348. [CrossRef]
21. Wang, Y.; Tian, Y.; Wu, M.; Geng, J. Two-part electricity price model for peak load regulation of natural gas power based on fuzzy clustering. *Proc. CSEE* **2017**, *6*, 38–46.
22. Kong, D.; Long, H.; Li, G.; Liu, Y. Research on equivalent smoothing strategy of PV output based on micro energy system. *Acta Energy Sol. Sin.* **2017**, *9*, 33–41.

23. Gil, M.; Duenas, P.; Reneses, J. Electricity and natural gas interdependency: Comparison of two methodologies for coupling large market models within the European regulatory framework. *IEEE Trans. Power Syst.* **2016**, *31*, 361–369. [CrossRef]

24. Xu, Z.; Zhang, Y.; Chen, Z.; Lin, X.; Chen, B. Bi-level optimal capacity configuration for power to gas facilities considering operation strategy and investment subject benefit. *Autom. Electr. Power Syst.* **2018**, *42*, 76–84.

25. Tan, Z.; Tan, Q.; Yang, S.; Ju, L.; De, G. A Robust Scheduling Optimization Model for an Integrated Energy System with P2G Based on Improved CVaR. *Energies* **2018**, *11*, 3437. [CrossRef]

26. Weiler, V.; Stave, J.; Eicker, U. Renewable Energy Generation Scenarios Using 3D Urban Modeling Tools Methodology for Heat Pump and Co-Generation Systems with Case Study Application [†]. *Energies* **2019**, *12*, 403. [CrossRef]

27. Cheng, J.; Choobineh, F. A Novel Wind Energy Conversion System with Storage for Spillage Recovery. *J. Power Energy Eng.* **2015**, *3*, 33–38. [CrossRef]

28. Wang, Y.; Lu, Y.; Ju, L.; Wang, T.; Tan, Q.; Wang, J.; Tan, Z. A Multi-objective Scheduling Optimization Model for Hybrid Energy System Connected with Wind-Photovoltaic-Conventional Gas Turbines, CHP Considering Heating Storage Mechanism. *Energies* **2019**, *12*, 425. [CrossRef]

29. Ju, L.; Zhang, Q.; Tan, Z.; Wang, W.; Xin, H.; Zhang, Z. Multi-agent-system-based coupling control optimization model for micro-grid group intelligent scheduling considering autonomy-cooperative operation strategy. *Energy* **2018**, *157*, 1035–1052. [CrossRef]

30. Andrea, M.; Ettore, B.; Gianfranco, C. Applications of power to gas technologies in emerging electrical systems. *Renew. Sustain. Energy Rev.* **2018**, *92*, 794–806.

31. Wei, Z.; Zhang, S.; Sun, G.; Zang, H.; Chen, S.; Chen, S. Power-to-gas Considered Peak Load Shifting Research for Integrated Electricity and Natural-gas Energy Systems. *Proc. CSEE* **2017**, *37*, 4601–4609.

32. Yang, S.; Tan, Z.; Ju, L.; Lin, H.; De, G.; Tan, Q.; Zhou, F. An Income Distributing Optimization Model for Cooperative Operation among Different Types of Power Sellers Considering Different Scenarios. *Energies* **2018**, *11*, 2895. [CrossRef]

33. Wang, J. Optimization of Power System Operation Based on GAMS. Master's Thesis, South China University of Technology, Guangzhou, China, 2014.

 *processes*

Article

Reliability Evaluation Method Considering Demand Response (DR) of Household Electrical Equipment in Distribution Networks

Hongzhong Chen [1], Jun Tang [1], Lei Sun [2], Jiawei Zhou [3], Xiaolei Wang [1,*] and Yeying Mao [1]

1   State Grid Suzhou Power Supply Company, Suzhou 215004, China; chenhongzhong@js.sgcc.com.cn (H.C.); tj_sz@js.sgcc.com.cn (J.T.); myy_sz@js.sgcc.com.cn (Y.M.)
2   School of Electrical and Automation Engineering, Hefei University of Technology, Hefei 230009, China; leisun@hfut.edu.cn
3   State Grid Suzhou Power Supply Company Suzhou Electric Power Design Institute Co., Ltd., Suzhou 215004, China; jwzhou1990@163.com
*   Correspondence: boboball.wang@hotmail.com

Received: 30 September 2019; Accepted: 23 October 2019; Published: 3 November 2019

Abstract: The load characteristic of typical household electrical equipment is elaborately analyzed. Considering the electric vehicles' (EVs') charging behavior and air conditioning's thermodynamic property, an electricity price-based demand response (DR) model and an incentive-based DR model for two kinds of typical high-power electrical equipment are proposed to obtain the load curve considering two different kinds of DR mechanisms. Afterwards, a load shedding strategy is introduced to improve the traditional reliability evaluation method for distribution networks, with the capacity constraints of tie lines taken into account. Subsequently, a reliability calculation method of distribution networks considering the shortage of power supply capacity and outages is presented. Finally, the Monte Carlo method is employed to calculate the reliability index of distribution networks with different load levels, and the impacts of different DR strategies on the reliability of distribution networks are analyzed. The results show that both DR strategies can improve the distribution system reliability.

Keywords: demand response; household electrical equipment; real-time electricity price; incentive mechanism; capacity constraint; reliability evaluation

---

1. Introduction

China's electricity generation and demands have been growing rapidly in recent years, but the annual utilization hours of power generation equipment are decreasing year-by-year, and the peak-valley difference of power system loads is gradually expanding. Statistically, China's total electricity consumption in 2018 was 6840 billion kW h, up 8.5% year-on-year, which is the highest growth rate since 2012. The maximum cooling loads in summer have reached 260 million kW, with a year-on-year growth of 10.5%, accounting for 27.8% of the maximum loads. However, the average utilization hours of power generation equipment in power plants of 6000 kW or above in China in 2019 were only 3862 h. In general, increasing generation capacity to meet peak load electricity demands will lead to an increase in investment cost and a decrease in equipment utilization hours, which cannot meet the requirements of economic operation of power grid and optimal allocation of resources.

As a prospective solution to the above issues, demand response (DR) has been widely proposed in many researches [1,2], where DR is regarded as an important measure to facilitate the penetration of renewable energy [3], realize friendly source-load interaction, balance the fluctuations of load profile, and consequently improve the system reliability and operation economy [4,5]. With the rapid development and application practice of ubiquitous power Internet of Things (IOT) technology, bi-directional communication and intelligent control between power grid and user side can be supported [6]. DR, normally including electricity price-based DR and incentive-based DR [7,8], could regulate consumers' electricity behaviors to reduce peak loads and improve load curve [9–12]. Real-time electricity price can reflect the relationship between power supply and demand, and thus a reasonable electricity price mechanism could efficiently guide users to participate in peak shaving and valley filling [13,14]. For incentive-based DR, users can be encouraged to adjust load demands by signing an agreement on condition of certain economic compensation [15,16].

Residential loads can be regarded as a resource that can be flexibly dispatched in the distribution network, which can be divided into temperature-controlled equipment, non-temperature-controlled equipment, and uncontrollable equipment according to the operation characteristics of household power loads [17]. Temperature-controlled equipment, such as air conditioning, can participate in DR by adjusting comfort interval [18]. Non-temperature-controlled equipment, such as electric vehicles (EVs), can change electricity consumption behavior by responding to real-time electricity price [19]. In [20], an optimal scheduling model, which considers the reliability of microgrid and the customers' electricity cost, is established to regulate the electricity consumption of home appliances and EVs. In [21], a home energy management system is proposed and an energy consumption optimization model is presented to minimize the electricity cost, and dynamic programming is employed to solve the real-time rescheduling model for determining the on/off status of home appliances. In [22], a customer utility function is defined and a novel economic model is presented to determine the customers' reaction to electricity price change. Reference [23] points out that the implementation of DR is limited by the accurate forecast of demand and price elasticity, and therefore presents a novel DR model based on consumers' information while avoiding predicting these two items. Both air conditioning loads and EV loads account for a large proportion in the daily load, and therefore enjoy a large potential for DR.

The reliability of distribution systems changes nonlinearly with the increase or decrease of the loads. By far, this issue has been studied by many researchers. A time-of-use electricity price-based DR model is established in [24,25] to identify the influence of electricity price-based DR on distribution network reliability. In [26], the incentive-based DR mechanism is systematically investigated to design the DR contract based on load transfer and load reduction, respectively, and then a bidding decision optimization model is presented to maximize the benefits of load aggregators. The influence of incentive-based DR on distribution network reliability is analyzed. However, the aforementioned works evaluate the reliability of distribution networks under a single DR mechanism. Actually, it is necessary to consider the comprehensive effects of two kinds of DR mechanism. In order to evaluate the performance of the implementation of DR, the customer baseline load is introduced in [27] with two different calculating methods, the day matching method and regression analysis. In [28], the concept of a virtual power plant is introduced, and the cost models considering the incentive- and electricity-based DR are respectively proposed to quantitatively analyze the influence of the uncertainty of DR on the expected losses of energy in distribution networks. It should be mentioned that the tie line capacity is generally assumed to be infinity when distribution network reliability is analyzed, which means the loads in non-fault areas could be entirely transferred after fault isolation, leading to overoptimistic reliability results. The line capacity constraint is considered in [29] when the outage is caused by a failure, assuming the power supply is sufficient. Nevertheless, the quantity of power users is rapidly increasing and the network structure is gradually becoming complex. The equipment in distribution networks that has not been upgraded, in time, may lead to electricity supply shortage in distribution systems. Therefore, these models may not perform well for calculation accuracy.

In this paper, two typical items of household equipment, EV and air conditioning, are taken into account. The DR models are presented based on different DR mechanisms, while considering load characteristic, charging behavior, and thermodynamic property. Also, it is necessary to present an improved distribution network reliability evaluation method considering tie line capacity constraint and power supply capacity shortage constraint, and therefore the accuracy of distribution network reliability index can be improved. Given the aforementioned reviewed literature, the contributions can be summarized as:

- The load characteristic of two typical items of household electrical equipment is elaborately analyzed.
- An electricity price-based DR model and an incentive-based DR model are proposed for two typical items of high-power electrical equipment, considering charging behavior and thermodynamic property.
- A load shedding strategy is introduced to improve the traditional reliability evaluation method for distribution networks, while taking into account the capacity constraints.
- A reliability calculation method of distribution networks with shortage of power supply capacity and faults taken into consideration is presented.

The remainder of this paper is given as follows. Section 2 present an electricity price-based DR model and an incentive-based DR model for two kinds of typical high-power electrical equipment. A load shedding strategy is proposed in Section 3, and the reliability index is detailed in Section 4. Section 5 proposes an improved reliability evaluation method of distribution networks, with shortage of power supply capacity and faults taken into account. The numerical results of the developed model are discussed in Section 6. The final section of the paper outlines conclusions based on this study.

2. DR Modeling

DR can be categorized into two types: Electricity price-based DR and incentive-based DR. The former guides users' electricity behavior by varying electricity price, while the latter adjusts electricity loads by contracts and compensation terms signed by customers. In this paper, two kinds of household electrical equipment, i.e., EVs and air conditioners, are chosen to establish DR models.

2.1. DR Modeling of EVs Based on Electricity Price

The charging loads of EVs depend on the user's habits, driving distance, status of the battery charge, the time when EVs plug in and out of distribution systems. Generally, EV owners use their cars during the day and charge them at night for the next day's trip. Real-time electricity price can reflect the relationship between power supply and demand, and therefore, real-time electricity price could be employed to guide users to adjust charging behavior of EVs. Given the development of communication technology, intelligent switch technology, and Internet of Things industry, the DR strategy for EV proposed in this paper can be realized in the near future.

2.1.1. Optimization Objective

When EVs plug into the grid, the owners are required to set the time of departure and the expected state of charge (SOC). The DR strategy is presented to minimize the charging cost of the owners by optimizing the charging power of EVs at each time interval according to the real-time electricity price. It should be mentioned that the owners are willing to participate in the DR because the cost of EV charging can be reduced while the owners' demands for the next day's travel are met. For a single EV, the objective function is defined as follows:

$$Min \sum_{t=1}^{T} P_t^{EV} C_t \Delta t \tag{1}$$

The objective Function (1) is devoted to minimize the charging cost of a single EV. P_t^{EV} and C_t represent the charging power of EV and time-of-use electricity price at time t, Δt is the length of one time interval, and T is the number of time intervals within the scheduling time.

2.1.2. Constraints

(1) SOC constraint

Considering the charging efficiency of the battery in an EV, the recursive formula of the SOC can be described as:

$$Soc_{t+1} = Soc_t + \frac{\varepsilon_{ch}P_t^{EV}\Delta t}{E_B} \tag{2}$$

$$Soc_{min} \leq Soc_t \leq Soc_{max} \tag{3}$$

where Soc_{t+1} and Soc_t denote the SOC of the battery at time $t+1$ and t, respectively; ε_{ch} represents the charging efficiency of the battery; E_B is the energy capacity of battery, and the unit is kW·h; Soc_{max} and Soc_{min} represent the upper and lower limit of SOC, respectively.

(2) Travel demand constraint

The SOC of the battery in an EV when it plugs out of the grid should be restricted to meet the owner's demand for travel.

$$Soc_{t_d} \geq Soc_{exp} \tag{4}$$

where Soc_{td} and Soc_{exp} respectively represent the actual and expected SOC of the battery at t_d; t_d denotes the time when EV plugs out, which is with strong uncertainty and can be approximately estimated according to the historical data of EVs.

(3) Battery safety constraint

SOC of the battery and the charging power in the charging process are described as:

$$0 \leq P_t^{EV} \leq P_{max}^{EV} \tag{5}$$

where $P_{max}{}^{EV}$ represents the maximum of charging power of EV.

(4) Undispatched time constraint

$$P_t^{EV} = 0 \quad t \notin [t_s, t_d] \tag{6}$$

where t_s represents the time when an EV plugs into the grid.

2.2. DR Modeling of Air Conditioners Based on Incentive

Air conditioning is chosen as the temperature-controlled load to participate in DR due to the widespread utilization of air conditioning and the large proportion of air conditioning loads in summer peak loads. Air conditioning could satisfy users' demands for temperature through cooling or heating equipment. The acceptable temperature range for a human body is relatively wide. Therefore, air conditioning can be regarded as a DR resource with great potential and flexibility.

2.2.1. Optimization Objective

The air conditioning load could be adjusted within the time period declared by consumers to smooth the load curve and realize peak load shifting. The air conditioning loads are controlled by adjusting the setting of the temperature. The scheduling model of air conditioning loads can be formulated as follows.

$$\min \frac{1}{T-1}\sum_{t=1}^{T}\left(P_t^{D} - \sum_{j=1}^{N}\Delta P_{k,t}^{air}x_{k,t} - \overline{P}^{D}\right)^2 \tag{7}$$

$$\overline{P}^{D} = \frac{1}{T}\sum_{t=1}^{T}\left(P_t^{D} - \sum_{j=1}^{N}\Delta P_{k,t}^{air}x_{k,t}\right) \tag{8}$$

The objective Function (7) aims to minimize the load variance, where P_t^D represents the base load of time t; $\Delta P_{k,t}^{air}$ denotes the load reduction of the k_{th} air conditioner in time t; $\overline{P}^D$ is the mean value of the total load in the dispatching period; N represents the total number of air conditioners; T is the number of time intervals within the scheduling time; $x_{k,t}$ is the binary decision variable indicating whether the k_{th} air conditioner participates in load reduction at time t; $x_{k,t}$ is equal to 1 if the air conditioner k participates in scheduling at time t, and 0 otherwise.

2.2.2. Constraints

(1) Electrical constraints of air conditioners

The energy efficiency ratio of air conditioners can be defined as the ratio of power and capacity of refrigerating. It should be noted that the energy efficiency ratio of inverter air conditioning cannot be simply modeled as a constant. The mathematical relationship among power, refrigerating capacity, and frequency in actual operation can be formulated as follows:

$$P_A = af_A + b \tag{9}$$

$$Q_A = mf_A^2 + nf_A + q \tag{10}$$

Equation (9) describes the relationship between the power and frequency of air conditioners, where P_A and f_A represent the operating power and frequency of air conditioners, respectively. a and b are constants. Equation (10) denotes the relationship between refrigerating capacity and frequency of air conditioners, where Q_A represents the refrigerating capacity of air conditioners. m, n, and q are coefficients.

(2) Thermodynamic model of air conditioners

The indoor temperature is generally chosen as the control variable, which is affected by many factors, such as the indoor and outdoor heat exchange, outdoor temperature, the heat transfer between the air conditioner and the indoor air, the air heat capacity, and the air conditioning refrigerating capacity. It should be mentioned that there are many other affecting factors; therefore, precisely modeling air conditioning thermodynamic characteristics can be quite complex. The thermodynamic model of air conditioning can be approximately described as follows:

$$\theta_{in,t+1} = \theta_{out,t+1} - Q_t R(1 - e^{-\Delta t/RC}) - (\theta_{out,t} - \theta_{in,t})e^{-\Delta t/RC} \tag{11}$$

where $\theta_{in,t+1}$ and $\theta_{in,t}$ represent the indoor temperature at time $t + 1$ and t, respectively; $\theta_{out,t+1}$ denotes the outdoor temperature at time $t + 1$; Q_t is the air conditioning refrigerating capacity at time t; R and C are the equivalent heat resistance and heat capacity of air conditioning, with the units of °C/kW kW h/°C, respectively.

(3) Inverter air conditioning operating frequency constraint

The frequency of the inverter air conditioning can be adjusted by the compressor, and the operating frequency should meet the following constraint:

$$f_{min,k} \leq f_k \leq f_{max,k} \tag{12}$$

where f_k represents the operating frequency of the k_{th} air conditioner; $f_{max,k}$ and $f_{min,k}$ denote the maximum and minimum operating frequency of the k_{th} air conditioner, respectively.

(4) Maximum dispatchable time constraints

The maximum dispatchable number of air conditioners in the scheduling period should be restricted in order to meet the users' comfort requirements.

$$\sum_{t=1}^{T} x_{k,t} \leq t_{c,k} \tag{13}$$

where $t_{c,k}$ is the maximum dispatchable number of the k_{th} air conditioner.

2.2.3. Control Method

The control methods of inverter air conditioning [30] can be divided into temperature control and frequency control. The changing curve of indoor temperature and frequency under the two control methods is shown in Figure 1. In the temperature control mode, it can be found that the frequency rapidly decreases to the lowest value if the set temperature of the inverter air conditioner raises. As the room temperature gradually increases, the frequency rises with fluctuations and tends to be stable. For the frequency control mode, the frequency is also rapidly reduced to the lowest value when the set temperature raises. The frequency continues to run at the lowest value until the temperature rises to the set temperature, and meanwhile, the frequency rises to a stable value. It can be found from Figure 1 that the temperature control method is simple but it is difficult to obtain an analytical solution of frequency deviation and control period, and therefore, the frequency control method is employed to control air conditioning loads; namely, when the user sets the comfort temperature range, i.e., from θ_1 to θ_2, the air conditioning would operate with the lowest frequency at the first time, and then the air conditioning frequency would be adjusted to stable value for θ_2.

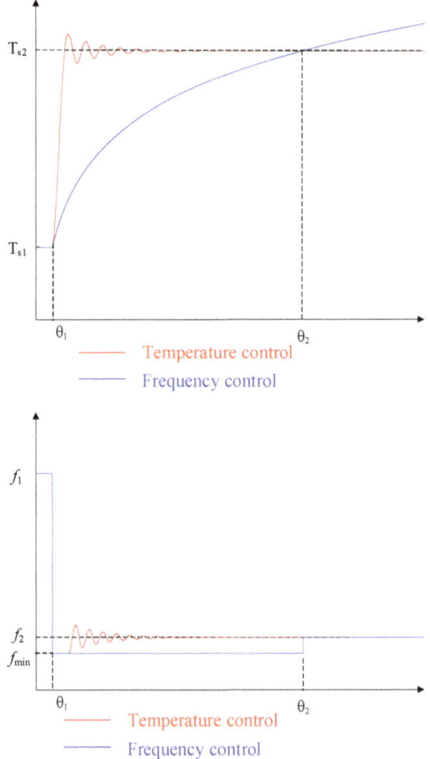

Figure 1. Schematic diagram of different control methods.

It is assumed that the indoor temperature at time $t + 1$ is equal to the indoor temperature at time t in stable operation. When the air conditioning runs stably, its operation power can be calculated by Equations (9)–(11).

$$P_t^{air} = \frac{-n + \sqrt{n^2 - 4m[q - (\theta_{out,t} - \theta_1)/R]}}{2m} a + b \tag{14}$$

Therefore, load reduction of air conditioning can be described as:

$$\Delta P_{k,t}^{air} = P_{k,t}^{air} - a f_{\min,k} - b \tag{15}$$

where $P_{k,t}^{air}$ denotes the power demand of the k_{th} air conditioner in stable operation at time t.

3. Reliability Evaluation of Distribution Networks Considering DR

3.1. Model of Load Transfer Capacity

When a failure occurs, non-fault areas can be supplied through tie lines after fault areas are isolated. Considering the maximum transmission capacity limit of the tie line, the transfer capacity of the tie line l can be calculated as follows [31]:

$$P_{lz} = \frac{P_{l\max} - P_l(1 + \beta_l)}{1 + \beta_l} \tag{16}$$

where P_{lz} represents the power supplied by the tie line; P_l is the load of a line itself; $P_{l\max}$ is the maximum transmission power of the line l; β_l is the line loss ratio of the line l.

If the tie line has already supplied heavy loads, the transfer power will not be enough to meet the power supply of all the loads in the downstream area of the fault. In this situation, the load shedding strategy described in Section 3.2 should be applied in order to meet the power supply requirements of important loads in the downstream area of the fault.

3.2. Load Shedding Strategy

The over-load operation of the feeder may occur during in the peak load period if the maximum transmission capacity limit of the feeder in the distribution network is taken into consideration. It is necessary to reduce the load amount of the feeder by cutting parts of the loads to ensure the secure and stable operation of the distribution networks. Therefore, the load shedding model is proposed and shown in Equations (17) and (18):

$$\max \sum_{i \in S} \omega_i P_i(t) \tag{17}$$

$$s.t.$$
$$\begin{cases} (1 + \beta) \sum_{i \in S} P_i(t) \leq P_{S\max} \\ U_{i,\min} \leq U_i \leq U_{i,\max} \end{cases} \tag{18}$$

The objective Function (17) should be maximized so as to supply as much load as possible, while taking the weight of the load points into consideration. ω_i is the importance coefficient of load point i for reliability requirement; S represents the collection set of all load points on the feeder; $P_i(t)$ is the power demand of load point i at time t. Equation (18) describes the power supply capacity constraint and node voltage constraint, where $P_{S\max}$ represents the upper limit of the power transmitted by the feeder and U_i, $U_{i,\max}$, and $U_{i,\min}$ are the voltage of node i and its maximum and minimum values, respectively.

4. Analysis of the Influence of DR on Distribution Network Reliability

The key of reliability evaluation is the selection and calculation of reliability index. In order to systematically study the influence of load changes caused by DR on the reliability of the distribution network, the reliability index influenced by load changes need to be firstly analyzed.

4.1. Load Point Reliability Index

(1) Load point average failure frequency index

The load point average failure frequency index is defined as the number of outages at load points in the statistical period. Long time overload operation of transformers will accelerate its aging, and then affect the average failure rate of the load point. DR can improve the load curve and reduce peak loads, thus reducing the failure frequency of load points.

(2) Load point average interruption frequency index

When the distribution transformer fails or the power supply is insufficient, load interruption occurs. The average interruption frequency index of load point i can be defined as:

$$P_{outage,i} = 1 - \frac{1}{T}\sum_{t=1}^{T} X(i,t) \tag{19}$$

where $P_{outage,i}$ denotes the outage probability of load point i; $X(I, t)$ is a binary decision variable for operation state of load point i at time t, which is equal to 1 in normally operating state, and 0 otherwise.

4.2. System Reliability Index

Three system reliability indexes are introduced in this paper as follows:

(1) Frequency index

Frequency index mainly refers to system average interruption frequency index (SAIFI).

(2) Time index

Time indexes mainly include customer average interruption duration index (CAIDI) and system average interruption duration index (SAIDI).

(3) Energy index

Energy not supplied (ENS) index mainly depends on the annual power outage time and load power of the system.

5. Improved Reliability Evaluation Method Based on Load Clustering

5.1. Improved Reliability Evaluation Method

According to the location of faults and their influence on other loads, the loads can be categorized into four types: (1) Type A: The loads in the fault area and their outage time depend on the time of fault isolation and repair; (2) Type B: The loads in the downstream of the fault area and their outage time depend on the load transfer time; (3) Type C: The loads in the upstream of the fault area, which can be supplied by the main transformer after fault isolation, and their outage time depend on the fault isolation time. It should be mentioned that if the supply capacity of the tie line is insufficient, the reliability indexes of Type B loads can be modified by applying the transferring capacity model and load shedding model described in Sections 3.1 and 3.2, and then the reliability index with and without DR implementation can be calculated, respectively.

The sequential Monte Carlo simulation method is applied to evaluate the reliability of distribution networks, and there are two situations that should be considered:

(1) When power supply capacity is insufficient in normal operation state, the load shedding strategy in Section 3.2 should be applied to supply power as much as possible with the feeder maximum capacity constraint respected. Calculate the system reliability indexes with and without DR, respectively.

(2) When a failure occurs in the distribution network, parts of the loads of Type B cannot get power supply due to restricted transfer capacity if the maximum capacity limit of feeders are considered.

5.2. Reliability Calculation Method of Distribution Networks Considering Load Clustering

The annual load peak is generally utilized to calculate the horizontal annual system reliability index in the traditional evaluation method. This method reflects the system reliability under the most severe situations, but ignores the impact of the load change on the system reliability. Calculating the reliability index only by the load peak value will greatly reduce the accuracy of evaluation results, since the annual load curve is composed of 8760 load points. However, the reliability calculation will be very time-consuming if each load point is substituted into the reliability evaluation. Therefore, it is necessary to cluster the annual load, then the reliability index can be calculated by employing the reliability calculation method in Section 5.1 based on the clustering results. The final system reliability index can be obtained according to the calculation results and weighted values of each cluster. In this way, the computational burden can be reduced under the premise of satisfying the accuracy.

6. Case Study

6.1. Case 1

A smart residential community in Suzhou is employed to illustrate the effectiveness of the proposed models. Residents' electricity consumption data at different time periods are collected by smart meters, and the residents' electricity consumption behavior is analyzed by the non-intrusive load monitoring, which could obtain the operation condition and loads of different types of electrical equipment by feature extraction and power decomposition technology.

6.1.1. Residential Electricity Load Analysis

The electricity consumption of a user in the smart community from 5–11 August in 2019 is shown in Figure 2, and the electricity consumption of each household equipment is depicted in Figure 3. It can be seen from Figure 2 that the electricity consumption of this user remained basically stable in a week, and the electricity consumption at the weekend increased compared to the working day. The maximum daily electricity consumption reached 23.44 kW h, since the residential community is mainly composed of villas. It can be concluded from Figure 3 that the electricity consumption of air conditioning accounts for up to 47% of the total electricity consumption, indicating that the air conditioning load has great potential for DR. Besides, there is little EV in this community, so only air conditioning is considered in this part.

6.1.2. Analysis on DR of Residential Load

In order to further analyze the DR potential of air conditioning load, the air conditioning load of 40 users in this smart community is selected for analysis. The air conditioning load curve can be roughly categorized into three classes, as depicted in Figure 4, and its daily load rate is shown in Table 1. Combined with Figure 4 and Table 1, it can be found that the air conditioning load characteristics of the three types of users are different from each other. The load curve of class A users have two peaks and two valleys, the peak of electricity consumption occurs at 14:00 and 22:00, and the daily load rate is relatively low, and therefore they have a great potential for DR. The power consumption of air conditioner of class B load in one day is almost zero, which has no potential of load control. Class C users are similar to class A, both of which show two peaks and two valleys, but the load rate of former one is relatively high, and they have a certain staggered peak response capability.

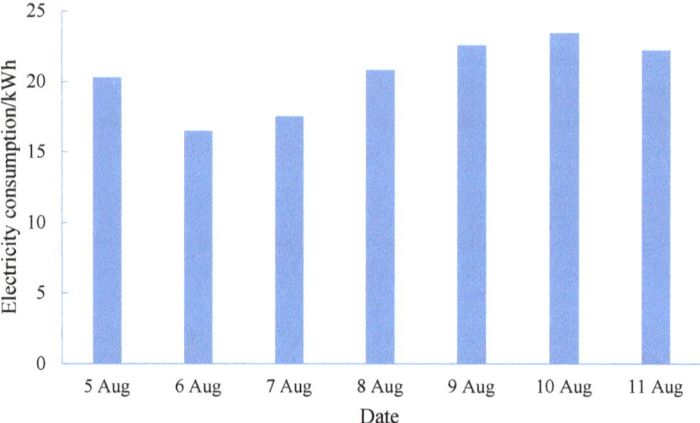

Figure 2. Residential electricity consumption in a week.

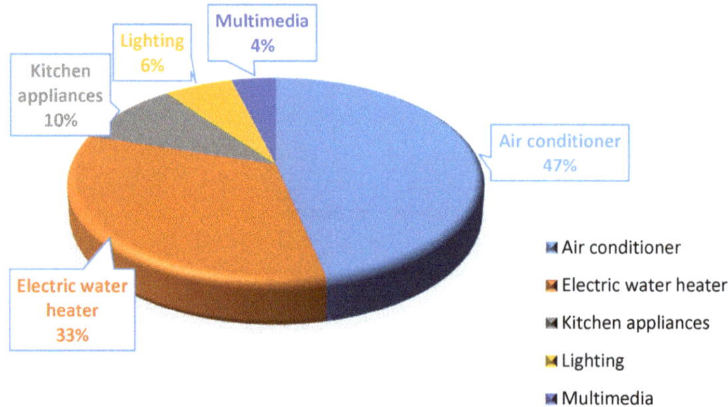

Figure 3. Household electricity composition analysis.

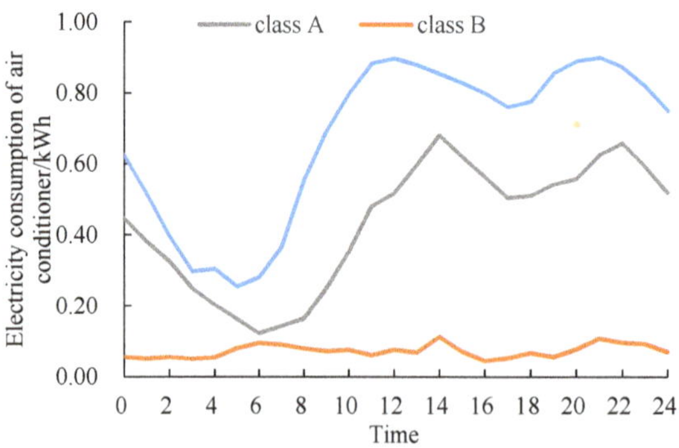

Figure 4. Air conditioning load curve of different classes.

Table 1. Daily load rate.

Type	Daily Load Rate	Percent
A	0.633253	56%
B	0.647609	11%
C	0.74984	33%

Among the 40 users in the smart community, the daily load curves of class A and class C users before and after participating in DR are shown in Figure 5. Both load peaks are reduced to a certain extent, the peak load decreases from 76.38 to 69.01 kW.

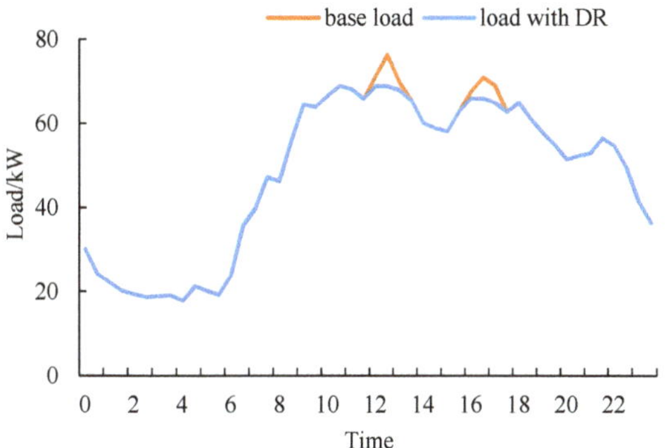

Figure 5. Comparison of load profile between base loads and loads with demand response (DR) taken into account.

6.2. Case 2

6.2.1. Simulation Settings

A modified RBTS-BUS 6 system is employed as the test system, in which the maximum power provided by the feeder F4 is 12 MW, and the transfer capacity of the tie line is 6 MW. As shown in Figure 6, the network frame includes 30 lines, 23 load points, 4 circuit breakers, and 21 load switches. The feeder length and load point data can be found in [32]. The circuit breakers, load switches, and fuses are assumed to be 100% reliable. The fault isolation time is 0.5 h, and the sum of fault isolation and load transfer time is 1 h. According to the current design requirements of distribution network capacity in residential areas, the capacity of each household is 8 kW, and the simultaneous factor is set at 0.5. The maximum power provided by the feeder F4 is 12 MW; therefore, there are about 3000 households in this area. The average EV inventory per household is set to 0.1, and all participate in DR. The air conditioning inventory per household is set to 2 and the proportion of air conditioning participating in DR is 0.2 considering the user's demand for comfort. The charging power and battery capacity of EVs is 5 kW and 40 kW·h, respectively. The energy consumption of EVs for traveling 100 km is 18 kW·h, and the battery charging efficiency is set to 0.9. It is assumed that the access time, departure time, and SOC of EVs are normally distributed, and could be obtained by employing Monte Carlo sampling. The highest and lowest running frequency of inverter air conditioning is 100 Hz and 20 Hz, respectively. R and C are specified as 5.56 °C/kW and 0.18 (kW h)/°C, respectively. The air conditioning power coefficient and the refrigerating capacity coefficient can be found in [33]. It is assumed that air conditioning cools in summer and heats in winter, and the comfort air conditioning interval is set to [22 °C, 28 °C]. The daily maximum dispatchable number is limited to 2 during the peak load period. The load data is shown in Appendix A, Table A1, and it is assumed that the load at each load point varies proportionally with the implementation of DR. The real-time electricity price curve is displayed in Appendix A, Figure A1. AMPL/CPLEX, an efficient commercial solver, is adopted to solve the proposed DR model, then the daily load curve considering DR of household equipment is obtained. Subsequently, the annual load curve considering DR could be determined according to the annual–weekly load curve and the weekly–daily load curve. Afterwards, the annual load curve is clustered, and the Monte Carlo simulation method is employed to calculate the reliability index of the distribution network [34].

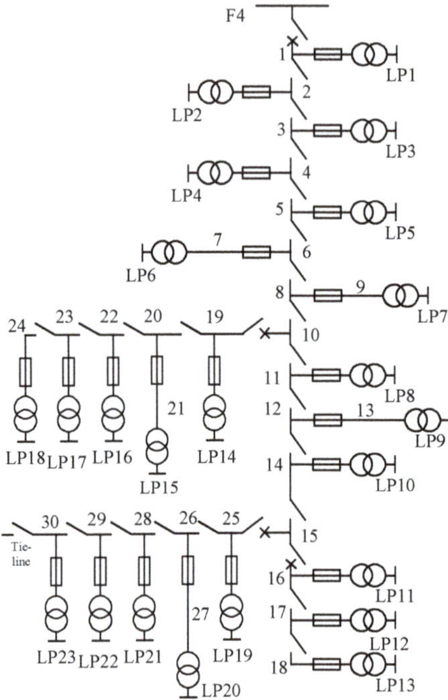

Figure 6. Topology of the test network.

6.2.2. Load Profile Considering DR

The daily load curve with and without EVs participating in DR is shown in Figure 7. It can be found that EV owners transfer the charging loads from the period of high electricity price to the period of low electricity price under the guidance of real-time electricity price, achieving peak load shaving and valley filling. The daily load curve with and without the air conditioning participating in DR is depicted in Figure 8. It can be found that the peak load is reduced and the load curve can be smoothed through the frequency reduction of air conditioning in the peak load period.

The daily load curve considering DR of EVs based on electricity price and air conditioning based on incentives is shown in Figure 9. It can be found that the load considering DR decreases during the peak period, and increases during the valley period compared to the base load, thus achieving peak load shaving and valley filling. Besides, the daily load peak without DR appears at 22:15, and the peak load is 12.27 MW. Considering the limitation of feeder capacity, part of the loads should be cut off. The daily load peak with DR appears at 22:30, and the maximum load is 11.57 MW, which satisfies the supply capacity constraints of feeders.

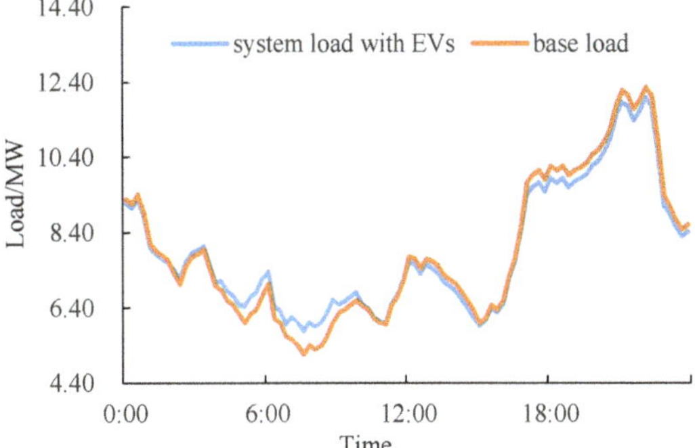

Figure 7. Comparison of load profile between base loads and loads with electric vehicles (EVs) taken into account

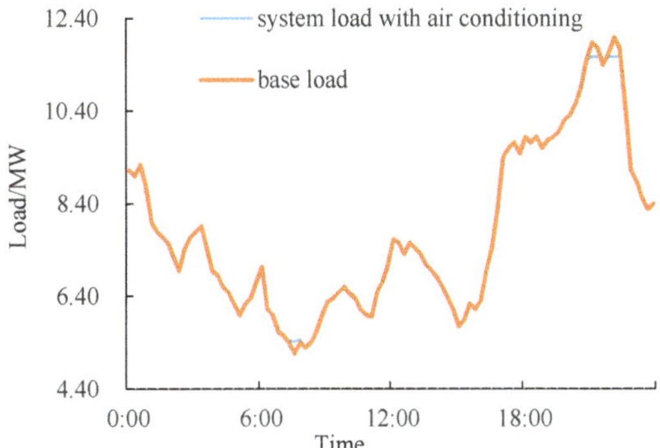

Figure 8. Comparison of load profile between base loads and loads with air conditioning taken into account.

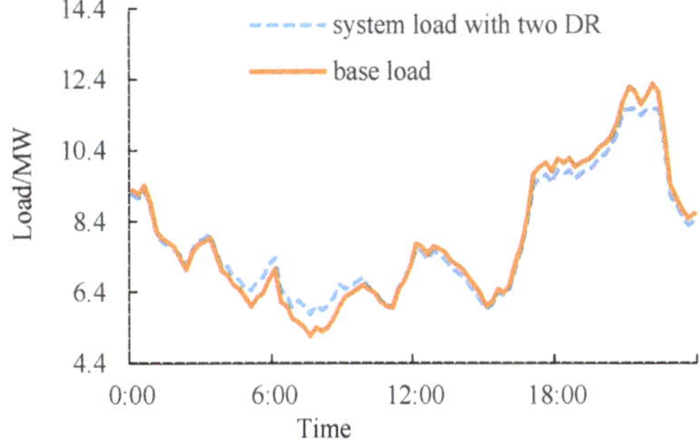

Figure 9. Comparison of load profile between base loads and loads with DR taken into account.

6.2.3. Influence of Real-Time Electricity Price on DR

The real-time electricity price, which has a certain randomness, is formulated according to the day-ahead load prediction, and plays the role of guiding electricity consumption. Considering the load uncertainty, a certain deviation between the actual electricity load and the predicted load may occur. Given this situation, applying the real-time electricity price shown in Appendix A, Figure A2, the daily load curve with and without EVs participating in the DR is shown in Figure 10. It can be found that the daily load peak appears at 21:00, because the real-time electricity in this period is relatively lower than others. Also, the daily load curve during the period of 17:00 to 18:45 is reduced, which is not the actual peak period needing to be clipped. Therefore, DR based on electricity price has a certain randomness and is greatly affected by real-time electricity price. Unreasonable real-time electricity price may lead to "peak-on-peak" or unnecessary load reduction.

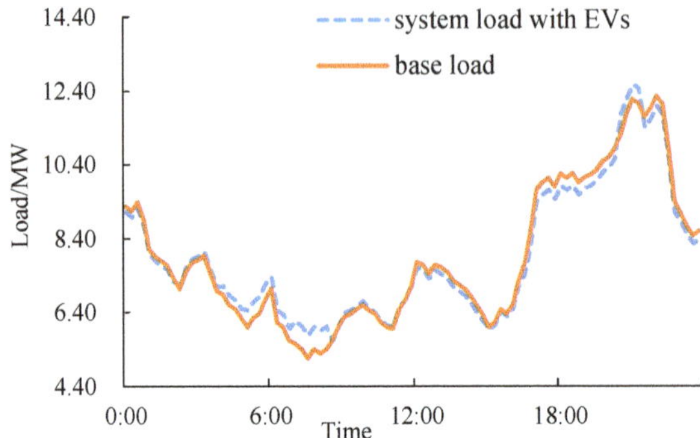

Figure 10. Comparison of load profile between base loads and loads with EVs taken into account.

6.2.4. Reliability Evaluation Considering DR

In this paper, four cases are conducted to evaluate the reliability of the distribution network considering DR. The annual load curve is composed of 8760 load points, which can be divided into 712 clusters, for there are 712 different points. The final system reliability evaluation indexes can be determined with the calculation results and weighted values of each cluster.

Case 1: DR is not taken into consideration.
Case 2: DR based on electricity price and incentive is considered.
Case 3: Case 1 with sufficient spare capacity of tie lines.
Case 4: Case 2 with sufficient spare capacity of tie lines.
The reliability index of the four cases is shown in Table 2.

Table 2. System reliability index (SAIFI, system average interruption frequency index; SAIDI, system average interruption duration index; CAIDI, customer average interruption duration index; ENS, energy not supplied).

Case	SAIFI/(times/a)	SAIDI/(h/a)	CAIDI/(h/times)	ENS/(MW h/a)
1	2.253	5.67	2.517	67.93
2	2.2052	5.2483	2.3798	59.31
3	2.2052	3.519	1.596	43.22
4	2.2052	3.519	1.596	40.76

By comparing the reliability indexes of cases 1 and 2, it can be concluded that DR can effectively improve the system reliability indexes. With DR, the peak load decreases 5.7%, the indexes of SAIDI, CAIDI, and ENS decrease 7.4%, 5.5%, and 12.7%, respectively. The main reason is that DR based on electricity price can transfer the loads from peak period to valley period to reduce the peak load of the system, and DR based on incentive can directly reduce part of the loads at the peak period, thus reducing the probability of power supply shortage of the distribution system. On the other hand, if a fault occurs during the peak period, the reduced loads considering DR can increase the transfer probability of the loads after fault isolation, thus reducing the average interruption frequency and interruption duration at the load point. Therefore the SAIFI, SAIDI, and ENS indexes are improved. By comparing cases 3 and 4, it can be found that when the capacity of the tie line is infinite, DR has no impact on the SAIFI and SAIDI indexes, and only the ENS index is improved due to the reduced peak load.

7. Conclusions

In this paper, two kinds of typical household electrical equipment are considered to participate in DR. DR models based on electricity price and incentive are proposed to obtain load curves with different DR mechanisms taken into account. Considering the transmission capacity limit of tie lines, a reliability evaluation method is proposed considering power supply capacity shortage. The following conclusions can be drawn from the simulation results:

(1) DR can improve the load curve and reduce the peak loads. DR based on electricity price has a certain randomness and is greatly affected by real-time electricity price. Unreasonable real-time electricity price may lead to "peak-on-peak" or unnecessary load reduction.

(2) Both DR based on electricity price and DR based on incentive can improve the reliability index of distribution networks. Compared with the reliability results attained by employing a single DR strategy, comprehensive DRs can improve reliability

(3) DR has no influence on the distribution network reliability index if transmission capacity of tie lines is assumed to be infinite. Under the premise of considering the tie line capacity limit, DR reduces the peak loads of the systems and decreases the probability of insufficient power supply capacity in normal operation, and meanwhile increases the possibility of the load point being transferred when a failure occurs.

Author Contributions: All the authors contributed to this work. H.C., J.T. and Y.M. conceived and structured the study, X.W. and J.Z. prepared the preliminary manuscript, L.S. reviewed and finalized the manuscript.

Funding: This research was funded by State Grid Suzhou Power Supply Company.

Acknowledgments: This work was supported by the Comprehensive Demonstration Project of Distribution Internet of Things in Suzhou Historic District.

Conflicts of Interest: The authors declare no conflicts of interest.

Appendix A

Table A1. Load data.

Load Number (Class)	Average Load (kW)	Load Number (Class)	Average Load (kW)	Load Number (Class)	Average Load (kW)	Load Number (Class)	Average Load (kW)
LP1(II)	165.9	LP7(I)	210.1	LP13(III)	250.1	LP19(III)	155.4
LP2(III)	180.8	LP8(III)	155.4	LP14(III)	155.4	LP20(II)	186.1
LP3(III)	250.1	LP9(I)	283.1	LP15(II)	186.1	LP21(I)	283.1
LP4(III)	243.1	LP10(II)	158.5	LP16(II)	158.5	LP22(II)	158.5
LP5(I)	207.0	LP11(III)	155.4	LP17(III)	250.1	LP23(I)	210.1
LP6(II)	165.9	LP12(II)	158.5	LP18(III)	243.1		

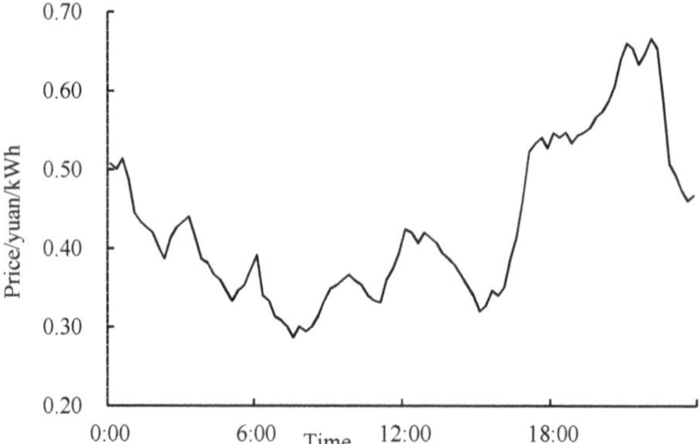

Figure A1. Real-time electricity price curve.

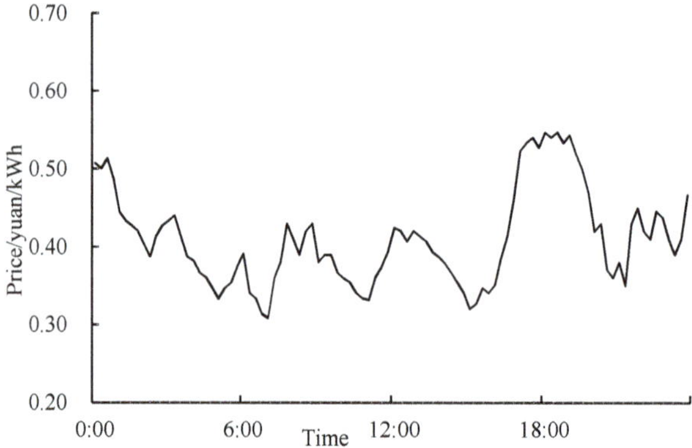

Figure A2. Real-time electricity price curve.

References

1. Eissa, M.M. Developing incentive demand response with commercial energy management system (CEMS) based on diffusion model, smart meters and new communication protocol. *Appl. Energy* **2019**, *236*, 273–292. [CrossRef]
2. Vahedipour-Dahraie, M.; Najafi, H.; Anvari-Moghaddam, A.; Guerrero, J. Study of the effect of time-based rate demand response programs on stochastic day-ahead energy and reserve scheduling in islanded residential microgrids. *Appl. Sci.* **2017**, *7*, 378. [CrossRef]
3. Pinson, P.; Madsen, H. Benefits and challenges of electrical demand response: A critical review. *Renew. Sustain. Energy Rev.* **2014**, *39*, 686–699.
4. Wang, F.; Zhou, L.; Ren, H.; Liu, X.; Talari, S.; Shafie-khah, M.; Catalão, J.P. Multi-objective optimization model of source-load-storage synergetic dispatch for building energy system based on TOU price demand response. *IEEE Trans. Ind. Appl.* **2018**, *54*, 1017–1028. [CrossRef]
5. Sharifi, R.; Anvari-Moghaddam, A.; Fathi, S.H.; Guerrero, J.M.; Vahidinasab, V. Economic demand response model in liberalised electricity markets with respect to flexibility of consumers. *IET Gener. Transm. Distrib.* **2017**, *11*, 4291–4298. [CrossRef]
6. Deng, R.; Yang, Z.; Chow, M.Y.; Chen, J.M. A survey on demand response in smart grids: Mathematical models and approaches. *IEEE Trans. Ind. Inform.* **2015**, *11*, 570–582. [CrossRef]
7. Yu, M.; Hong, S.H.; Ding, Y.; Ye, H. An incentive-based demand response (DR) model considering composited DR resources. *IEEE Trans. Ind. Electro.* **2018**, *66*, 1488–1498. [CrossRef]

8. Zhou, B.R.; Huang, Y.C.; Zhang, Y.J. Reliability analysis on microgrid considering incentive demand response. *Autom. Electr. Power Syst.* **2017**, *41*, 70–78.

9. Brahman, F.; Honarmand, M.; Jadid, S. Optimal electrical and thermal energy management of a residential energy hub, integrating demand response and energy storage system. *Energy Build.* **2015**, *90*, 65–75. [CrossRef]

10. Sharifi, R.; Moghaddam, A.A.; Fathi, S.H.; Vahidinasab, V. A flexible responsive load economic model for industrial demands. *Processes* **2019**, *7*, 147. [CrossRef]

11. Sharifi, R.; Fathi, S.H.; Vahidinasab, V. A review on demand-side tools in electricity market. *Renew. Sustain. Energy Rev.* **2017**, *72*, 565–572. [CrossRef]

12. Aalami, H.A.; Moghaddam, M.P.; Yousefi, G.R. Modelling and prioritizing demand response programs in power markets. *Electr. Power Syst. Res.* **2010**, *80*, 426–435. [CrossRef]

13. Smith, R.; Meng, K.; Dong, Z.; Simpson, R. Demand response: A strategy to address residential air-conditioning peak load in Australia. *J. Mod. Power Syst. Clean Energy* **2013**, *1*, 223–230. [CrossRef]

14. Reynolds, S.P.; Creighton, T.E. Time-of-use rates for very large customers on the pacific gas and electric company system. *IEEE Trans. Power Appar. Syst.* **1980**, *99*, 147–151. [CrossRef]

15. Kai, M.; Hu, G.; Spanos, C.J. A cooperative demand response scheme using punishment mechanism and application to industrial refrigerated warehouses. *IEEE Trans. Ind. Inform.* **2015**, *11*, 1520–1531.

16. Chen, C.; Kishore, S.; Wang, Z.; Alizadeh, M.; Scaglione, A. How will Demand Response aggregators affect electricity markets ?—A Cournot game analysis. In Proceedings of the 5th International Symposium on Communications Control and Signal Processing (ISCCSP), Roma, Italy, 2–4 May 2012; pp. 1–6.

17. Zeng, B.; Jiang, W.Q.; Yang, Z.; Li, G. Optimal load dispatching method based on chance-constrained programming for household electrical equipment. *Autom. Electr. Power Syst.* **2018**, *38*, 27–33.

18. Chen, Z.; Li, Y.Q.; Leng, Z.Y.; Lu, G.X. Refined modeling and energy management strategy of typical household high-power loads. *Autom. Electr. Power Syst.* **2018**, *42*, 135–143.

19. Sharifi, R.; Anvari-Moghaddam, A.; Fathi, S.H.; Guerrero, J.M.; Vahidinasab, V. An optimal market-oriented demand response model for price-responsive residential consumers. *Energy Eff.* **2019**, *12*, 803–815. [CrossRef]

20. Tushar, M.H.K.; Assi, C.; Maier, M.; Uddin, M.F. Smart microgrids: Optimal joint scheduling for electric vehicles and home appliances. *IEEE Trans. Smart Grid* **2014**, *5*, 239–250. [CrossRef]

21. Khalid, A.; Javaid, N.; Guizani, M.; Alhussein, M.; Aurangzeb, K.; Ilahi, M. Towards dynamic coordination among home appliances using multi-objective energy optimization for demand side management in smart buildings. *IEEE Access* **2018**, *6*, 19509–19529. [CrossRef]

22. Mohajeryami, S.; Moghaddam, I.N.; Doostan, M.; Vatani, B.; Schwarz, P. A novel economic model for price-based demand response. *Electr. Power Syst. Res.* **2016**, *135*, 1–9. [CrossRef]

23. Mnatsakanyan, A.; Kennedy, S.W. A novel demand response model with an application for a virtual power plant. *IEEE Trans. Smart Grid* **2015**, *6*, 230–237. [CrossRef]

24. Zhao, H.S.; Wang, Y.Y.; Chen, S. Impact of DR on distribution system reliability. *Autom. Electr. Power Syst.* **2015**, *39*, 49–55.

25. Zhang, Y.B.; Ren, S.J.; Yang, X.D.; Bao, K.K.; Xie, L.Y.; Qi, J. Optimal configuration considering price-based demand response for stand-alone microgrid. *Electr. Power Autom. Equip.* **2017**, *37*, 55–62.

26. Qi, X.J.; Cheng, Q.; Wu, H.B.; Yang, S.H.; Li, Z.X. Impact of incentive-based DR on operational reliability of distribution network. *Trans. China Electrotech. Soc.* **2018**, *33*, 5319–5326.

27. Sharifi, R.; Fathi, S.H.; Vahidinasab, V. Customer baseline load models for residential sector in a smart-grid environment. *Energy Rep.* **2016**, *2*, 74–81. [CrossRef]

28. Niu, W.J.; Li, Y.; Wang, B.B. Demand Response Based Virtual Power Plant Modeling Considering Uncertainty. *Proc. CSEE* **2014**, *34*, 3630–3637.

29. Lei, M.; Wei, W.; Zeng, J.H.; Mo, S.Y. Effect of load control on power supply reliability considering demand response. *Autom. Electr. Power Syst.* **2018**, *42*, 53–59.

30. Yang, J.R.; Shi, K.; Cui, X.Q.; Gao, C.W.; Cui, G.Y.; Yang, J.L. Peak Load Reduction Method of Inverter Air-conditioning Group Under DR. *Autom. Electr. Power Syst.* **2018**, *42*, 44–56.

31. Kwac, J.; Flora, J.; Rajagopal, R. Household energy consumption segmentation using hourly data. *IEEE Trans. Smart Grid* **2014**, *5*, 420–430. [CrossRef]

32. Billinton, R.; Jonnavithula, S. A test system for teaching overall power system reliability assessment. *IEEE Trans. Power Syst.* **1996**, *11*, 1670–1676. [CrossRef]

33. Cao, X.L.; Yu, S.X.; Li, X.L.; Wang, W.; Liao, S.M. Theoretic and experimental study on domestic air-conditioner with R410A as refrigerant. *J. Cent. South Univ. (Sci. Technol.)* **2010**, *41*, 759–763.
34. Transmission and Distribution Committee. *IEEE Guide for Electric Power Distribution Reliability Indices Redline*; IEEE: NY, USA, 2012.

Article

A Time-Sequence Simulation Method for Power Unit's Monthly Energy-Trade Scheduling with Multiple Energy Sources

Liang Sun [1,2], Qi Zhang [1], Na Zhang [3], Zhuoran Song [4], Xinglong Liu [5] and Weidong Li [1,*]

[1] School of Electrical Engineering, Dalian University of Technology, Dalian 116024, China; sunliang_3333@sina.com (L.S.); zhangqizq@mail.dlut.edu.cn (Q.Z.)
[2] State Grid Shenyang Electric Power Supply Company, Shenyang 110081, China
[3] State Grid Liaoning Economic Research Institute, Shenyang 110015, China; zhangnamoumou@126.com
[4] State Grid Liaoning Electric Power Company Limited, Shenyang 110006, China; zrsong_sg@126.com
[5] China Energy Engineering Group Liaoning Electric Power Survey & Design Institute Co. Ltd., Shenyang 110179, China; liuxinglong_01@126.com
* Correspondence: wdli@dlut.edu.cn; Tel.: +86-0411-8470-8923

Received: 11 September 2019; Accepted: 17 October 2019; Published: 21 October 2019

Abstract: The uncertainty of new energy output from wind power is rarely considered in the monthly energy-trade scheduling. This causes many problems since the new energy penetration level increases. The fairness of the scheduled energy for the power suppliers is difficult to guarantee. Because the actual power system operation is far away from scheduling when the monthly energy-trade schedule is carried out, unnecessary wind curtailment might occur, and even the feasibility of monthly energy-trade schedule might not be guaranteed. This affects the security and reliability of the power system operation. In this paper, a new time-sequence simulation method for the monthly energy-trade scheduling is proposed, which considers the new energy power forecasting characteristic and the computational load problem of hourly energy-trade simulation in the remaining months. The proposed method is based on a segment modelling strategy. The power generation in the scheduling month is optimized hourly, and the energy generation is optimized in the subsequent months on a monthly basis. For the scheduling month, accurate cost function is applied in the objective function, and detailed short-term operation constraints and the new energy forecasting results are considered, which can guarantee the feasibility of the new monthly energy-trade scheduling and lay a solid foundation for daily dispatching. For the subsequent months, since the load forecast accuracy is lower and no wind power forecasting results could be used, the rough cost function is applied, and only monthly constraints are considered. To ensure a balance in the execution progress of each power generating entity, the simulation time-scale is set as the remainder of the months in the study year. The new approach ensures the fairness of power execution progress and improves the new energy consumption level. A case study was used to verify the feasibility and effectiveness of the proposed method, which provides a theoretical reference for the monthly electrical energy-trade scheduling.

Keywords: monthly energy-trade scheduling; time-sequence simulation method; feasibility; fairness; consumption of renewable energy

---

1. Introduction

Among the generation scheduling processes in China, the annual contract energy planning is carried out according to the contract energy of the generating companies and the national scheduled energy. Then, the monthly energy-trade scheduling and daily generation dispatching are performed by shortening the time scale [1]. Monthly energy-trade scheduling is the intermediate link between the annual contract planning and daily dispatching. If the monthly energy-trade scheduling is improper, the annual contract energy may become difficult to complete. Therefore, the daily dispatching feasibility, operation economy, and security is impacted.

At present, most of the provincial power trading centers in China apply the average decomposition method for monthly energy-trade scheduling. In past decades, the average decomposition method was sufficient to solve the principal contradiction of the power balance. However, this method may be very simple and easy to be implement, but its predominant factors such as the operation condition are not comprehensive enough [2]. Moreover, since the end of the 20th century, with the increasing energy gap and deteriorating ecology, the other drawbacks of the average decomposition method, i.e., lack of energy conservation and emission reduction techniques, have become increasingly prominent. Realizing these drawbacks, some studies [3,4] optimized the formulation of monthly energy-trade scheduling with respect to energy-saving power generation. For instance, a study [3] proposed the combination scheme and calculation method for the non-heating thermal power units of monthly energy-trade scheduling with regard to coal consumption and pollutant emission. Another study [4] expanded the optimization space of energy savings and emission reduction benefits from the time dimension, by optimizing the monthly unit commitment and the current electrical energy distribution. Some improved methods of the conventional units based on the average decomposition method can be found in Reference [5–7]. In the monthly trade scheduling method proposed in Reference [5], the deviation of the load ratio was adjusted moderately, according to the comprehensive cost ranking results of the generation unit cost. A comprehensive consumption cost optimization method is proposed in Reference [6], which considers the impact of the monthly electric energy nonlinear fluctuation on the relational consumption cost. On the basis of the unit electric energy integrative cost diversity of different generation units, a new monthly energy-trade scheduling method is proposed in Reference [7], in which the unit electric energy integrative cost of the generation unit was weight-modified. The balance between the economic benefits, energy conservation, and emission reduction can be achieved using the method in Reference [7]. These research works considered the interests of generation units and power grid companies, as well as the social environment. As a result, the benefits of energy saving and emission reduction could be obtained, to some extent, in monthly energy-trade scheduling. However, these research works did not consider the environmental benefits of integrating renewable energy resources into the monthly power trade scheduling process.

In the existing research, the renewable energy units and the conventional units were mostly considered respectively and serially in the monthly energy-trade scheduling. In other words, the remaining load power was reserved for the traditional units in the monthly energy-trade scheduling once the renewable energy power generation is deducted, according to the total load power. This method is feasible when the penetration level of renewable energy is low. However, with the increasing penetration of renewable energy, generation scheduling of thermal power units is influenced by the high volatility, randomness, and intermittence of the renewable energy. A study [8] proposed a market clearing model for energy and reserve products, in coordinated power and gas networks with the integration of compressed air energy storage and wind energy sources (WES), based on two-stage stochastic network-constrained unit commitment. In Reference [9], a sustainable day-ahead scheduling of the grid-connected home-type micro-grids with the integration of distributed energy resources and responsive load demand was co-investigated, and an efficient energy management system optimization algorithm was studied. A novel control algorithm for joint demand response management and thermal comfort optimization in micro-grids equipped with renewable energy sources and energy storage units was presented in Reference [10]. In Reference [11], a simulation-based optimization approach for the design of an EMS in grid-connected photovoltaic-equipped micro-grids with a heterogeneous occupancy schedule was presented. The research studies above could effectively cope with the energy-optimizing problems with large-scale integration of renewable energy into the grid. However, these research studies were all for the micro-grids, which were very different from the regional power grid studied in this paper, and the optimizing time scale is also different from this paper. For the monthly energy-trade scheduling problem with the large-scale integration of renewable energy studied in this paper, the risk of the thermal power units not completing the annual contract energy planning is much higher. The fairness requirement of the deviation from the annual base power completion rate is also difficult to meet. At the same time, because the previous methods of monthly energy-trade scheduling focus on the economics of electricity decomposition, and do not consider the difficulty in implementing the subsequent daily dispatching, it might result in unnecessary water and wind curtailment when the clean energy output fluctuation is large. What is more, the operation feasibility might be difficult to guarantee, which may affect the security and reliability of the operation [12–15]. Therefore, with the large-scale integration of renewable energy into the grid, the following new formulating principles of monthly energy-trade scheduling should be followed.

- The feasibility of power generation scheduling should be ensured [16].
- Impartial and open dispatching requirements should be met [17].
- The benefits of energy conservation and emission reduction should be considered, which means priority should be given to renewable energy units [18–20].

As mentioned above, existing methods such as the average decomposition method, the deviation of load ratio method, etc. cannot meet all the above principles. Therefore, a new method tailored around three principles is needed. The time-sequence simulation method is an effective method to solve optimizing problems with multiple variables. With this method, the various complex constraints, including fairness and operation, can be comprehensively considered and higher resolutions can be achieved. In recent years, the time-sequence simulation method has been widely used in the power system area, such as in the modelling of the annual wind power trade scheduling and the low-carbon benefit evaluation [21,22]. A previous study [23] applied the time-sequence simulation method to the daily dispatching and ultra-short-term scheduling including wind power. Another study [24] optimized the monthly unit commitment model, including wind turbines with the time-sequence simulation method. The literature [25] attempted to apply the time-sequence simulation method to model monthly energy-trade scheduling. However, Reference [25] focused on optimizing the distribution of electricity during the scheduling month, whereas the economics and fairness of electricity distribution in other months of the year were not considered.

Based on the above research, this paper proposes a new time-sequence simulation method for monthly energy-trade scheduling, based on the segment-formulating strategy. The simulation time-scale is set as the remaining months of the year, so as to ensure a balance in the execution progress of each power generation entity. Optimizing convergence and efficiency can be ensured, by applying the strategy of optimizing the generation power hourly in the scheduling month and optimizing the generation energy monthly in subsequent months. The proposed method considers the power generation characteristics of various energy units and improve the consumption capability of renewable energy. Thus, it enhances the efficiency of energy conservation and emission reduction of the power grid by considering several operational constraints in the scheduling month (the following month). It can maintain the balance of the annual base electrical energy completion rate of each thermal power unit more efficiently. Thus, the fairness of execution of the generation schedule is realized. The comprehensive consideration of the system operation constraints significantly improves the feasibility of the generation schedule and provides a good foundation for subsequent daily dispatching.

2. Modelling Concept and Method

As stated previously, methods such as average decomposition and the deviation of the load ratio cannot effectively solve the problem of formulating the monthly energy-trade scheduling, when the renewable energy is absorbed by the power grid on a large scale. However, the time-sequence simulation method is a dynamic optimization method simulated by setting up an objective function and the relevant constraints. It is characterized by both long-time and short-time scales. In the long-time scale, the optimal values at each time point can be accumulated and the sum can be macroscopically restricted by the fairness constraints. In the short time scale, the optimal solutions can be obtained considering the detailed operation constraints at each time interval. The time-sequence simulation method matches with the modelling demand of monthly energy-trade scheduling. Therefore, the time-sequence simulation method is very suitable for the modelling of the monthly energy-trade scheduling.

To improve the operation economy and utilization rate of renewable energy, the objective function of the model is to minimize the combined costs including the coal consumption cost, peaking cost, start-up and shutdown cost, hydropower abandoning cost, wind power curtailment cost, and nuclear power peak shaving cost in the remaining months. To better consider multiple factors during the actual power operation, the annual constraints, monthly constraints, and short-term constraints are all considered. In the annual constraints, the energy completion rate deviation constraints are introduced to meet the power generation fairness requirement. The monthly power energy balance constraints for each generation unit are considered in the monthly constraints, to ensure reasonable allocation of the monthly scheduling energy. Operation constraints including the generator characteristic constraints, power balance constraints, and spinning reserve constraints are considered in the hourly short-term constraints, to reduce the deviation between monthly scheduling and daily dispatching, and improve the schedule feasibility. According to the typical time-sequence simulation method, the monthly energy-trade scheduling problem is modelled and simulated on an hourly basis, from the beginning of the decision point to the end of the whole year. The detailed process is shown in Figure 1.

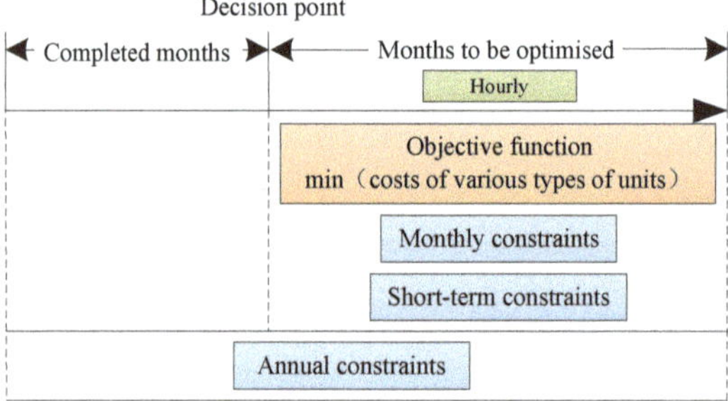

Figure 1. The monthly energy-trade scheduling model according to the typical time-sequence simulation.

However, some problems are faced when using the typical time-sequence simulation method mentioned above in the modelling of monthly energy-trade scheduling.

- During the optimizing months, the method is optimized on an hourly basis. A system operation with up to 8760 time intervals must be simulated. For large-scale power systems with hundreds of generation units, such a demand presents a massive scale optimization problem, which is difficult to solve.
- Considering the low accuracy of long-term forecasting of wind power, water inflow, and the load, the simulation results in the months farther from the decision point might be vastly different from the actual operation, which may increase the wind and water curtailment level.

Therefore, it is neither necessary nor feasible to simulate the system operation in the whole period on an hourly basis. Some works in the literature proposed preliminary solutions to this problem. For example, Reference [26] proposed a segmented idea to simplify the calculation, but adopted the method of average segmentation throughout the process, which is too simplistic and does not conform to the physical characteristics of the accuracy of the predicted value decay with time. In another example, Reference [27] divided each day into three sections—peak, flat, and valley—which did not meet the law of change in prediction. Based on these ideas, a segment modelling strategy for the monthly energy-trade scheduling is presented. The system simulation is former-accurate and after-rough. The remaining months are decomposed into the scheduling month and the subsequent months. The scheduling month is the next month after the decision point, and the subsequent months include the months after the scheduling month to the end of the year. For the scheduling month, since the load forecast accuracy is higher and wind power forecasting data could be used, the system operation is still simulated hourly, and, thus, the cost function (defined as the accurate cost function), the monthly constraints, and the short-term constraints are the same as in the typical method introduced above. In addition, for the subsequent months, since the load forecast accuracy is lower and no wind power forecasting results could be used, the system operation is simulated monthly, and the cost function (defined as the rough cost function) only contains the coal consumption cost of thermal power units, to ensure the economy of the model, and only the monthly constraints are considered. The short-term constraints, which require the prediction data with high accuracy and large-scale calculation, are abandoned in the subsequent months. The electric energy completion situation in the completed month involves the formulation for the next month as an input. Additionally, it is constrained by the annual constraints with the electrical energy of the scheduling month and the subsequent months. The process is shown in Figure 2.

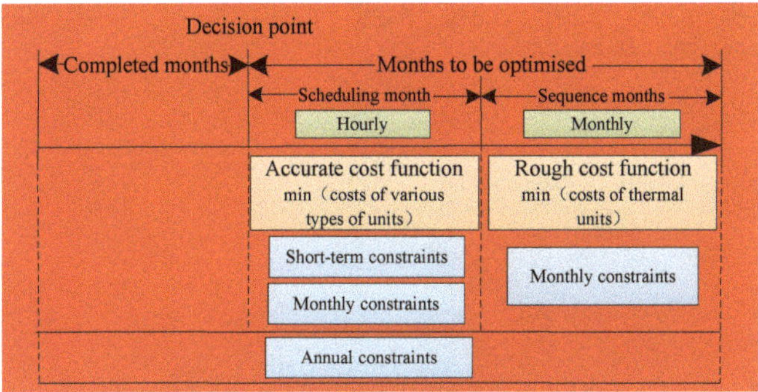

Figure 2. The monthly energy-trade scheduling model according to the improved time series simulation.

In the actual application, the monthly electrical energy-trade scheduling is rolling when scheduled from the start of the year to November of that year. During each decision period, the completed energy of each generation unit is applied to correct the energy target of the remaining months.

3. Mathematical Model

3.1. Objective Function

The objective function of the monthly energy-trade scheduling model is to minimize the sum of total costs for the remaining months. As mentioned above, to improve the feasibility of the model and the solving efficiency, the accurate cost function F_1 is used in the scheduling month, and the rough cost function F_2 is used in subsequent months.

$$minF = min(F_1 + F_2) \tag{1}$$

3.1.1. Accurate Cost Function for the Scheduling Month

The accurate cost function for the scheduling month is the total costs of all the different types of units in the scheduling month, which consist of the operational costs of the condensing thermal power units and operational cost, the steam extraction thermal power units, the curtailment cost of the wind power units, the abandoning cost of the hydropower units, and the peak shaving cost of the nuclear power units.

$$F_1 = \sum_{t=1}^{T} \left\{ \sum_{i=1}^{N_{CN}} f_{CN}(i,t) + \sum_{k=1}^{N_{CQ}} f_{CQ}(k,t) + \sum_{l=1}^{N_w} f_w(l,t) + \sum_{s=1}^{N_m} f_m(s,t) + \sum_{v=1}^{N_n} f_n(v,t) \right\} \tag{2}$$

The nomenclature of variables can be found in Appendix A.

The operation costs of the condensing thermal power units are composed of the coal consumption cost, deep-peak-regulation cost, and start-up and shut-down costs.

$$f_{CN}(i,t) = a_i (P^t_{CN_i})^2 + b_i (P^t_{CN_i}) + c_i + M \times \Delta P^t_{CN_i} + S_{CN_i} u^t_i (1 - u^t_i) \tag{3}$$

The operation costs of the steam extraction thermal power units include the coal consumption cost, deep peak-regulation cost, and the start-up and shut-down costs.

$$\begin{aligned} f_{CQ}(k,t) &= a_k (P^t_{CQ_k} + C_{v1} \times H^t_{CQ_k})^2 + b_k (P^t_{CQ_k} + C_{v1} \times H^t_{CQ_k}) + c_k \\ &\quad + M \times \Delta P^t_{CQ_k} + S_{CQ_k} u^t_k (1 - u^t_k) \end{aligned} \tag{4}$$

The wind power curtailment cost is:

$$f_w(l,t) = C_w (P^t_{w_l} - \tilde{P}^t_{w_l}) \tag{5}$$

The hydropower abandoning cost is:

$$f_m(s,t) = C_m P^t_{m,g} \tag{6}$$

The nuclear power peak-regulation cost is:

$$f_n(v,t) = C_n (\bar{P}_{n_v} - P^t_{n_v}) \tag{7}$$

3.1.2. Rough Cost Function for Subsequent Months

The rough cost function is reduced to the sum costs of thermal units in the sequence months, including the condensing thermal power units and steam extraction thermal power units.

$$F_2 = \sum_{e=m+1}^{12} \left(\sum_{i=1}^{N_{CN}} W^e_{CN_i,RE} \times C_{CN_i} + \sum_{k=1}^{N_{CQ}} W^e_{CQ_k,RE} \times C_{CQ_k} \right) \tag{8}$$

3.2. Constraint Conditions

3.2.1. Short-Term Constraints

To ensure the feasibility of the monthly energy-trade scheduling, the short-term constraints limit the operation status of the units in each time interval of the scheduling month. The operation constraints of various units, and the power balance constraint and spinning reserve constraints are considered.

The short-term constraints of the condensing thermal power units include the maximum and minimum output constraints, ramp rate constraints, and minimum start-off time constraints. The constraints considered are as follows.

$$\begin{cases} u_i^t P_{CN_i,min} \leq P_{CN_i}^t \leq u_i^t P_{CN_i,max} \\ -\Delta P_{CN_i,down} \leq P_{CN_i}^t - P_{CN_i}^{t-1} \leq \Delta P_{CN_i,up} \\ (u_i^t - u_i^{t-1})(X_{CN_i,on}^{t-1} - T_{CN_i,on}) \leq 0 \\ (u_i^{t-1} - u_i^t)(X_{CN_i,off}^{t-1} - T_{CN_i,off}) \leq 0 \end{cases} \tag{9}$$

The short-term constraints of the steam extraction thermal power units include the maximum and minimum output constraints, ramp rate constraints, minimum start-off time constraints, thermoelectric relationship constraint, reserve capacity constraint, and thermal balance constraint. The constraints considered are below.

$$\begin{cases} u_k^t P_{CQ_k,min} \leq P_{CQ_k}^t \leq u_k^t P_{CQ_k,max} \\ -\Delta P_{CQ_k,down} \leq P_{CQ_k}^t - P_{CQ_k}^{t-1} \leq \Delta P_{CQ_k,up} \\ (u_k^{t-1} - u_k^t)(X_{CQ_k,off}^{t-1} - T_{CQ_k,off}) \leq 0 \\ (u_k^t - u_k^{t-1})(X_{CQ_k,on}^{t-1} - T_{CQ_k,on}) \leq 0 \\ \max(C_{m_k} \times H_{CQ_k}^t + K_k, P_{CQ_k,min} - C_{v2_k} \times H_{CQ_k}^t) \leq P_{CQ_k}^t \leq P_{CQ_k,max} - C_{v1_k} \times H_{CQ_k}^t \\ 0 \leq H_{CQ_k}^t \leq H_{CQ_k,max} \\ \sum_{k=1}^{N_{CQ}} H_{CQ_k}^t = H_L^t \end{cases} \tag{10}$$

In the short-term constraints of wind turbines, the wind power output should be equal to or less than the forecasted wind power. The constraints considered are below.

$$0 \leq P_{w_l}^t \leq \overline{P}_{w_l}^t \tag{11}$$

The short-term constraints of the hydropower units are composed of the upper and lower limits of the reserve capacity, the upper limit constraint of the water flow for generating power, and the maximum output constraint. The constraints considered are below.

$$\begin{cases} V_m^{t+1} = V_m^t + f_m^t - \sum_{s=1}^{N_m} q_{m_s}^t - g_m^t \\ V_m^{min} \leq V_{m_s}^t \leq V_m^{max} \\ 0 \leq q_{m_s}^t \leq q_{m_s,max} \\ P_{m_s}^t = a q_{m_s}^t h \\ 0 \leq P_{m_s}^t \leq \min\{P_{m_s,max}, a q_{m_s}^t h_{m_s}^t\} \end{cases} \tag{12}$$

The nuclear power units perform the 15-1-7-1 power generation mode, which means that the nuclear power units maintain the state of full power generation for 15 h, and then decrease to the state of low power generation within 1 h. Then the nuclear power units maintain the low power generation state for 7 h, and return to the full power generation state in 1 h. The constraints considered are below.

$$\begin{cases} u_{n_v,y_1}^t + u_{n_v,y_2}^t = 1 \\ P_{n_v,y_1}^t = P_{n_v,max} \times u_{n_v,y_1}^t \\ P_{n_v,y_2}^t = \alpha P_{n_v max} \times u_{n_v,y_2}^t \\ P_{n_v}^t = \max\{P_{n_v,y_1}^t, P_{n_v,y_2}^t\} \end{cases} \tag{13}$$

The power balance constraint of the system is:

$$\sum_{i=1}^{N_{CN}} P_{CN_i}^t + \sum_{k=1}^{N_{CQ}} P_{CQ_k}^t + \sum_{l=1}^{N_w} P_{w_l}^t + \sum_{s=1}^{N_m} P_{m_s}^t + \sum_{v=1}^{N_n} P_{n_v}^t = P_L^t \tag{14}$$

The up and down spinning reserve constraints of the system are:

$$\sum_{i=1}^{N_{CN}} P^t_{CN_i,\max} + \sum_{k=1}^{N_{CQ}} P^t_{CQ_k,\max} + \sum_{s=1}^{N_m} P^t_{m_s,\max} + \sum_{v=1}^{N_n} P^t_{n_v,\max} \geq 1.05 P^t_L - 1.2 P^t_w + 0.5 \beta P^t_w \tag{15}$$

$$\sum_{s=1}^{N_m} P^t_{m_s,\min} + \sum_{v=1}^{N_n} P^t_{n_v,\min} \leq 0.95 P^t_L - 0.8 P^t_w - 0.5 \beta P^t_w \tag{16}$$

3.2.2. Monthly Constraints

The monthly constraints limit the monthly scheduled energy in the scheduling month and the subsequent months, to ensure that the energy allocated in each month can be used to reasonably formulate the daily dispatching. The constraints include the upper and lower limits of the monthly power generation energy of various units and the monthly electricity balance constraint.

The monthly operation rate of the units in the monthly constraints of the condensing thermal power units is defined as:

$$\frac{W^{year}_{CN_i} - W^{ywc}_{CN_i}}{T_{wwc} \times P_{CN_i,\max} \times \eta} \leq \delta^e_{CN_i} \leq \min\left(1, \frac{W^{year}_{CN_i} - W^{ywc}_{CN_i}}{T_{wwc} \times P_{CN_i,\min}}\right) \tag{17}$$

Then, the upper and lower limit constraints of electrical energy of the condensing power units are as follows:

$$P_{CN_i,\min} \times t^e \times \delta^e_{CN_i} \leq W^e_{CN_i,RE} \leq P_{CN_i,\max} \times \eta \times t^e \times \delta^e_{CN_i} \tag{18}$$

The electrical energy balance constraints of the steam extraction thermal power units are:

$$\begin{cases} \dfrac{W^{year}_{CQ_k} - W^{ywc}_{CQ_k}}{T_{wwc} \times P_{CQ_k,\max} \times \eta} \leq \delta^e_{CQ_k} \leq \min\left(1, \dfrac{W^{year}_{CQ_k} - W^{ywc}_{CQ_k}}{T_{wwc} \times P_{CQ_k,\min}}\right) \\ P_{CQ_k,\min} \times t^e \times \delta^e_{CQ_k} \leq W^e_{CQ_k,RE} \leq P_{CQ_k,\max} \times \eta \times t^e \times \delta^e_{CQ_k} \end{cases} \tag{19}$$

The monthly electrical energy constraints of the wind turbine, hydropower unit, and nuclear power unit are:

$$0 \leq W^e_{w_l,RE} \leq W^e_{w_l,F} \tag{20}$$

$$0 \leq W^e_{m_s,RE} \leq W^e_{m_s,F} \tag{21}$$

$$0 \leq W^e_{n_v,RE} \leq W^e_{n_v,F} \tag{22}$$

The balance constraint of the monthly electrical energy is:

$$\sum_{i=1}^{N_{CN}} W^e_{CN_i,RE} + \sum_{k=1}^{N_{CQ}} W^e_{CQ_k,RE} + \sum_{l=1}^{N_w} W^e_{w_l,RE} + \sum_{m=1}^{N_m} W^e_{m_l,RE} + \sum_{v=1}^{N_n} W^e_{n_v,RE} = W^e_{L,RE} \tag{23}$$

3.2.3. Annual Constraints

The energy generated in the completed months, the scheduled energy in the scheduling month, and the scheduled energy in the subsequent months constitute the total annual energy of each thermal power unit.

$$W_{T_i,A} = \sum_{e=1}^{m-1} W^e_{T_i,H} + \sum_{t=1}^{T} P^t_{T_i} + \sum_{e=m+1}^{12} W^e_{T_i,RE}, \tag{24}$$

where the electrical energy of the completed month is the input, and the amount of electricity in the scheduling month and the subsequent months is constrained by the annual fairness constraints as the optimization value. The deviation of electrical energy in the completed month can be corrected when formulating the monthly energy-trade scheduling.

The total annual energy of the thermal power units is composed of the annual base electrical energy and the annual transaction energy.

$$W_{T_i,A} = W_{T_i,B} + W_{T_i,D} \tag{25}$$

The annual transaction energy is composed of the tie-line energy, the generation right transfer trading energy, and the trading energy of large consumers.

$$W_{T_i,D} = W_{T_i,N} + W_{T_i,O} + W_{T_i,Y} \tag{26}$$

The completion rate of the annual base electrical energy of the total thermal power units, ρ_T is defined as the ratio of the optimized annual base electrical energy of all thermal power units and the expected annual base electrical energy of all thermal power units.

$$\rho_T = \frac{\sum\limits_{i=1}^{N_T} W_{T_i,B}}{\sum\limits_{i=1}^{N_T} W_{T_i,BF}} \tag{27}$$

According to the National Development and Reform Committee of China's guide on strengthening and improving the regulation of power generation operation, the deviation threshold of the completion rate of the annual base electrical energy of each thermal unit is defined as $\lambda\%$. The allocated fairness constraint for the annual base electrical energy of thermal power units is as follows:

$$(1 - \lambda\%)\rho_T \leq \frac{W_{T_i,B}}{W_{T_i,BF}} \leq (1 + \lambda\%)\rho_T \tag{28}$$

4. Simulation and Analysis

4.1. Simulation Condition

For the solving methods in this paper, since it is actually a large-scale mixed integer quadratic programming model, the Branch Bound (BB) methods, for instance, could be applied. In the case studies, we used the IBM CPLEX Business optimization software to solve the monthly energy-trade scheduling problem in the test system.

The test system consists of four condensing thermal power units, two steam-extraction thermal power units, one 600 MW hydropower station, one 250 MW wind farm, and one 60 MW nuclear power station. The simulation process of monthly energy-trade scheduling is shown in Figure 3.

April is assumed to be the scheduling month. The electrical energy and operation parameters and the operation cost coefficient of the thermal power units are shown in Tables 1–3, respectively. The thermoelectric relationship coefficient data of the thermoelectric units are shown in Table 4. The characteristic parameters of the hydropower station and the reservoir are shown in Tables 5 and 6, respectively. The monthly predicted hydropower electrical energy (PHEQ) is shown in Table 7. The characteristic parameters of the reservoir are shown in Table 6. The cost for deep-peak regulation is 500 ¥/MWh.

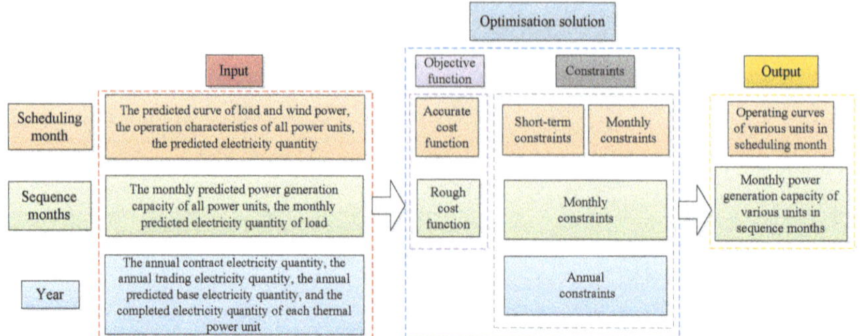

Figure 3. Simulation process of the monthly energy-trade scheduling.

Table 1. Electrical energy data of thermal power units.

Units	Annual Contract Electricity Quantity (MWh)	Annual Trading Electricity Quantity (MWh)	Annual Predicted Base Electricity Quantity (MWh)	The Completer Electricity Quantity in 1~3 Months (MWh)
1	3,139,000	2,511,200	627,800	876,740.1
2	1,569,500	1,255,600	313,900	452,014.1
3	1,307,900	1,046,320	261,580	322,918.3
4	784,800	627,840	156,960	197,038.9
5 (CHP)	1,689,800	1,351,840	337,960	441,868.7
6 (CHP)	1,109,100	887,280	221,820	325,208.2

Table 2. The operation parameters of thermal power units.

Units	Pmax (MW)	Pmin (MW)	Ton, min/Toff, min (h)	Pup (MW/h)	Pdown (MW/h)
1	600	280	8	168	168
2	350	140	5	80	80
3	250	100	5	80	80
4	150	70	6	42	42
5 (CHP)	323	150	6	90	90
6 (CHP)	212	100	6	60	60

Table 3. The operation cost coefficient of the thermal power units.

Units	S (¥/MWh)	A (¥/MW^{2}h)	B (¥/MWh)	C (¥/h)	Average Coal Consumption Cost (¥/h)
1	1,200,000	0.06	157.8	6300	203.0
2	650,000	0.048	112.8	13,440	174.6
3	500,000	0.045	130.8	8640	182.8
4	260,000	0.04	164.4	3240	195.8
5 (CHP)	600,000	0.046	163.0	11,293	218.5
6 (CHP)	500,000	0.103	162.3	6922	221.4

Table 4. The thermoelectric relationship coefficient of thermoelectric units.

Units	C_{v1}	C_{v2}	C_m	K
5 (CHP)	0.23	0.23	0.45	80.7
6 (CHP)	0.21	0.21	0.45	45.4

Table 5. The characteristic parameters of the hydropower station.

Unit	Pmax (MW)	qmax (m^3/s)	a	Annual Contract Electricity Quantity (MWh)
1	600	705.9	8.5	2,607,169

Table 6. The characteristic parameters of the reservoir.

Reservoir	Vmax (m^3)	Vmin (m^3)	V$_0$ (m^3)	h (m)
1	90.18×10^8	40.09×10^8	60×10^8	100

Table 7. The monthly predicted hydropower electrical energy.

Month	1	2	3	4	5	6
PHEQ (MWh)	44,847	81,109	102,817	138,739	174,379	300,159
Month	7	8	9	10	11	12
PHEQ (MWh)	359,829	429,932	350,262	307,388	161,367	156,341

Load and wind power prediction methods have always been research interests in recent decades. A lot of research results have been obtained and many methods have been proposed. The widely used prediction methods include the Depth Ridgelet Neural Network method, support vector machine, etc. This paper focuses on the formulation of monthly energy-trade scheduling. Before it is formulated, the load and wind power forecasting results have been obtained by using existing prediction methods. Therefore, the load and wind power prediction methods are not the research contents of this paper. In the case studies, the predicted wind power electrical energy (PWEE) is randomly generated based on the historical wind power data. As shown in Table 8, the error range of monthly wind electricity energy prediction in Reference [28,29] was adopted for the calculation. The actual wind generation energy in April was 61,654 MWh, which was 6.04% deviation from the predicted wind power electric energy. The actual wind power generation curve is shown in Figure 4.

Table 8. The monthly predicted wind-power electrical energy.

Month	1	2	3	4	5	6
PWEQ (MWh)	20,772	20,772	34,620	58,162	50,546	46,391
Month	7	8	9	10	11	12
PWEQ (MWh)	29,842	26,721	39,744	48,884	51,792	49,720

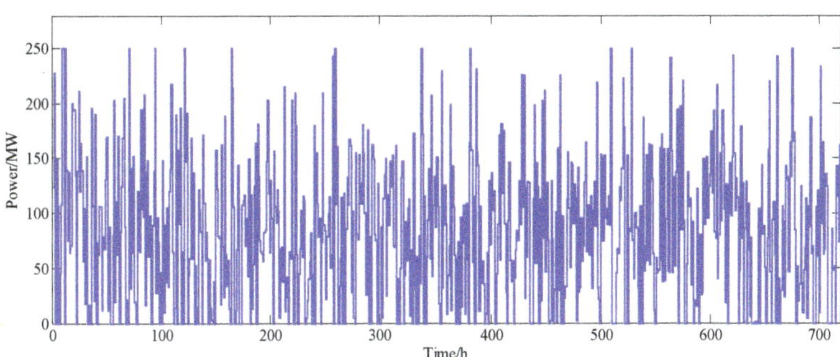

Figure 4. The actual wind power generation curve.

The annual load electrical energy is assumed to be 13.123323×10^6 MWh and the monthly load coefficients are shown in Table 9. The load curve in April (the scheduling month) is shown in Figure 5.

Table 9. The monthly load coefficients.

Month	1	2	3	4	5	6
Load coefficients	0.0949	0.0682	0.0702	0.0732	0.0752	0.0772
Month	**7**	**8**	**9**	**10**	**11**	**12**
Load coefficients	0.0992	0.0972	0.0912	0.0832	0.0722	0.0982

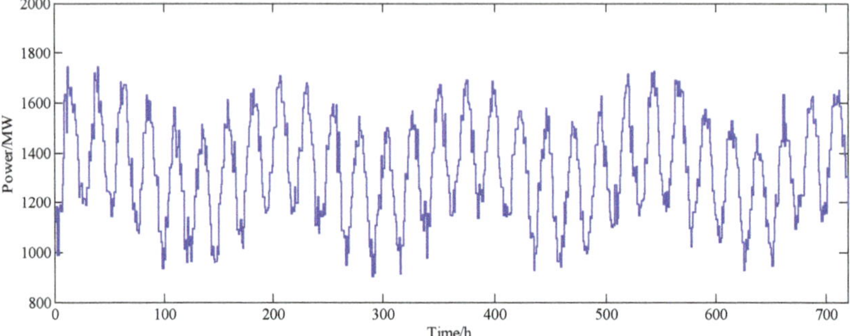

Figure 5. The load curve in April.

4.2. Simulation Results

4.2.1. Simulation Results of April (the Scheduling Month)

1. Computational Performance Analysis

To verify the computational performance of the proposed approach, a simplified test system only including the thermal power units was simulated with simple constraints. The running time of the simplified traditional model on an hourly simulation was 1024.92 s. The optimization of the traditional model did not converge because the calculation amount was excessive. However, the computational time using the proposed approach and model was 28.15 s. The computational speed was increased by 97.25%, and the computational volume was reduced effectively. The convergence of optimizing could be ensured when applying the proposed method and model.

2. Economy, Energy Conservation, and Emission Reduction Effect Analysis

The monthly scheduled energy of the thermal power units from April to December, according to the proposed method, are shown in Table 10. For comparison, the monthly scheduled energy values of the thermal power units, according to the load rate deviation method, are shown in Table 11.

Table 10. The scheduled energy of thermal power units using the proposed method.

Units	Month								
	4	**5**	**6**	**7**	**8**	**9**	**10**	**11**	**12**
1	206,840.2	204,913.7	143,514.6	148,298.4	357,120.0	345,600.0	344,914.7	187,073.0	316,955.0
2	102,170.2	63,372.9	182,371.2	208,320.0	63,372.9	67,450.3	63,372.9	201,600.0	172,327.0
3	87,763.6	55,116.8	139,164.2	148,800.0	55,116.8	144,000.0	55,116.8	144,000.0	148,800.0
4	63,404.6	89,280.0	37,705.2	60,493.8	89,280.0	86,400.0	38,962.1	37,705.2	89,280.0
5 (CHP)	177,520.1	185,531.3	77,667.0	192,249.6	165,640.7	77,667.0	80,255.9	77,667.0	192,249.6
6 (CHP)	88,819.6	126,182.4	49,856.7	117,176.2	51,518.6	49,856.7	115,656.6	49,856.7	126,182.4

Table 11. The scheduled energy of thermal power units using the load rate deviation method.

Units	Month								
	4	5	6	7	8	9	10	11	12
1	235,580.1	243,661.8	205,164.9	294,433.0	263,053.9	250,963.2	234,876.6	227,177.2	351,768.5
2	120,967.0	120,210.3	101,217.9	145,258.2	129,777.4	123,812.4	115,876.1	112,077.6	173,544.6
3	105,224.8	104,566.5	88,045.8	126,354.8	112,888.6	107,699.9	100,796.4	97,492.2	150,960.1
4	67,181.1	63,367.4	53,355.8	76,571.2	68,410.6	65,266.3	61,082.7	59,080.4	91,482.0
5 (CHP)	128,855.4	133,275.8	112,219.2	161,046.2	143,882.7	137,269.5	128,470.6	124,259.3	192,407.0
6 (CHP)	81,473.6	84,268.5	70,954.7	101,827.4	90,975.1	86,793.7	81,230.2	78,567.5	121,656.4

The thermal power generation cost was calculated according to the average coal consumption cost coefficient in Table 3. During the period from April to December, the thermal power generation costs, according to the proposed method and the load rate deviation method, were ¥ 1,388,520,214.3 (¥ 1.388 billion) ¥ and ¥ 1,397,646,068.9 (¥ 1.397 billion), respectively. Compared with the load rate deviation method, the proposed method could save ¥ 9,125,854.6 (¥9.126 million).

The total scheduled energy values of thermal power units in the scheduling month were 61,654.1 MWh when the presented time-sequence simulation method was applied in the simulation and 58,162.0 MWh when the load rate deviation method was applied. The energy of renewable energy units generated by the time-sequence simulation method and the benefits of energy conservation and emission reduction were verified.

If the operating cost of various types of units, the deep peak-regulation cost, and the start-up cost of thermal power units were considered in the scheduling month, then the total comprehensive generation cost from April to December with the time-sequence simulation method was ¥ 1,389,168,214.3 (or ¥ 1.389 billion).

3. Feasibility Analysis

The hourly power output curve of each thermal power unit in April (the scheduling month), according to the presented method is shown in Figure 6. The hourly output of each power unit in the scheduling month with the time-sequence simulation method can be obtained under the premise that the reliability of the generator and system operation is guaranteed. With the time-sequence simulation method, all the constraints together could ensure that the power output of each power unit is close to the actual operating conditions, which means a higher feasibility could be ensured.

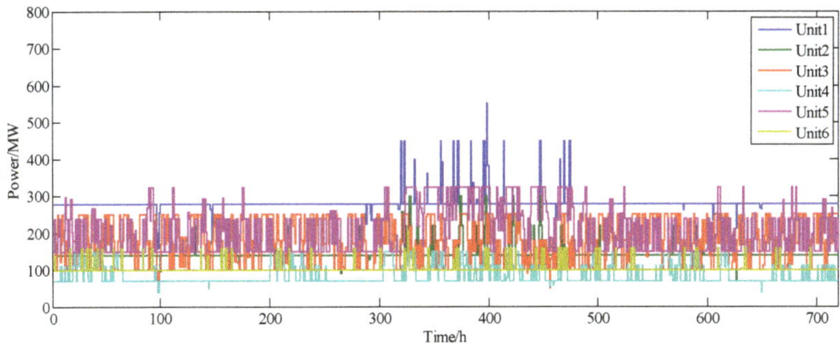

Figure 6. The hourly power output curve of each thermal power unit in April.

4. Fairness Analysis

The monthly energy-trade scheduling is simulated, according to the presented method, when considering the completed rate thresholds (CRTs) for different annual base electrical energies. The total costs based on different CRTs are shown in Table 12.

Table 12. Total cost based on different CRTs.

CRT	The Total Cost (¥)
2%	1,389,737,549.8
3%	1,389,168,214.3
5%	1,389,087,840.1
7%	1,388,881,156.7
10%	1,387,398,273.3

From Table 12, it can be seen that the relationship between the total cost and CRT are negatively correlated. With the increase in CRT, the relative optimization space of the monthly energy-trade scheduling became larger, so that there could be a more economical way to revise the monthly energy-trade scheduling. However, due to the increase in CRT, the fairness of generation scheduling was weaker. Therefore, the contradiction between the power generation units and the dispatching department was sharper. According to the current national regulations in China, the CRT was set to 3%. With the gradual improvement of market mechanism for electricity, the CRT could be adjusted more reasonably, according to the proportion of fairness and economy of the monthly energy-trade scheduling.

4.2.2. Rolling Correction Results during the Whole Year

The scheduling month was moved from January to December. The monthly energy-trade scheduling results for the whole year after the rolling correction are shown in Table 13.

Table 13. The monthly energy-trade scheduling results for the whole year.

Units		Month 1	2	3	4	5	6
Thermal Power plants	1	387,134.0	242,411.3	247,194.7	206,840.2	232,986.9	202,367.6
	2	198,550.9	131,918.9	121,544.3	102,170.2	140,135.5	101,553.7
	3	148,287.9	90,969.3	83,661.1	87,763.6	92,291.6	74,537.9
	4	80,500.3	58,160.3	58,378.4	63,404.6	56,976.4	57,423.3
	5 (CHP)	198,032.3	122,431.6	121,404.8	177,520.1	122,722.2	112,861.3
	6 (CHP)	118,291.4	100,518.6	106,398.2	88,819.6	78,201.0	75,727.2
Wind unit		36,659.6	32,809.7	46,058.9	61,654.1	56,428.5	54,982.0
Hydropower station		44,847.0	81,109.0	102,817.0	138,739.0	174,379.0	300,159.0
Nuclear plant		37,200.0	33,600.0	37,200.0	36,000.0	37,200.0	36,000.0

Units		Month 7	8	9	10	11	12
Thermal Power plants	1	280,140.4	248,503.1	232,828.2	217,412.6	248,327.6	357,120.0
	2	151,354.0	117,253.8	108,857.5	114,130.4	73,712.6	205,244.2
	3	109,850.7	85,535.5	95,751.3	113,900.4	122,991.0	148,800.0
	4	71,886.7	64,416.1	57,709.4	53,655.8	64,151.3	89,280.0
	5 (CHP)	150,314.1	154,141.0	163,445.3	122,121.1	108,187.5	163,484.7
	6 (CHP)	102,852.1	104,726.4	107,678.0	76,745.4	79,663.5	81,865.1
Wind unit		42,685.4	40,271.0	48,806.3	52,327.0	57,692.1	49,720.0
Hydropower station		359,829.0	429,932.0	350,262.0	307,388.0	161,367.0	156,341.0
Nuclear plant		37,200.0	37,200.0	36,000.0	37,200.0	36,000.0	37,200.0

5. Conclusions

A new time-sequence simulation method for monthly energy-trade scheduling is presented in this paper. The feasibility of the presented method is validated at the theoretical level using a case study. The segment modeling strategy is applied to simulate on an hourly basis for the scheduling month and for the sequence months on a monthly basis. In the power systems integrated with large-scale new energy power, multiple factors such as the generation coordinate among various types of power units, equitable distribution of electrical energy, and system operation reliability could be comprehensively considered in the monthly energy-trade scheduling process.

The results of the case study verified the feasibility and effectiveness of the proposed approach.

- The characteristics of wind power, nuclear power, hydropower, thermal power, and combined heat and power (CHP) generators were comprehensively considered. Therefore, the consumption capability of renewable energy power can be improved, according to the presented monthly energy-trade scheduling method. Thus, the energy saving and emission reduction benefits can be improved.
- By efficiently managing the balance of the annual base electrical energy completion rate of each thermal power unit, the monthly energy trade scheduling fairness can be ensured in a better way.
- Because the necessary operating constraints in the short-term time-scale could be easily introduced into the mathematical model for the scheduling month, the feasibility of the monthly energy-trade scheduling could be improved significantly. This improvement can lay a good foundation for daily dispatching.

The limitation of this study is that it does not involve an experimental study. Therefore, in future studies, this method might need to be modified according to the actual operating conditions. These conditions may be the number of power plants in a regional power grid, calculation time limit, etc., which may improve the practicality of the proposed method.

Author Contributions: Conceptualization, L.S. and W.L. Methodology, N.Z. and Z.S. Software, Q.Z. and X.L. Validation, L.S., W.L., and N.Z. Formal Analysis, Q.Z. Data Curation, Q.Z. Writing—Original Draft Preparation, Q.Z. and N.Z. Writing—Review and Editing, L.S. Visualization, X.L. Supervision, W.L. Project Administration, L.S.

Funding: The National Basic Research Program (China), grant number 51607021, funded this research.

Conflicts of Interest: The authors declare no conflict of interest.

Appendix A NOMENCLATURE

Table A1. Variable comparison table.

F_1	Precise objective function for the scheduling month
F_2	Rough objective function for the subsequent months
T	Time intervals in the scheduling month
i, k, l, s, v	Sequence numbers of pure condensing thermal power units, extraction steam thermal power units, wind units, hydropower stations, and nuclear plants
$N_{CN}, N_{CQ}, N_w, N_m, N_n$	The numbers of pure condensing thermal power units, extraction steam thermal power units, wind units, hydropower stations, and nuclear plants
$f_{CN}(i,t)$	Operating cost of pure condensing thermal power unit i at time t
$f_{CQ}(k,t)$	Operating cost of extraction steam thermal power unit k at time t
$f_w(l,t)$	Operating cost of wind unit l at time t
$f_m(s,t)$	Operating cost of hydropower station s at time t
$f_n(v,t)$	Operating cost of nuclear plant v at time t
a_i, b_i, c_i	Fuel cost coefficients of pure condensing thermal power unit i
$P^t_{CN_i}$	Power by the pure condensing thermal power unit i at time t
M	Average cost of deep peak regulation of thermal power units
$\Delta P^t_{CN_i}$	Power for deep peak regulation by pure condensing thermal power unit I at time t
S_{CN_i}	Start-up cost of pure condensing thermal power unit i
u^t_i	States of the pure condensing thermal power unit I at time t
a_k, b_k, c_k	Fuel cost coefficients of extraction steam thermal power unit k
$P^t_{CQ_k}$	Power by extraction steam thermal power unit k at time t
$H^t_{CQ_k}$	Thermal power by extraction-steam thermal power unit k at time t

Table A1. *Cont.*

$\Delta P^t_{CQ_k}$	Power for deep peak regulation of extraction-steam thermal power unit k at time t
S_{CQ_k}	Start-up cost of extraction steam thermal power unit k
u^t_k	States of extraction steam thermal power unit k at time t
C_w	Average cost of wind power curtailment
$P^t_{w_l}$	Consumptive power of wind unit l at time t
$\bar{P}^t_{w_l}$	Prediction power of wind unit l at time t
C_m	Average cost of hydropower curtailment
$P^t_{m,g}$	Theoretical power generated by curtail water at time t
C_n	Average cost of peak regulation by nuclear plants
$\bar{P}_{n_v}$	Rated power of nuclear plant v
$P^t_{n_v}$	Power by nuclear plant v at time t
m	Serial number of the scheduling month
e	Serial numbers of the subsequent months
C_{CN_i}	Average fuel cost coefficient of pure condensing thermal power unit i
$W^e_{CN_i,RE}$	Planned generation energy of pure condensing thermal power unit i in month e
C_{CQ_k}	Average fuel cost coefficient of extraction steam thermal power unit k
$W^e_{CQ_k,RE}$	Planned generation energy of the extraction steam thermal power unit k in month e
$P_{CN_i,min}$	The maximum output power of the pure condensing thermal power unit i
$P_{CN_i,max}$	The minimum output power of the pure condensing thermal power unit i
$\Delta P_{CN_i,down}, \Delta P_{CN_i,up}$	The maximum rate of downward ramping / upward ramping of the pure condensing thermal power unit i
$X^{t-1}_{CN_i,on}, X^{t-1}_{CN_i,off}$	The continuous starting time / downtime of the pure condensing thermal power unit I until time t-1
$T_{CN_i,on}, T_{CN_i,off}$	The minimum starting time/downtime of the pure condensing thermal power unit i
$C_{m_k}, K_k. C_{v2_k}, C_{v1_k}$	The heat-electric coefficients of the extraction-steam thermal power unit k
$H_{CQ_k,max}$	The upper output thermal power limit of the extraction steam thermal power unit k
H^t_L	The thermal load at time t
$P_{w_l,max}$	The rated power of wind unit l
V^t_m	Volume of water in reservoir at time t
f^t_m	Volume of water entering the reservoir at time t
$q^t_{m_s}$	Volume of water for power generation of hydropower station s at time t
g^t_m	Volume of abandoned water at time t
V^{min}_m	The minimum volume for saving reservoir water
V^{max}_m	The maximum volume for saving reservoir water
$q_{m_s,max}$	The acceptable maximum water flow of hydropower unit s
a	The power coefficient of the hydropower unit
h	The head of the reservoir
$P_{m_s,max}$	The rated power of the hydropower unit s
$u^t_{n_v,y_1}, u^t_{n_v,y_2}$	The states of nuclear plant v at time t
$P_{n_v,max}$	The rated power of nuclear plant v
$P^t_{n_v,y_1}$	The power of nuclear plant v at time t corresponding to '$u^t_{n_v,y_1}$'
$P^t_{n_v,y_2}$	Power of nuclear plant v at time t corresponding to '$u^t_{n_v,y_2}$'
α	The ratio of '$P^t_{n_v,y_2}$' to '$P_{n_v,max}$'
P^t_L	Load at time t
β	The confidence coefficient
$\delta^e_{CN_i}$	The operating rate of the pure condensing thermal power unit i in month e
$W^{year}_{CN_i}$	The annual contract electricity energy of the pure condensing thermal power unit i

Table A1. *Cont.*

$W_{CN_i}^{ywc}$	The generation energy that the pure condensing thermal power unit i has generated until decision time
T_{wwc}	The sum of scheduling month's number of hours and the subsequent months' number of hours
η	The empirical value from the annual operating rate of the thermal power unit
t^e	The number of hours in month e
$W_{w_l,RE}^e$	The generation energy of wind unit l in month e
$W_{w_l,F}^e$	The maximum generation energy of wind unit l in month e
$W_{m_s,RE}^e$	The generation energy of hydropower station s in month e
$W_{m_s,F}^e$	The maximum generation energy of the hydropower station s in month e
$W_{n_v,RE}^e$	The generation energy of the nuclear plant v in month e
$W_{n_v,F}^e$	The maximum generation energy of the nuclear plant v in month e
$W_{L,RE}^e$	The power load energy in month e
$W_{T_i,A}$	The annual planned generation energy of the thermal power unit i
$W_{T_i,H}^e$	Before the scheduling month, the generation energy that the thermal power unit i generated in month e
$P_{T_i}^t$	In the scheduling month, the power of thermal power unit i at time t
$W_{T_i,RE}^e$	In the subsequent month, the generation energy that the thermal power unit i generates in month e
$W_{T_i,B}$	The annual basic generation energy of thermal power unit i
$W_{T_i,D}$	The annual transactional generation energy of thermal power unit i
ρ_T	The completion rate of all thermal power units' annual basic generation energy
$W_{T_i,BF}$	The specified annual basic generation energy of the thermal power unit i
λ	The percentage of the annual base generation energy completion rate deviation threshold

References

1. Han, J.H. The Formulation Method of Annual Contract Energy Planning for Power Generation Unit under the Constraint of Energy Saving and Emission Reduction. Master's Thesis, Harbin Institute of Technology, Harbin, China, 2011.
2. Tang, W. Research on the Method of Compiling Monthly Energy Trade Scheduling. Ph.D. Thesis, Harbin Institute of Technology, Harbin, China, 2009.
3. Zhang, H.Q.; Chang, Y.J.; Tang, D.Y.; Wang, S.Y.; Chen, Q.; Li, Z.G.; Wang, Y.Q.; Yu, J.L. A monthly electric energy plan making method of thermal power generation unit in Energy-Saving generation dispatching mode. *Power Syst. Prot. Control* **2011**, *39*, 84–89. [CrossRef]
4. Xia, Q.; Chen, Y.G.; Chen, L. Energy-Saving power generation scheduling mode and method considering monthly unit combination. *Power Syst. Technol.* **2011**, *6*, 27–33. [CrossRef]
5. Wang, Y.; Tang, W.; Luo, H.H.; Yu, F.; Liu, Z.Y.; Jin, Z.H.; Liu, J.; Guo, Y.F.; Yu, J.L.; Liu, Z. Load Rate Deviation Method for Preparing Monthly Energy Trade Scheduling for Thermal Power Generation Units. *Power Syst. Prot. Control* **2009**, *22*, 134–140. [CrossRef]
6. Tang, W.; Wang, Y.; Yu, F.; Liu, Z.Y.; Luo, H.H.; Jin, Z.H.; Liu, J.; Guo, Y.F.; Yu, J.L.; Liu, Z. Comprehensive Cost-weighting Method for Preparing Monthly Energy Trade Scheduling for Thermal Power Units. *Power Syst. Technol.* **2009**, *33*, 167–173. [CrossRef]
7. Tang, W.; Wang, Y.; Yu, J.L.; Min, D.J.; Luo, H.H.; Guo, Y.F.; Jin, Z.H.; Liu, J.; Liu, Z. Comprehensive Consumption Optimization Method for Preparing Monthly energy trade scheduling for Thermal Power Units. *Proc. CSEE* **2009**, *29*, 64–70. [CrossRef]
8. Mohammad, A.M.; Ahmad, S.Y.; Behnam, M.; Mousa, M.; Miadreza, S.-K.; Catalão, J.P.S. Stochastic network-constrained Co-Optimization of energy and reserve products in renewable energy integrated power and gas networks with energy storage system. *J. Clean. Prod.* **2019**, *223*, 747–758. [CrossRef]

9. Mousa, M.; Hamed, A.; Seyedeh, S.G.; Hasan, U.; Terrence, F. Optimal energy management system based on stochastic approach fora home Microgrid with integrated responsive load demand and energy storage. *Sustain. Cities Soc.* **2017**, *28*, 256–264. [CrossRef]
10. Korkas, C.D.; Baldi, S.; Michailidis, L.; Kosmatopoulos, E.B. Occupancy-based demand response and thermal comfort optimization in microgrids with renewable energy sources and energy storage. *Appl. Energy* **2016**, *163*, 93–104. [CrossRef]
11. Korkas, C.D.; Baldi, S.; Michailidis, L.; Kosmatopoulos, E.B. Intelligent energy and thermal comfort management in grid-connected microgrids with heterogeneous occupancy schedule. *Appl. Energy* **2015**, *149*, 194–203. [CrossRef]
12. Yu, J.; Ji, F.; Zhang, L.; Chen, Y.S. An over painted oriental arts: Evaluation of the development of the Chinese renewable energy market using the wind power market as a model. *Energy Policy* **2009**, *37*, 5221–5225. [CrossRef]
13. Chen, Q.; Fan-Chao, G.U.; Jin, Y.Q.; Qin, C.; Chen, X.Y.; Li, Z.R. Energy-Saving Generation Dispatch of Power System Including Large-scale Wind Farms. *Proc. CSU-EPSA* **2014**. [CrossRef]
14. Chi, Y.N.; Liu, Y.H.; Wang, W.S.; Chen, M.Z.; Dai, H.Z. Study on Impact of Wind Power Integration on Power System. *Power Syst. Technol.* **2007**, *31*, 77–81. [CrossRef]
15. Rao, P.; Peng, C.H. A Research on Power Dispatch of Energy-saving and Emission-reduction Generation Based on the Improved Differential Evolution Algorithm. *J. East China Jiaotong Univ.* **2010**, *2010*. [CrossRef]
16. Ting, L.I.; Zhang, H.L.; Shao, P.; Guo, S.Q.; Guang, H.H. Economic evaluation method of monthly power generation plan considering the direct power purchase' influence. *Technol. Dev. Enterp.* **2016**. [CrossRef]
17. Wei, X.H.; Hu, Z.Y.; Yang, L. Thoughts and Suggestions about The Current Evaluation Indicators of the San Gong Dispatching. *Autom. Electr. Power Syst.* **2012**, *36*, 109–112. [CrossRef]
18. Tian, K.; Yao, J.; Wenhui, L.I. Power System Optimal Dispatching Model with Low-Carbon under Clean Energy Generation Integration. *Shaanxi Electr. Power* **2016**. [CrossRef]
19. Zhong, H.; Xia, Q.; Chen, Y.; Kang, C.Q. Energy-Saving generation dispatch toward a sustainable electric power industry in China. *Energy Policy* **2015**, *83*, 14–25. [CrossRef]
20. Lin, R.M.; Jin, H.G.; Cai, R.X. Researching Direction and Development for New Generation Energy Power System. *Power Eng.* **2003**. [CrossRef]
21. Liu, C.; Cao, Y.; Huang, Y.H.; Li, P.; Sun, Y.; Yuan, Y. The Method for Preparing Annual Wind Power Wind Power Plan Based on Time Series Simulation. *Autom. Electr. Power Syst.* **2014**, *38*, 13–19. [CrossRef]
22. Cao, Y.; Li, P.; Yuan, Y.; Zhang, X.S.; Guo, S.Q.; Zhang, C.F. New Energy Consumption Capacity Analysis and Low Carbon Benefit Evaluation Based on Time Series Simulation. *Autom. Electr. Power Syst.* **2014**, *38*, 60–66. [CrossRef]
23. Bai, L.P. The Economic Constraints and Economic Dispatch of Power Systems Considering New Energy Sources. Master's Thesis, Harbin Engineering University, Harbin, China, 2016.
24. Ji, F.; Cai, X.G. Monthly unit combination model with wind power system. *J. Harbin Inst. Technol.* **2017**, *3*. [CrossRef]
25. Sun, L.; Xu, J.; Zhang, Q.; Shen, J.K.; Liu, X.L.; Li, W.D. A time sequential simulation method for monthly energy trade planning in the power system with multiple energy resources. *J. Phys. Conf. Ser. IOP Publ.* **2018**, *1053*, 012008. [CrossRef]
26. Li, X. Research on Optimization Method of Power System Monthly Power Purchase Scheduling. Master's Thesis, Chongqing University, Chongqing, China, 2013.
27. Liang, Z.F.; Xia, Q.; Xu, H.Q.; Zhu, M.X.; Zhang, J.; Yang, M.H. Provincial grid monthly power generation plan based on multi-objective optimization model. *Power Syst. Technol.* **2009**, *13*, 90–95. [CrossRef]
28. Zhang, X.Q.; Bai, X.; Wan, Y.Z.; Peng, M.Q.; Wang, P.; Xiang, Y. Application of Grey Model in Wind Power Generation Forecast of Northwest Power Grid. *Power Syst. Clean Energy* **2011**, *4*, 66–70. [CrossRef]
29. Guo, D. Research on High Proportion of Wind Power Grid-Connected to Improve Consumption Capacity. Ph.D. Thesis, Shenyang Agricultural University, Shenyang, China, 2017.

 *processes*

Article

Simulation-Based Design and Economic Evaluation of a Novel Internally Circulating Fluidized Bed Reactor for Power Production with Integrated CO_2 Capture

Jan Hendrik Cloete [1,*], Mohammed N. Khan [2], Schalk Cloete [1] and Shahriar Amini [1,2,*]

[1] Flow Technology Research Group, SINTEF Industry, 7465 Trondheim, Norway; schalk.cloete@sintef.no

[2] Department of Energy and Process Engineering, Norwegian University of Science and Technology, 7491 Trondheim, Norway; mkhan@ntnu.no

* Correspondence: henri.cloete@sintef.no (J.H.C.); shahriar.amini@sintef.no (S.A.); Tel.: +47-930-02214 (J.H.C.); +47-466-39721 (S.A.)

Received: 10 September 2019; Accepted: 30 September 2019; Published: 11 October 2019

Abstract: Limiting global temperature rise to well below 2 °C according to the Paris climate accord will require accelerated development, scale-up, and commercialization of innovative and environmentally friendly reactor concepts. Simulation-based design can play a central role in achieving this goal by decreasing the number of costly and time-consuming experimental scale-up steps. To illustrate this approach, a multiscale computational fluid dynamics (CFD) approach was utilized in this study to simulate a novel internally circulating fluidized bed reactor (ICR) for power production with integrated CO_2 capture on an industrial scale. These simulations were made computationally feasible by using closures in a filtered two-fluid model (fTFM) to model the effects of important subgrid multiphase structures. The CFD simulations provided valuable insight regarding ICR behavior, predicting that CO_2 capture efficiencies and purities above 95% can be achieved, and proposing a reasonable reactor size. The results from the reactor simulations were then used as input for an economic evaluation of an ICR-based natural gas combined cycle power plant. The economic performance results showed that the ICR plant can achieve a CO_2 avoidance cost as low as $58/ton. Future work will investigate additional firing after the ICR to reach the high inlet temperatures of modern gas turbines.

Keywords: chemical looping combustion; power production; carbon capture; internally circulating reactor; reactor design; fluidization; techno-economics; computational fluid dynamics; filtered two-fluid model; coarse-grid simulations

---

1. Introduction

Several high-profile studies have shown that carbon capture and storage must play a central role in the future energy mix to reach the goal of limiting the global temperature increase to well below 2 C above preindustrial limits at a reasonable cost [1–3]. A low-cost pathway to limiting global CO_2 emissions will be essential to prevent the negative consequences of climate change, while allowing for continued development in developing nations where billions of people still live in poverty.

Many different technologies have been proposed to capture CO_2 from fossil-fuel power plants, after which the CO_2 can either be stored or utilized in other industrial processes. However, a major challenge of such processes is the energy penalty associated with CO_2 capture. An increased energy penalty requires more fuel to be used to achieve the same power output, increasing operating and capital costs, but also increasing the amount of CO_2 that must be dealt with.

A promising group of technologies for capturing CO_2 are those based on chemical looping combustion (CLC) [4], as they can essentially eliminate the energy penalty of CO_2 and potentially even offer efficiency improvements in comparison to unabated plants [5]. Traditionally, the CLC process is performed in a dual circulating fluidized bed (CFB) configuration. In the oxidation reactor, a metallic oxygen carrier is oxidized, providing large amounts of heat. The thermal energy in the gas phase is used for power production, whereas the hot particles are transported to the fuel reactor where the oxidized particles are reduced by a fuel, producing CO_2 and steam. The CLC process therefore keeps the CO_2 stream separate from the nitrogen-containing air stream, allowing an almost pure CO_2 stream for storage to be obtained simply by knocking out the water.

A drawback of the dual circulating fluidized bed CLC approach is that efficient power production with CO_2 storage requires high pressure operation. However, progress on the scale-up of pressurized CLC systems has been limited [6] due to the complexity of pressurizing the two reactors, loop seals, and cyclones, and in maintaining the required solids circulation between the reactors. Consequently, several alternative CLC configurations have been proposed to overcome the challenges of the pressurized dual CFB CLC system. These include gas switching technologies [7,8], rotating bed reactors [9,10], packed bed chemical looping [11,12], and internally circulating reactors (ICRs) [13–15], which will be the focus of the present study.

The internally circulating reactor concept replaces the loop seals and cyclones that separate reactors in the dual CFB with simple ports connecting two sections of a reactor vessel, allowing the oxygen carrier to circulate between the reducing and oxidizing sections. This allows the CLC process to take place within a single unit, significantly simplifying pressurization and scale-up. The disadvantage is that gas will leak through the ports along with the circulating solids, reducing the CO_2 capture efficiency and the purity of the captured CO_2. However, it has been shown that the detrimental effect of gas leakage can be limited by controlling the fluidization velocity ratio of the two sections and the bed loading [13], achieving CO_2 capture efficiencies greater than 95% and purities greater than 92%.

Academia has been prolific in proposing novel processes and reactors for CO_2 capture. Unfortunately, implementation of new technologies in the process and energy industry has traditionally been slow, requiring several decades from process conception to commercial reality. The urgency of climate change will require very rapid scale-up and industrialization of these novel CO_2 capture technologies, starting once governments start imposing strong policies to reduce carbon emissions (the IEA Sustainable Development Scenario assumes CO_2 prices of \$63/ton and \$140/ton in 2025 and 2040, respectively [16]). This is also valid for other industries—rapid innovation and implementation of new process technologies will be necessary in a world that is increasingly environmentally and resource-constrained.

Simulation-based engineering will be an essential tool in enabling such rapid innovation by decreasing the number of costly and time-consuming experimental scale-up steps, and computational fluid dynamics (CFD) is the most suitable tool for investigating the chemical reactors common in the process and energy industries. However, although CFD has proven extremely useful in better understanding flow processes on the lab-scale, a common challenge to industrial simulation is the fact that important phenomena may occur on time- and length-scales that are several orders of magnitude smaller than those associated with the industrial processes [17]. This is especially relevant in multiphase processes and for the fluidized beds used in the ICR reactor studied here, where gas bubbles and particle clusters of length-scales in the order of ten particle diameters play an important role on the overall fluidized bed behavior. Using small enough grid cells and time steps to resolve these small-scale phenomena remains impossible for parametric studies of industrial-scale devices, even with large, modern computational clusters.

Multiscale methods are necessary to overcome this challenge—allowing the use of coarse computational grids to achieve reasonable computational times by using closures for unresolved subgrid effects to maintain acceptable accuracy. The filtered two-fluid model (fTFM) [18] is a common approach for multiscale modeling of fluidized beds. In the fTFM, the governing equations of the two-fluid model (TFM) closed by the kinetic theory of granular flow, where the solids phase is assumed to behave as a continuum and closures capture the effects of random particle collisions and translation, is spatially averaged, revealing subgrid terms that require closure. Several groups have strived to develop such closures. Most of the work has focused on the subgrid correction to the drag [19–26], which substantially reduces the drag compared to the drag law evaluated at the resolved quantities, although several other closures are necessary for fluidized bed hydrodynamics [27]. Research on closures for reactive flow has been limited. Most studies have investigated the influence of subgrid effects on the effective reaction rate of first-order solids-catalyzed reactions [28–30], where mass-transfer limitations imposed by the bubbles and clusters drastically reduce the effective reaction rates. Closures are also required for the dispersion of scalars (such as species and enthalpy) due to subgrid velocity fluctuations and for the effective interphase heat transfer rate [31,32].

The present study aimed to demonstrate how multiscale CFD simulations can be used to assist the evaluation of novel reactor concepts on an industrial scale, focusing on an internally circulating fluidized bed reactor for power production with CO_2 capture. Firstly, some improvements were proposed for existing fTFM closures, improving the accuracy and simplicity of existing hydrodynamics closures [27,33] and, most importantly, proposing a generalized reactive fTFM closure. The latter is important, since existing closures [28–30] are only valid for simple first-order solids-catalyzed reaction equations. Next, an fTFM accounting for all important subgrid effects in reactive flows was used to evaluate the effect of several design and operating parameters on the ICR behavior. It can be noted that, to the best of the authors' knowledge, this is the most complete implementation—in terms of the number of subgrid effects accounted for—of a reactive fTFM to date. Then, results from the reactor simulations were combined with previously published power plant simulations by the same authors [34] to conduct an economic assessment of the ICR concept for low carbon power production from natural gas. Finally, the results are used to discuss the future of virtual prototyping of novel reactors using multiscale CFD simulations, as well as the potential of the ICR to combat climate change.

2. Materials and Methods

The present study utilizes both multiscale CFD reactor modeling and process modeling to inform the economic evaluation of power production with CO_2 capture using the ICR concept. Figure 1 shows how information flows between these three parts of the study, and the subsequent sections describe each part in detail.

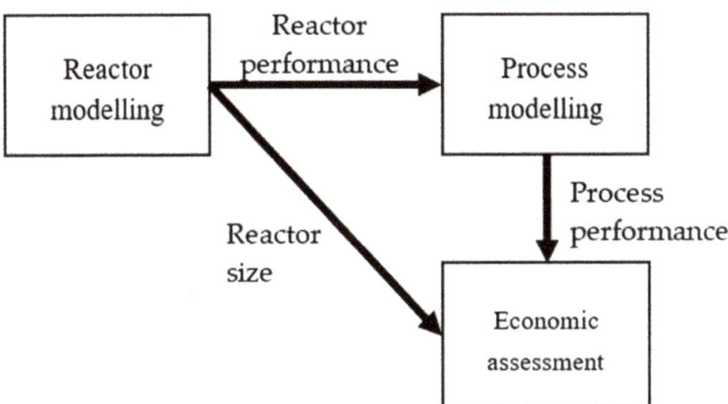

Figure 1. Information flow between different parts of the present study.

2.1. Reactor Modeling

2.1.1. The Filtered Two-Fluid Model (fTFM)

The fTFM solves the spatially-averaged (or filtered) form of the governing equations for the two-fluid model closed by the kinetic theory of granular flow [35,36]. This section briefly presents the filtered governing equations, as well as the closures that are used for the subgrid terms. The interested reader may find a more complete discussion of the derivation of the filtered equations in earlier studies [33,37].

The filtered continuity equations are given below. S_R is a source term due to the mass transfer during the reduction and oxidation of the oxygen carrier. Closures for the filtered reaction rates, which can be used to calculate the source term, are discussed in Section 2.1.2.

$$\frac{\partial}{\partial t}\left(\overline{\alpha_g \rho_g}\right) + \nabla \cdot \left(\overline{\alpha_g \rho_g} \widetilde{\vec{v}}_g\right) = S_R, \tag{1}$$

$$\frac{\partial}{\partial t}\left(\overline{\alpha_s \rho_s}\right) + \nabla \cdot \left(\overline{\alpha_s \rho_s} \widetilde{\vec{v}}_s\right) = -S_R. \tag{2}$$

Next, the filtered momentum equations are shown in Equations (3) and (4):

$$\frac{\partial}{\partial t}\left(\rho_g \overline{\alpha_g} \widetilde{\vec{v}}_g\right) + \nabla \cdot \left(\rho_g \overline{\alpha_g} \widetilde{\vec{v}}_g \widetilde{\vec{v}}_g\right) = -\overline{\alpha_g} \nabla \widetilde{p} - \nabla \cdot \left(\rho_g \overline{\alpha_g \vec{v}_g'' \vec{v}_g''}\right) + \nabla \cdot \overline{\overline{\overline{\tau}}}_g + \overline{\alpha_g \rho_g} \vec{g} + \overline{K_{sg}\left(\vec{v}_s - \vec{v}_g\right)} - \overline{\alpha_g' \nabla p'}, \tag{3}$$

$$\frac{\partial}{\partial t}\left(\rho_s \overline{\alpha_s} \widetilde{\vec{v}}_s\right) + \nabla \cdot \left(\rho_s \overline{\alpha_s} \widetilde{\vec{v}}_s \widetilde{\vec{v}}_s\right) = -\overline{\alpha_s} \nabla \widetilde{p} - \nabla \cdot \left(\rho_s \overline{\alpha_s \vec{v}_s'' \vec{v}_s''}\right) - \nabla \overline{p_s} + \nabla \cdot \overline{\overline{\overline{\tau}}}_s + \overline{\alpha_s \rho_s} \vec{g} + \overline{K_{gs}\left(\vec{v}_g - \vec{v}_s\right)} - \overline{\alpha_s' \nabla p'}. \tag{4}$$

In both filtered momentum equations, the second term on the right-hand side represents the stresses due to subgrid velocity fluctuations (arising from gas bubbles and solids clusters), which add to the diffusive momentum transport. The gas-phase subgrid stress is usually relatively small compared to the solids stresses due to the large difference in the phase densities [38] and can safely be neglected. However, the solids subgrid stresses are accounted for by means of an anisotropic stress closure [39], which has been shown to offer significant improvements compared to Boussinesq approximation-based closures using isotropic independent variables [39,40]. In the filtered solids momentum equation, the filtered kinetic theory stresses (third and fourth terms on the right-hand side) are small at the grid sizes that are relevant for industrial-scale fluidized beds [39], which was used in the present study. Therefore, the filtered kinetic theory stresses were estimated on the basis of the unfiltered granular temperature equation, as it was previously shown to be sufficient [41].

The second-to-last term on the right-hand side of both momentum equations represents the filtered drag force, where subgrid effects generally reduce the drag compared to that in a homogenous suspension. This is due to the tendency of fluidized particles to form solids clusters and gas bubbles, which are not resolved on a coarse grid. These meso-scale structures vary in size and shape due to local flow conditions. Gas will tend to pass through dilute regions, reducing the effective drag on the solids clusters, the effect of which must be accounted for in a closure. A modified version of a 3-marker anisotropic closure published previously [27] was used to close the filtered drag force. It was found that the 3-marker closure could be simplified significantly, while maintaining similar accuracy, by eliminating the filtered slip velocity as a marker. More information about the development and verification of this new closure can be found in the Supplementary Material.

Finally, the last term on the right-hand side of both momentum equations is due to subgrid pressure gradient fluctuations and is referred to here as the meso-scale interphase force. This contribution arises from the redistribution of the pressure gradient over subgrid gas bubbles and solids clusters [42] and tends to add to the effective drag force [33]. For the present study, an older anisotropic closure [33] was improved on by drawing an analogy to the closure for the meso-scale solids stresses [39], where it was found that a filtered co-variance term can be accurately closed as a function of the relevant gradients. The Supplementary Material also details the development and verification of this new closure.

Next, Equation (5) gives the filtered species transport equation for reactant A, which is consumed in an nth order reaction.

$$\frac{\partial}{\partial t}\left(\rho_g \overline{\alpha_g} \widetilde{X_A}\right) + \nabla \cdot \left(\rho_g \overline{\alpha_g} \widetilde{X_A} \widetilde{\vec{v}}_g\right) = \nabla \cdot \left(\rho_g \overline{D \alpha_g \nabla X_A}\right) - \nabla \cdot \left(\rho_g \overline{\alpha_g X_A'' \vec{v}_g''}\right) - \overline{k_A \alpha_s C_A^n M_A}. \tag{5}$$

The species dispersion due to the filtered microscopic diffusion (the first term on the right-hand side) was expected to be small relative to meso-scale dispersion, as well as convective transport. Therefore, in line with previous work regarding scalar dispersion in fTFMs [31], it was simply evaluated as $\nabla \cdot \left(\rho_g D \overline{\alpha_g} \nabla \widetilde{X}_A \right)$ in the present study. The subgrid species dispersion rate (the second term on the right-hand side) tends to disperse the species due to sub-grid velocities arising from unresolved gas bubbles and solids clusters. This effect was accounted for using the closures of Agrawal et al. [31] but has been shown to only have a minor effect on the overall reaction rate [28]. The filtered reaction rate (third term on the right-hand side) is typically substantially reduced by subgrid bubbles and clusters and is essential to model [28]. This is because, for gas-solid reactions, the reactant will be consumed faster inside dense regions, creating a mass transfer limitation due to the finite rate at which reactants are transported to these dense regions. A limited number of studies have investigated reactive fTFM closures [28,30,41], but they have all focused on reactions that are solids catalyzed and first-order with respect to the gaseous reactant. The next section of the present study therefore proposes a novel, simplified approach for accounting for different reaction orders and for reactions where the solids phase participate in the reaction. Finally, it can be noted that the filtered solids species equations are similar to those of the gas-phase species and are thus treated in a similar way. Consequently, they are not discussed separately.

Finally, Equation (6) shows the filtered enthalpy transport equation for the gas-phase.

$$\frac{\partial}{\partial t}\left(\rho_g \overline{\alpha_g} \widetilde{h}_g\right) + \nabla \cdot \left(\rho_s \overline{\alpha_g} \widetilde{h}_g \overrightarrow{\widetilde{v}}_g\right) = \nabla \cdot \left(\overline{\kappa_g \alpha_g \nabla T_g}\right) - \nabla \cdot \left(\overline{\rho_s \alpha_s h_g'' \overrightarrow{v}_g''}\right) + \overline{\gamma\left(T_s - T_g\right)} + \overline{k_A \alpha_s C_A^n M_A \Delta H_{r,A}}. \tag{6}$$

Here, as with the species diffusion, the filtered microscopic conductivity (first term on the right-hand side) is small compared to the enthalpy dispersion from sub-grid velocity fluctuations and is simply approximated as $\overline{\kappa_g \alpha_g \nabla T_g}$. The subgrid enthalpy dispersion rate (second term on the right-hand side) and the filtered heat transfer rate (third term on the right-hand side) were modeled using the closures of Agrawal et al. [31]. The physical behavior of these contributions is analogous to that of the species dispersion rate and filtered reaction rate, due to the similarity of mass and heat transfer. The enthalpy source term due to reaction (fourth term on the right-hand side) was evaluated at the filtered reaction rate modeled in Equation (5), assuming that the heat of reaction is uniform for each cell in the coarse-grid simulations. This is a reasonable assumption based on the good mixing and fast heat transfer in fluidized beds. Finally, it can be noted that the solids-phase filtered enthalpy equation was treated in a similar way, and it is therefore not discussed here separately.

2.1.2. Reaction Modeling

In the present study, Ni/NiO (supported on Al_2O_3) was used as oxygen carrier due to its high reactivity [43] and its ability to tolerate high operating temperatures [44]. In the fuel section, the oxygen carrier was reduced by the fuel according to:

$$4NiO(s) + CH_4(g) \rightarrow 4Ni(s) + 2H_2O(g) + CO_2(g). \tag{7}$$

In the air side of the ICR, the oxygen carrier re-oxidizes as:

$$2Ni(s) + O_2(g) \rightarrow 2NiO(s). \tag{8}$$

The reactions are implemented in the filtered species conservation equations, as follows, for species i taking part in reaction k, where v_i is the stoichiometric constant:

$$\frac{\partial}{\partial t}\left(\rho_g \overline{\alpha_g} \widetilde{X}_i\right) + \nabla \cdot \left(\rho_g \overline{\alpha_g} \widetilde{X}_i \overrightarrow{\widetilde{v}}_g\right) = \nabla \cdot \left(\rho_g D \overline{\alpha_g} \nabla \widetilde{X}_i\right) - \nabla \cdot \left(\overline{\rho_g \alpha_g X_i'' \overrightarrow{v}_g''}\right) + \sum_{k=0}^{n} v_i R_k^H M_i. \tag{9}$$

The effective reaction rate, R_k^H, in units of $mol/(m^3s)$ was calculated as shown in Equation (10), where gas species A reacts with solids species B (here, B represents NiO in reduction and Ni in oxidation).

$$R_k^H = \eta \overline{\alpha_s} \rho_s a_0 \left(\widetilde{X_{Ni}} + \widetilde{X_{NiO}} \right) k_s \widetilde{C_A}^n \left(1 - \frac{\widetilde{X_B}}{\widetilde{X_{Ni}} + \widetilde{X_{NiO}}} \right)^{\frac{2}{3}}. \tag{10}$$

The solid particles are porous, and the reaction can be considered to be kinetically controlled following the shrinking core model applied to microscopic grains inside the porous particle [45]. Application of the shrinking core model with reaction rate control [46] is evident in the final factor of Equation (10). The reaction rate constant, k_s, is expressed as follows, where the detrimental effect of increasing pressure is accounted for in the pre-exponential factor:

$$k_s = \frac{k_0}{\overline{p}^q} e^{-\frac{E_a}{RT}}. \tag{11}$$

The kinetic parameters for the reduction and oxidation reactions, as well as the oxygen carrier properties, were obtained from the experimental work of Abad et al. [45]. It can be noted that the aforementioned study found no intraparticle mass transfer limitations, as may be expected for such small, porous particles.

In the fTFM, the subgrid bubbles and clusters impose an additional mass transfer limitation on the reactions, since the gaseous reactants have to be transported into the dense solid clusters for the reactions to occur. This effect is modeled in Equation (10) by means of an effectiveness factor, η. In the present study, η was first modeled for a reference first-order solids catalyzed reaction with a fixed reaction rate constant, as in previous studies [28,41]. It was then found that the reference closure can be effectively scaled to different reaction rate constants and reaction orders by drawing an analogy with packed bed theory and defining a cluster-scale Thiele modulus, as follows:

$$\phi = \sqrt{\frac{n+1}{2} \frac{k'(d_p \mathcal{L})^2}{D}}. \tag{12}$$

Here, $\mathcal{L}$ is the average ratio of the cluster diameter to the particle diameter, which requires closure, and ϕ is the Thiele modulus [47]. The effective reaction rate constant, k', was obtained by re-writing the reaction equation as first-order with respect to the gaseous reactant and the solids volume fraction. This approach has previously been shown to be useful to extend effectiveness factors from intraparticle mass transfer theory to various reaction orders [48]. For the example of Equation (10), the effective reaction rate constant becomes:

$$k' = \rho_s a_0 \left(\widetilde{X_{Ni}} + \widetilde{X_{NiO}} \right) k_s \widetilde{C_A}^{n-1} \left(1 - \frac{\widetilde{X_B}}{\widetilde{X_{Ni}} + \widetilde{X_{NiO}}} \right)^{\frac{2}{3}}. \tag{13}$$

The effectiveness factor for a spherical particle can then be written as follows [49]:

$$\eta = \frac{1}{\phi} \left(\frac{1}{\tanh(3\phi)} - \frac{1}{3\phi} \right). \tag{14}$$

This relation is exact for a first-order reaction in a porous particle with no convective transport. Relatively small discrepancies arise for reactions of different order, but the largest uncertainty in this application is the constant deformation of the clusters in the fluidized bed, as well as the convective species transport taking place inside the cluster.

The basic premise of the approach proposed in the present study is that the effectiveness factor in Equation (14) is analogous to the effectiveness factor of a particle cluster at the largest achievable mass transfer resistance (smallest effectiveness factor). This will typically occur at intermediate filtered solids volume fractions when maximum phase segregation is achieved and clusters are relatively large. As the filtered solids volume fraction tends to the limits of zero or maximum packing, clustering disappears, and the mass transfer resistance tends to zero ($\eta = 1$).

A hypothesis can then be formulated that, for different cluster-scale Thiele moduli, the minimum effectiveness factor (η_{min}) can be scaled by using Equation (15) when a filtered effectiveness factor closure (η_{ref}) is derived from resolved simulation data for a reference Thiele modulus (ϕ_{ref}).

$$\eta_{new,min} = \eta_{ref,min} \frac{\frac{1}{\phi_{new}}\left(\frac{1}{\tanh(3\phi_{new})} - \frac{1}{3\phi_{new}}\right)}{\frac{1}{\phi_{ref}}\left(\frac{1}{\tanh\left(3\phi_{ref}\right)} - \frac{1}{3\phi_{ref}}\right)}. \tag{15}$$

Then, the new effectiveness factor (η_{new}) can be calculated as follows, assuming that the tendency towards $\eta = 1$ will be proportional to the tendency of η_{min} to unity (no subgrid correction):

$$\eta_{new} = 1 - \left(1 - \eta_{ref}\right)\frac{\left(1 - \eta_{new,min}\right)}{\left(1 - \eta_{ref,min}\right)}. \tag{16}$$

It was found that the suggested hypothesis holds well and that this approach is essential to accurately model reactions that are not simple first-order solids catalyzed reactions in the fTFM. Consequently, the proposed approach was used to model the reactions in the present study. The complete development and verification of the generalized reactive fTFM closure is presented in the Appendix A.

Finally, it can be noted that the effectiveness factor closure presented here does not account for the Stefan flow (one mole of methane produces three moles of gas products) occurring in the fuel section of the ICR and investigation of this topic is recommended for future work. However, considering that the reduction reactions are extremely fast (see Section 3.1.1) and occur only near the inlet, it is not expected to have a large impact on the overall reactor behavior.

2.1.3. Simulation Geometry and Mesh

Figure 2 shows the reactor geometry that was considered for the ICR. In the base case, the reactor consists of a cylinder with a height of 6.92 m and a diameter of 3.46 m. These sizes were selected to yield a fluidization velocity of roughly 1 m/s in the freeboard, which is a typical value for vigorous bubbling fluidization. An aspect ratio of 2, typical of fluidized beds, was chosen. A thin wall separates the reactor into the reduction and oxidation sections, consisting of a 2 m high vertical section at the center of the bed and a section sloping at an angle of 30° with the vertical axis to the reactor wall to ensure that solids will not deposit on this surface. Two ports allow the oxygen carrier to circulate between the sections. Reduced oxygen carrier travels through the bottom port (height of 0.6 m) to the air section, whereas oxidized oxygen carrier is carried through the top port (terminating 0.7 m above the bottom of the reactor) to the reduction section. In the base geometry, the width of the square ports (see Figure 2b) is 20 cm. The gas outlet from each section was sized to yield a velocity of roughly 50 m/s, accounting for the much larger flow rate in the air section, which is a typical value for gas transport.

A cut-cell mesh was used to mesh the complex ICR geometry. Long simulations, in the order of 2500 simulated seconds, were necessary to achieve steady reactor behavior; therefore, the average grid size was chosen to yield a coarse mesh of approximately 50,000 cells. A minimum of five cells across the gaps in the ports and the outlets were specified to resolve the most important flow gradients.

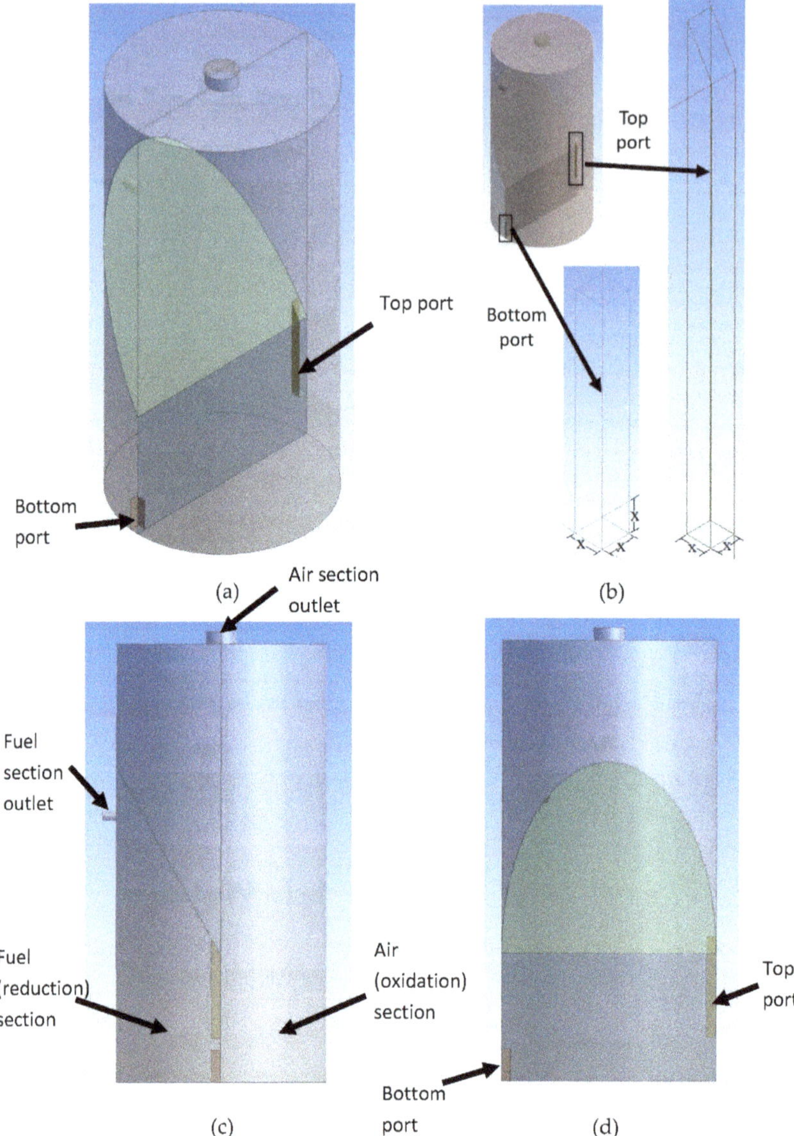

Figure 2. Simulated geometry for the internally circulating reactor (ICR): (**a**) isometric view; (**b**) details of the ports connecting the reactor sections, where x is the port width; (**c**) front view; (**d**) side view.

2.1.4. Reactor Operating Conditions

NiO particles supported on Al_2O_3 were used as oxygen carrier. The oxygen carrier particles were considered to have a diameter of 150 µm, a typical value for bubbling fluidization, a density of 3446 kg/m^3, and an active mass fraction of 0.4 [45]. It can be noted that the reactor model assumes monodisperse particles, due to the complexity of accounting for particle size distributions in fTFMs and due to the limited state of development of subgrid closures accounting for polydispersity [50]. Additionally, the simulations assume the particle density to be constant during the reactor operation, since the density changes will be small (mostly less than 5%) due to the high inert content of the oxygen carrier. The loading of the bed corresponds to an initial bed height of 1 m at a solids volume fraction of 0.6.

Fluidization gas was added uniformly at the bottom of the reactor to the two reactor sections, assuming a perfect gas distributor. Additionally, the effect of the inlet conditions on the subgrid behavior were not accounted for in the fTFM closures. These simplifications are necessary, since none of the state-of-the-art fTFMs have thus far included these effects in their closures. However, considering the large dimensions of the reactors considered, inlet effects are expected to have a relatively small influence on the overall reactor behavior, thereby minimizing the error associated with these simplifications. The other inlet boundary conditions are listed in Table 1 (note that the natural gas used in the process simulations was replaced with an equivalent amount of methane in the reactor simulations).

Table 1. Summary of the conditions for the inlets of the two reactor sections.

Inlet	Oxidation	Reduction
Mass flow rate (kg/s)	41.15	0.698
Temperature (°C)	422	434
Composition	Air	Methane

Uniform pressure outlet boundary conditions were considered for the fuel and air section outlets. For the air section outlet, a pressure of 18 bar (absolute) was considered, which results from the air compressor pressure ratio of 18 [51] employed in the process simulations, which is a typical value for standard, large-scale, F-class gas turbines [52]. For the fuel section outlet, a relatively small overpressure relative to the air section outlet was employed to achieve a target flow rate. This is discussed in more detail in Section 3.1.2.

A no-slip boundary condition was specified for the gas at the walls, whereas partial slip boundary conditions with a specularity coefficient of 0.1 was employed at the walls, based on the model of Johnson and Jackson [53]. It can be noted that, technically, a subgrid closure is required for the particle–wall interaction. However, such closures have not yet been developed in the fTFM community and were therefore neglected in the present study. The effect of the particle–wall boundary condition is expected to be small for the large reactor dimensions considered in the present study; therefore, neglecting the sub-grid effects is a reasonable assumption.

2.1.5. Solver

The reactor simulations were performed in the commercial CFD solver, ANSYS FLUENT 19.2, using user defined functions to implement the subgrid closures of the fTFM. The phase-coupled SIMPLE algorithm [54] was used for pressure-velocity coupling, and all other equations were discretized based on the QUICK scheme [55].

2.2. Process Modeling

This study conducted an economic assessment of the ICR integrated into a natural gas combined cycle (NGCC) power plant, as recently evaluated for CO_2 capture using CLC [34]. The interested reader is referred to that study for details about the process modeling methodology. One important change from this previous work was the inclusion of the gas leakage between reactor sections in the ICR. The mixing between the outlet streams of the oxidation and reduction reactors was adjusted to yield 95% CO_2 capture and purity (molar percentage and dry basis), based on the reactor simulations (see Section 3.1.3) for the conditions considered.

The layout of the simulated plant is shown in Figure 3, where 20 parallel ICR reactors were needed to accommodate the required air throughput. Natural gas is pre-heated and fed to the fuel section of the ICR reactors where it is converted to CO_2 and H_2O, which is expanded to generate some power.

After H_2O is condensed out, the remaining CO_2 is compressed and pumped to 110 bar for transport and storage. The air section of the ICR replaces the combustor for the main gas turbine. Air from the main compressor reacts with the reduced oxygen carrier in a highly exothermic reaction and is heated to 1150 °C in the base case. This temperature was selected based on material limitations, and a sensitivity analysis of this value was performed in the economic evaluation in Section 3.2. The hot depleted air stream is then expanded in the main gas turbine before being sent to a heat recovery steam generator for extra power production using a steam cycle. The results of this plant were compared to the reference NGCC plant detailed in Khan et al. [34].

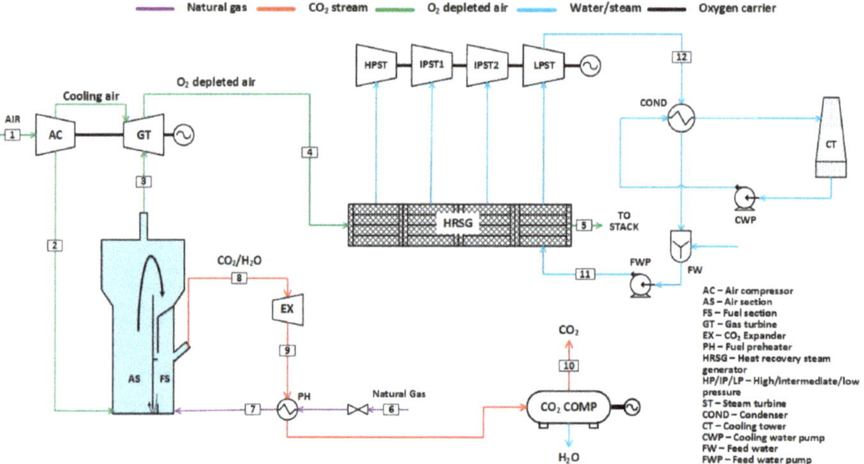

Figure 3. Process flowsheet of the ICR integrated into a combined cycle. It can be noted that the reactor in the flowsheet represents a cluster of ICR reactors and that the outputs from these reactors were combined in stream 3.

2.3. Economic Assessment

Capital costs: The total cost of the combined power cycle was obtained directly from the PEACE component in Thermoflex. This includes direct component costs and several additional cost components accounting for construction, engineering, contingencies, and other cost components.

Costs related to the CO_2 compressors and intercoolers were estimated using installed cost data from Aspen Plus. This cost was increased by approximately 74% to account for engineering, contingencies, and owner's costs, based on the methodology of Gerdes et al. [56].

The ICR capital costs were estimated based on cost correlations for process vessels from Turton et al. [57]. Each ICR was composed of two process vessels: (1) an inner vessel to carry the temperature, attrition, and corrosive loads constructed from an expensive Ni-alloy, and (2) a thick pressure shell carrying the pressure load constructed from carbon steel. An insulation layer of 0.4 m thickness was inserted between these two vessels. To account for the relatively complex ICR geometry, the cost of the inner vessel was increased by a factor of three. This was a somewhat arbitrary adjustment, and a sensitivity of total plant economics to ICR cost is therefore presented later. Costs for auxiliaries and contingencies were subsequently added according to Turton et al. [57] to yield the total reactor cost. A breakdown of the different components of the cost of the 20 ICR units required for the base case is shown in Figure 4. It can be noted that the capital cost associated with the oxygen carrier was only for the initial loading (replacement of the oxygen carrier was considered under operating and maintenance costs). Further, the number of ICR reactors selected to deliver the required process throughput at the reactor operating conditions is specified in Section 2.1.4.

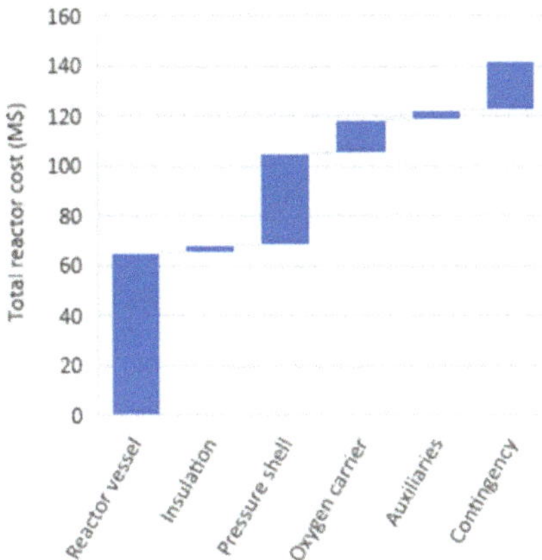

Figure 4. Breakdown of the cost of 20 ICR units.

The total costs of these three main components were then added together to yield the total plant cost. All costs are reported in 2019 US dollars using the chemical engineering plant cost index (CEPCI).

Operating and maintenance (O&M) costs: Fixed O&M labor costs were calculated assuming 11 personnel per shift for the NGCC reference plant and 13 personnel per shift for the ICR plant using the methodology of Peters and Timmerhaus [58]. A $45/hour rate was used with appropriate increases for benefits, maintenance labor, and overheads. In addition, 1.5% of total plant costs per year was added for insurance and property taxes. The key assumptions for variable O&M costs are summarized in Table 2. Costs for the oxygen carrier [59] and water [60] were taken from the literature, whereas the oxygen carrier lifetime was specified based on discussions with catalyst suppliers. CO_2 transport and storage costs can vary widely based on the transport distance and the type of transport and storage, but a reasonable average value was selected based on costs provided in two IEA reports [61,62]. Natural gas prices are known to vary widely, and a value representative of Europe was assumed for this study.

Table 2. Variable O&M cost assumptions.

Natural Gas	8 $/GJ
Oxygen carrier (OC)	15 $/kg
OC replacement period	2 years
Process water	2 $/m^3
Cooling water	0.35 $/m^3
CO_2 transport and storage costs	12 $/ton CO_2

Capital and O&M costs were then used to calculate the levelized cost of electricity (Equation (17)) and the CO_2 avoidance costs (Equation (18)) using a discount rate of 8%, a plant economic lifetime of 30 years, and a construction period of 2 years for NGCC and 3 years for the ICR plants (investment is assumed to be linear over the construction period).

$$\text{LCOE (\$/MWh)} = \frac{\sum_{t=1}^{m} \frac{I_t + M_t + F_t}{(1+r)^t}}{\sum_{t=1}^{m} \frac{E_t}{(1+r)^t}}, \tag{17}$$

$$\text{CAC (\$/ton)} = \frac{LCOE_{CCS} - LCOE_{ref}}{e_{ref} - e_{CCS}}. \tag{18}$$

Here, the summations are done for each year during construction and operation (t) up to the end of the plant economic lifetime (m). I is the investment expenditures, M is the O&M expenditures, F is the fuel expenditure, E is the electricity generation, r is the discount rate, and e is the plant-specific emissions (ton/MWh).

3. Results

This section outlines how the fTFM described in Section 2.1 was first used to optimize and size an industrial-scale ICR reactor for power production with integrated CO_2 capture. Subsequently, the ICR process was then evaluated economically using the reactor size and performance suggested by the simulations.

3.1. Reactor Optimization

3.1.1. Characteristics of ICR Operation

In this study, plots and animations from the reactor simulations were used to introduce important characteristics of ICR operation for a typical case. Firstly, Figure 5 (as well as the associated Video S1 in the Supplementary Material) demonstrates the circulation of the oxygen carrier between the two reactor sections. In the reduction section, the relatively low fluidization velocity from the fuel feed results in a dense bubbling bed. Due to the very fast reaction of the oxygen carrier with the methane, most of the conversion takes place near the inlet, leading to a highly reduced oxygen carrier. However, the mixing is very fast in the fluidized bed and a relatively uniform distribution of the oxygen carrier is rapidly attained in the rest of the bed on the fuel side. Oxygen carrier particles, reduced by the fuel, pass through the bottom port to the air (oxidation) section, where they are rapidly oxidized and mixed into the rest of the particles. Owing to the much larger molar flow rate on the air side, a more vigorous fluidization occurs, lifting the particles to the freeboard, including a diameter expansion (which helps to reduce particle elutriation), and allowing them to pass back to the fuel section through the top port. Again, the oxidized particles mix rapidly into the reduction side bed, where they are reduced by the fuel.

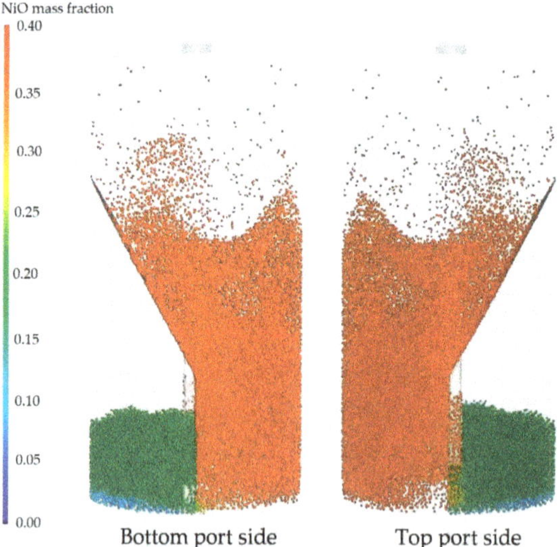

Figure 5. Particle plot of the instantaneous NiO mass fraction. A corresponding animation is provided in Video S1. Note that the particles pictured are tracers following the continuous solids flow for visualization purposes and do not influence the simulation solution.

Some further comments can be made about the nature of the solids flow through the ports. In the bottom port, the flow is quite dense, with a time-average solids volume fraction of about 0.5. The flow through the top port is more dilute, with a time-averaged solids volume fraction between 0.3 and 0.4. The animations show some transient fluctuations of solids in both ports; therefore, the flow in not completely steady and there is a risk of backflow, which might reduce the reactor performance. It may be noted that no problems with blockage of the ports have been experienced during extensive experimental evaluations of the ICR concept [14,15].

Figure 6 (and Video S2) shows that the solids circulation between the reactor sections is associated with undesired gas leakage—CO_2 leaks from the fuel section to the air section, reducing the CO_2 capture efficiency of the reactor, and N_2 leaks from the air section to the fuel section, reducing the purity of the CO_2. One of the most important criteria for designing and operating the ICR is therefore to minimize the amount of gas leakage between the reactor sections, while maintaining sufficient oxygen carrier circulation to ensure that the fuel is completely converted in the reduction sector.

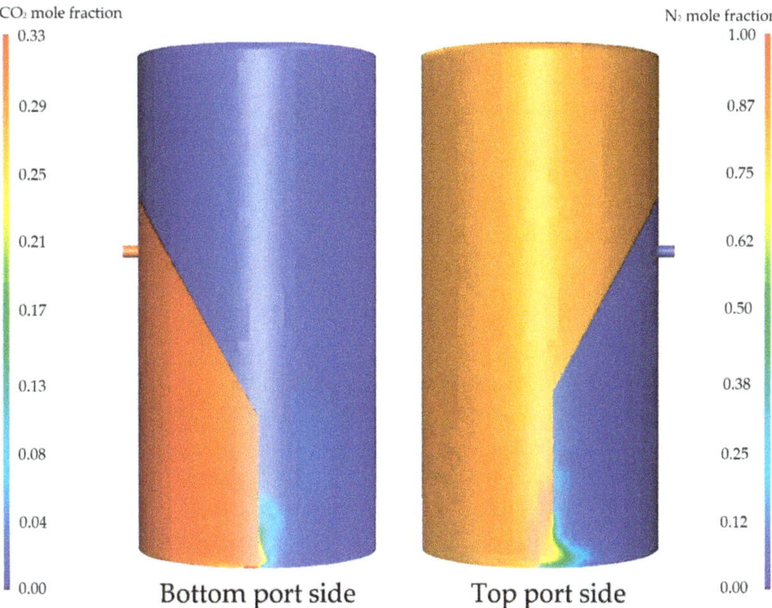

Figure 6. Contour plot of the instantaneous CO_2 and N_2 mole fractions at the outer wall of the ICR showing the undesired gas leakage between the two sections of the reactor. A corresponding animation is provided in Video S2.

Many design and operating parameters can influence the ICR performance. These include, but are not limited to, the solids loading, the particle size and distribution, the gas flow rates to the reactor sections, the operating pressure and temperature, and several dimensions of the reactor and internals. Due to the complexity of simultaneously optimizing these parameters, the scope of the present study was limited to three important factors that will be investigated in the subsequent sections:

- The pressure difference between the two reactor outlets, which can be used to control the solids distribution between the reactor sections, as well as the solids circulation.
- The size of the ports connecting the reactor sections, which can be sized to allow sufficient solids circulation while limiting undesired gas leakage.
- The overall reactor size, which primarily determines the amount of solids elutriation from the reactor.

3.1.2. Reduction Section Overpressure

During ICR operation, the most practical way to control the reactor behavior would be to tune the pressure difference between the reduction and oxidation section outlets. For example, applying an overpressure at the reduction side outlet will lead to more gas exiting the reactor on the oxidation side. To satisfy the mass balance of the reactor, this means that relatively more gas must pass through the bottom port to reach the oxidation section. As is shown in this section, this gas flow influences the solids circulation between the sections, as well as the distribution of solids in the two sections. Both these factors have a critical effect on the reactor performance.

To better understand this behavior, ICR simulations were performed at different overpressures applied at the reduction section outlet. Specifically, reduction outlet flow ratios (ROFR) from 0.96 to 1 were investigated. The ROFR is defined as the ratio of the reduction outlet molar flow rate to the ideal outlet molar flow rate that would occur in case of no gas leakage between the reactor sections and complete fuel conversion. A lower ROFR implies a higher overpressure in the fuel section, with the ROFR = 0.96 case corresponding to an overpressure of 0.34 bar. It can also be noted that the figures and animations presented in the previous section are for the case of ROFR = 0.98.

Figure 7 (and Video S3) shows the effect of the reduction section outlet overpressure on the ICR behavior. Firstly, it is interesting to note from the solids volume fraction values that, despite the large grid sizes employed, a substantial amount of phase segregation is still resolved in the more vigorously fluidized air section. In the slowly fluidized fuel section, the resolved solids distribution is nearly homogenous, and the effects of particle clusters and bubbles are therefore nearly completely accounted for in the subgrid closures of the fTFM.

Increasing the overpressure (corresponding to lower ROFR values), more gas will pass through the bottom port to the air section, which also increases the solids flow rate through the bottom port. This causes the bed loading on the oxidation side to increase, which creates the hydrostatic pressure buildup required to achieve solids flow through the top port against the overpressure imposed in the fuel section.

Therefore, cases with a lower ROFR will reach a pseudo-steady state (where the time-averaged solids flow rates through the top and bottom ports are equal) with a larger fraction of the oxygen carrier on the oxidation side, as shown in Figure 8b. The solids elutriation rate (Figure 8a), which occurs almost entirely from the oxidation side, is therefore greatest at lower ROFR values.

Furthermore, Figure 8a shows that a maximum solids circulation rate occurs at an ROFR of 0.98. At first, when decreasing the ROFR from 1, the solids circulation rate increases due to more gas flow through the bottom port, thereby entraining more solids, as well as more solids flowing through the top port as a result of the higher bed loading on the air side. However, as the ROFR further decreases, the bed on the fuel side becomes so low that the solids barely reach the top of the bottom port, thus limiting the achievable circulation rate through the bottom port. Regular backflow through the bottom port also starts to happen during these cases, since the low bed height on the fuel side does not provide sufficient hydrostatic pressure to maintain a steady flow through the bottom port. If the solids circulation rate becomes too low, significant fuel slip, that is, incomplete fuel conversion, starts to occur (Figure 8c), since not enough oxygen is available in the fuel section to convert all of the fuel and the low bed height reduces the gas residence time in the bed.

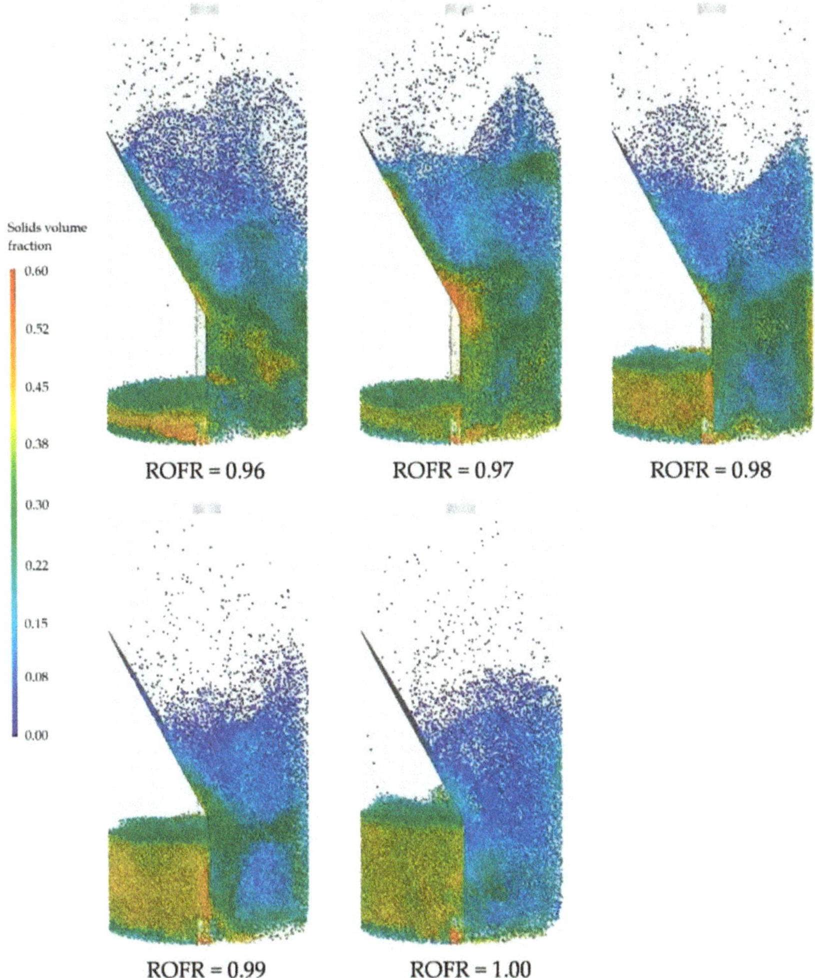

Figure 7. Instantaneous particle plot of the cases with varying reduction outlet flow ratios, where the particles are colored by the particle volume fraction. A corresponding animation is provided in Video S3. Note that the particles pictured are tracers following the continuous solids flow for visualization purposes and do not influence the simulation solution.

Figure 8c also shows that the CO_2 capture percentage decreases as the ROFR decreases, due to the increased gas flow through the bottom port, allowing more CO_2 to exit with the depleted air at the oxidation section outlet. The CO_2 purity behavior is more complex—it remains relatively constant when lowering the ROFR from 1 to 0.98, despite the increasing solids flow rate through the top port. This indicates that the gas-to-solids leakage ratio through the top port increases at high ROFR values, resulting in more gas leakage per unit of solids circulation. The purity increases in the ROFR = 0.97 case due to the lowering solids flow rate but decreases in the ROFR = 0.96 case due to backflow through the bottom port.

Based on the results in this section, the ROFR = 0.98 case (corresponding to a reduction outlet overpressure of 0.26 bar) was chosen for further investigation, primarily due to the high solids circulation rate that was obtained. Furthermore, this case showed a good compromise between decreasing solids elutriation and decreasing CO$_2$ capture with decreasing ROFR.

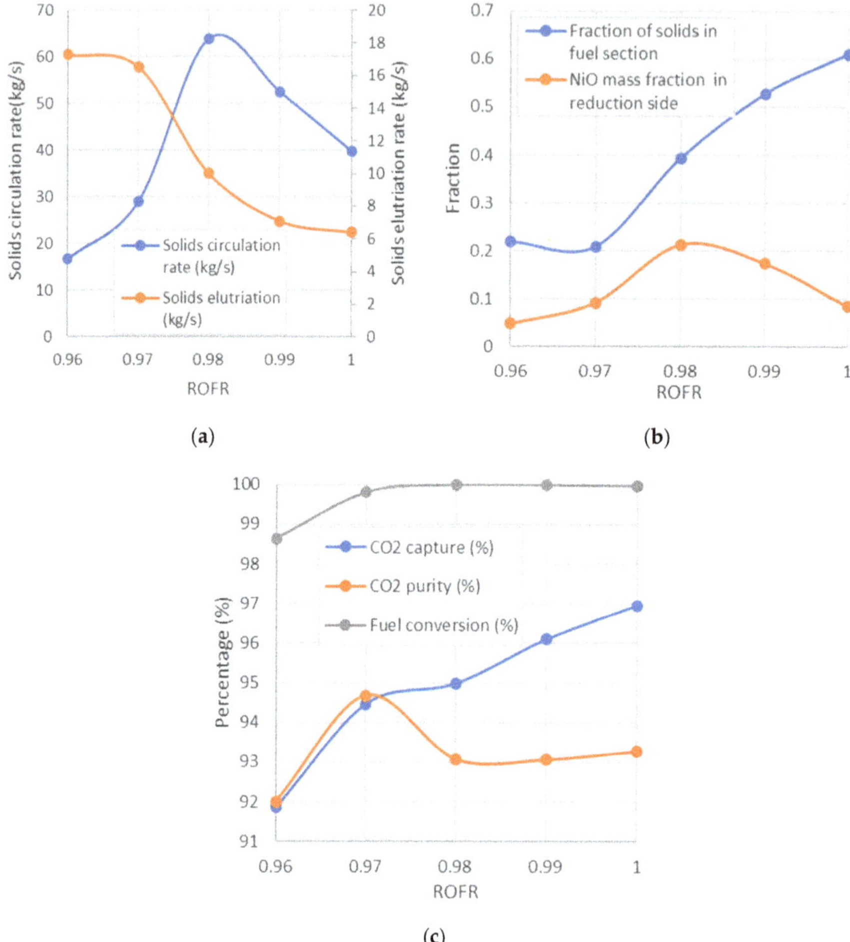

Figure 8. Summary of important time-averaged properties of the ICR as a function of the reduction outlet flow ratio: (**a**) solids circulation and elutriation rates; (**b**) fraction of solids and NiO mass fraction in the reduction side; (**c**) CO$_2$ capture, CO$_2$ purity (molar percentage on a dry basis), and fuel conversion.

3.1.3. Port Size

The previous section revealed that the primary criterion for achieving complete fuel conversion is a sufficient solids circulation rate to transport enough oxygen to the fuel reactor to oxidize the methane. However, since the ROFR = 0.98 case from the previous section had a much higher than necessary solids circulation rate, there is the potential to further increase the reactor performance by decreasing the size of the ports (dimension x in Figure 2). This will decrease the solids circulation rate between the reactor sections, but also decrease the associated undesired gas leakage.

Simulations were therefore performed at an ROFR of 0.98 while decreasing the port size dimensions, as shown in Figure 9. As expected, the results showed a decreasing solids circulation rate with decreasing port size. Consequently, the average NiO mass fraction in the fuel section was also reduced. In the case with a port size of 16 cm, not enough NiO was present in the reduction section to fully oxidize the fuel, leading to a large fuel slip, which would be unacceptable in practice.

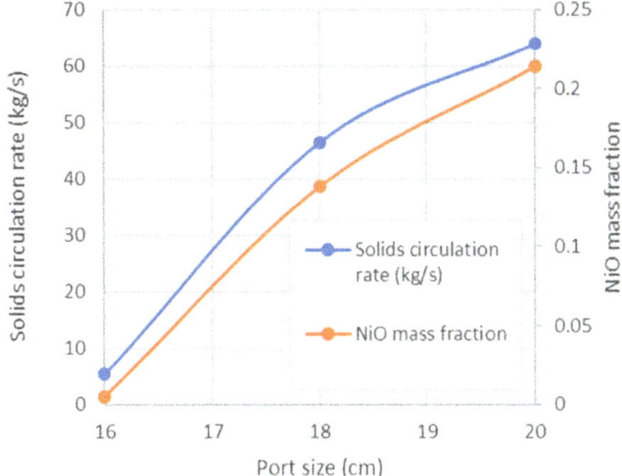

Figure 9. Plot of the solids circulation rate and the NiO mass fraction in the fuel section with changing port size.

The case with ROFR = 0.98 and a port size of 18 cm was consequently chosen as the best ICR case. Compared to the case with ROFR = 0.98 and a port size of 20 cm, the CO_2 capture efficiency increased from 95.0% to 95.7%, and the CO_2 purity increased from 93.1% to 95.2% while still preserving complete fuel conversion. Consequently, CO_2 capture efficiencies and purities of 95% are assumed as reasonable for the economic assessment in Section 3.2.

3.1.4. Reactor Size

One of the main criteria for reactor sizing is usually to ensure complete reactant conversion, with the increasing reactor diameter and height both serving to increase the gas residence time (the latter by decreasing the superficial velocity of a fixed mass flow rate of gas) for complete conversion. However, as previously noted, the oxidation reaction of methane with NiO is very fast for the temperatures considered for ICR operation. Additionally, the height of the bed on the fuel side is limited to a minimum due to the height of the bottom port. Consequently, the solids circulation rate between the reactor sections is primarily responsible for ensuring complete fuel conversion, and the overall reactor size is of lesser importance.

Nonetheless, the overall reactor size has an important effect on the particle elutriation. In the best case investigated so far (ROFR = 0.98 and a port size of 18 cm), the predicted solids elutriation rate was 11 kg/s, corresponding to 2.0% of the total solids loading per hour. Consequently, a rather large amount of elutriated solids will have to be separated using cyclones and filters downstream of the reactor. The amount of solids elutriation can be reduced by increasing the reactor height (more space for solids to fall back down) and diameter (lower superficial gas velocities).

To investigate the effect of increasing the reactor size, all the reactor dimensions were multiplied by a scaling factor (SF) from 1 to 1.2 (compared to the base case geometry discussed in Section 2.1.3. with a port size of 18 cm). Figure 10 (and Video S4) visualize the change in reactor behavior with increasing scaling factor, showing clearly that a larger reactor leads to a slower fluidization and less solids elutriation on the air side.

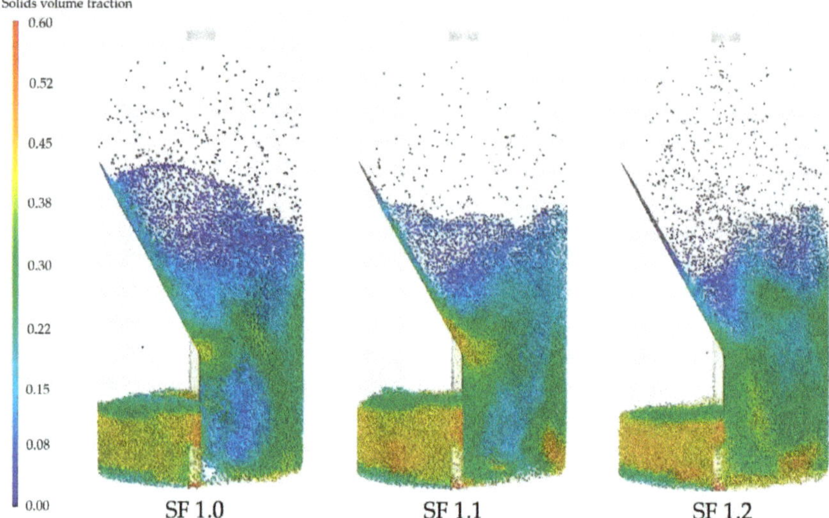

Figure 10. Instantaneous particle plot of the cases with reactor scaling factors ranging from 1 to 1.2, where the particles are colored by the particle volume fraction. A corresponding animation is provided in Video S4. Note that the particles pictured are tracers following the continuous solids flow for visualization purposes and do not influence the simulation solution.

The decreasing elutriation rate is quantified in Figure 11a, with the solids elutriation (in percentage of solids loading per hour) decreasing from 2.0% to 0.49% at a scaling factor of 1.1, and 0.20% at a scaling factor of 1.2. However, it was also found that the CO_2 capture and purity decreases with increasing scaling factor (Figure 11c). This is because the port sizes were also scaled along with the other reactor dimensions, leading to an increase in the solids circulation rate (Figure 11a) and the associated gas leakage. However, the results also showed an increase in the average NiO fraction in the fuel section (Figure 11b), indicating higher than necessary solids circulation rate. As a result, it is expected that CO_2 capture efficiencies and purities of more than 95% can be obtained by again decreasing the port size to achieve a solids circulation rate close to the minimum required for complete fuel conversion, as was done in Section 3.1.3.

In summary, the preferred ICR reactor size will in practice be determined by the trade-off in increasing capital cost of the reactor and the decreasing cost of handling elutriated solids with increasing reactor size. The cost of the former is quantified in the subsequent economic assessment, whereas quantification of the latter is left for a future study.

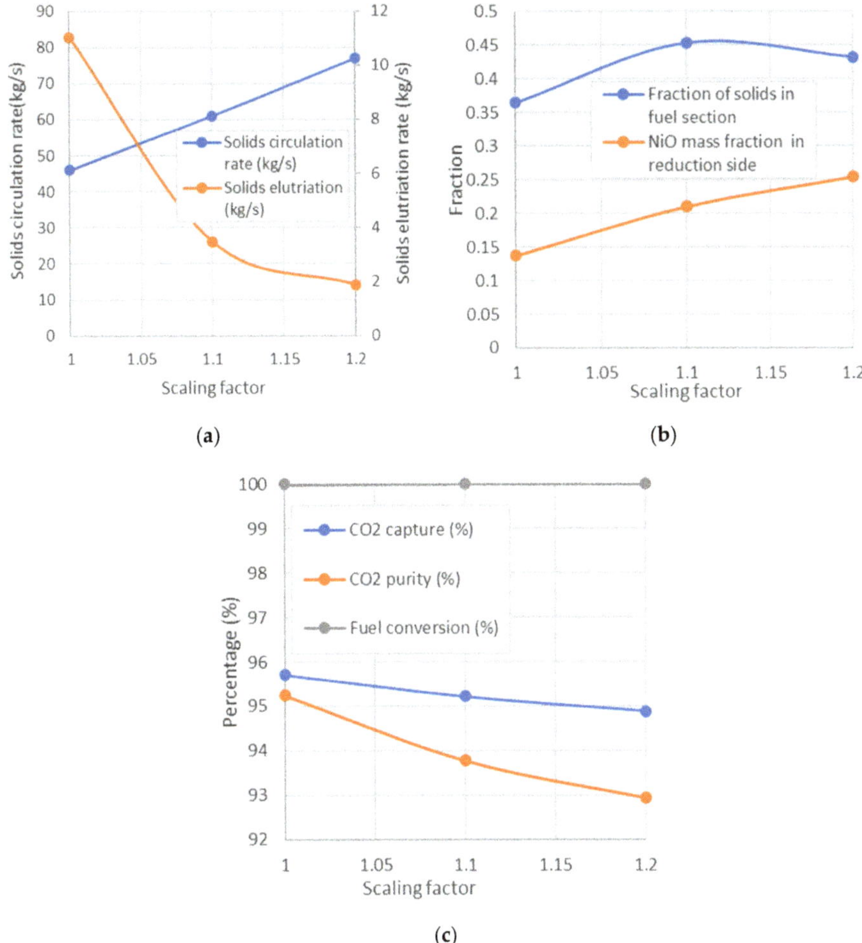

Figure 11. Summary of important time-averaged properties of the ICR as a function of the reactor size scaling factor: (**a**) solids circulation and elutriation rates; (**b**) fraction of solids and NiO mass fraction in the reduction side; (**c**) CO_2 capture, CO_2 purity, and fuel conversion.

3.2. Economic Evaluation

In this section, an ICR-based NGCC plant is evaluated economically based on the CO_2 capture and purity estimated in the reactor simulations, as well as the size of reactor that was found to be adequate (SF = 1). In addition, an unabated NGCC plant is assessed as a reference.

Three ICR cases were investigated with different assumptions about the maximum reactor temperature. In combined cycles, it is crucial to maximize the turbine inlet temperature to maximize net electric efficiency and minimize the required air flow rate through the system. However, the reactor, oxygen carrier, and downstream filters will impose constraints on the maximum allowable temperature. The oxygen carrier material may be the most important constraint. To date, the highest successfully demonstrated operating temperature has been a NiO-based oxygen carrier successfully operated at 1185 °C [44]. However, modern gas turbines can operate with combustor outlet temperatures well over 1600 °C, implying that this is a very important constraint. The economic impact of this constraint will be illustrated in this work.

To this end, the cases listed in Table 3 were evaluated in this study. It can be noted that the reactor simulations were performed only for the ICR-1150 case, since it was expected that the gas leakage would not vary much between the cases at different temperatures.

It is immediately evident from the results in Table 3 that the achieved plant efficiency is strongly dependent on the ICR outlet temperature. When the temperature can reach up to 1300 °C, an attractively low energy penalty of 3.6% points is attained, but this increases to an unacceptable 12.3% points in the case with an ICR outlet temperature of 1000 °C. Since fuel costs are typically the major cost component in a natural gas fired plant, this is expected to have a major impact on the economic performance.

Table 3. Performance of the different plants evaluated in this study.

Plant	NGCC	ICR-1000	ICR-1150	ICR-1300
Combustor/reactor outlet temperature (°C)	1416	1000	1150	1300
Thermal input (MW)	765.0	697.5	697.5	697.5
Gas turbine (MW)	292.5	220.9	223.0	235.5
Steam turbine (MW)	161.6	84.3	106.9	121.5
CO_2 expander (MW)		36.3	41.4	46.4
CO_2 compressors (MW)		−15.9	−15.7	−15.5
Auxiliaries (MW)	−8.5	−4.6	−6.0	−6.6
Net power (MW)	445.6	321.0	349.7	381.3
Net electric efficiency (%)	58.3	46.0	50.1	54.7
CO_2 intensity (kg/MWh)	352.2	22.3	20.5	18.8

The levelized cost of electricity and CO_2 avoidance cost for the different cases are shown in Figure 12. Clearly, the ICR outlet temperature has a large impact on the economic performance of the plant. Interestingly, the relative increase in capital costs from the ICR-1300 case to the ICR-1000 case (50%) is greater than the relative increase in fuel costs (25%). This is due to the large increase in air flowrate required by the cases with lower ICR outlet temperatures. The larger air flowrate increases the costs of ICR reactors, gas turbines, and heat recovery steam generators, adding to the increase in specific capital costs caused by a reduction in net electric efficiency.

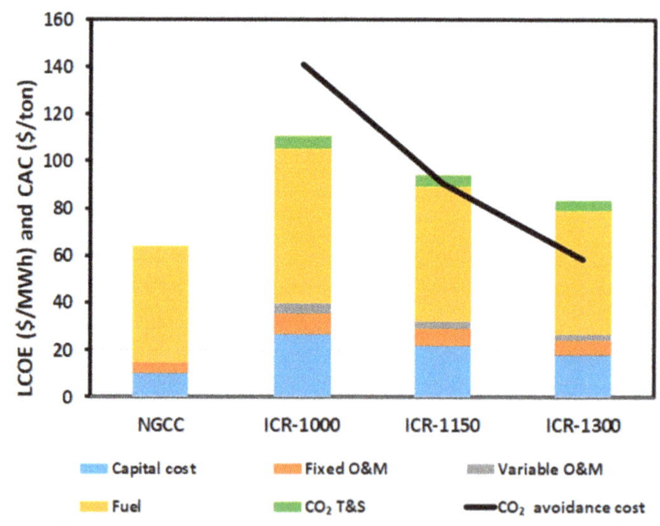

Figure 12. Levelized cost of electricity (LCOE) and CO_2 avoidance costs (CAC).

CO_2 avoidance costs range from $58/ton to $141/ton. For comparison, a thorough review by Rubin et al. [63] found the representative CO_2 avoidance cost for NGCC plants with post-combustion capture to be \$107/ton (after adjusting to 2019 currency and adding CO_2 transport and storage costs). The ICR-1300 case, if it is technically possible, therefore appears attractive relative to this benchmark, whereas the \$91/ton CO_2 avoidance cost of the more realistic ICR-1150 case offers marginal benefits.

As outlined earlier, all the reactor sizes investigated performed well in terms of fuel conversion, and the primary difference was the amount of particle elutriation. Preventing particle elutriation will require the addition of a cyclone (possibly an internal cyclone placed in the ICR freeboard), including smaller reactor sizes needing cyclones with a higher separation efficiency to keep escaped fines constant. Inclusion of such a cyclone was not studied here, but the effect of the change in reactor size was investigated.

Figure 13 shows that higher reactor costs have a relatively small impact on overall economic performance of the plant. An increase in reactor costs from 108 to 233 M\$ only increased the levelized cost of electricity by \$5.5/MWh and the CO_2 avoidance cost by \$16.5/ton. The uncertainty in the reactor cost estimation was also highlighted in the methodology description, particularly the factor 3 that was used to account for the cost increase of including the ICR internal structure in a Ni-alloy process vessel. For perspective, the range of reactor costs shown in Figure 13 is equivalent to a wide range in this tuning factor of 1.7–6.4, indicating that the overall plant performance is not overly sensitive to uncertainties in the ICR cost assessment.

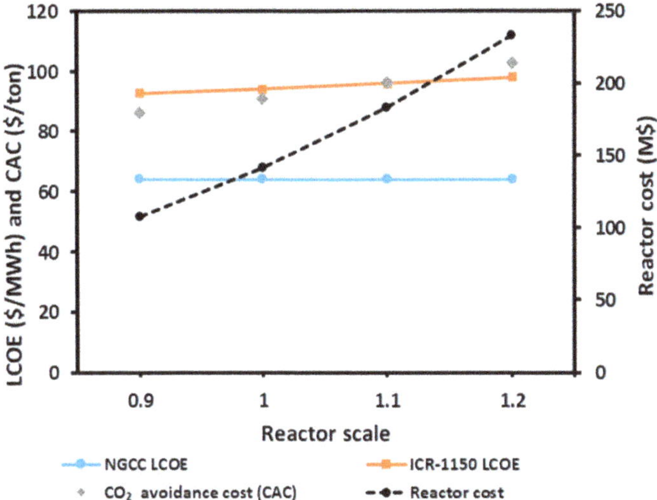

Figure 13. The effect of reactor scale factor (reactor diameter and height relative to the base case) on economic performance indicators.

4. Discussion

The energy and process industry is in a state of flux due to climate change concerns, increasingly stringent environmental regulations, and constraints on raw material availability. However, a swift response to this changing market environment is challenging due to the multidecade scale-up and demonstration timeframes typical in the industry. This study applied a simulation-based approach that could alleviate this challenge to the design and economic evaluation of a novel reactor concept for clean power production from natural gas.

Specifically, filtered two-fluid model (fTFM) simulations were used to design and further investigate a relatively complex internally circulating reactor (ICR) at an industrial scale. Naturally, there are many uncertainties with modeling such a complex multiphase reactor at industrial scales and operating conditions, but the simulation results served to increase confidence in the ICR technology and offered several insights on the practical operation of a large-scale ICR. Such insights would be very expensive and time-consuming to gain via gradual experimental scale-up and demonstration.

The ICR has many different design and operating parameters that can be adjusted to optimize performance. Simulations can allow for cost-effective completion of the very large number of design iterations that will be required to successfully optimize within this large parameter space. Model accuracy is the main uncertainty in this strategy, making validation against large-scale experimental data a high priority. Unfortunately, such data is scarce, often not publicly available, and generally not detailed enough for proper model validation. Given the large fundamental benefits of such a simulation-based design approach, efforts to collect such data from large-scale reactors for further fTFM development and validation is highly recommended. Furthermore, there is substantial scope for further development of fTFM closures for improved accuracy and generality, which would reduce the uncertainty of simulation-based investigations of industrial-scale fluidized beds. Accounting for polydispersity in fTFMs is an especially rich area for research, considering the importance of polydispersity to many industrial applications, the complexity of its subgrid modeling, and the limited amount of research on the topic to date.

Sizing of the ICR via simulations allowed for an economic assessment of the ICR integrated into a natural gas combined cycle (NGCC) power plant for clean power production using the chemical looping combustion (CLC) principle. The primary challenge of using CLC in combined cycles is the maximum temperature achievable in the reactor, which limits the turbine inlet temperature and, hence, the power cycle efficiency. Results from the economic assessment illustrated this with a decrease in CO_2 avoidance costs from \$141/ton to \$58/ton if the ICR operating temperature could be increased from 1000 to 1300 °C. These results emphasize that one of the main focusses of oxygen carrier development effort should be the high temperature durability of the materials.

Since it is doubtful that ICR operating temperatures can increase far beyond 1200 °C and gas turbine technology keeps progressing beyond 1600 °C to allow for higher efficiencies, the maximum reactor temperature is a serious limitation. One potential solution is to combust additional fuel after the ICR to further increase the gas temperature, as outlined in Khan et al. [34]. The use of such a combustor can eliminate most of the energy penalty, although this is at the expense of increased CO_2 emissions if natural gas is used for the extra firing, or at the expense of added costs if hydrogen is used. The economic implications of this trade-off will be explored in future work involving detailed modeling of a modern gas turbine with a combustor outlet temperature exceeding 1600 °C.

Supplementary Materials: The following are available online at http://www.mdpi.com/2227-9717/7/10/723/s1, Document: Hydrodynamic closure development; Video S1: Oxygen carrier circulation; Video S2: Gas leakage; Video S3: Effect of reduction section outlet overpressure; Video S4: Effect of reactor size.

Author Contributions: Conceptualization, J.H.C. and S.C.; data curation, J.H.C. and M.N.K.; formal analysis, J.H.C., M.N.K., and S.C.; funding acquisition, S.A.; investigation, J.H.C. and M.N.K.; methodology, J.H.C., M.N.K., and S.C.; project administration, S.A.; software, J.H.C.; supervision, S.C. and S.A.; validation, J.H.C. and M.N.K.; visualization, J.H.C. and M.N.K.; writing—original draft, J.H.C., M.N.K., and S.C.; writing—review and editing, J.H.C., S.C., and S.A.

Funding: This research was funded by the Research Council of Norway under the CLIMIT program, grant number: 255462.

Acknowledgments: The resolved simulations used for closure development and verification in this study were performed on computing resources provided by UNINETT Sigma2—the National Infrastructure for High Performance Computing and Data Storage in Norway.

Conflicts of Interest: The authors declare no conflict of interest.

Nomenclature

Acronym Definitions

CCS	Carbon capture and storage	IEA	International Energy Agency
CEPCI	Chemical engineering plant cost index	LCOE	Levelized cost of electricity
CFB	Circulating fluidized bed	NGCC	Natural gas combined cycle
CFD	Computation fluid dynamics	OC	Oxygen carrier
CLC	Chemical looping combustion	O&M	Operating and maintenance
CAC	CO_2 avoidance cost	ROFR	Reduction outlet flow ratio
fTFM	Filtered two-fluid model	T&S	Transport and storage
GPM	Gradient product marker	TFM	Two-fluid model
ICR	Internally circulating reactor		

Main Symbol Definitions

a_0	Specific surface area (m^2/kg)	q	Pressure exponent
C	Molar concentration (mol/m^3)	R	Universal gas constant (J/(mol. K))
D	Diffusion coefficient (m^2/s)	R^H	Heterogenous reaction rate (mol/(m^3.s))
d_p	Particle diameter (m)	S_R	Mass transfer source term (kg/(m^3.s))
E_a	Activation energy (J/mol)	r	Discount rate
e	Plant specific emissions (ton/MWh)	T	Temperature (K)
F	Fuel expenditure ($/year)	t	Time (s)
$\vec{g}$	Gravitational acceleration (m/s^2)	v	Stoichiometric constant
h	Specific enthalpy (J/kg)	X	Species mass fraction
I	Investment expenditures ($/year)	α	Volume fraction
K_{gs}	Interphase momentum exchange coefficient (kg/(m.s))	γ	Heat transfer coefficient (W/(m^3.K))
k_0	Pre-exponential factor	ΔH_r	Heat of reaction (J/kg)
k	Reaction rate constant (m^{3n-3}/mol^{n-1}s)	Δ_f	Filter size (m)
k_s	Reaction rate constant (m^{3n-2}/mol^{n-1}s)	η	Effectiveness factor
k'	Effective reaction rate constant (1/s)	κ	Thermal conductivity (W/(m.K))
M	Molecular weight (kg/mol)	ρ	Density (kg/m^3)
M	O&M expenditures ($/year)	$\bar{\bar{\tau}}$	Stress tensor (Pa)
m	Plant economic lifetime (years)	υ	Velocity (m/s)
n	Reaction order	ϕ	Thiele modulus
p	Pressure (Pa)	$\mathcal{L}$	Dimensionless cluster length-scale

Sub- and Superscript Definitions

f	Filter	min	Minimum
g	Gas	p	Particle
i	Species index	r	Reaction
k	Reaction index	ref	Reference
max	Maximum	s	Solids

Sub- and Superscript Definitions

$\bar{x}$	Algebraic volume average	$\tilde{x}$	Phase-weighted volume average
x'	Fluctuation from mean (algebraic)	x^*	Scaled value
x''	Fluctuation from mean (phase-weighted)	$\vec{x}$	Vector quantity
$\hat{x}$	Dimensionless value		

Appendix A. Reactive Closure Development

Appendix A.1. Resolved Two-Fluid Model (TFM) Simulations

The subgrid reactive closure was developed from resolved TFM simulations using a similar setup, material properties, and methodology to what has been done in previous work [41]; therefore, the detailed description of the simulation setup and statistical analysis of the data is not repeated here.

However, new resolved simulations were performed with some changes in place compared to the aforementioned study in order to collect better data for reaction modeling. Firstly, the height of the periodic simulation domain was doubled to 1.28 m to increase the region for collecting data. Secondly, the species boundary conditions for the reactants were set to a fixed value at the bottom boundary. This ensured that the reactants were never depleted in the periodic simulations, allowing data to be collected continuously. However, imposing this fixed boundary condition influenced the effectiveness factors predicted near the boundary. Therefore, data was only collected from heights between 0.4 and 0.88 m, where the effectiveness factor was found to be independent of the domain height. Four simulations were performed with domain-averaged solids volume fractions of 0.05, 0.1, 0.2, and 0.4 to provide continuous data over the entire range of filtered solids volume fractions.

In the present study, isothermal conditions were assumed, and five solids-catalyzed reactions were considered. In each of the reactions, a single product gas phase species was converted to another product species. All reactions were independent from each other and each species only took part in one reaction. Three first-order reactions were considered with different rate constants, as well as a 0.5th order and a 2nd order reaction. The different reactions are summarized in Table A1. Consequently, the species transport equation can be written as follows for the reactants and products, respectively.

$$\frac{\partial}{\partial t}\left(\alpha_g \rho_g X_i\right) + \nabla \cdot \left(\alpha_g \rho_g \vec{v}_g X_i\right) = \nabla \cdot \left(\alpha_g \rho_g D_i \nabla X_i\right) - k_i \alpha_s C_i^{n_i} M_i, \tag{A1}$$

$$\frac{\partial}{\partial t}\left(\alpha_g \rho_g X_i\right) + \nabla \cdot \left(\alpha_g \rho_g \vec{v}_g X_i\right) = \nabla \cdot \left(\alpha_g \rho_g D_i \nabla X_i\right) + k_i \alpha_s C_i^{n_i} M_i. \tag{A2}$$

Table A1. Summary of the reactions that were considered.

Abbreviation	i—Reactant	i—Product	k_i ($m^{3n-3}/mol^{n-1}s$)	n
1 slow	A	B	15.8	1
1 mid	C	D	63.0	1
1 fast	E	F	252	1
0.5 mid	G	H	63.0	0.5
2 mid	I	J	63.0	2

Appendix A.2. Closure Development

It was previously established that a simple effectiveness factor closure is sufficient for use in reactive fTFMs [41]. Therefore, a simple one-marker closure, similar to the one described previously [41], was developed for the reference reaction, 1 mid. It was found that the following expression can accurately predict the effectiveness factor for the reference reaction as a function of the filter size and the filtered solids volume fraction:

$$-\log(\eta) = \left(\frac{2}{\pi}\right)^3 \mathrm{atan}\left(x_1 \Delta_f^{*\,x_2} \overline{\alpha}_s\right) \mathrm{atan}\left(x_3 \Delta_f^{*\,x_2} \max(\alpha_{s,\max} - \overline{\alpha}_s, 0)\right) \mathrm{atan}\left(x_4 \Delta_f^*\right) x_5. \tag{A3}$$

An excellent fit ($R^2 = 0.991$) was obtained against the binned (conditionally-averaged) data from the resolved TFM simulations, using the following model coefficient values: $x_1 = 2.162$, $x_2 = 0.3204$, $x_3 = 5.504$, $x_4 = 1.163$, $x_5 = 1.281$, and $\alpha_{s,\max} = 0.5621$.

Next, it was found that the length factor, $\mathcal{L}$, in Equation (12) can be closed as a function of the filter size. This is understandable, as the average size of the subgrid clusters will increase as the grid size increases in the coarse-grid simulations. Using the methodology described in Section 2.1.2, the best fit to the data from all five reactions was obtained with the following length factor closure:

$$\mathcal{L} = \left(\frac{2}{\pi}\right) \mathrm{atan}\left(x_1 \Delta_f^*\right) x_2. \tag{A4}$$

A reasonably good fit of $R^2 = 0.864$ was obtained over all the reaction data with the coefficients $x_1 = 0.08768$ and $x_2 = 71.24$, and when evaluating the minimum effectiveness factor of the reference closure (used in Equations (15) and (16)) at a filtered solids volume fraction of 0.3403.

The good fit against the binned data for such a large variety of reactions implies that applying an analogy to intraparticle mass transfer works sufficiently well for generalizing the reactive fTFM closure. This is further shown in Figure A1, which compares the binned reactions rates from the resolved simulations to the model predictions as a function of the reactant concentration. Both axes are shown on a base-10 log scale to show the entire range of data collected. It is clearly seen from Figure A1a that the importance of the effectiveness factor due to sub-grid effects increases as the reaction rate increases. Furthermore, the generalized effectiveness factor closure performs well in capturing this effect.

Next, Figure A1b shows the importance of accounting for the reaction rate order in the effectiveness factor closure, especially for the 0.5 order reaction. Although some deviations exist in the 0.5 order case, the intraparticle mass transfer analogy generally captures the effect of changing reaction order well. Furthermore, the generalized closure clearly presents a substantial improvement over simply using the closure for a first order reaction, noting that all previous fTFM closures for the reaction effectiveness factor only considered first order reactions.

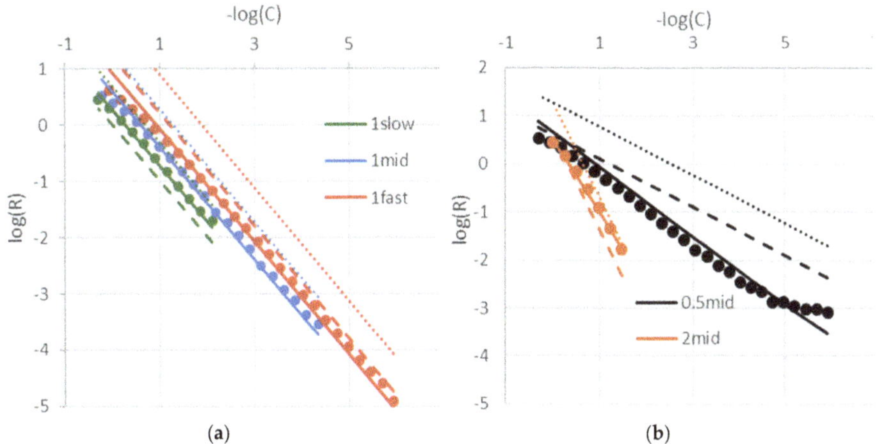

Figure A1. The reaction rate plotted against the concentration, comparing the binned data from resolved simulations (symbols) to model predictions (solid lines) for (**a**) first order and (**b**) other order reactions. The dashed lines show the predictions when using the effectiveness factor from the reference reaction (1 mid) and the dotted line the predictions without an effectiveness factor. Data is shown for the largest filter size considered ($\hat{\Delta}_f = 10.4$) and for an intermediate filtered solids volume fraction ($\overline{\alpha_s} = 0.30$).

Appendix A.3. Closure Verification

The generalized reactive effectiveness factor closure was verified by comparing coarse-grid simulation predictions to that of resolved simulations for the fast bubbling case discussed in the Supplementary Material. Figure A2a shows the importance of subgrid modeling for the reactions, showcasing large overpredictions of the scaled conversion occurring when a closure is neglected, especially for large grid sizes and for the case with large subgrid corrections (fast reaction rate and/or low reaction order). Figure A2b shows that even though both the reference and generalized effectiveness factor closures offer a significant improvement over the case with no closure, the generalized closure consistently outperforms the reference closure for all reactions (except for the reference reaction, C, where the closures are identical and the performance is the same). It is interesting to note that an excellent prediction of the conversion could be obtained despite larger inaccuracies in the hydrodynamics prediction (Figure S4b in the Supplementary Material), suggesting that the effectiveness factor closure is robust.

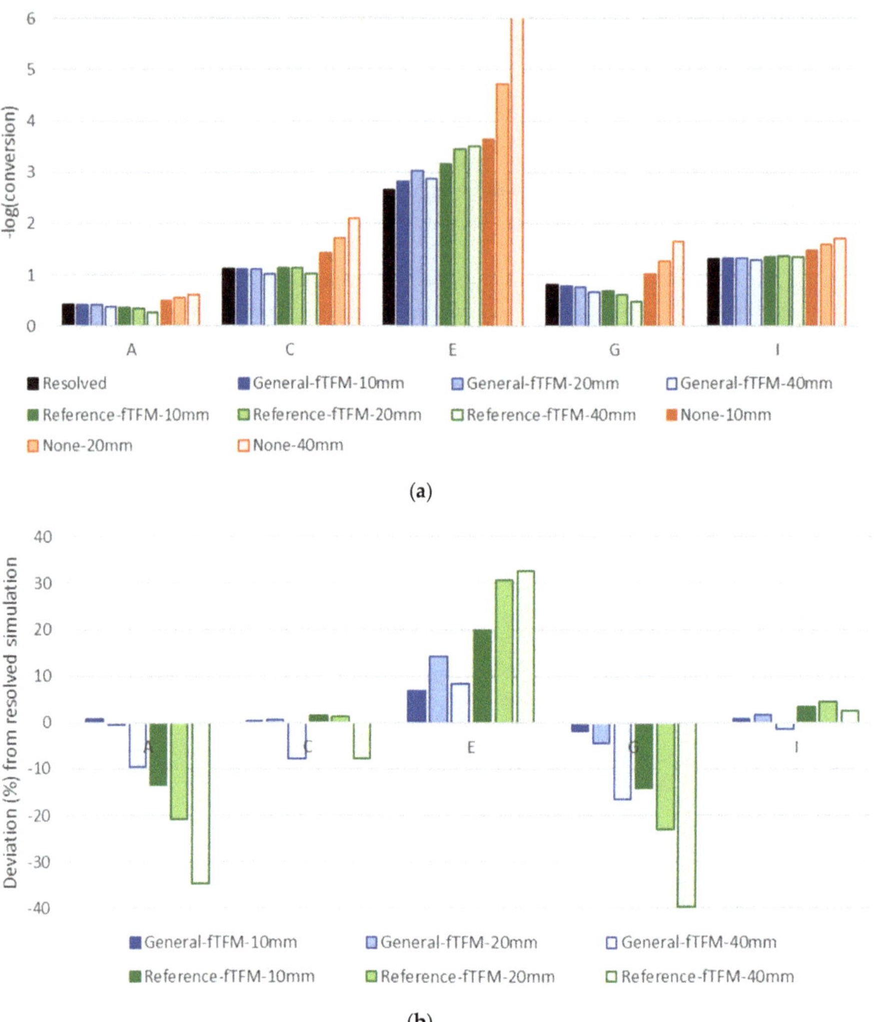

(a)

(b)

Figure A2. Evaluation of different approaches to modeling the filtered reaction rate in coarse-grid simulations of fast bubbling flow: neglecting subgrid effects on the filtered reaction rate (None), using an effectiveness factor for a reference first-order reaction (Reference), and using the generalized effectiveness factor closure described in this paper (General). Three different grid sizes were considered in the coarse-grid simulations. (**a**) Comparison of the scaled conversion for the different cases; (**b**) comparison of the percentage error of the generalized and reference closure predictions when compared to the benchmark resolved simulations.

The generalized closure is reasonably accurate in predicting the scaled conversion for all reactions, although some discrepancies occur, especially at large grid sizes. However, it is likely that these discrepancies are mainly due to the grid size being too coarse to resolve important macro-scale hydrodynamic flow structures [27], and may therefore not be primarily due to shortcomings of the reactive closure. Future studies will focus on a more detailed verification of reactive fTFM closures.

The results of this section emphasize the benefit of the generalized effectiveness factor closure; therefore, such a closure is highly recommended for future studies of industrial-scale fluidized bed reactors using reactive fTFM simulations.

References

1. IPCC. *Fifth Assessment Report: Mitigation of Climate Change*; Intergovernmental Panel on Climate Change: Cambridge University Press: Cambridge, UK; New York, NY, USA, 2014.
2. Bauer, N.; Calvin, K.; Emmerling, J.; Fricko, O.; Fujimori, S.; Hilaire, J.; Eom, J.; Krey, V.; Kriegler, E.; Mouratiadou, I.; et al. Shared Socio-Economic Pathways of the Energy Sector–Quantifying the Narratives. *Glob. Environ. Chang.* **2017**, *42*, 316–330. [CrossRef]
3. IEAGHG. *CCS in Energy and Climate Scenarios*; IEA Greenhouse Gas R & D Programme: Cheltenham, UK, 2019.
4. Ishida, M.; Zheng, D.; Akehata, T. Evaluation of a chemical-looping-combustion power-generation system by graphic exergy analysis. *Energy* **1987**, *12*, 147–154. [CrossRef]
5. Arnaiz del Pozo, C.; Cloete, S.; Cloete, J.H.; Jiménez Álvaro, Á.; Amini, S. The potential of chemical looping combustion using the gas switching concept to eliminate the energy penalty of CO_2 capture. *Int. J. Greenh. Gas Control* **2019**, *83*, 265–281. [CrossRef]
6. Mattisson, T.; Keller, M.; Linderholm, C.; Moldenhauer, P.; Rydén, M.; Leion, H.; Lyngfelt, A. Chemical-looping technologies using circulating fluidized bed systems: Status of development. *Fuel Process. Technol.* **2018**, *172*, 1–12. [CrossRef]
7. Zaabout, A.; Cloete, S.; Johansen, S.T.; van Sint Annaland, M.; Gallucci, F.; Amini, S. Experimental Demonstration of a Novel Gas Switching Combustion Reactor for Power Production with Integrated CO_2 Capture. *Ind. Eng. Chem. Res.* **2013**, *52*, 14241–14250. [CrossRef]
8. Cloete, S.; Romano, M.C.; Chiesa, P.; Lozza, G.; Amini, S. Integration of a Gas Switching Combustion (GSC) system in integrated gasification combined cycles. *Int. J. Greenh. Gas Control* **2015**, *42*, 340–356. [CrossRef]
9. Håkonsen, S.F.; Blom, R. Chemical Looping Combustion in a Rotating Bed Reactor–Finding Optimal Process Conditions for Prototype Reactor. *Environ. Sci. Technol.* **2011**, *45*, 9619–9626. [CrossRef]
10. Håkonsen, S.F.; Grande, C.A.; Blom, R. Rotating bed reactor for CLC: Bed characteristics dependencies on internal gas mixing. *Appl. Energy* **2014**, *113*, 1952–1957. [CrossRef]
11. Noorman, S.; Gallucci, F.; van Sint Annaland, M.; Kuipers, J.A.M. Experimental Investigation of Chemical-Looping Combustion in Packed Beds: A Parametric Study. *Ind. Eng. Chem. Res.* **2011**, *50*, 1968–1980. [CrossRef]
12. Spallina, V.; Gallucci, F.; Romano, M.C.; Chiesa, P.; Lozza, G.; van Sint Annaland, M. Investigation of heat management for CLC of syngas in packed bed reactors. *Chem. Eng. J.* **2013**, *225*, 174–191. [CrossRef]
13. Zaabout, A.; Cloete, S.; Amini, S. Innovative Internally Circulating Reactor Concept for Chemical Looping-Based CO_2 Capture Processes: Hydrodynamic Investigation. *Chem. Eng. Technol.* **2016**, *39*, 1413–1424. [CrossRef]
14. Osman, M.; Zaabout, A.; Cloete, S.; Amini, S. Internally circulating fluidized-bed reactor for syngas production using chemical looping reforming. *Chem. Eng. J.* **2018**. [CrossRef]
15. Osman, M.; Zaabout, A.; Cloete, S.; Amini, S. Mapping the operating performance of a novel internally circulating fluidized bed reactor applied to chemical looping combustion. *Fuel Process. Technol.* **2020**, *197*, 106183. [CrossRef]
16. IEA. *World Energy Outlook*; International Energy Agency: Paris, France, 2018.
17. Ge, W.; Chang, Q.; Li, C.; Wang, J. Multiscale structures in particle–fluid systems: Characterization, modeling, and simulation. *Chem. Eng. Sci.* **2019**, *198*, 198–223. [CrossRef]
18. Igci, Y.; Andrews, A.T.; Sundaresan, S.; Pannala, S.; O'Brien, T. Filtered two-fluid models for fluidized gas-particle suspensions. *AIChE J.* **2008**, *54*, 1431–1448. [CrossRef]
19. Igci, Y.; Sundaresan, S. Constitutive Models for Filtered Two-Fluid Models of Fluidized Gas–Particle Flows. *Ind. Eng. Chem. Res.* **2011**, *50*, 13190–13201. [CrossRef]
20. Sarkar, A.; Milioli, F.E.; Ozarkar, S.; Li, T.; Sun, X.; Sundaresan, S. Filtered sub-grid constitutive models for fluidized gas-particle flows constructed from 3-D simulations. *Chem. Eng. Sci.* **2016**, *152*, 443–456. [CrossRef]

21. Ozel, A.; Fede, P.; Simonin, O. Development of filtered Euler–Euler two-phase model for circulating fluidised bed: High resolution simulation, formulation and a priori analyses. *Int. J. Multiph. Flow* **2013**, *55*, 43–63. [CrossRef]

22. Ozel, A.; Gu, Y.; Milioli, C.C.; Kolehmainen, J.; Sundaresan, S. Towards filtered drag force model for non-cohesive and cohesive particle-gas flows. *Phys. Fluids* **2017**, *29*, 103308. [CrossRef]

23. Jiang, Y.; Kolehmainen, J.; Gu, Y.; Kevrekidis, Y.G.; Ozel, A.; Sundaresan, S. Neural-network-based filtered drag model for gas-particle flows. *Powder Technol.* **2018**. [CrossRef]

24. Schneiderbauer, S. A spatially-averaged two-fluid model for dense large-scale gas-solid flows. *AIChE J.* **2017**, *63*, 3544–3562. [CrossRef]

25. Schneiderbauer, S.; Saeedipour, M. Approximate deconvolution model for the simulation of turbulent gas-solid flows: An a priori analysis. *Phys. Fluids* **2018**, *30*, 023301. [CrossRef]

26. Gao, X.; Li, T.; Sarkar, A.; Lu, L.; Rogers, W.A. Development and validation of an enhanced filtered drag model for simulating gas-solid fluidization of Geldart A particles in all flow regimes. *Chem. Eng. Sci.* **2018**, *184*, 33–51. [CrossRef]

27. Cloete, J.H.; Cloete, S.; Radl, S.; Amini, S. On the choice of closure complexity in anisotropic drag closures for filtered Two Fluid Models. *Chem. Eng. Sci.* **2019**, *207*, 379–396. [CrossRef]

28. Cloete, J.H.; Cloete, S.; Radl, S.; Amini, S. Verification of filtered Two Fluid Models for reactive gas-solid flows. In Proceedings of the CFD 2017, Trondheim, Norway, 30 May–1 June 2017.

29. Cloete, J.H.; Cloete, S.; Radl, S.; Amini, S. Verification study of anisotropic filtered Two Fluid Model Closures. In Proceedings of the AIChE Annual Meeting, Minneapolis, MN, USA, 29 October–3 November 2017.

30. Holloway, W.; Sundaresan, S. Filtered models for reacting gas–particle flows. *Chem. Eng. Sci.* **2012**, *82*, 132–143. [CrossRef]

31. Agrawal, K.; Holloway, W.; Milioli, C.C.; Milioli, F.E.; Sundaresan, S. Filtered models for scalar transport in gas–particle flows. *Chem. Eng. Sci.* **2013**, *95*, 291–300. [CrossRef]

32. Huang, Z.; Zhang, C.; Jiang, M.; Zhou, Q. Development of a Filtered Interphase Heat Transfer Model Based on Fine-Grid Simulations of Gas-Solid Flows. *AIChE J.* **2019**. [CrossRef]

33. Cloete, J.H.; Cloete, S.; Municchi, F.; Radl, S.; Amini, S. Development and verification of anisotropic drag closures for filtered Two Fluid Models. *Chem. Eng. Sci.* **2018**, *192*, 930–954. [CrossRef]

34. Khan, M.N.; Cloete, S.; Amini, S. Efficiency Improvement of Chemical Looping Combustion Combined Cycle Power Plants. *Energy Technol.* **2019**. in Press. [CrossRef]

35. Gidaspow, D.; Bezburuah, R.; Ding, J. Hydrodynamics of Circulating Fluidized Beds, Kinetic Theory Approach. In Proceedings of the 7th Engineering Foundation Conference on Fluidization, Brisbane, Australia, 3–8 May 1992; pp. 75–82.

36. Lun, C.K.K.; Savage, S.B.; Jeffrey, D.J.; Chepurniy, N. Kinetic Theories for Granular Flow: Inelastic Particles in Couette Flow and Slightly Inelastic Particles in a General Flow Field. *J. Fluid Mech.* **1984**, *140*, 223–256. [CrossRef]

37. Cloete, J.H.; Cloete, S.; Municchi, F.; Radl, S.; Amini, S. The sensitivity of filtered Two Fluid Model to the underlying resolved simulation setup. *Powder Technol.* **2017**, *316*, 265–277. [CrossRef]

38. Milioli, C.C.; Milioli, F.E.; Holloway, W.; Agrawal, K.; Sundaresan, S. Filtered two-fluid models of fluidized gas-particle flows: New constitutive relations. *AIChE J.* **2013**, *59*, 3265–3275. [CrossRef]

39. Cloete, J.H.; Cloete, S.; Radl, S.; Amini, S. Development and verification of anisotropic solids stress closures for filtered Two Fluid Models. *Chem. Eng. Sci.* **2018**, *192*, 906–929. [CrossRef]

40. Cloete, S.; Cloete, J.H.; Amini, S. Hydrodynamic validation study of filtered Two Fluid Models. *Chem. Eng. Sci.* **2018**, *182*, 93–107. [CrossRef]

41. Cloete, J.H. Development of Anisotropic Filtered Two Fluid Model Closures. Ph.D. Thesis, Norwegian University of Science and Technology, Trondheim, Norway, 2018.

42. Zhang, D.Z.; VanderHeyden, W.B. The effects of mesoscale structures on the macroscopic momentum equations for two-phase flows. *Int. J. Multiph. Flow* **2002**, *28*, 805–822. [CrossRef]

43. Zaabout, A.; Cloete, S.; Amini, S. Autothermal operation of a pressurized Gas Switching Combustion with ilmenite ore. *Int. J. Greenh. Gas Control* **2017**, *63*, 175–183. [CrossRef]

44. Kuusik, R.; Trikkel, A.; Lyngfelt, A.; Mattisson, T. High temperature behavior of NiO-based oxygen carriers for Chemical Looping Combustion. *Energy Procedia* **2009**, *1*, 3885–3892. [CrossRef]

45. Abad, A.; Adánez, J.; García-Labiano, F.; de Diego, L.F.; Gayán, P.; Celaya, J. Mapping of the range of operational conditions for Cu-, Fe-, and Ni-based oxygen carriers in chemical-looping combustion. *Chem. Eng. Sci.* **2007**, *62*, 533–549. [CrossRef]
46. Levenspiel, O. *Chemical Reaction Engineering*, 3rd ed.; John Wiley & Sons: Hoboken, NJ, USA, 1999.
47. Thiele, E.W. Relation between Catalytic Activity and Size of Particle. *Ind. Eng. Chem.* **1939**, *31*, 916–920. [CrossRef]
48. Yang, W.; Cloete, S.; Morud, J.; Amini, S. An Effective Reaction Rate Model for Gas-Solid Reactions with High Intra-Particle Diffusion Resistance. *Int. J. Chem. React. Eng.* **2016**, *14*, 331. [CrossRef]
49. Rawlings, J.B.; Ekerdt, J.G. Chapter 7. In *Chemical Reactor Analysis and Design Fundamentals*; Nob Hill Publishing: Madison, WI, USA, 2002.
50. Chevrier, S. Development of Subgrid Models for a Periodic Circulating Fluidized Bed of Binary Mixture of Particles. Ph.D. Thesis, Université de Toulouse, Toulouse, France, 2017.
51. Naqvi, R.; Wolf, J.; Bolland, O. Part-load analysis of a chemical looping combustion (CLC) combined cycle with CO_2 capture. *Energy* **2007**, *32*, 360–370. [CrossRef]
52. EBTF. *European Best Practice Guide for Assessment of CO_2 Capture Technologies*; European Benchmark Task Force, European Commission: Brussels, Belgium, 2011.
53. Johnson, P.C.; Jackson, R. Frictional-Collisional Constitutive Relations for Granular Materials, with Application to Plane Shearing. *J. Fluid Mech.* **1987**, *176*, 67–93. [CrossRef]
54. Patankar, S. *Numerical Heat Transfer and Fluid Flow*; Hemisphere Publishing Corporation: New York, NY, USA, 1980.
55. Leonard, B.P.; Mokhtari, S. *ULTRA-SHARP Nonoscillatory Convection Schemes for High-Speed Steady Multidimensional Flow*; NASA Lewis Research Center: Cleveland, OH, USA, 1990.
56. Gerdes, K.; Summers, W.M.; Wimer. *Quality Guidelines for Energy System Studies: Cost Estimation Methodology for NETL Assessments of Power Plant Performance*; DOE/NETL-2011/1455 United States 10.2172/1513278 NETL-IR English; NETL: Pittsburgh, PA, USA, 2011; p. Medium: ED.
57. Turton, R.; Bailie, R.C.; Whiting, W.B.; Shaeiwitz, J.A.; Bhattacharyya, D. Appendix A. In *Analysis, Synthesis and Design of Chemical Processes*; Pearson Education: Upper Saddle River, NJ, USA, 2008.
58. Peters, M.S.; Timmerhaus, K.D. *Plant Design and Economics for Chemical Engineers*; McGraw-Hill: New York, NY, USA, 1991. [CrossRef]
59. Adanez, J.; Abad, A.; Garcia-Labiano, F.; Gayan, P.; de Diego, L.F. Progress in Chemical-Looping Combustion and Reforming technologies. *Prog. Energy Combust. Sci.* **2012**, *38*, 215–282. [CrossRef]
60. Spallina, V.; Pandolfo, D.; Battistella, A.; Romano, M.C.; Van Sint Annaland, M.; Gallucci, F. Techno-economic assessment of membrane assisted fluidized bed reactors for pure H2 production with CO_2 capture. *Energy Convers. Manag.* **2016**, *120*, 257–273. [CrossRef]
61. IEAGHG. *The Costs of CO_2 Transport: Post-Demonstration CCS in the EU*; European Technology Platform for Zero Emission Fossil Fuel Power Plants: Brussels, Belgium, 2011.
62. IEAGHG. *The Costs of CO_2 Storage: Post-Demonstration CCS in the EU*; European Technology Platform for Zero Emission Fossil Fuel Power Plants: Brussels, Belgium, 2011.
63. Rubin, E.S.; Davison, J.E.; Herzog, H.J. The cost of CO_2 capture and storage. *Int. J. Greenh. Gas Control* **2015**, *40*, 378–400. [CrossRef]

Article

Numerical and Experimental Study of a Vortex Structure and Energy Loss in a Novel Self-Priming Pump

Hao Chang [1,2], Ramesh K. Agarwal [2,*], Wei Li [1,2], Ling Zhou [1,2] and Weidong Shi [3,*]

[1] Research Center of Fluid Machinery Engineering and Technology, Jiangsu University, Zhenjiang 212013, China; haochang@wustl.edu (H.C.); 1000003263@ujs.edu.cn (W.L.); lingzhoo@hotmail.com (L.Z.)

[2] Department of Mechanical Engineering and Materials Science, Washington University in St. Louis, St. Louis, MO 63130, USA

[3] School of Mechanical Engineering, Nantong University, Nantong 226019, China

[*] Correspondence: rka@wustl.edu (R.K.A.); wdshi@ujs.edu.cn (W.S.)

Received: 28 August 2019; Accepted: 27 September 2019; Published: 4 October 2019

Abstract: The self-priming pump as an essential energy conversion equipment is widely used in hydropower and thermal power plants. The energy losses in the internal flow passage of the pump directly affect its work efficiency. Therefore, it is important to improve the internal flow characteristic of the pump. In the present work, a novel self-priming pump which starts without water is proposed; this pump can reduce the energy consumption as well as the time needed to start its operation. The spatial structure of the vortices in the pump is investigated by employing the Q criterion with the numerical solution of the vorticity transport equation. Based on the morphology, the vortices can be separated into three categories: Trailing Edge Vortex (TEV), Leading Edge Vortex (LEV) and Gap Leakage Vortex (GLV). Generally, the morphology of the TEV is more disorderly than that of LEV and GLV, and the intensity of TEV is significantly higher than that of the other two vortices. To determine the magnitude and distribution of energy loss in the pump, entropy production analysis is employed to study the influence of blade thickness on energy characteristics of the pump. It is found that with an increase in the flow rate, the location of energy loss transfers from the trailing edge to the leading edge of the blade, and viscous entropy production (VEP) and turbulence entropy production (TEP) are the dominant factors which influence the energy conversion in the pump. More importantly, employing the blade with a thin leading edge and a thick trailing edge can not only significantly reduce the impact of incoming flow under over-load condition (flow rate higher than the design condition) but can also increase the efficiency of the pump. Thus, an increase in thickness of the blade from the leading edge to the trailing edge is beneficial for improving the pump performance. The results of this paper can be helpful in providing guidelines for reducing the energy loss and in improving the performance of a self-priming pump.

Keywords: vortex structure; energy loss; entropy production; self-priming pump

---

1. Introduction

The self-priming pump as an essential form of energy conversion equipment is widely used in hydropower and thermal power plants. The energy loss in the internal flow passage of the pump directly affects its work efficiency. Therefore, it is important to improve its internal flow characteristics by improving the design of its blades [1]. In recent years, considerable research has been conducted on improving the energy characteristics of the pump by modifying its blade profile. Wang et al. [2] proposed an innovative impeller with forward-curving blade by numerically and experimentally investigating the relationship between the energy loss in the impeller and the blade inlet angle, and obtained an optimal range for the inlet angles. Elyamin et al. [3] investigated the influence of the number of blades on the energy performance by using numerical simulations and concluded that the increase in number of blades can weaken the secondary flow, resulting in a reduction in the energy loss of the impeller. Shao et al. [4] used high-speed photography to study the internal characteristics of the pump with an impeller with different geometrical structures; they found that the complex vortex structures at low flow rates (partial-load conditions) can be eliminated by decreasing the outlet angle and width of the blade. Nejad and Riasi [5] conducted an experimental test to analyze the relationship between the blade profile and the energy performance; the blades were manufactured from straight to curved profiles with uniform inlet and outlet angles. They showed that reducing the blade angle is beneficial for improving the pump efficiency. Han et al. [6] selected the blade wrap angle and blade exit angle as an analysis variable in their numerical and experimental investigation to optimize the pump performance; they considered the effect of blade parameters on the distribution of turbulent kinetic energy, which affects the performance of the pump. Jeon et al. [7] designed a novel S-shape blade by employing the design of experiment (DOE) and the response surface method (RSM), and showed that the blade's efficiency was enhanced by 3% compared to under the original model. Nejad at el. [8] modified the blade profile into a bucket shape and studied the influence of blade angle, chord, and other parameters on the hydraulic performance; they concluded that the maximum efficiency of the pump can be improved by employing an optimal combination of these parameters. Wang et al. [9–11] pointed out the factors influencing the energy loss in a multistage centrifugal pump and analyzed the hydraulic loss caused by surface roughness of the impeller shroud experimentally.

However, while most of the investigations noted above studied the effect of varying the number of blades, inlet angle, outlet width, wrap angle, and riding position on the energy efficiency of a pump; on the effect of modification of blade thickness on pump performance has hardly been studied. Furthermore, the influence of unstable vortex structures on the internal flow passage of the pump and their contribution to energy loss has not been studied. In addition, the overall energy loss analysis has only been conducted using the conventional approach by determining the hydraulic efficiency or turbulent kinetic energy, which does not provide the main details of energy loss, namely its location, sources, and type. In the present work, entropy production analysis is introduced to identify the sources, locations, and reasons for energy loss/dissipation. Wang [12] investigated the cavitation performance of the LNG cryogenic pump; they showed that the evolution and degradation that occur due to cavitation can be determined by applying the entropy production model, and then verified the results of their analysis through an experiment. Hou et al. [13] combined the orthogonal design with entropy production analysis to optimize the geometric parameters of the pump; the optimal design was obtained which improved the energy performance of the pump. Gu et al. [14] analyzed the influence of the clocking positions on the hydraulic loss and obtained the energy loss distribution in the pump by employing entropy production analysis. Li et al. [15] introduced the concept of wall entropy production in their investigation; they pointed out that a large deviation in energy loss occurs with respect to experiments without considering the losses that occur due to the wall effect. Thus, entropy production analysis is used in this paper to accurately evaluate the energy loss that occurs in the pump.

In this paper, a novel self-priming pump start without water has been proposed and studied by using the Q criterion with vorticity transport equation and employing the entropy production analysis for quantification of energy losses. The vortex structures inside and energy characteristics of the pump with different blade thicknesses are systematically investigated by using numerical simulations. Numerical simulations are validated for one of the most promising blade thickness distribution in terms of pump head and efficiency by conducting an experiment. Finally, guidelines for obtaining a good thickness distribution for the blades are provided for improving the flow characteristics and performance of the pump.

2. Structure of Novel Self-Priming Pump

Traditional centrifugal pumps must fill the pump with water before starting, which takes lot of time and results in energy loss. Therefore, a novel self-priming pump is proposed in this paper which can start without water. Figure 1 shows the structure of the proposed novel self-priming pump; it shows the inlet, self-priming system, compressor, pump, outlet and outlet valve of the novel self-priming pump. The specifications of the pump are listed in Table 1.

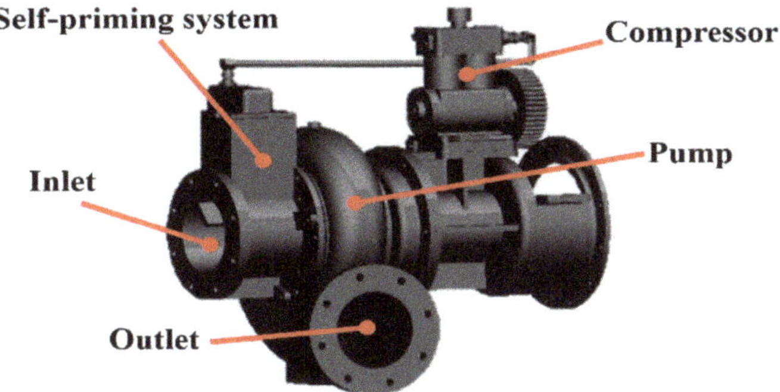

Figure 1. The structure and components of the pump.

Table 1. Specifications of the pump.

Parameter	Symbol	Value
Head	H	45 m
Design flow condition	Q_d	500 m^3/h
Rotation speed	n	2200 r/min
Specific speed	n_s	172
Impeller inlet diameter	D_1	200 mm
Impeller outlet diameter	D_2	280 mm
Blade numbers	Z	6
Blade wrap angles	ϕ	120°
Volute inlet width	b_1	80 mm
Volute outlet diameter	D_4	200 mm

As shown in Figure 1, the self-priming system is installed immediately after the inlet and consists of an induction chamber, filter screen, valve seat, ball valve, plate, valve chamber, intake tube, nozzle 1, joint, and nozzle 2, as shown in Figure 2. The inlet tube extends into the water before the pump starts, and then the compressor starts running with the operation of the pump. The jet-stream created in the compressor is discharged into nozzle 1 through the intake tube. The inner structure of nozzle 1 first shrinks and then smoothly expands, which results in a high-velocity jet-stream being created through nozzle 1.

A joint is installed between the nozzle 1 and nozzle 2 and is connected with the valve chamber. Hence, the high-velocity jet-stream passes through the joint and nozzle 2 exhausts directly into the atmosphere. At the same time, the internal pressure of the joint is continuously decreasing which leads to pushing the air in the valve box into the joint and eventually discharging it from nozzle 2. Accordingly, the pressure in the valve box gradually reduces. When the gravity of the ball valve is less than the pressure difference experienced by it, the ball valve is lifted from the valve seat and the air in the induction chamber flows into the valve box and eventually drains into the atmosphere. Furthermore, since the induction chamber is installed at the inlet, the air in the pump exhausts into the atmosphere. Ultimately, water flows into the pump along the inlet tube and the pump is filled with water, which achieves the pump start without water. When the process of self-priming is finished, the ball valve is merged in the water. Under the effect of gravity, it returns to the valve seat and closes the connection between the induction chamber and the valve box, avoiding the energy loss resulting from the water discharging into the atmosphere.

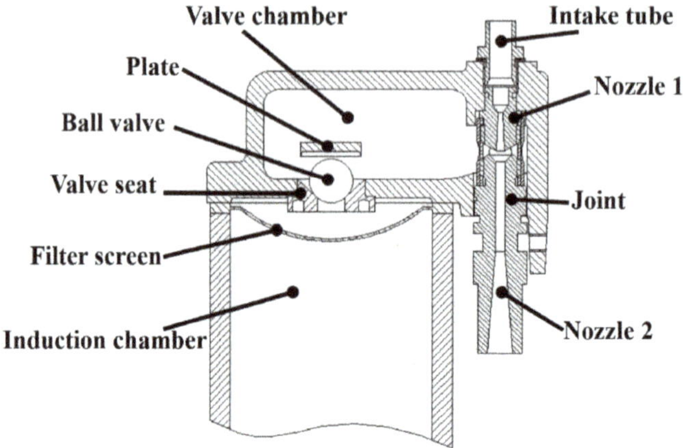

Figure 2. Structure of the self-priming system.

The outlet valve is installed on the outlet and is connected to the atmosphere in order to prevent the air flow into the pump. A rubber plate is pressed against the inclined exit of the outlet tube. When the air inside the pump is drained, the pump is filled with water and under the effect of pressure, the rubber plate separates from the exit of the outlet tube, and thus the self-priming of the pump is completed. The pump thus accomplishes the waterless start and can deliver water normally. The structure of the outlet valve as shown in the Figure 3.

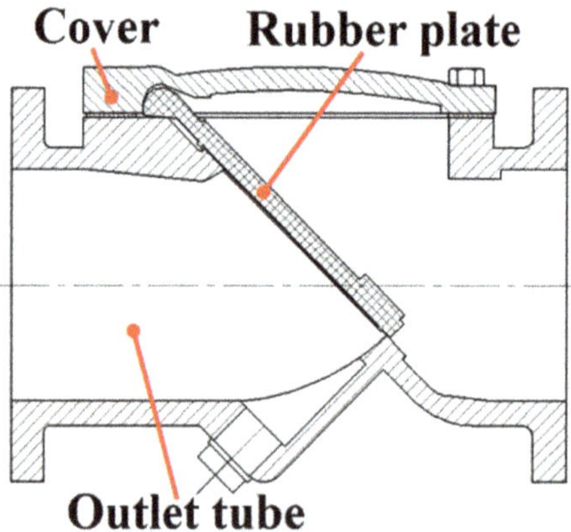

Figure 3. Structure of the outlet valve.

The geometric parameters of the blade cannot only decide the work capability of the pump, but can also affect the internal flow characteristic In line with our previous investigation, four different blade thickness distributions are proposed [16]; these are shown in Table 2. The 3D models of four blade thickness distribution schemes is shown in the Figure 4. Scheme 1 and scheme 2 employ the uniform thin and thick blade respectively, while the scheme 3 and scheme 4 employ the non-uniform increasing and decreasing blade thickness distribution respectively.

Table 2. Blades with different thickness distributions.

Scheme	Thickness of Leading Edge (mm)	Thickness of Middle Part (mm)	Thickness of Trailing Edge (mm)
1	3	3	3
2	6	6	6
3	3	6	6
4	6	6	3

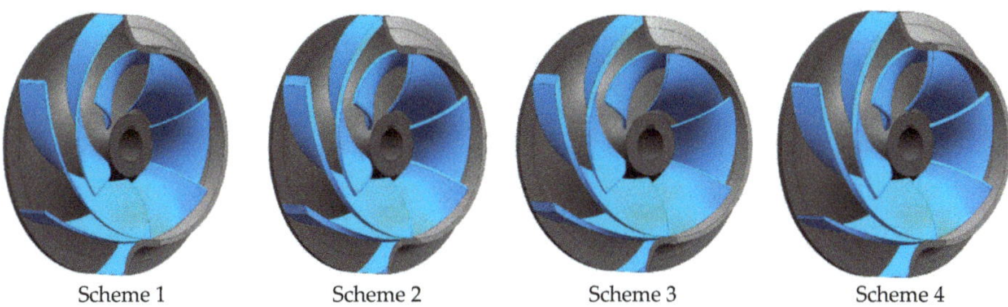

Figure 4. The 3D models based on four blade thickness distribution schemes given in Table 2.

3. Computational Model and Numerical Method

3.1. Computational Domain

Figure 5 shows the computational domains of s novel self-priming pump including the outlet, volute, impeller, self-priming system, and inlet. In addition, the stable inflow and outflow can be obtained by extending the inlet and outlet, respectively.

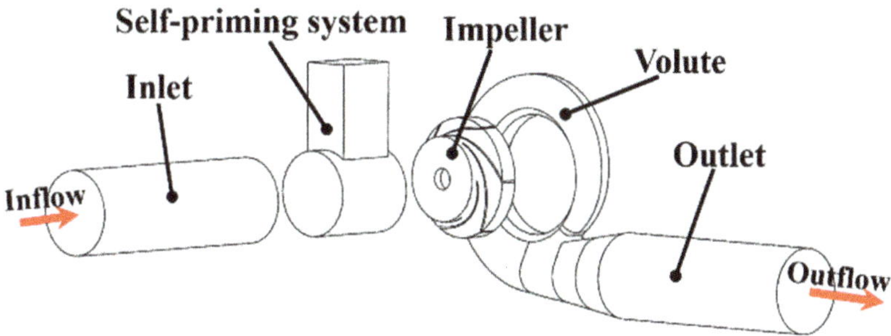

Figure 5. The computational domains.

3.2. Grid Generation

The computational domains of the self-priming pump are discretized into a hexahedral structure by employing the ICEM-CFD (The Integrated Computer Engineering and Manufacturing code for Computational Fluid Dynamics). The leading edge of the blade and tongue of the volute are refined to capture the details of the internal flow characters, as shown in Figures 6 and 7.

Since the grid resolution can significantly affect the accuracy of the simulation, grid sensitivity analysis was conducted. As shown in Figure 8, the number of grid elements was varied from 1.2×10^6 to 5.3×10^6 to evaluate the grid independence of the solution. When the number of grid elements exceeds 2.6×10^6, the deviation in pump efficiency and pump head are reduced to less than 1% by further increasing the number of grid elements. Therefore, considering the computational efficiency of the simulation, the grid elements are set as 2.6×10^6 in all simulations reported in this paper and y + is less than 60. The number of grid elements for the impeller, volute, outlet, inlet, and self-priming system are 478,998, 720,480 504,000, 696,000 and 272,232, respectively.

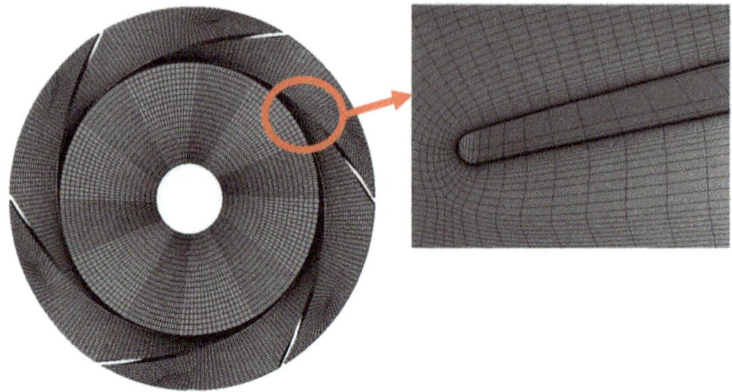

Figure 6. Grid distribution in an impeller.

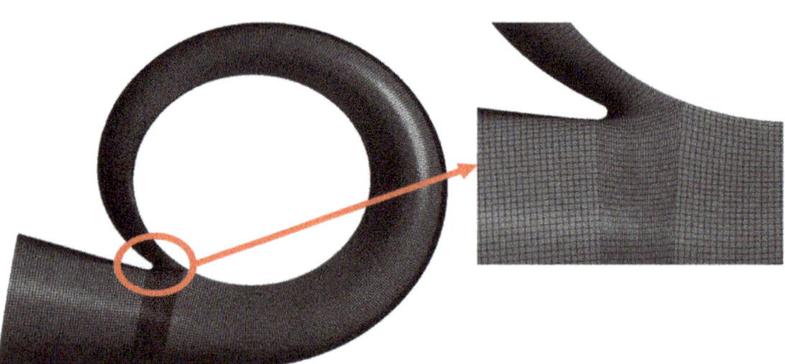

Figure 7. Grid distribution in the volute.

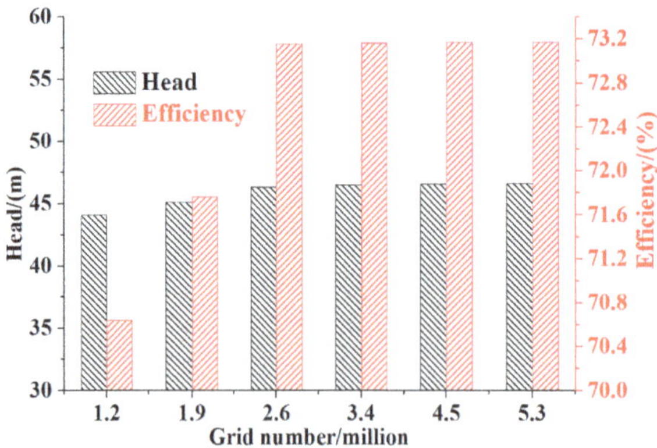

Figure 8. Sensitivity analysis of the number of grid elements used for the solution.

3.3. Calculation Method

In this paper, the SST k–ω turbulence model with Reynolds-Averaged Navier-Stokes (RANS) equations is employed for developing the numerical solution using the commercial CFD solver ANSYS CFX 17.1. The fluid medium is incompressible water at 25 °C. Considering the impeller as a rotating domain, the surfaces at the inlet and outlet of the impeller are set as the frozen rotor interfaces. Figure 9 shows the sensitivity of the calculations to time steps. The step sizes of 1/120 T (2.27×10^{-4} s), 1/60 T (4.54×10^{-4} s) and 1/30 T (9.08×10^{-4} s) are selected as time steps in the investigation. By comparing the pressure at the outlet with different time steps, it was found that small variations exist among the results when using different time steps. Based on this study on the influence of time steps on the solution, 1/60 T (4.54×10^{-4} s) is selected as the time step for computational efficiency without compromising accuracy [17,18]. The numerical setup is described in Table 3.

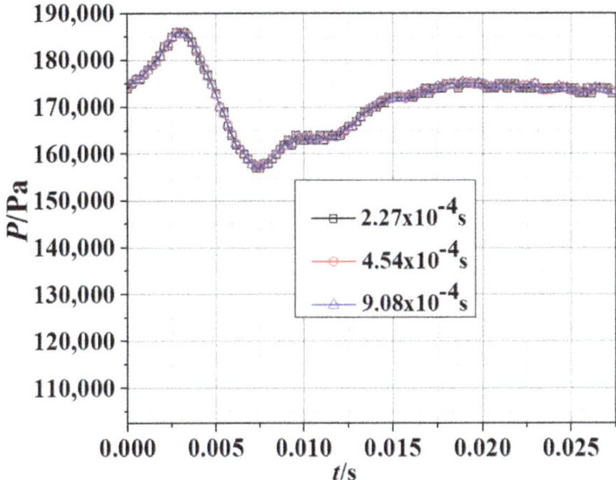

Figure 9. Analysis of the sensitivity of time steps in the numerical simulation.

Table 3. The numerical setup.

CFD Software	ANSYS CFX 17.1
Turbulence model	SST k–ω
Flow medium	Water at 25 °C
Inlet boundary condition	Total pressure (1 atm)
Outlet boundary condition	Mass flow
Wall roughness	50 µm
Steady state	Frozen rotor
Transient state	Transient rotor-stator
Advection scheme	High resolution
RMS residual	0.00001
Transient calculation time	0.2 s
Time step	4.54×10^{-4} s
Turbulence intensity	5%

4. Results and Discussion

4.1. Analysis of the Vortex Structure

Due to the complex structure of the novel self-priming pump, several types of vortices are generated in the internal flow field, namely the secondary flow vortex, reflux vortex, and flow separation vortex. Therefore, to reduce the influence of the vortices on the hydraulic performance of the pump, the Q criterion is employed to investigate the morphology and structure of vortices. Figure 10 shows the vortices in the impeller and the pressure distributions on the impeller for the four different blade thickness distributions given in Table 2. The equations of the Q criterion can be written as follows [19]:

$$\lambda^3 + P\lambda^2 + Q\lambda + R = 0. \tag{1}$$

Considering water as an incompressible medium, the equivalent equations are:

$$P = S_{ii} = 0 \tag{2}$$

$$Q = \left(\Omega_{ij}\Omega_{ji} - S_{ij}S_{ji}\right)/2 \tag{3}$$

$$R = -\left(S_{ij}S_{jk}S_{ki} + 3\Omega_{ij}\Omega_{jk}S_{ki}\right)/3 \tag{4}$$

and

$$S_{ij} = \left(\partial u_i/\partial x_j + \partial u_j/\partial x_i\right)/2 \tag{5}$$

$$\Omega_{ij} = \left(\partial u_i/\partial x_j - \partial u_j/\partial x_i\right)/2. \tag{6}$$

In Equation (1), P, Q, and R represent the invariants of velocity gradient tensor. The Q criterion was proposed based on the second invariant of velocity gradient tensor. S_{ij} and Ω_{ij} represent tensors of strain rate and vorticity, respectively and reflect the deformation and rotational motion. Figure 10 depicts the three-dimensional (3D) vortex structures with blade thickness distribution given in Table 2 under various flow conditions. According to the morphology and structure of vortices, they can be separated into three categories, namely the leading edge vortex (LEV), trailing edge vortex (TEV), and gap leakage vortex (GLV).

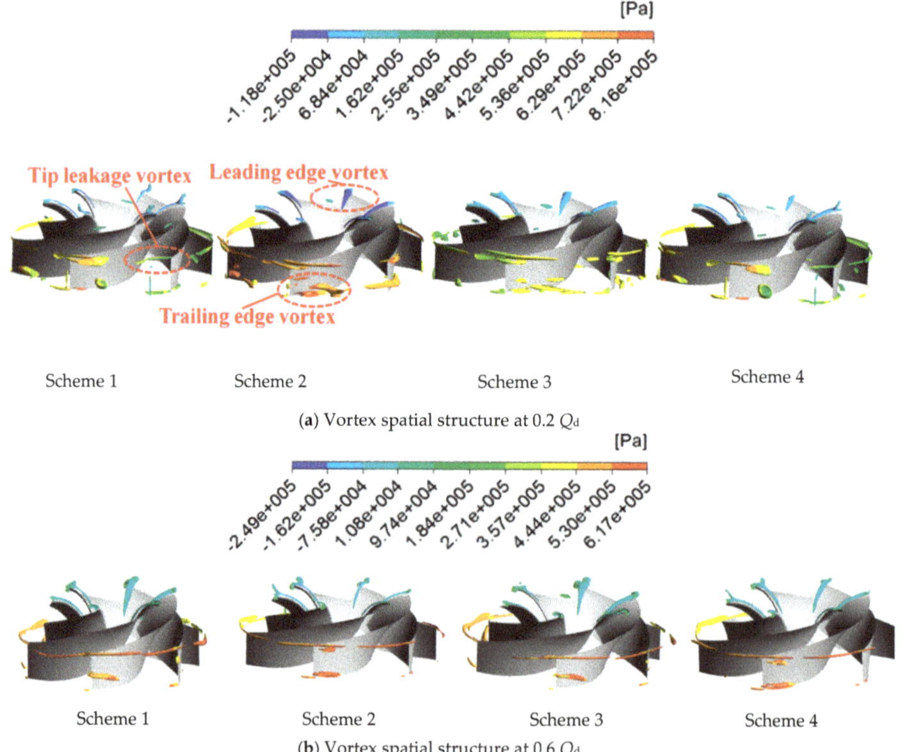

(**a**) Vortex spatial structure at 0.2 Q_d

(**b**) Vortex spatial structure at 0.6 Q_d

Figure 10. *Cont.*

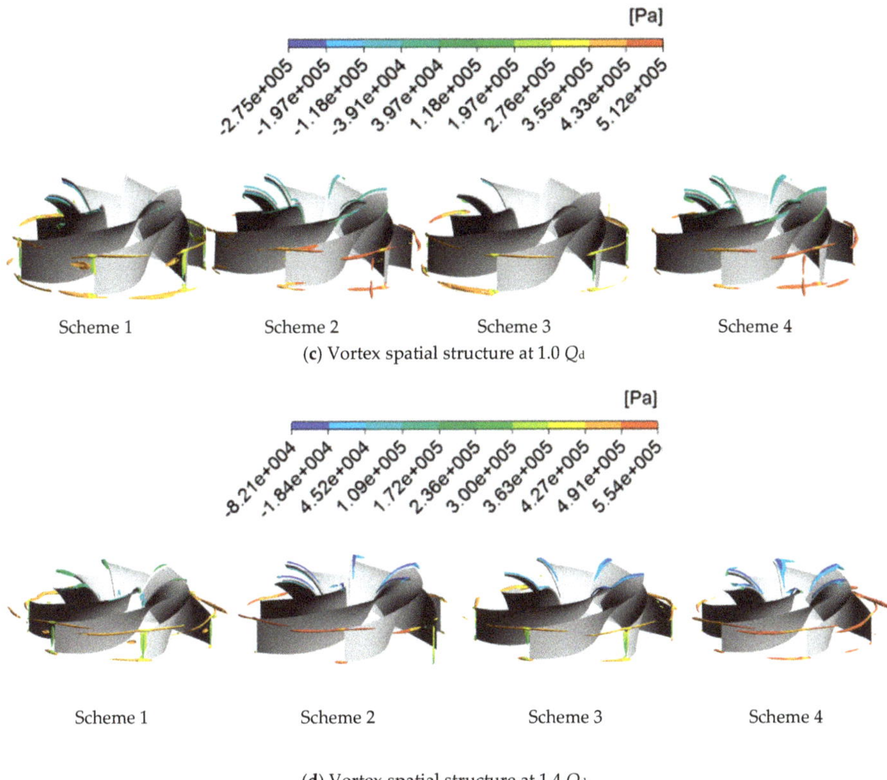

(c) Vortex spatial structure at 1.0 Q_d

(d) Vortex spatial structure at 1.4 Q_d

Figure 10. 3D vortex structure for various flow conditions for four different blade thickness distributions.

As shown in Figure 10, GLV is generated at the gap between the blade and volute, which looks like a silk ribbon surrounding the entire impeller. This vortex structure is generated due to pressure difference between the pressure surface of the blade (PSB) and the suction surface of the blade (SSB), which results in water flowing from PSB to SSB through the gap and interacting with the mainstream in the impeller passage. The rotation direction of the vortex is the same as the impeller. It can also be noted that with the increase in the flow condition, the pressure difference between the PSB and SSB gradually decreases, which results in a corresponding decline in the size and strength of GLV.

Meanwhile, due to the impact of the incoming flow on the impeller, LEV can be observed at the leading edge of the blade. By comparing the LEV at various flow conditions, it can be concluded that the size and strength of LEV has a negative correlation with the flow condition; in other words, the size and strength of LEV reaches the maximum at the partial-load flow condition ($0.2Q_d$) and reaches the minimum at $1.0Q_d$. The incoming flow on the impeller at the design flow condition is considered as the non-impact flow, that is, the internal flow field is smooth without any vortices. However, when the flow condition decreases from $1.0Q_d$ to the $0.2Q_d$, the axial velocity remains the same, but the meridional velocity gradually decreases, which results in the increase of the impeller incidence angle. Therefore, the impact of incoming flow is felt at the leading edge of the blade. When the flow condition increases from $1.0Q_d$ to the $1.4Q_d$, the incidence angle is larger than the design incidence angle which also leads to generation of LEV at leading edge of the blade. In addition, it can be noted that the negative pressure is created at the leading edges of the blades which may induce cavitation that can deteriorate the impeller. The incoming flow at the inlet of the impeller is squeezed, which a causes large change in the velocity direction and thus the pressure reduces sharply at the leading edge of the blade. In addition, since the expelling coefficient of the blade with thicker leading edge is larger than that with thinner leading edge, the magnitude and area of negative pressure on the thicker leading edge blade (Schemes 2 and 4) is significantly larger than that on the thinner leading edge blade (Schemes 1 and 3). Furthermore, as the flow condition increases, the area of the negative pressure gradually decreases.

At the partial-load conditions (0.2Q_d–0.6Q_d), the TEV can be seen at the trailing edge of the blade, as shown in Figure 10a,b, where the shape of the TEV is like a bubble attached to the PSB. Generally, the morphology of the TEV is more disordered than that of the LEV and GLV, and the pressure and intensity of TEV are significantly higher than for the other two vortices. As the flow condition continues to increase, the intensity of the TEV gradually declines. Furthermore, it can be observed that the TEV of the Scheme 2 is the most disorderly one, which has a twisted shape. This not only effects the flow movement of the impeller passage, but also decreases the hydraulic performance. Therefore, based on these observations on vortex structures, it can be concluded that the characteristics of the TEV are more unstable than those of LEV and GLV.

To further investigate the formation principles of the unsteady vortex in detail, the vorticity transport equation [20,21] is used in this research:

$$\frac{D\vec{\omega}}{Dt} = \left(\vec{\omega}\cdot\nabla\right)\vec{V} - \vec{\omega}(\nabla\cdot\vec{V}) + \frac{\nabla\rho_m \times \nabla p}{\rho_m^2} + v\nabla^2\vec{\omega}. \tag{7}$$

In Equation (7), $D\vec{\omega}/Dt$ is the rate of the vorticity variation, $\left(\vec{\omega}\cdot\nabla\right)\vec{V}$ is the vortex stretching variable (VSV) caused by the velocity gradients, and $\vec{\omega}(\nabla\cdot\vec{V})$ represents the vortex dilation, which reveals the influence of the medium compressibility on the vorticity. The term $\nabla\rho_m \times \nabla p/\rho_m^2$ is the baroclinic torque, which is induced by the gradient of both the pressure and density. $v\nabla^2\vec{\omega}$ is the viscidity diffusion variable (VDV), which determines the effect of viscidity diffusion on vorticity. Considering the fact that the medium is water which is incompressible, the terms $\vec{\omega}(\nabla\cdot\vec{V})$ and $\nabla\rho_m \times \nabla p/\rho_m^2$ are neglected in this paper.

Figure 11 shows the distribution of the viscidity diffusion variable (VDV) on the PSB for various flow conditions. It can be observed that an oval shape VDV is generated near the trailing edge of the blade, and the magnitude of the VDV gradually decreases with an increase in the flow condition. The VDV distribution reaches maximum at the trailing edge at a flow condition of 100 m³/h and it moves towards the leading edge for high flow condition of 700 m³/h. Since the blade passage is full of uneven flow and axial eddies at the partial-load conditions, the induced relative velocity on the PSB is smaller than the ideal design relative velocity. The separation of the flow in the boundary layer occurs due to the effect of the positive incidence angle. The conclusion can be drawn that the thicker the blade is, the greater the flow separation becomes. Furthermore, when the flow condition reaches the design value, the VDV distribution reduces to its minimum. And at an over-load condition (700 m³/h), a larger VDV is generated at the leading edge.

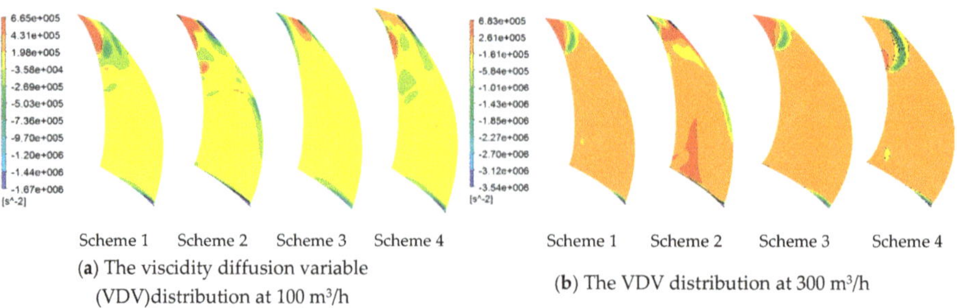

(**a**) The viscidity diffusion variable (VDV)distribution at 100 m³/h

(**b**) The VDV distribution at 300 m³/h

Figure 11. *Cont.*

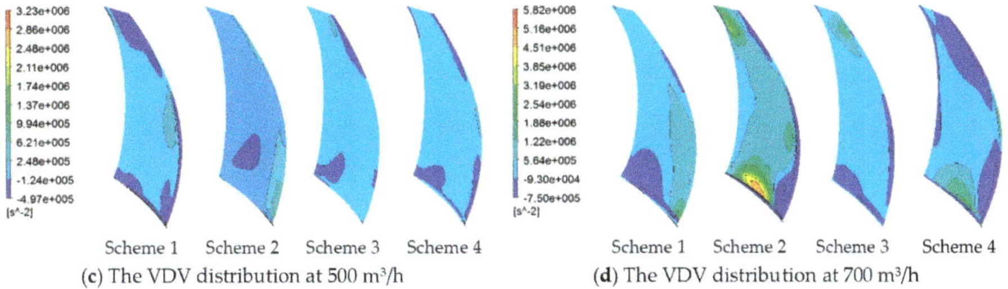

Scheme 1   Scheme 2   Scheme 3   Scheme 4   Scheme 1   Scheme 2   Scheme 3   Scheme 4

(**c**) The VDV distribution at 500 m³/h   (**d**) The VDV distribution at 700 m³/h

Figure 11. The VDV distribution at various flow conditions.

The vortex Stretching Variable (VSV) distribution is similar to the VDV distribution, as shown in Figure 12. The VSV distribution is based on the angular momentum conservation principle [22]. VSV first reduces and then gradually increases; the lowest value of VSV is obtained at the design condition. Based on this discussion, it can be concluded that the VSV plays an essential role in the formation of the vortex, and more importantly that increasing the thickness of the blade is an effective way to decrease the VSV.

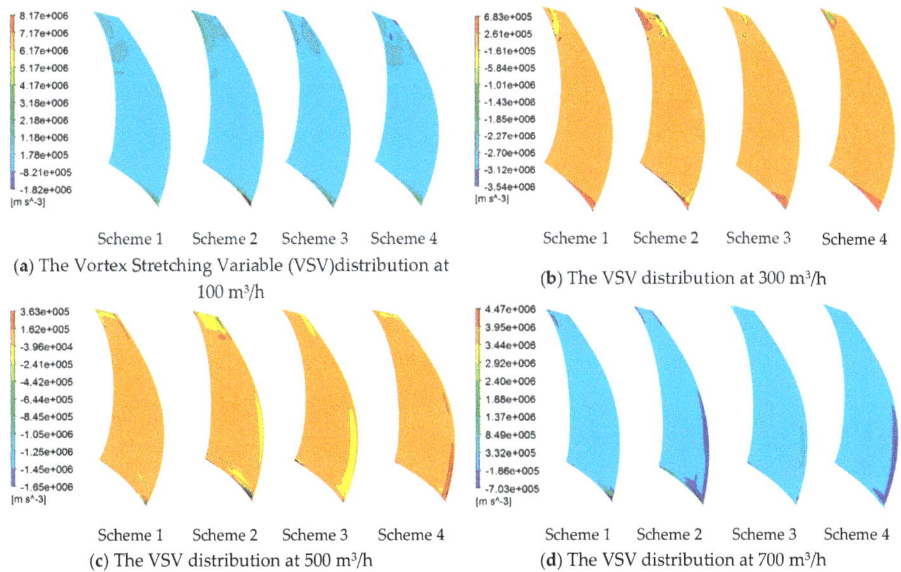

Scheme 1   Scheme 2   Scheme 3   Scheme 4   Scheme 1   Scheme 2   Scheme 3   Scheme 4

(**a**) The Vortex Stretching Variable (VSV)distribution at 100 m³/h   (**b**) The VSV distribution at 300 m³/h

Scheme 1   Scheme 2   Scheme 3   Scheme 4   Scheme 1   Scheme 2   Scheme 3   Scheme 4

(**c**) The VSV distribution at 500 m³/h   (**d**) The VSV distribution at 700 m³/h

Figure 12. The VSV distribution at various flow conditions.

4.2. Energy Characteristic of the Pump

The self-priming pump consumes and converts energy during its operation. Noting that the entropy always increases in the irreversible thermal processes, the entropy production is calculated to evaluate the energy loss of the pump. The total entropy production can be obtained as follows [15]:

$$S_{\text{Total}} = S_H + S_V + S_T + S_W \tag{8}$$

In Equation (8), S_H represents the entropy production caused by the heat conversion. Since the self-priming pump operates under the condition of a constant temperature, this term is neglected in this study. S_W denotes the entropy production due to the dissipation of kinetic energy and potential energy on the wall (WEP). S_V is the entropy production from the viscidity diffusion (VEP) and S_T is

117

the entropy production generated by the turbulence which can determined by the loss in turbulent kinetic energy (TEP). The rate of the entropy production is obtained as follows: [12]:

The rate of entropy production from VEP can be expressed as:

$$s_v = \frac{\mu}{T}\left[2\left\{\left(\frac{\partial \overline{u}}{\partial x}\right)^2 + \left(\frac{\partial \overline{v}}{\partial y}\right)^2 + \left(\frac{\partial \overline{w}}{\partial z}\right)^2\right\} + \left(\frac{\partial \overline{u}}{\partial y} + \frac{\partial \overline{v}}{\partial x}\right)^2 + \left(\frac{\partial \overline{u}}{\partial z} + \frac{\partial \overline{w}}{\partial x}\right)^2 + \left(\frac{\partial \overline{v}}{\partial z} + \frac{\partial \overline{w}}{\partial y}\right)^2\right] \tag{9}$$

where μ is the dynamic viscosity and T is the temperature.

The rate of entropy production from VEP can be expressed as:

$$s_t = \frac{\mu}{T}\left[2\left\{\overline{\left(\frac{\partial u'}{\partial x}\right)^2} + \overline{\left(\frac{\partial v'}{\partial y}\right)^2} + \overline{\left(\frac{\partial w'}{\partial z}\right)^2}\right\} + \overline{\left(\frac{\partial u'}{\partial y} + \frac{\partial v'}{\partial x}\right)^2} + \overline{\left(\frac{\partial u'}{\partial z} + \frac{\partial w'}{\partial x}\right)^2} + \overline{\left(\frac{\partial v'}{\partial z} + \frac{\partial w'}{\partial y}\right)^2}\right] \tag{10}$$

Since S_T cannot be solved directly, the rate of the TEP can obtained as follows [13]:

$$s_t = \frac{\rho \varepsilon}{T} \tag{11}$$

where ρ is the density of water (1000 kg/m^3) and ε is the turbulence dissipation.

It should be noted that due to the velocity gradient on the wall, and the dissipation of kinetic and potential energy occurs during the water flow between the turbulent layer and the laminar layer. Most of the previous work in this area has neglected the energy loss of the WEP. In this paper, the rate of WEP is calculated by the expression [23]:

$$s_w = \frac{\tau_w v_w}{T} \tag{12}$$

The WEP on the wall is obtained as:

$$S_W = \int_A \frac{\tau_w v_w}{T} dA \tag{13}$$

where τ_w represents the shear stress of the wall, while v_w is the speed of the first fluid layer near the wall. Thus, the VEP and TEP generated in the fluid region can be obtained as follows:

$$S_T = \int_V s_t dV \tag{14}$$

$$S_V = \int_V s_v dV. \tag{15}$$

Figure 13 shows the entropy production of the impeller for various flow conditions. It can be observed that the TEP of the blades with a thick leading edge (Schemes 2 and 4) is larger than that of the blades with a thin leading edge (Schemes 1 and 3). Furthermore, the TEP of the blade with increasing thickness from the leading edge to the middle section and trailing edge (Scheme 3) is lower than that of the uniformly thin blade (Scheme 1). As the flow condition gradually increases, the TEP first decreases drastically but then smoothly increases at the design flow condition. In addition, the decrement rate of TEP at partial-load condition is larger than the growth rate under the over-load condition, and the TEP at design flow condition drops down to the minimum value. By comparing the VEP and WEP for different blade thicknesses, it can be observed that their variation trends are consistent with TEP; however, both the VEP and WEP of Scheme 2 are higher than those of other schemes. Furthermore, as the flow condition increases, both the VEP and WEP continue to rise but the growth rate of VEP is larger than that of WEP. More importantly, comparing the magnitudes of TEP, VEP, and WEP, it can be concluded that the effects of TEP and VEP on energy loss are stronger than that of WEP. In other words, the energy loss caused by VEP and TEP are the dominant factors influencing the energy conversion in the novel self-priming pump.

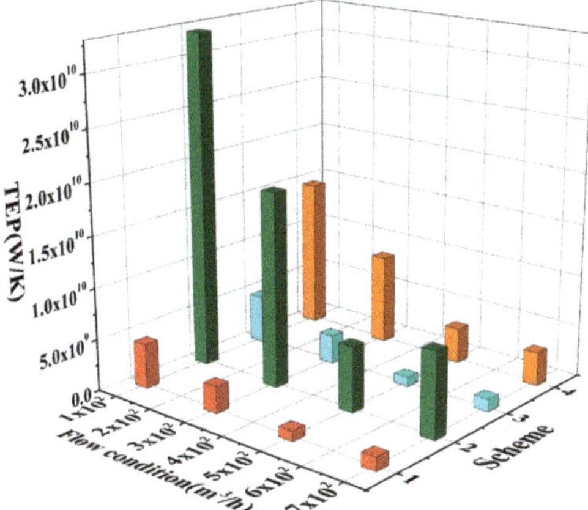

(**a**) The turbulence entropy production (TEP) of the impeller at various flow conditions

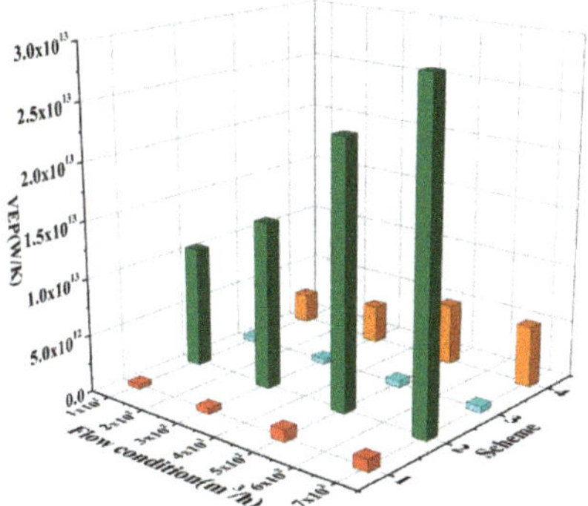

(**b**) The viscous entropy production (VEP) of the impeller at various flow conditions

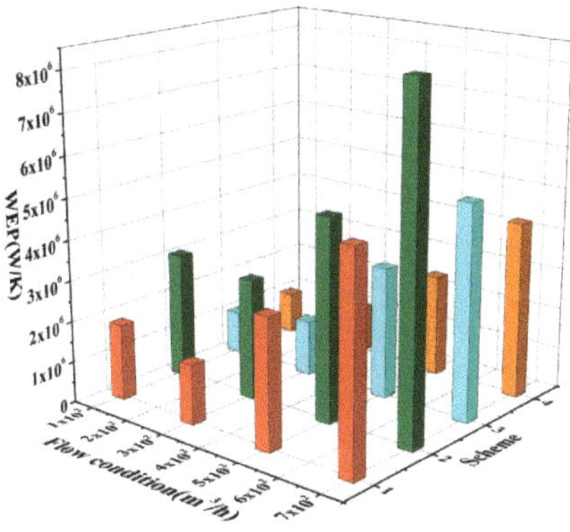

(**c**) The entropy production due to the dissipation of kinetic energy and potential energy on the wall (WEP) of the impeller at various flow conditions.

Figure 13. Impeller entropy production at various flow conditions.

As shown in Figure 14, the distribution of the VEP rate coincides with the VDV distribution rate. It can be noticed that the energy loss caused by the VEP rate is mainly concentrated near the trailing edge for a partial-load condition. There is a negative correlation between the VEP and flow conditions in the range $0.2Q_d$ to $1.0Q_d$. This means the magnitude of the VEP rate gradually decreases when the flow condition increases to the design flow condition. On the other hand, as the flow condition further increases beyond the design flow condition, the VEP rate shows an increasing trend and the location of energy loss changes from the trailing edge to the leading edge of the blade. These results can be attributed to the change in VEP due to viscous diffusion; greater viscous diffusion can be found at the leading edge under partial-load conditions, which leads to separation of the boundary layer. Moreover, the area and magnitude of viscous diffusion in Scheme 3 are obviously lower than those in Scheme 2; in other words, utilizing the blade with increasing thickness from the leading edge (Scheme 3) is a good solution to reduce the boundary layer separation and its effect.

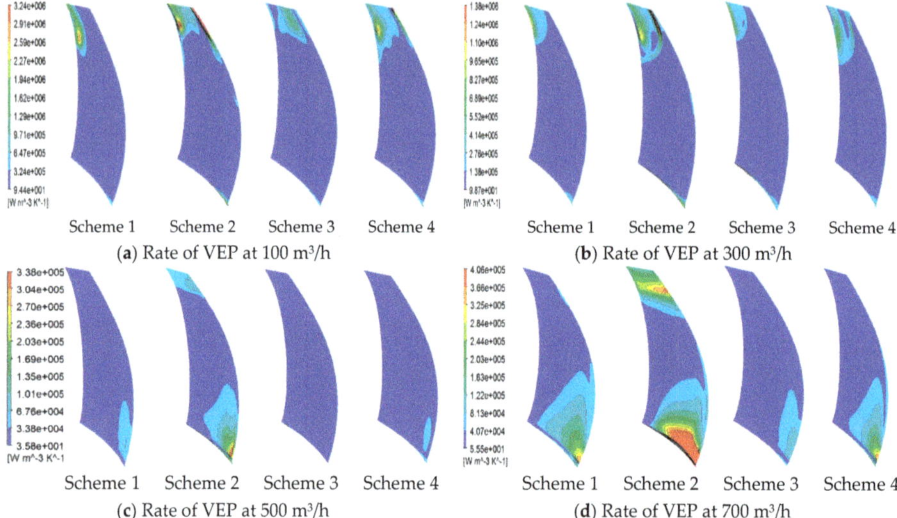

Figure 14. Rate of VEP at various flow conditions [24].

Figure 15 presents the distribution of the TEP rate at various flow conditions. It can be noticed that the TEP rate distribution has a consistent variation trend with the VEP rate, but its intensity is relatively smaller. Since the TEV is created under partial-load conditions, it can not only reduce the flow area of the impeller but also can induce turbulent dissipation at the trailing edge. In addition, by comparing the TEP rates of each schemes, it can be found that a larger TEP rate occurs at the trailing edge of the thick blade (Schemes 2 and 4), and that the TEP of Scheme 3 is closer to the ideal flow field.

Thus, based on the above observations, it can be concluded that the flow condition creates negative effect on the TEP and VEP rate in the range $0.2Q_d$ to $1.0Q_d$, and the lowest TEP and VEP rates are obtained for $1.0Q_d$. Furthermore, as the flow condition increases to the over-load condition, the TEP and VEP rates gradually increase and the energy loss caused by the VEP and TEP decrease to the minimum of the design condition. In addition, the entropy production from the thick blade is larger than that from the thin blade. Therefore, it can be concluded that the blade thickness distribution plays an essential role in the energy loss of the self-priming pump.

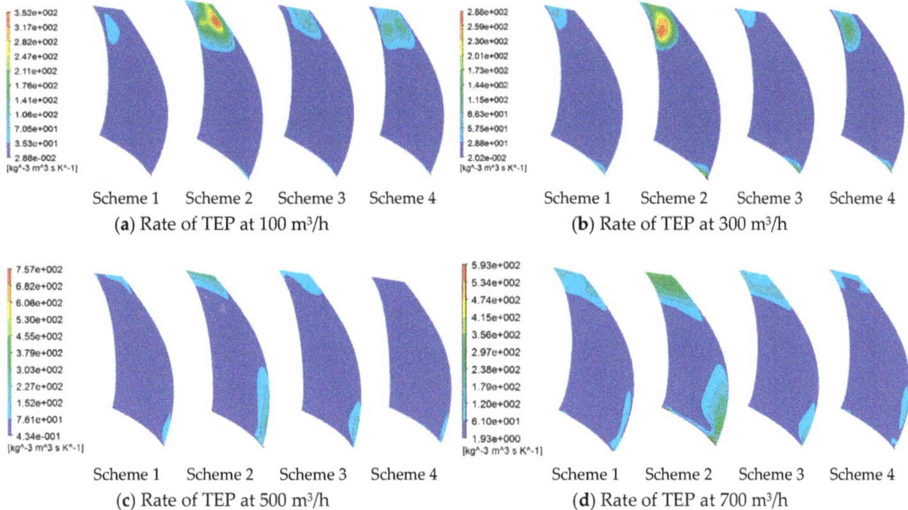

Scheme 1   Scheme 2   Scheme 3   Scheme 4   Scheme 1   Scheme 2   Scheme 3   Scheme 4

(**a**) Rate of TEP at 100 m³/h   (**b**) Rate of TEP at 300 m³/h

Scheme 1   Scheme 2   Scheme 3   Scheme 4   Scheme 1   Scheme 2   Scheme 3   Scheme 4

(**c**) Rate of TEP at 500 m³/h   (**d**) Rate of TEP at 700 m³/h

Figure 15. Rate of TEP at various flow conditions [24].

5. Hydraulic Performance of the Pump

To compare the hydraulic performance of the pump with different thicknesses of the blade (Schemes 1–4), the head and efficiency are selected as the two key quantities for evaluating its hydraulic characteristics. The head indicates the hoisting height of water when it receives energy from the pump, while the efficiency expresses the energy conversion ability of the self-priming pump. The head and efficiency are defined as:

$$H = \frac{P_O - P_I}{\rho g} + \frac{V_O^2 - V_I^2}{2g} + (Z_O - Z_I) \tag{16}$$

and

$$\eta = \frac{60gHQ_F}{\pi Mn} \tag{17}$$

where $P_O - P_I$, $V_O^2 - V_I^2$ and $Z_O - Z_I$ denote the pressure difference, velocity difference, and height difference obtained when the water flow through the pump, M is the torque, and Q_F is the flow conditions of the self-priming pump. In the analysis and presentation of results, the head and flow conditions are converted into the dimensionless parameters given below [24]:

$$\Phi = \frac{60Q_F}{\pi^2 D_2 b_2 n} \tag{18}$$

$$\Psi = \frac{7200Hg}{\pi^2 D_2^2 n^2}. \tag{19}$$

Figure 16 shows the differences in hydraulic performance of different schemes. It shows the formation of energy loss in the partial-load condition and over-load condition, which becomes minimal in the design condition. This trend in Figure 16 can be mainly attributed to the generation of TEV under a part-load condition and LEV under an over-load condition; however, the energy loss in the partial-load condition is more severe than in the over-load condition. Furthermore, it can be noted that the head of the Scheme 2 is lower than that of Scheme 1 overall except under the partial-load condition. Since a large TEV is generated in Scheme 2, the flow area reduces sharply, which results in large changes in the internal flow. Due to the effect of incoming flow, the head increases but results in large energy dissipation as shown in Figure 16b. Thus, the efficiency of Scheme 2 is obviously lower than that of

Scheme 1 for flow condition of $0.6Q_d$. Although the energy loss of the thinner blades (Scheme 1) is lower, its energy conversion ability is weaker than that of the blade with thicker trailing edge. More importantly, by comparing the efficiency of each scheme, it can be concluded that the energy conversion ability of Scheme 3 is superior to that of other schemes. Due to the advantages of the thin leading edge and thick trailing edge, Scheme 3 can not only significantly reduce the impact of the incoming flow in over-load condition, but also can promote the work capability of the blade on the fluid.

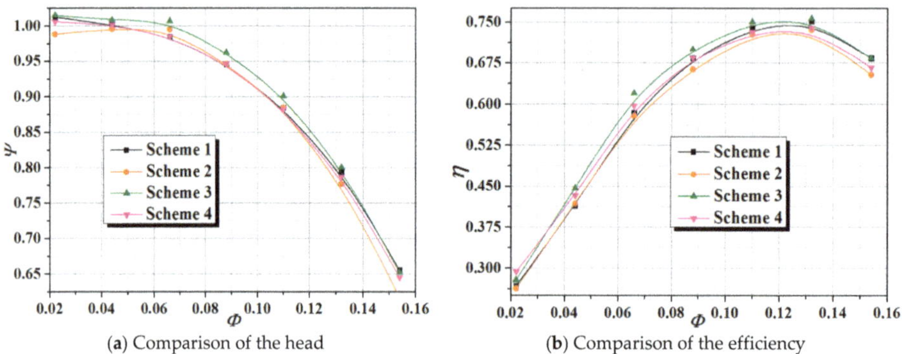

(a) Comparison of the head   (b) Comparison of the efficiency

Figure 16. Head and efficiency under various flow conditions.

6. Experiment Results

To verify the accuracy of the calculations reported in Section 5, the hydraulic performance experiment for Scheme 3 (since it was found to be the best scheme in Section 5) was conducted at Jiangsu University in China. The measurement deviation of the pressure transducer (D_P) is ± 0.5 %. Considering that the monitoring points were set on the inlet and outlet tank and the measurement deviation of dynamic head can be ignored (D_D = 0), the measurement deviation of head (D_H) is obtained as follows:

$$D_H = \sqrt[2]{D_P^2 + D_D^2} \tag{20}$$

Furthermore, the deviation of the transducer turbine flowmeter (D_Q) is ± 1.0 %, the measurement deviations of torque (D_M) and speed transducer (D_N) are ± 0.5 % and ± 0.1 %, respectively. Therefore, the measurement deviation of shaft power (D_T) can be determined by:

$$D_T = \sqrt[2]{D_N^2 + D_M^2} \tag{21}$$

Finally, the measurement deviation of experimental system (D_S) can be expressed as follows:

$$D_S = \sqrt[2]{D_H^2 + D_Q^2 + D_T^2} \tag{22}$$

Therefore, the measurement deviation of the experiment is within ±1.22 % which quantifies the uncertainty in experimental measurements. The experimental results are shown in Figure 17. It can be seen from this figure that overall the experimental and numerical results for pump head and efficiency are in reasonable agreement with minor discrepancies. It should be noted that in the calculations, the internal structure of the pump was simplified. This meant the pump cavity, seal, and bearing were not considered in the numerical model. As a result, the pump performance results obtained in the calculations are slightly larger than those in the experiment. However, the maximum discrepancy between the numerical results and experimental results is within 3 %, which is in the acceptable range. It can therefore be concluded that the numerical model and computations reported in this paper are accurate and the results are reliable.

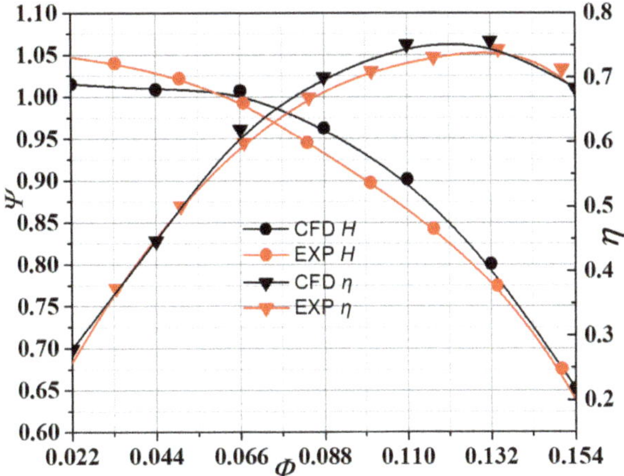

Figure 17. Comparison of experimental and numerical results for pump head and efficiency for Scheme 3.

In addition, to validate the capability of a novel self-priming pump, the experiments with waterless start were conducted at different altitudes at Jiangsu University in China.

Figure 18 shows the results of self-priming pump experiment at 5.3 m, 6 m, 7 m, and 8 m altitudes. It should be noted that the self-priming time reaches 62 s at 5.3 m altitude, which is remarkably shorter than the standard Chinese self-priming pump time of 120 s. Furthermore, it can be noted from Figure 18 that the increase rate in self-priming time R with the altitude can be separated into two stages, namely the altitude from 0 m to 6 m and from 6 m to 8 m. The first stage shows the non-linear variation while the second stage shows the linear variation in start-up time with the altitude. When the self-priming altitude is below 6 m, with increase in the altitude more resistance of water needs to be overcome, therefore negative pressure is generated in the pump which induces the air in the water to separate and leads to the air in the pump increasing. Thus, the line of the increase rate R in Figure 18 shows a nonlinear increase. On the other hand, as the altitude continues to increase, the air in the water is no longer discharged. Therefore, the relationship between air volume in the pump and altitude shows a linear increase, which is the reason for the linear growth in the second stages.

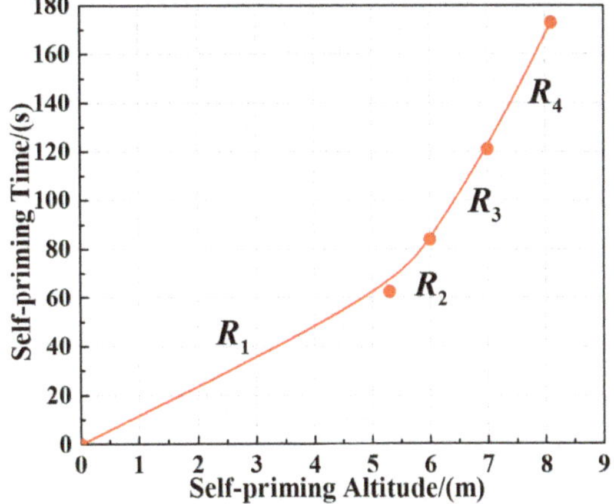

Figure 18. Self-priming time at different altitudes.

7. Conclusions

In this paper, by using the Q criterion with vorticity transport equation and analyzing the entropy production in the flow, the vortex structures and energy characteristics of a novel self-priming pump with different blade thicknesses were systematically investigated by numerical simulations which are validated for one of the most promising blade thickness distributions in terms of the pump head and efficiency. The following conclusions can be drawn.

A new self-priming pump has been designed which can achieve the pump start without water. Under the effect of the self-priming system, negative pressure is generated in the pump which results in rapid suction of the water in the pump. It is demonstrated that the self-priming capability of the pump is remarkably superior to the traditional pump.

The spatial structure of vortices in the pump is investigated by employing the Q criterion with the numerical solution of the vorticity transport equation. Based on the morphology, the vortices be separated into three categories: Trailing Edge Vortex (TEV), Leading Edge Vortex (LEV), and Gap Leakage Vortex (GLV). Generally, the morphology of the TEV is more disorderly than that of LEV and GLV, and the pressure and intensity of TEV are significantly higher than the other two vortex structures. TEV cannot only block the passage of the impeller but can also increase instability in the internal flow. Analysis of the formation mechanism of the vortices using the numerical solution of the vorticity transport equation shows that the Viscous Diffusion Variable (VDV) plays an essential role in the separation of the boundary layer which results in the generation of Tip Leakage Vortex (TLV). In addition, the thicker the blade is, the more severe the flow separation becomes.

To determine the magnitude and distribution of energy loss in the pump, entropy production analysis is applied to study the influence of blade thickness on energy characteristics. Based on the analysis, greater energy loss is generated in both the partial-load condition and the over-load condition relative to design condition; however, the energy loss in partial-load condition is larger than that in the over-load condition due to differences in the size and intensity of TLV. Furthermore, in other words, the energy loss caused by viscous entropy production (VEP) and turbulence entropy production (TEP) are the dominant factors influencing the energy loss in the novel self-priming pump. The energy loss of the thicker blade is larger than that of the thinner blade; however, employing the blade with a thin leading edge and thick trailing edge can not only significantly reduce the impact of incoming flow under over-load condition but can also increase the work capability of the blade on the fluid. Thus, increase in thickness of the blade from the leading edge to the trailing edge is beneficial for pump performance.

Finally, the hydraulic test confirmed the accuracy of the simulations. Therefore, the numerical results presented in this paper can be considered as accurate and reliable.

Author Contributions: This is a joint work and the authors contributed according to their expertise and capability: H.C. conducted the investigation and analysis, and writing and revisions of the paper; R.K.A. ontributed to methodology and revisions of the paper; W.L. Performed the validation and revision; L.Z. Performed the data analysis and W.S. was involved in revision of the manuscript.

Funding: This work was sponsored by the National Natural Science Foundation of China (No. 51979138, No. 51679111, No. 51409127 and No. 51579118), National Key R&D Program Project (No. 2017YFC0403703), PAPD, and Six Talents Peak Project of Jiangsu Province (No. JNHB-CXTD-005), Key R&D Program Project in Jiangsu Province (BE2015119, BE2015001-4, BE2016319, BE2017126), Natural Science Foundation of Jiangsu Province (No. BK20161472), Science and Technology Support Program of Changzhou (No. CE20162004) and Graduate Student Scientific Research Innovation Projects of Jiangsu Province (Grant No. SJKY19_2547). The first author would like to thank Chinese Scholarship Council (CSC) for the financial support and Washington University in St. Louis for providing the opportunity to spend a year as a visiting PhD student (201808320259).

Nomenclature

H	head m
Q_d	design flow condition, m^3/h
n	rotation speed, r/min

n_s	specific speed
D_1	impeller inlet diameter, mm
D_2	impeller outlet diameter, mm
Z	blade number
φ	blade wrap angles, °
b_1	volute inlet width, mm
D_4	volute outlet diameter, mm
SST	shear stress transition
t	rotation period, s
LEV	leading edge vortex
TEV	trailing edge vortex
GLV	gap leakage vortex
PSB	suction surface of the blade
VDV	viscidity diffusion variable
VSV	vortex stretching variable
WEP	wall entropy production
VEP	viscidity entropy production
TEP	turbulence entropy production
ρ	density of water, kg/m^3
ε	turbulence dissipation
M	torque, Nm
Q_F	flow condition, m^3/h
D_P	deviation of the pressure transducer
D_D	measurement deviation of dynamic head
D_H	deviation of transducer turbine flowmeter
D_M	measurement deviation of torque
D_N	deviation of speed transducer
D_T	measurement deviation of shaft power
D_S	measurement deviation of experiment system

References

1. Turkyilmazoglu, M. Performance of direct absorption solar collector with nanofluid mixture. *Energy Convers. Manag.* **2016**, *114*, 1–10. [CrossRef]
2. Wang, T.; Kong, F.; Xia, B.; Bai, Y.; Wang, C. The method for determining blade inlet angle of special impeller using in turbine mode of centrifugal pump as turbine. *Renew. Energy* **2017**, *109*, 518–528. [CrossRef]
3. Elyamin, G.R.H.A.; Bassily, M.A.; Khalil, K.Y.; Mohamed, S.G. Effect of impeller blades number on the performance of a centrifugal pump. *Alex. Eng. J.* **2019**, *58*, 39–48. [CrossRef]
4. Shao, C.; Zhou, J.; Cheng, W. Effect of viscosity on the external characteristics and flow field of a molten salt pump in the view of energy loss. *Heat Mass Transf.* **2019**, *55*, 711–722. [CrossRef]
5. Nejad, J.; Riasi, A.; Nourbakhsh, A. Efficiency improvement of regenerative pump using blade profile modification: Experimental study. *Proc. Inst. Mech. Eng. Part E J. Process Mech. Eng.* **2019**, *233*, 448–455. [CrossRef]
6. Han, X.; Kang, Y.; Li, D.; Zhao, W. Impeller optimized design of the centrifugal pump: A numerical and experimental investigation. *Energies* **2018**, *11*, 1444. [CrossRef]
7. Jeon, S.Y.; Yoon, J.Y.; Jang, C.M. Optimal design of a novel 'S-shape' impeller blade for a microbubble pump. *Energies* **2019**, *12*, 179. [CrossRef]
8. Nejad, J.; Riasi, A.; Nourbakhsh, A. Parametric study and performance improvement of regenerative flow pump considering the modification in blade and casing geometry. *Int. J. Numer. Methods Heat Fluid Flow* **2017**, *27*, 1887–1906. [CrossRef]
9. Wang, C.; Hu, B.; Zhu, Y.; Wang, X.; Luo, C.; Cheng, L. Numerical study on the gas-water two-phase flow in the self-priming process of self-priming centrifugal pump. *Processes* **2019**, *7*, 330. [CrossRef]
10. Wang, C.; He, X.; Zhang, D.; Hu, B.; Shi, W. Numerical and experimental study of the self-priming process of a multistage self-priming centrifugal pump. *Int. J. Energy Res.* **2019**, 1–19. [CrossRef]

11. Wang, C.; He, X.; Shi, W.; Wang, X.; Wang, X.; Qiu, N. Numerical study on pressure fluctuation of a multistage centrifugal pump based on whole flow field. *AIP Adv.* **2019**, *9*, 035118. [CrossRef]

12. Wang, C.; Zhang, Y.; Hou, H.; Zhang, J.; Xu, C. Entropy production diagnostic analysis of energy consumption for cavitation flow in a two-stage LNG cryogenic submerged pump. *Int. J. Heat Mass Transf.* **2019**, *129*, 342–356. [CrossRef]

13. Hou, H.; Zhang, Y.; Zhou, X.; Zuo, Z.; Chen, H. Optimal hydraulic design of an ultra-low specific speed centrifugal pump based on the local entropy production theory. *Proc. Inst. Mech. Eng. Part A J. Power Energy* **2019**, *233*, 715–726. [CrossRef]

14. Gu, Y.; Pei, J.; Yuan, S.; Wang, W.; Zhang, F.; Wang, P.; Appiah, D.; Liu, Y. Clocking effect of vaned diffuser on hydraulic performance of high-power pump by using the numerical flow loss visualization method. *Energy* **2019**, *170*, 986–997. [CrossRef]

15. Li, D.; Wang, H.; Qin, Y.; Han, L.; Wei, X.; Qin, D. Entropy production analysis of hysteresis characteristic of a pump-turbine model. *Energy Convers. Manag.* **2017**, *149*, 175–191. [CrossRef]

16. Chang, H.; Li, W.; Shi, W.; Liu, J. Effect of blade profile with different thickness distribution on the pressure characteristics of novel self-priming pump. *J. Braz. Soc. Mech. Sci. Eng.* **2018**, *40*, 518. [CrossRef]

17. Bai, L.; Zhou, L.; Han, C.; Zhu, Y.; Shi, W. Numerical study of pressure fluctuation and unsteady flow in a centrifugal pump. *Processes* **2019**, *7*, 354. [CrossRef]

18. Bai, L.; Zhou, L.; Jiang, X.; Pang, Q.; Ye, D. Vibration in a multistage centrifugal pump under varied conditions. *Shock Vib.* **2019**, 2057031. [CrossRef]

19. Alfonsi, G.; Primavera, L. Temporal evolution of vortical structures in the wall region of turbulent channel flow. *Flow Turbul. Combust.* **2009**, *83*, 61–79. [CrossRef]

20. Ji, B.; Luo, X.; Arndt, R.E.A.; Wu, Y. Numerical simulation of three-dimensional cavitation shedding dynamics with special emphasis on cavitation–vortex interaction. *Ocean Eng.* **2014**, *87*, 64–77. [CrossRef]

21. Liu, Y.; Tan, L. Tip clearance on pressure fluctuation intensity and vortex characteristic of a mixed flow pump as turbine at pump mode. *Renew. Energy* **2018**, *129*, 606–615. [CrossRef]

22. Wu, J.Z.; Ma, H.Y.; Zhou, M.D. *Vorticity and Vortex Dynamic*; Springer Science & Business Media: Berlin/Heidelberg, Germany, 2007.

23. Hucan, H.; Zhang, Y.; Li, Z. A numerical research on energy loss evaluation in a centrifugal pump system based on local entropy production method. *Therm. Sci.* **2017**, *21*, 1287–1299.

24. Chang, H.; Shi, W.; Li, W.; Liu, J. Energy loss analysis of novel self-priming pump based on the entropy production theory. *J. Therm. Sci.* **2019**, *28*, 306–318. [CrossRef]

 *processes*

Article

Multi-Agent Consensus Algorithm-Based Optimal Power Dispatch for Islanded Multi-Microgrids

Xingli Zhai [1] **and Ning Wang** [2,*]

[1] Jinan Power Supply Company in Shandong Provincial Electric Power Company of State Grid, Jinan 250000, China; 15169199001@163.com

[2] College of Electrical Engineering, Zhejiang University, Hangzhou 310027, China

* Correspondence: 11610045@zju.edu.cn; Tel.: +86-1776-7069-885

Received: 25 August 2019; Accepted: 19 September 2019; Published: 1 October 2019

Abstract: Islanded multi-microgrids formed by interconnections of microgrids will be conducive to the improvement of system economic efficiency and supply reliability. Due to the lack of support from a main grid, the requirement of real-time power balance of the islanded multi-microgrid is relatively high. In order to solve real-time dispatch problems in an island multi-microgrid system, a real-time cooperative power dispatch framework is proposed by using the multi-agent consensus algorithm. On this basis, a regulation cost model for the microgrid is developed. Then a consensus algorithm of power dispatch is designed by selecting the regulation cost of each microgrid as the consensus variable to make all microgrids share the power unbalance, thus reducing the total regulation cost. Simulation results show that the proposed consensus algorithm can effectively solve the real-time power dispatch problem for islanded multi-microgrids.

Keywords: islanded multi-microgrids; real-time power dispatch; multi-agent; consensus algorithm

---

1. Introduction

The emergence of microgrids (MG) provides a new technical means for the comprehensive utilization of renewable energy [1–3]. According to whether there is an electrical connection with the main grid, microgrids feature two typical operation modes, i.e., grid-connected and islanded. In general, microgrids can operate in grid-connected mode to exchange power with the main grid. On the other hand, when being used in a remote area or in an emergency situation, they can also be transferred to islanded mode to guarantee local grid services [4]. Therefore, an islanded microgrid is more suitable for remote areas without grid coverage, which can improve the utilization efficiency of local renewable energy and reduce the cost of power supply in remote areas. Recently, the multi-microgrids (MMGs) system has become an integrated, flexible network that incorporates multiple individual microgrids (MGs) [5,6], which are often geographically close and connected to a distribution bus. If the neighboring islanded microgrid in a remote area can realize the cluster operation through interconnection, the mutual energy between the microgrids not only helps to absorb excess power energy, but also supports each other as a backup power supply [7,8], which is beneficial to improving the overall power supply reliability and economy.

The islanded microgrids plays an important role in renewable energy applications and power sharing among different loads connected to multi-microgrid systems [9,10]. The key issue with islanded microgrids is how to ensure a power balance between generation and demand in a cost-effective way. Hence, problems of microgrid real-time dispatching and operations have received considerable attention in the literature. There are some achievements on the dispatching and operations of individual microgrids that mainly focus on reducing the regulation cost and the coordination of various devices in the microgrid. Due to the intermittency and variability of renewables-based distributed generation (DG), the methods for uncertainty power dispatch of individual microgrid are mainly classified into three categories: stochastic power dispatch [11], robust power dispatch [12,13], and rolling power dispatch [14]. In order to deal with the uncertainty of demand response and renewable energy, the authors of [11] presented a stochastic programming framework for 24-h optimal scheduling of combined heat and power (CHP) systems-based MG. Wang R. and Luo Z. of [12] and [13] adopted a robust optimization approach to accommodate the uncertainties of demand response and renewable energy; it performs better than deterministic optimization in terms of the expected operational costs. For a real renewable-based microgrid in the north of Chile, the authors of [14] proposed a moving horizon optimization strategy to eliminate the forecasting errors caused by renewable energy. The optimal dispatch of an individual microgrid owes more to the coordination of the controllable units in a microgrid. However, for multi-microgrids, the power dispatch between each microgrid is a critical issue, so the optimal dispatch of multi-microgrids is more complex.

Due to the frequent fluctuation of power supply and load on both sides, offline optimal dispatch methods have not been suitable for the multi-microgrids, especially for working in islanded mode, which lacks support from the main grid [15]. Therefore, the power dispatch of the islanded multi-microgrids should focus on real-time optimal dispatch capability in order to maintain a real-time power balance between power generation and load demand, ensure the stable operation of the multi-microgrids system, and take into account the economic operation of the system. As long as the power command calculation process is required for centralized optimization, it will cause a certain computational complexity. In the case of high real-time requirements, the requirements for the generation and delivery speed of dispatching command also increase. With the advantages of distributed control architecture in smart grids, the multi-agent theory [16] and distributed control method [17,18] are applied to the microgrid dispatching model. The authors of [19] developed a new hybrid intelligent algorithm called imperialist competitive algorithm-genetic algorithm (ICA-GA) to determine both the optimal location and operation of an islanded MG. In order to minimize the islanded microgrids' operational losses, the authors of [20] adopted the glow-worm swarm optimization (GSO) algorithm to solve an optimal power flow problem. However, these approaches are a centralized optimization method that collects the whole network's information via a central controller and uses an intelligent optimization algorithm, such as glow-worm swarm optimization, particle swarm optimization, ant colony optimization, etc. Although the centralized optimization method has high regulation accuracy, when the number of network nodes is large, the communication volume is too large, the communication line is required to be high, and the scalability is poor. In addition, the intelligent optimization algorithm is unstable and cannot guarantee convergence to the optimal solution, which could cause a decline in the control performance. A new distributed reinforcement learning approach based on the multi-agent systems algorithm was applied to minimize the power losses under given operational constraints in [21]. The multi-agent system based consensus method [22] provides a new way of solving the real-time power dispatch problem of islanded multi-microgrids. The main issue with a consensus problem is achieving agreement regarding certain quantities of interest associated with agents in multi-agent systems by utilizing a local information exchange [23]. The traditional consensus algorithm is a very simple local coordination rule, which results in agreement at the group level, and no centralized task planner or global information is required by the algorithm. Due to its distributed implementation, robustness, and scalability, multi-agent consensus algorithms have been widely applied in many coordination problems, such as power system economic dispatch [24,25], power allocation [26], optimal control [27,28], etc. Compared with the traditional centralized optimization algorithm, the consensus method only requires each agent to obtain the information on the local and neighboring agents in real time. Hence, it can obtain the ideal convergence value with less transmission information and a shorter optimization time.

Motivated by these works, this paper provides a multi-agent system-based consensus algorithm to solve the real-time power dispatch problem of islanded multi-microgrids, which have a lower communication burden and better dynamic performance. In order to ensure the overall real-time power balance of the islanded multi-microgrids and reduce the power regulation costs, the real-time cooperative power dispatch framework of the islanded multi-microgrids is built by using a multi-agent system consensus algorithm. Simultaneously, the real-time dispatch of power imbalance is optimized to reduce the overall regulation costs of the islanded multi-microgrids, so as to ensure that each microgrid is responsible for the corresponding power regulation tasks according to its own situation. At the same time, the speed of dispatching command generation and release is accelerated because of avoiding centralized optimization, so the system can better adapt to the dynamic requirements of real-time power dispatch of the islanded multi-microgrids. Based on the consensus method, the real-time power dispatch strategy works in a fully distributed manner without a central coordinator; communication occurs only between the device and its neighbors. The main contributions of this paper are as follows:

(1)   A real-time cooperative dispatch framework for islanded multi-microgrids based on multi-agent consensus method is built that can ensure the overall real-time power balance and minimize the power regulation costs.

(2)   The consensus method only needs a small amount of information from the local and neighboring microgrids, which reduces the communications burden and increases the reliability compared with the traditional centralized optimization method.

The remainder of this paper is organized as follows. In Section 2, we establish a cooperative power dispatch framework for islanded multi-microgrids based on the consensus algorithm. We then model the regulation cost of each controllable unit in the microgrid to quantify the regulation costs of each controllable unit participating in the real-time control process in Section 3. The power dispatch consensus algorithm is designed for an islanded multi-microgrid in Section 4. Several numerical simulations are conducted and analyzed in Section 5, and Section 6 concludes this paper.

2. Cooperative Power Dispatch Framework of Islanded Multi-Microgrids

2.1. Cooperative Power Dispatch Framework Based on Consensus Algorithm

The microgrid is equipped with a microgrid controller (MGC) according to the requirements of the control. The MGC is responsible for ensuring the stable real-time operation of the microgrid. When many microgrids, through interconnection, constitute a multi-microgrids system, the traditional centralized dispatch framework requires the upper system to gather real-time information on each microgrid. This information is distributed and dispatched to each unit after centralized optimization. Although this method can more fully acquire system information and adapt to various optimization algorithms, it increases the burden on the communication network and the controller. At the same time, it cannot adapt to the requirements of plug and play and real-time control.

The number of controllable units and amount of data in multi-microgrids is large. The multi-microgrids' interconnection makes the operation mode diversified and requires the control mode to be easily extended, which need to satisfies the requirement of "plug and play." Therefore, the architecture and operational characteristics of the islanded multi-microgrids determine that it is suitable to adopt a decentralized control architecture [29]. This paper establishes a real-time cooperative dispatch framework for islanded multi-microgrids based on the multi-agent consensus method, which is shown in Figure 1.

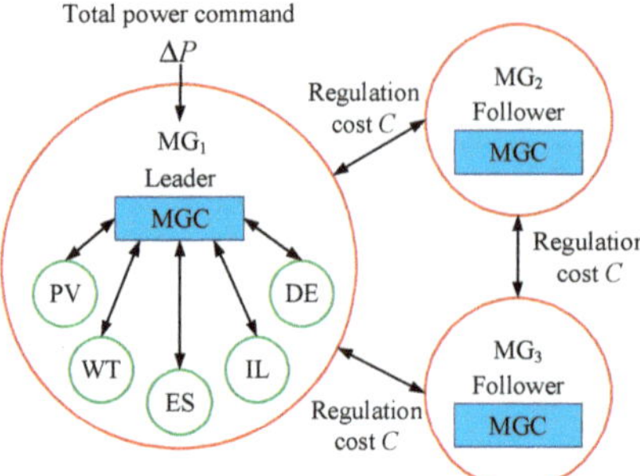

Figure 1. Real-time cooperative dispatch framework for islanded multi-microgrids.

The control objective is to ensure the overall real-time power balance of islanded multi-microgrids and reduce the power regulation costs, which is equivalent to the problem of converting the economic dispatch problem into the consensus of the regulation costs in the power allocation process [30]. Therefore, the role of each MGC is as the agent in the network and the regulation cost of each microgrid is the concerned consensus variable. Under the premise that interconnected multi-microgrids system belongs to the same community of interests, the regulation of each microgrid is guided by the economic signal of the whole system, i.e., the regulation cost signal. Thus the economy of islanded multi-microgrids operation can be improved by reasonably allocating the regulation power. The islanded multi-microgrids adopt the "leader-follower" model [31,32]. The leader MGC is a communication center of all microgrids, which calculates the total power command, communicates and cooperates with other MGs, and balances the power of the entire control area. The follower exchanges the regulation costs with its adjacent MGs. After the communication iteration of each microgrid, the total power command is delivered to each microgrid and the MGC delegates the power commands of each control unit in the microgrid according to the established strategy. Because each MGC does not need to acquire the global information of the system, the information needed by each microgrid in the iteration is only local information, and the regulation cost information transmitted by the neighbors, so the communication burden is small.

2.2. Power Dispatch Strategy of Microgrid

After receiving the allocated power command, each microgrid needs to decentralize the power commands of the controllable units according to the power allocation strategy. It is assumed that each microgrid includes the following units: photovoltaic (PV), wind turbine (WT), energy storage device (ES), and controllable micro-power sources, such as diesel engine (DE) or interruptible load (IL).

Considering the regulation costs and dispatching priority of various regulation methods, the decision order of power allocation strategy for positive and negative power command is specified as follows.

(1) Power allocation strategy for positive power command. When the microgrid receives a positive power command, it requires the microgrid to increase the power generation or reduce the power consumption. The energy storage discharge is preferred. If the energy storage reaches the discharge limit but still does not meet the demand, the diesel generator output power is increased. If the output of the diesel generator reaches the limit but still does not meet the demand, then only the load shedding method can be adopted to regulation the interruptible load.

(2) Power allocation strategy for negative power command. When the microgrid receives a negative power command, it requires the microgrid to reduce the power generation. The energy storage charge is preferred. If the energy storage reaches the charge limit but still does not meet the demand, the output of the photovoltaic system or wind turbine will be reduced and part of the power generation will be abandoned.

3. Microgrid Regulation Cost Modeling

In this section, the regulation cost of each controllable unit in the microgrid is modeled to quantify the regulation cost of each controllable unit participating in the real-time control process, thus providing a basis for calculating the consensus power allocation.

Considering that the regulation cost of most controllable units is more suitable to be measured by electricity, the state duration window T_S is introduced here. When calculating the regulation cost of each controllable unit at time t, the cumulated regulation power of each controllable unit in T_S period starting from time t is calculated by using real-time power commands, so that the real-time regulation cost of each microgrid at time t can be quantified according to the amount of electricity. The significance of the regulation cost described in this paper is to calculate the regulation cost of running each microgrid in the state specified by the current power command within a given time window, and to quantify the regulation cost by converting the power into electricity.

3.1. Regulation Cost of Energy Storage Battery

The regulation cost of an energy storage battery includes the damage to the battery life caused by the charge and discharge behavior and the corresponding maintenance cost, which is related to the charge and discharge power and the state of charge (SOC). According to the working requirements of an energy storage battery, four reference limits are selected to divide the battery capacity of an energy storage battery into five areas, as shown in Figure 2.

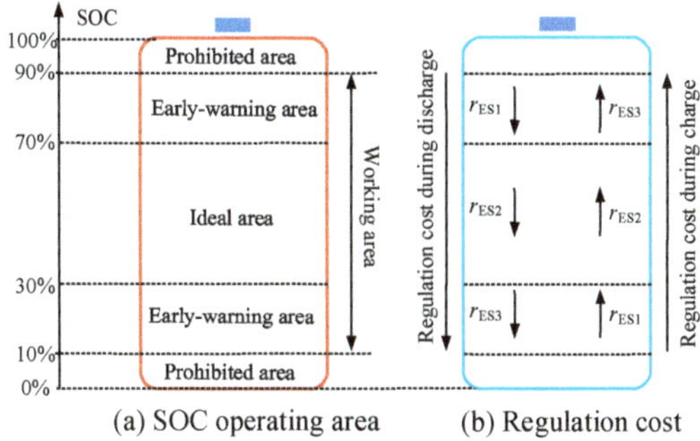

Figure 2. Schematic diagram for energy storage system SOC subarea: (**a**) Description of the SOC operating area; (**b**) Description of the regulation cost.

The four reference limits are SOC_{max}, SOC_{high}, SOC_{low}, and SOC_{min}, which are 90%, 70%, 30%, and 10%, respectively. Between SOC_{min} and SOC_{max} is the working area. It is required that the SOC of an energy storage battery be confined to within this range. The ideal working area (30-70%) is the optimal range of SOC. When the SOC enters the early-warning working area, it should not further approach the prohibited working area. In order to enable the dispatching energy storage battery to conform to the above basic idea, the unit regulation cost of each area is set according to the charge and discharge process in the working area, and it satisfies $r_{ES1} < r_{ES2} < r_{ES3}$. For example, in the early-warning area with high SOC, if the charging continues, the unit regulation cost will be higher. It can be considered that the damage to the battery will be greater, however. Conversely, the discharging will help the SOC approach the ideal working area. It can be considered that this is beneficial to the life of the battery and will lead to lower regulation costs. Therefore, the real-time maximum charge and discharge power of the energy storage system are given as follows:

$$P_{\text{ES}+}^{\max} = \begin{cases} \dfrac{\eta(SOC(t-\Delta t)-SOC_{\min})V_{\text{ES}}}{\Delta t}, & SOC(t) < SOC_{\min} \\ \eta P_{\text{ES}}^{\max}, & SOC(t) \geq SOC_{\min} \end{cases}$$
$$SOC(t) = SOC(t - \Delta t) - \dfrac{P_{\text{ES}}^{\max}\Delta t}{V_{\text{ES}}} \tag{1}$$

$$P_{\text{ES}-}^{\max} = \begin{cases} P_{\text{ES}}^{\max}, & SOC(t) \leq SOC_{\max} \\ \dfrac{(SOC_{\max}-SOC(t-\Delta t))V_{\text{ES}}}{\eta\Delta t}, & SOC(t) > SOC_{\max} \end{cases}$$
$$SOC(t) = SOC(t - \Delta t) + \dfrac{\eta P_{\text{ES}}^{\max}\Delta t}{V_{\text{ES}}} \tag{2}$$

where Δt is the time interval of power command; $SOC(t)$ and $SOC(t - \Delta t)$ represent the current SOC calculated value and the actual SOC value of the previous time point, respectively; and V_{ES}, $P_{\text{ES}}^{\max}$, and η are the energy storage system capacity, power limits, and efficiency, respectively.

The real-time regulation cost of the energy storage system can be expressed as follows:

$$\begin{cases} C_{\text{ES}}(\Delta P_{\text{ES}}) = V_{\text{ES}}T_{\text{S}} \displaystyle\sum_{i=1}^{3} r_{\text{ES}i}\Delta S_i \\ \Delta P_{\text{ES}}\Delta P \geq 0 \\ -P_{\text{ES}-}^{\max} \leq \Delta P_{\text{ES}} \leq P_{\text{ES}+}^{\max} \end{cases}, \tag{3}$$

where ΔP_{ES} is the power command of the energy storage device, $C_{\text{ES}}(\Delta P_{\text{ES}})$ is the regulation cost generated when the energy storage system is adjusted to the power command ΔP_{ES} and the regulation cost in the other adjustment modes below is the same, and ΔP is the total power imbalance of the multi-microgrids, that is, the real-time power imbalance. ΔS_i is the SOC change caused by the SOC area with a regulation cost of $r_{\text{ES}i}$ after running the T_{S} period with the current power command. The interval that the SOC may actually traverse during the calculation of the real-time power command is also related to the selection of the state duration window T_{S}. $P_{\text{ES}-}^{\max}$ and $P_{\text{ES}+}^{\max}$ are the maximum charge and discharge power of the energy storage system, respectively.

3.2. Regulation Cost of Diesel Generator

The generation cost of the diesel generator is determined by its fuel consumption coefficient and fuel unit price. In this paper, the power commands of diesel generators in the power allocation process refer to the output value within the range of its adjustable output force, i.e.,

$$\begin{cases} C_{\text{DE}}(\Delta P_{\text{DE}}) = r_{\text{DE}}\Delta P_{\text{DE}}T_{\text{S}} = \lambda r_{\text{oil}}\Delta P_{\text{DE}}T \\ \Delta P_{\text{DE}} = P_{\text{DE}}(t) - P_{\text{DE}}^{\min} \\ \Delta P_{\text{DE}}^{\max} = P_{\text{DE}}^{\max} - P_{\text{DE}}^{\min} \\ P_{\text{DE}}^{\min} = \max\{0.3P_{\text{DE}}^{\text{N}}, P_{\text{DE}}(t - \Delta t) - R_{\text{d}}\Delta t\} \\ P_{\text{DE}}^{\max} = \min\{P_{\text{DE}}^{\text{N}}, P_{\text{DE}}(t - \Delta t) + R_{u}\Delta t\} \\ P_{\text{DE}}^{\min} \leq P_{\text{DE}}(t) \leq P_{\text{DE}}^{\max} \\ 0 \leq \Delta P_{\text{DE}} \leq \Delta P_{\text{DE}}^{\max} \end{cases}, \tag{4}$$

where ΔP_{DE} is the power command of the diesel generator; r_{DE}, λ, and r_{oil} are the unit generation cost, unit fuel coefficient, and fuel unit cost of the diesel generator, respectively. P_{DE}^{N}, $P_{\text{DE}}^{\max}$, and $P_{\text{DE}}^{\min}$ are the rated power and real-time output upper and lower limit of diesel generators, respectively. $\Delta P_{\text{DE}}^{\max}$ is the upper limit of diesel generator power command. R_u and R_{d} are the rates of increasing and decreasing output of diesel generators, respectively.

Here ΔP_{DE} is the excess part of the current real-time output $P_{\text{DE}}(t)$, relative to the lower limit of output $P_{\text{DE}}^{\min}$ at this time. That is, the $P_{\text{DE}}^{\min}$ output part is regarded as the unadjustable part in the power allocation calculation process, which is recorded as the forced output. The operation cost generated by $P_{\text{DE}}^{\min}$ is recorded as the forced cost, which is not included in the regulation cost of the microgrid.

In this paper, the starting and stopping of the diesel generator are determined by setting the start and stop thresholds, given the start and stop coefficients k_{DE_on} and k_{DE_off}. When the power shortage of the i-th microgrid equipped with diesel generator exceeds $k_{DE\_on}P^N_{DE}$ and the diesel generator satisfies the other operational constraints, the diesel generator is started. If the power shortage of the i-th microgrid is lower than $k_{DE\_off}P^N_{DE}$, and the diesel generator has satisfied the other operation constraints, the diesel generator is shut down. When the diesel generator enters the start-stop state, it runs according to the increase/decrease output rate until the state transition is completed.

3.3. Regulation Cost of Other Units

For interruptible load and DG, the following assumptions are made in this paper: (1) when a part of the interruptible load needs to be cut off, the current calculation point calculates the current interruptible load amount and the next calculation point restores part of the load under the necessary conditions by default. If the next calculation point is still unable to restore the load, that part will be delayed according to the actual demand; (2) photovoltaic or wind turbines operate in the maximum power point tracking (MPPT) mode. When the output of distributed generation needs to be reduced, the current calculation point calculates the reduction command at the current maximum output, and the next calculation point is still calculated by the maximum output by default, so there is no need to consider the increase of the distributed generation output.

The regulation costs of load shedding and distributed generation abandoned generation are as follows:

$$\begin{cases} C_{IL}(\Delta P_{IL}) = r_{IL}\Delta P_{IL}T_S \\ 0 \le \Delta P_{IL} \le P^{max}_{IL} \\ P^{max}_{IL} = P_{IL}(t) \end{cases} \tag{5}$$

$$\begin{cases} C_{DG}(\Delta P_{DG-}) = r_{DG}|\Delta P_{DG-}|T_S \\ -P^{max}_{DG-} \le \Delta P_{DG-} \le 0 \\ P^{max}_{DG} = P_{PV}(t) + P_{WT}(t) \end{cases}, \tag{6}$$

where ΔP_{IL} and ΔP_{DG-} are the power commands of the interruptible load and the distributed generation output reduction, respectively. r_{IL} and r_{DG} are the unit regulation costs of the interruptible load and the distributed generation output reduction, respectively. P^{max}_{IL} and P^{max}_{DG-} are the upper limit of interruptible load and the distributed generation output reduction, respectively. P^{max}_{IL} is the real-time maximum load $P_{IL}(t)$ of the current interruptible load, i.e., P^{max}_{DG-} is the sum of the real-time output of PV ($P_{PV}(t)$) and the real-time output of ($P_{WT}(t)$).

3.4. Regulation Cost Function of Microgrid

Combining with the microgrid power allocation strategy and the regulation cost model of each controllable unit, the regulation cost function of each microgrid can be constructed as shown in Figure 3.

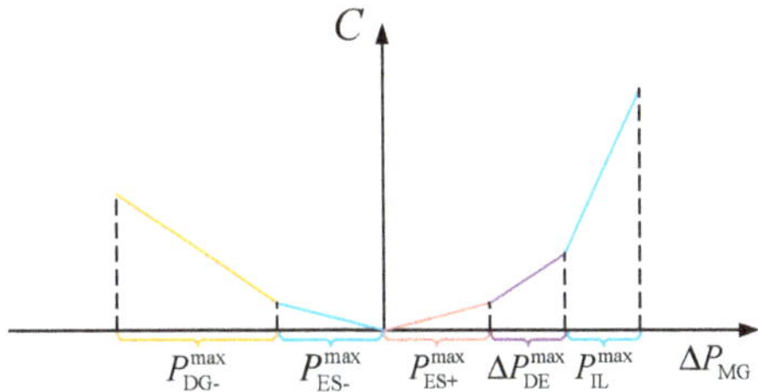

Figure 3. Regulation cost function of microgrids.

From the previous analysis, it can be seen that the regulation cost of each controllable unit can be approximated as a linear or piecewise linear function (e.g., the SOC variation in the charge and discharge process of energy storage system spans two areas of different unit regulation cost). It can be seen from Figure 3 that, when considering the dispatch priority, this paper can combine the regulation cost model of each controllable unit with the segmental linear function according to the power allocation strategy decision order of positive and negative power commands, that is, the microgrid regulation cost function.

This profile has the following salient features:

(1)   Because the power allocation strategy determines the dispatching order according to the regulation cost, the unit with low regulation cost takes priority in the task of power regulation. Therefore, the curve along the increasing direction of positive and negative power command is steeper, that is, the slope value increases.

(2)   The positive and negative power command parts of the regulation cost function are single-valued functions. In other words, for any positive and negative power command, as long as the microgrid power command ΔP_{MG} is determined, the corresponding power allocation scheme and regulation cost can be uniquely determined. Conversely, a uniquely determined power command can be obtained when the regulation cost is given.

(3)   For the positive power command, P_{ES+}^{max}, ΔP_{DE}^{max}, and P_{IL}^{max} jointly determine the upper limit of the microgrid power regulation amount, i.e., $\Delta P_{MG}^{max} = P_{ES+}^{max} + \Delta P_{DE}^{max} + P_{IL}^{max}$. For the negative power command, P_{ES-}^{max} and P_{DG-}^{max} jointly determine the lower limit of the microgrid power regulation amount, i.e., $\Delta P_{MG}^{min} = P_{ES-}^{max} + P_{DG-}^{max}$.

(4)   Due to the above assumptions in this paper, it is the case that the load-shedding operation and the distributed generation increasing the output after reducing the output can be regarded as the default cost-free adjustment strategy with the highest priority, which is not reflected in the regulation cost function.

4. Consensus Algorithm Design for Power Dispatch

4.1. Cooperative Power Dispatch Model for Islanded Multi-Microgrids

In this paper, a power dispatch consensus algorithm is designed for islanded multi-microgrids, which take the regulation cost of each microgrid as a consensus variable and use the discrete-time first-order consensus algorithm [33] to solve the problem iteratively. Hence, the total regulation cost of the multi-microgrids is chosen as the objective function. The mathematical model is constructed as follows:

$$
\begin{cases}
\mathrm{min}F = \sum\limits_{i=1}^{n} f_C(\Delta P_{MGi}) \\
s.t. \ \Delta P = \sum\limits_{i=1}^{n} \Delta P_{MGi} \\
\qquad = \sum\limits_{i=1}^{n}\left(P_{Loadi} - P_{PVi} - P_{WTi} - P_{DEi}^{min}\right) \\
\Delta P_{MGi} = \Delta P_{ESi} + \Delta P_{DEi} + \Delta P_{ILi} + \Delta P_{DG-i} \\
\Delta P \Delta P_{MGi} > 0 \\
\Delta P_{MGi}^{min} \le \Delta P_{MGi} \le \Delta P_{MGi}^{max}
\end{cases}
\tag{7}
$$

where the total power command ΔP of the islanded multi-microgrids is the difference between the total load and the total output of the wind power, PV, and the forced output of the energy storage device. ΔP_{MGi} is the power command of the i-th microgrid, that is, the power regulation amount of the i-th microgrid. $f_C(\Delta P_{MGi})$ is the regulation cost of the i-th microgrid, which is a function of the microgrid power regulation. P_{Loadi} is the total load of the i-th microgrid. ΔP_{MGi}^{max} and ΔP_{MGi}^{min} are the upper and lower limits of the power regulation of the i-th microgrid, respectively, which depend on the real-time status of each control unit in the microgrid. n is the number of microgrids.

Therefore, when a power imbalance occurs in the islanded multi-microgrids, it will be jointly undertaken by all the microgrids that can participate in the regulation. Selecting the regulation cost of the microgrid as the consensus variable can enable each microgrid to participate in the regulation according to its own resources. Its essence is to allocate the power regulation amount to each microgrid according to the slope value of each segment of each microgrid regulation cost function, which aims at reducing the overall regulation cost of the multi-microgrids.

4.2. Microgrid Regulation Cost Consensus

The regulation cost of each microgrid is selected as the consensus variable, which is abbreviated as C. Based on the discrete-time first-order consensus algorithm [33], the formula for updating the consensus variables of each agent is as follows:

$$C_i^{(k+1)} = \sum_{j=1}^{n} d_{ij}^{(k)} C_j^{(k)}, \tag{8}$$

where $C_i^{(k)}$ is the regulation cost calculated by the k-th iteration of the i-th microgrid. $d_{ij}^{(k)}$ is the i-th row and j-th column element of the row-stochastic matrix D when iterating at the k-th step. The row-stochastic matrix D is obtained from the Laplacian matrix L of the communication topology and is related to the structure of the communication topology, which is defined by the following:

$$d_{ij} = \frac{|l_{ij}|}{\sum\limits_{j=1}^{n} |l_{ij}|}, \quad i = 1, \cdots, n \text{ with } \begin{cases} l_{ii} = \sum\limits_{i \neq j} a_{ij} \\ l_{ij} = -a_{ij} \end{cases}, \tag{9}$$

where a_{ij} is the (i, j) entry of the adjacency matrix A. In this paper, the adjacency matrix A is

$$A = \begin{bmatrix} 0 & 1 & 1 \\ 1 & 0 & 1 \\ 1 & 1 & 0 \end{bmatrix}. \tag{10}$$

It can be known from Equation (8) that each agent obtains the state information of the previous iteration of the neighboring agent by means of the row stochastic matrix D, which is related to the communication topology to update its state. In order to ensure the power balance, the leader guides the regulation direction and magnitude of the regulation cost. As a leader, the microgrid regulation cost update formula is given as follows:

$$C_i^{(k+1)} = \begin{cases} \sum\limits_{j=1}^{n} d_{ij}^{(k)} C_j^{(k)} + \mu \Delta P_{error} & \Delta P > 0 \\ \sum\limits_{j=1}^{n} d_{ij}^{(k)} C_j^{(k)} - \mu \Delta P_{error} & \Delta P < 0 \end{cases}, \tag{11}$$

where μ is the error adjustment step size; ΔP_{error} is the deviation between the total power command and the sum of the microgrid power commands, which ignores the line loss of the islanded multi-microgrids system. The expression of ΔP_{error} is:

$$\Delta P_{error} = \Delta P - \sum_{i=1}^{n} \Delta P_{MGi}. \tag{12}$$

The error adjustment step size μ can be artificially given an appropriate parameter or adaptively adjusted by detecting ΔP_{error}. The meaning of Equation (10) can be explained as follows: taking the current total power command $\Delta P > 0$ as an example, if $\Delta P_{error} > 0$, the sum of the power commands of each microgrid is still insufficient to balance the current power imbalance, and the regulation cost needs to be increased accordingly; if $\Delta P_{error} < 0$, the regulation cost can be reduced.

In principle, the selection of leaders should be determined by the regulation capability. Selecting the microgrid with the largest regulation capability as the leader can reduce the need to replace leaders. The so-called regulation capability can be reflected by the parameters that reflect the regulatable resource of microgrid, such as energy storage capacity, diesel generator capacity, interruptible load capacity, etc.

In summary, the regulation cost of each microgrid in the iterative process is the weighted average of the regulation cost of the local and neighbors' previous iterations, so the communication burden on the network is small. When some microgrids have reached the limit of the power regulation, they should quit the communication topology and stop updating. The adjacent microgrids should also modify the corresponding row random matrix elements according to the new communication topology. In the convergence process of the consensus algorithm, $|\Delta P_{\text{error}}| \leq \varepsilon$ is taken as the convergence condition, where ε is the convergence error.

When the regulation cost is updated by communication interaction between microgrids, the power command of each unit needs to be inversely solved by the variable of regulation cost and the corresponding regulation cost function. Because the regulation cost function is a piecewise linear function and has a unique solution, the calculation process is not complicated, and each MGC can quickly solve and calculate the microgrid total power command for the next iteration. The flowchart of the proposed consensus algorithm for real-time cooperative power dispatch of islanded multi-microgrids is shown in Figure 4.

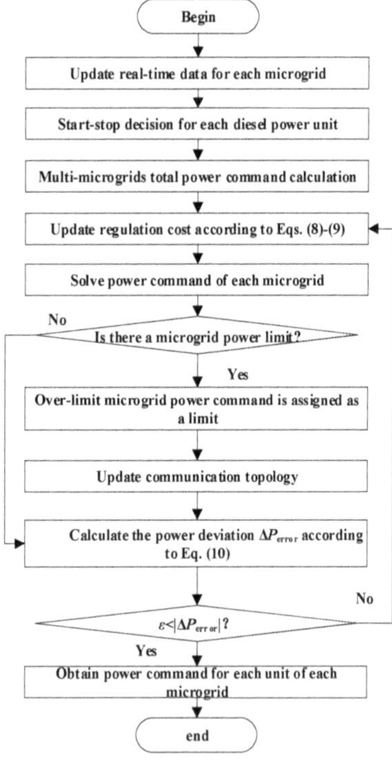

Figure 4. Flowchart of consensus algorithm for real-time cooperative power dispatch of islanded multi-microgrids.

5. Numerical Simulations

5.1. Simulation Parameter Setting

In this paper, an islanded multi-microgrids model formed by three interconnected microgrids is established. The main program is carried out in a Matlab 2015b (USA mathworks company) environment. The communication topology is shown in Figure 1. In the absence of power overrun in microgrids, there is communication between any two microgrids. Considering the regulatable resources of each microgrid, MG1 is the leader, and MG2 and MG3 are followers.

The equipment parameters of the multi-microgrids are given in Table 1. The minimum output of the diesel generator is 30% of the rated output, the minimum running time is 60 min, the minimum stop time is 30 min, and the start and stop coefficients k_{DE_on} and k_{DE_off} are set at 1.5 and 0.5, respectively. The regulation cost parameters are given in Table 2.

Table 1. Equipment parameters of multi-microgrids.

Equipment Parameters	MG1	MG2	MG3
PV capacity/kW	180	100	120
WT capacity/kW	40	80	80
ES capacity/kWh	200	200	300
ES rated power/kW	50	50	100
ES efficiency η	0.9	0.9	0.9
ES initial SOC	0.34	0.25	0.21
DE rated power/kW	50	50	-
DE increase output rate/(kW/min)	2	2	-
DE decrease output rate/(kW/min)	2	2	-
Maximum load/kW	130	180	140

Table 2. Parameters of regulating cost.

Parameters of Regulating Cost ($/kWh)	MG1	MG2	MG3
r_{ES1}	0.05	0.05	0.05
r_{ES2}	0.1	0.1	0.1
r_{ES3}	0.25	0.25	0.25
r_{DE}	1.4	1.4	1.4
r_{DG}	1.6	1.6	1.6
r_{IL}	1.6	1.9	1.8

5.2. Simulation of Multi-Microgrids Power Allocation

The simulation analysis of continuous real-time power allocation is performed at intervals of 1 min. The total duration is 24 h. The shorter interval is also applicable in practical applications. The time window is taken as $T_S = 30$ min or 0.5 h, the error adjustment step size of consensus algorithm is $\mu = 0.01$, and the convergence error is $\Delta P_{error} = 0.1$ kW.

The load curve of the multi-microgrids, the total output curve of wind power and photovoltaics, and the corresponding power imbalance curve are shown in Figure 5, whose data source is an actual islanded multi-microgrid [34]. It can be seen that the power imbalance mainly occurs at 09:00–22:00. On the one hand, due to the large PV output at noon, there is a situation of excess power, and then a power shortage occurs after the peak load arrives at night.

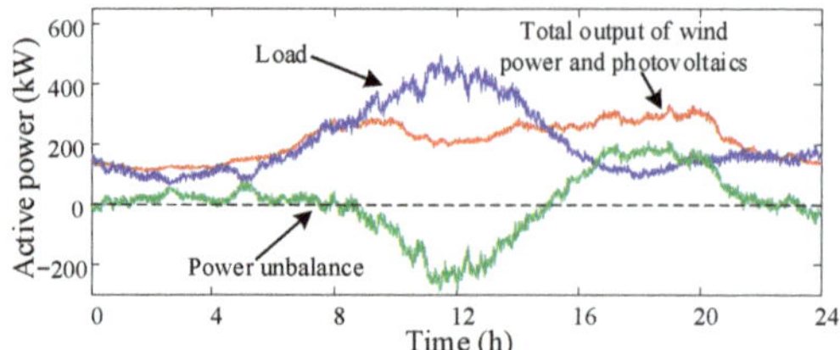

Figure 5. Power unbalance of multi-microgrids.

After simulation, the power command of each microgrid basically follows the trend of the total power command. The regulation power of each unit throughout the day and the corresponding microgrid regulation cost are given in Table 3. The total regulation cost of MG1 and MG2 includes the forced cost corresponding to the unadjustable part of the output of the energy storage device. Since MG1 and MG2 are equipped with energy storage devices, the actual regulation costs of MG1 and MG2 are higher than that of MG3. The power command curves of each microgrid and the output curves of various distributed generations are shown in Figures 6 and 7, respectively.

Table 3. Regulation power and cost.

	MG1	**MG2**	**MG3**	**Total**
Diesel power generation (kWh)	248.07	260.88	0	508.95
Load shedding (kWh)	9.63	8.00	6.18	23.81
Abandoned power generation (kWh)	120.77	116.38	97.09	334.25
Regulation cost ($)	484.06	498.88	140.76	1123.70

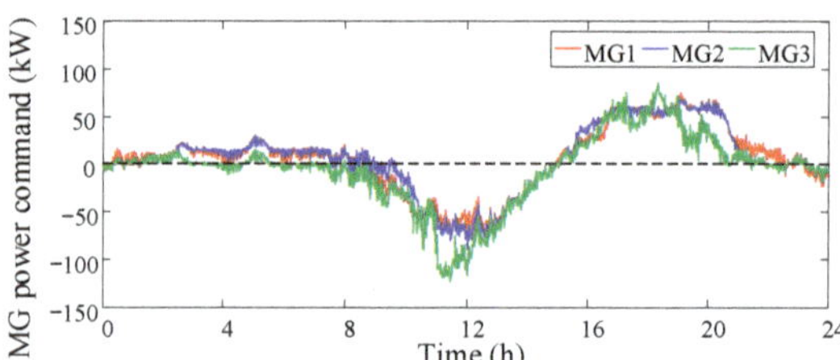

Figure 6. Power command curve of each microgrid.

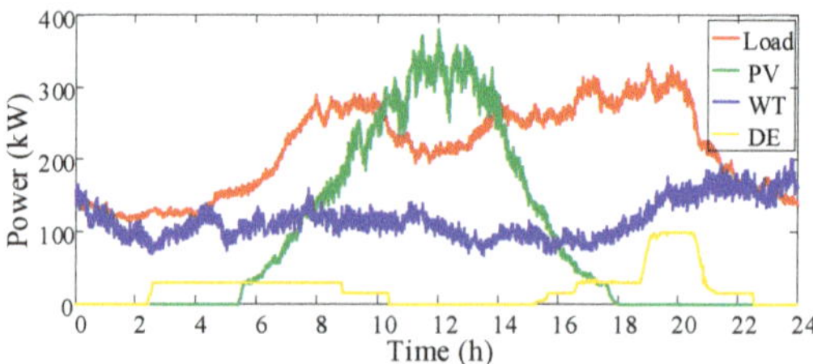

Figure 7. Power curve of DGs.

The results of charge and discharge power control of each microgrid energy storage unit are analyzed. The real-time power curve and SOC curve of each microgrid energy storage unit are shown in Figures 8 and 9. It can be seen that the energy storage units of each microgrid are charged during the period of excessive photovoltaic output during the daytime, and discharge is completed during the peak load period, which is in accordance with the scheduling rules. The SOC change trend of each energy storage is basically the same, the capacity can be effectively utilized, and the power command allocation is reasonable.

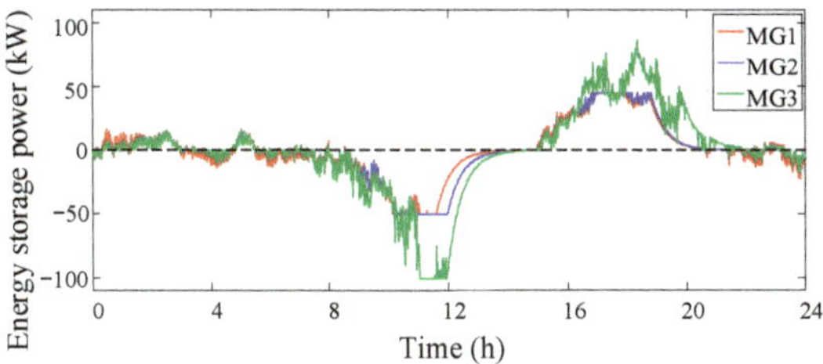

Figure 8. Power curve of each ES.

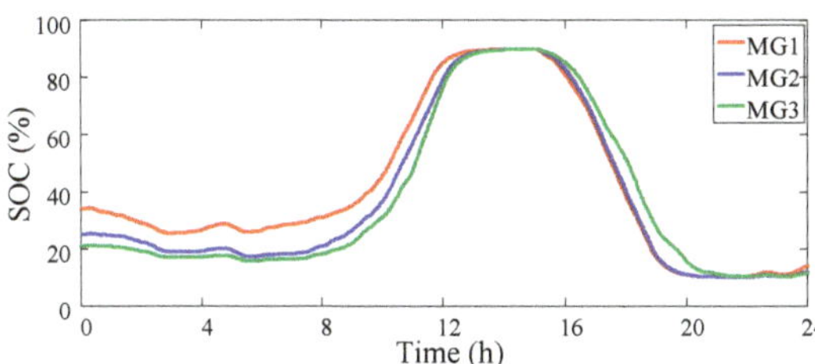

Figure 9. SOC curve of each ES.

The consensus convergence process of the two sections in the continuous simulation process is selected. The power command is $\Delta P = -205$ kW, $\Delta P = +210$ kW, respectively. The convergence process curves are shown in Figures 10 and 11.

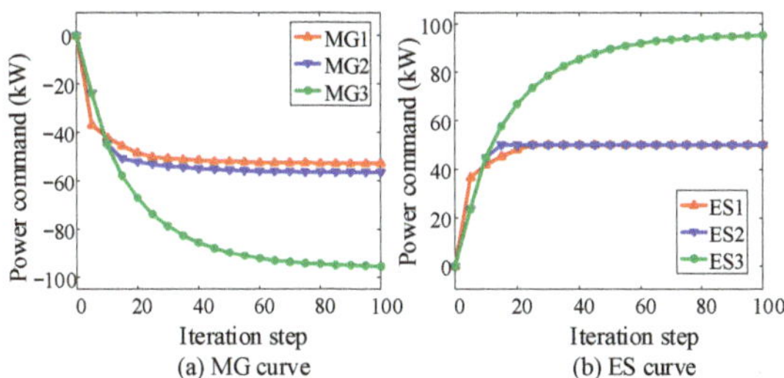

Figure 10. Consensus convergence process ($\Delta P = -205$ kW): (**a**) Description of the three MG iteration process; (**b**) Description of the three ES iteration process.

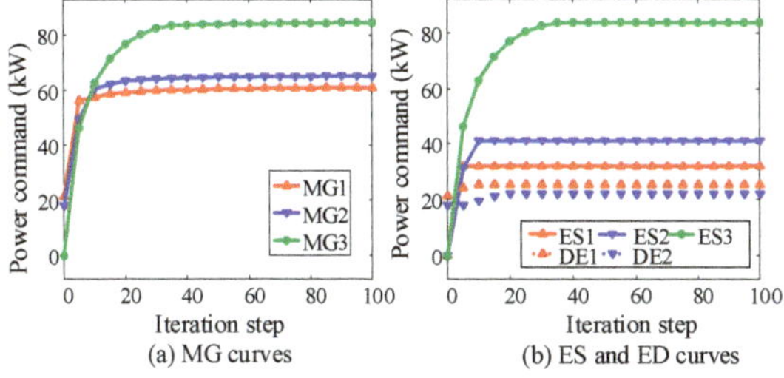

Figure 11. Consensus convergence process ($\Delta P = +210$ kW): (**a**) Description of the three MG iteration processes; (**b**) Description of the three ES and two DE iteration processed.

As can be seen from Figure 10, the current total power command is negative; the energy storage device in each microgrid charges to balance the corresponding unbalance amount, of which MG3 bears the largest regulation power. The reason is that the energy storage device of MG3 has the maximum capacity and rated power. When the energy storage devices of MG1 and MG2 reach the charge limit in the iteration process, it is necessary to adopt a regulation approach with a higher regulation cost. The energy storage device bears the low cost of power regulation. When each microgrid consumes a consensus regulation cost, MG3 would bear more power regulation.

Similarly, as shown in Figure 11, when the current total power command is positive, each microgrid should increase its power generation to balance the corresponding unbalance. Each energy storage device discharges and two diesel generators have been turned on. When the output of ES1 and ES2 reaches the limit, the corresponding DE1 and DE2 gradually increase their output in the iteration process to promote the consensus regulation cost of each microgrid.

5.3. Comparative Analysis under Different Operation Modes

In order to reflect the advantages of the interconnected operation of microgrid and consensus power allocation model, the different operation modes of microgrid are compared and analyzed in this section, including the following three cases.

Mode 1: each microgrid operates independently [35].

Mode 2: the multi-microgrids are interconnected to form a microgrid cluster and each microgrid gives priority to autonomous operation. When its own power commands are different from the total power command of the microgrid cluster, it actively exchanges power with the microgrid cluster system [36].

Mode 3: the multi-microgrids are interconnected to form a microgrid cluster and the power allocation is based on the real-time cooperative power dispatch model established in this paper.

The three models are simulated continuously throughout the day and the regulation costs of each microgrid and microgrid cluster are given in Table 4. From the perspective of total regulation costs, the order is mode 1 > mode 2 > mode 3. Under mode 1, each microgrid operates independently and can only rely on its own supply for balance. When the generation and load are not balanced, there will be more abandoned power generation or load shedding, which would lead to the regulation costs being higher. The interconnection of the microgrids in mode 2 will reduce the regulation costs, which reflects the advantages of cluster operation of adjacent microgrids to a certain extent. However, the benefits of clustering operations have not been fully explored due to the priority of autonomous balance of microgrids. Mode 3 adopts the cooperative power dispatch strategy in which three microgrids in total are dispatched, and the energy transfer between each microgrid is mutually beneficial. Each microgrid allocates the regulation power reasonably according to the signal of regulation costs. Overall, the regulation capability of the microgrid cluster system has been optimized and the economy has been improved.

Table 4. Regulation costs under the three modes.

Microgrid	Regulation Cost ($)		
	Mode 1	Mode 2	Mode 3
MG1	597.86	560.29	484.06
MG2	694.19	644.66	498.88
MG3	317.22	338.23	140.76
Total	1609.27	1543.18	1123.70

Figure 12 shows the MG1 power command under the three modes; the PV configuration of MG1 is the largest. It can be seen that the negative power command of MG1 at noon is smaller than that of the other two modes. Modes 1 and 2 decrease obviously at noon due to the bias towards autonomy.

Under mode 3, due to the interconnection of each microgrid, the surplus output of MG1 can partly support the other two microgrids, thus reducing the negative power command in this period.

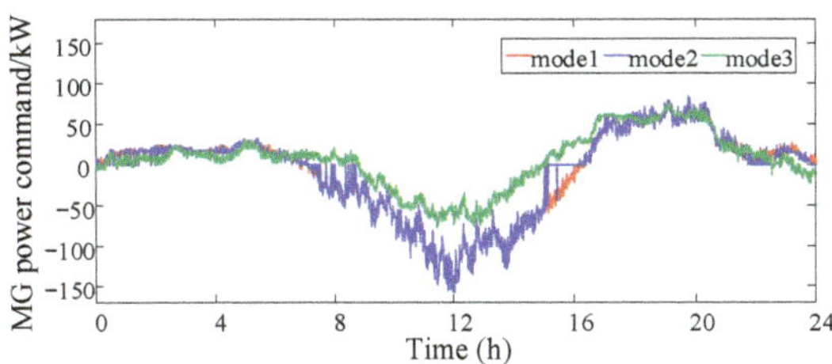

Figure 12. Power command of MG1 under the three modes.

5.4. Comparison with Centralized Optimization

From the perspective of engineering applications, this paper compares the optimization of performance with the centralized optimization method. Aiming at the time section of power command $P = +210$ kW, a multiple population genetic algorithm (MPGA) [37] is selected for centralized optimization, in which the MPGA runs 10,000 times and the results are compared with those of the proposed consensus algorithm. The CPU (central processing unit) used in the simulation is 3.2 GHz, with 2 GB memory. A performance comparison of the two optimization approaches is given in Table 5.

Table 5. Comparison of optimization performance.

	MPGA		**Consensus Algorithmic**
Regulation cost ($)	optimal	20.35	21.16
	mean	23.44	
Optimization time (s)	mean	1.66	0.02
Optimization approach	centralized, static		distributed, dynamic

From Table 5, it can be seen that the consensus algorithm belongs to distributed optimization technology, and its convergence time is greatly reduced compared with centralized optimization. As the scale of the microgrid clusters further increases, its advantages will become more obvious. MPGA is optimized 10,000 times and the optimal regulation cost is better than that of the consensus algorithm. It can be seen that the optimization result of the consensus algorithm is not the global optimal solution. The mean value of 10,000 optimization results obtained by MPGA is slightly lower than that of the consensus algorithm. According to statistics, the probability that the MPGA optimization result will be better than the consensus algorithm is about 30.69%. For the comparison of the optimization results shown in Figure 13, most of the optimization algorithms belong to static optimization, and it is difficult to solve the real-time dynamic problem.

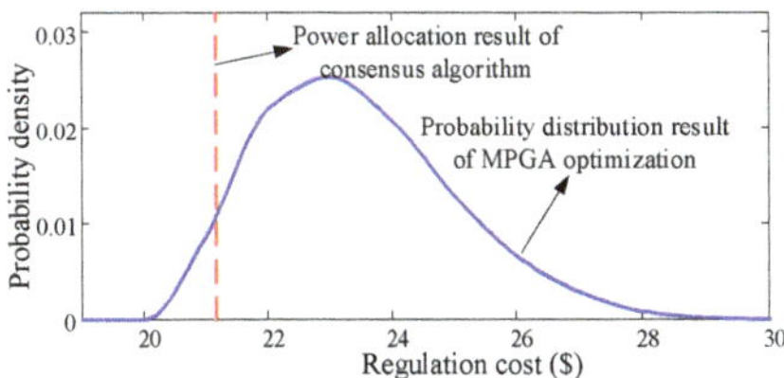

Figure 13. Comparison of optimization results.

5.5. Impact Analysis of Different T_S

The value of time window T_S will affect the results, as reflected in the power control of the energy storage. The reason is that the SOC electricity calculation of the energy storage is related to the length of the time window. This section compares the impact of different T_S.

Taking ES3 as an example, the power curve and SOC curve of ES3 are shown in Figures 14 and 15, respectively. Figure 14a shows the overall situation throughout the day and Figure 14b,c shows that the time period has a significant impact under different T_S. Between 11:00 and 14:00, because of the high photovoltaic output at noon, the charge of the energy storage device is close to the SOC upper limit. When T_S is small (such as $T_S = 10/60$ h), the power command of the energy storage device will quickly drop to 0 and stop charging until the SOC reaches the upper limit. If T_S is large (such as $T_S = 90/60$ h), the energy storage device will start earlier to reduce the charge power. This proves that the control of the TS energy storage device tends to be conservative when T_S is large, i.e., a larger T_S can control the SOC of the energy storage device to avoid approaching the limit value.

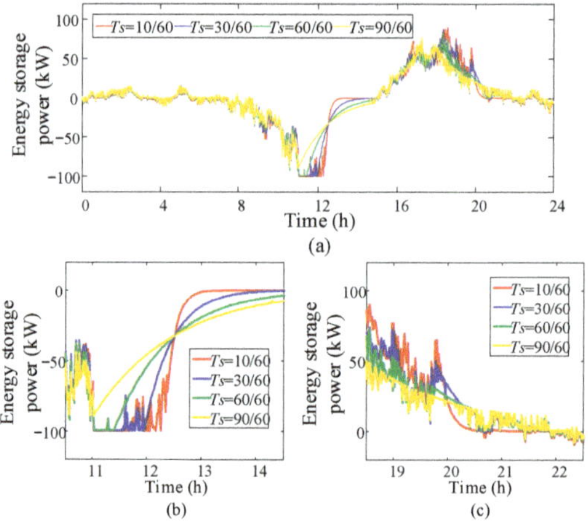

Figure 14. Power curve of ES3 with different Ts: (**a**) Description of the Power curve of ES3 with different Ts from 0:00 to 24: 00; (**b**) Description of the Power curve of ES3 with different Ts from 10:30 to 14: 20; (**c**) Description of the Power curve of ES3 with different Ts from 18:40 to 22: 20.

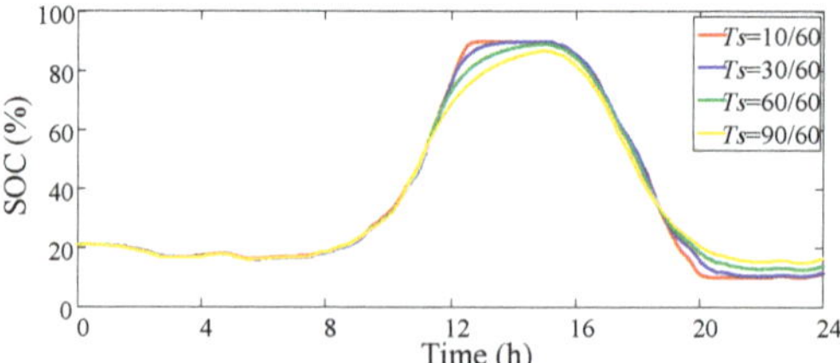

Figure 15. SOC curve of ES3 with different Ts.

However, the larger T_S will increase the regulation cost of other controllable units, that is, the power regulation capability of the energy storage device is restricted in some time periods and the corresponding parts need to be borne by other controllable units. Table 6 gives the regulation cost under different values. Generally speaking, T_S should not be too large or too small. The value of T_S can be determined by reference to the type of energy storage battery and the consideration of the life of the energy storage battery, as well as the actual application requirements. The recommended value is 0.5–1.5 h.

Table 6. Regulation cost with different T_S.

T_S (h)	MG1 Regulation Cost ($)	MG2 Regulation Cost ($)	MG3 Regulation Cost ($)	Total ($)
10/60	472.14	493.30	146.29	1111.73
30/60	484.06	498.88	140.76	1123.70
60/60	493.24	512.18	141.82	1147.24
90/60	515.62	532.72	147.83	1196.17

6. Conclusions

In recent years, the development of microgrids has attracted interest to its related topics. This paper studies the real-time power allocation problem of islanded multi-microgrids. Considering the architecture and operational characteristics of islanded multi-microgrids, it is suitable to adopt a decentralized control architecture. A real-time cooperative power dispatch framework for islanded multi-microgrids is proposed. The microgrid regulation cost is modeled and a consensus algorithm is introduced to realize the power allocation. The conclusions are as follows:

(1) The consensus power allocation algorithm proposed in this paper can allocate power reasonably and ensure the real-time power balance of the islanded multi-microgrids.
(2) Each microgrid is guided by the regulation cost, and rationally optimizes the regulation power allocated by each microgrid, thereby improving the overall economics of the microgrid cluster system.
(3) The communication burden of the control architecture is small. Each microgrid only needs a small amount of information from the local and neighboring microgrids, which can satisfy the real-time dynamic power allocation requirements of islanded multi-microgrids.

Author Contributions: All the authors have contributed to this paper in different aspects. X.Z. proposed the original concept and wrote the original draft. N.W. acted as supervisor and gave suggestions on paper improvement.

Funding: This research received no external funding.

Conflicts of Interest: The authors declare no conflict of interest.

Abbreviations and Nomenclature

MG	microgrid
MMGs	multi-microgrids
CHP	combined heat and power
MGC	microgrid controller
ICA-GA	imperialist competitive algorithm-genetic algorithm
GSO	glowworm swarm optimization
PV	photovoltaic
WT	wind turbine
ES	energy storage device
DE	diesel engine
IL	interruptible load
SOC	state of charge
DG	distributed generation
MPPT	maximum power point tracking
MPGA	multiple population genetic algorithm
Δt	time interval of power command
$SOC(t)$	current SOC calculated value
$SOC(t - \Delta t)$	previous actual SOC value
V_{ES}	energy storage system capacity
P_{ES}^{max}	energy storage system power limits
η	energy storage system efficiency

T_S	state duration window
ΔP_{ES}	power command of the energy storage device
$C_{ES}(\Delta P_{ES})$	regulation cost generated by power command ΔP_{ES}
ΔP	total power imbalance
ΔS_i	SOC change caused by the SOC area with the
r_{ESi}	unit regulation cost of ES
P_{ES-}^{max}	maximum charge power
P_{ES+}^{max}	maximum discharge power
ΔP_{DE}	power command of the diesel generator
r_{DE}	unit generation cost of the diesel generator
λ	unit fuel coefficient of the diesel generator
r_{oil}	fuel unit cost of the diesel generator
P_{DE}^{N}	rated power of diesel generators
P_{DE}^{max}	upper limit of diesel generators output
P_{DE}^{min}	lower limit of diesel generators output
ΔP_{DE}^{max}	upper limit of diesel generator power command
R_u	rates of decreasing output of diesel generators
R_d	rates of increasing output of diesel generators
ΔP_{DE}	excess part of the current real-time output
k_{DE_on}	start coefficients of the diesel generator
k_{DE_off}	stop coefficients of the diesel generator
ΔP_{IL}	power commands of interruptible load
ΔP_{DG-}	power command of distributed generation output reduction
r_{IL}	unit regulation cost of interruptible load
r_{DG}	unit regulation cost of distributed generation output reduction
P_{IL}^{max}	upper limit of interruptible load
P_{DG-}^{max}	upper limit of distributed generation output reduction
ΔP_{MG}	microgrid power command
ΔP_{MGi}	power command of the i-th microgrid
$f_C(\Delta P_{MGi})$	regulation cost of the i-th microgrid
P_{Loadi}	total load of the i-th microgrid
ΔP_{MGi}^{max}	upper limit of the power regulation of the i-th microgrid
ΔP_{MGi}^{min}	lower limit of the power regulation of the i-th microgrid
n	number of microgrid
μ	error adjustment step size
ΔP_{error}	power deviation commands
ε	convergence error

References

1. Hatziargyriou, N.; Asano, H.; Iravani, R.; Marnay, C. Microgrids. *IEEE Power Energy Mag.* **2007**, *5*, 78–94. [CrossRef]
2. Katiraei, F.; Iravani, R.; Hatziargyriou, N.; Dimeas, A. Microgrids management. *IEEE Power Energy Mag.* **2008**, *6*, 54–65. [CrossRef]
3. Che, L. Microgrids and distributed generation systems: Control, operation, coordination and planning. *Diss. Theses-Gradworks* **2015**, *63*, 33–39.
4. Katiraei, F.; Iravani, M.R.; Lehn, P.W. Micro-grid autonomous operation during and subsequent to islanding process. *IEEE Trans. Power Deliv.* **2005**, *20*, 248–257. [CrossRef]

5. Amoateng, D.O.; Al Hosani, M.; El Moursi, M.S.; Turitsyn, K.; Kirtley, J.L. Adaptive voltage and frequency control of islanded multi-microgrids. *IEEE Trans. Power Syst.* **2018**, *33*, 4454–4465. [CrossRef]

6. Hualei, Z.; Shiwen, M.; Yu, W.; Zhang, F.; Chen, X.; Cheng, L. A survey of energy management in interconnected multi-microgrids. *IEEE Access* **2019**, *7*, 72158–72169.

7. John, T.; Ping Lam, S. Voltage and frequency control during microgrid islanding in a multi-area multi-microgrid system. *IET Gener. Transm. Distrib.* **2017**, *11*, 1502–1512. [CrossRef]

8. Farrokhabadi, M.; Canizares, C.A.; Bhattacharya, K. Frequency control in isolated/islanded microgrids through voltage regulation. *IEEE Trans. Smart Grid* **2017**, *8*, 1185–1194. [CrossRef]

9. Chowdhury, D.; Khalid Hasan, A.; Rahman Khan, M.Z. Scalable DC microgrid architecture with phase shifted full bridge converter based power management unit. In Proceedings of the 2018 10th International Conference on Electrical and Computer Engineering (ICECE), Dhaka, Bangladesh, 20–22 December 2018; pp. 22–25.

10. Khalid Hasan, A.; Chowdhury, D.; Rahman Khan, M. Scalable DC microgrid architecture with a one-way communication based control interface. In Proceedings of the 2018 10th International Conference on Electrical and Computer Engineering (ICECE), Dhaka, Bangladesh, 20–22 December 2018; pp. 265–268.

11. Alipour, M.; Mohammadi-Ivatloo, B.; Zare, K. Stochastic scheduling of renewable and CHP based microgrids. *IEEE Trans. Ind. Inform.* **2015**, *11*, 1049–1058. [CrossRef]

12. Wang, R.; Wang, P.; Xiao, G. A robust optimization approach for energy generation scheduling in microgrids. *Energy Convers. Manag.* **2015**, *106*, 597–607. [CrossRef]

13. Luo, Z.; Wei, G.U.; Zhi, W.U.; Wang, Z.; Tan, Y. A robust optimization method for energy management of CCHP microgrid. *J. Mod. Power Syst. Clean Energy* **2018**, *6*, 132–144. [CrossRef]

14. Palma-Behnke, R.; Benavides, C.; Lanas, F.; Severino, B.; Reyes, L.; Llanos, J.; Sáez, D. A microgrid energy management system based on the rolling horizon strategy. *IEEE Trans. Smart Grid* **2013**, *4*, 996–1006. [CrossRef]

15. Solanki, B.V.; Canizares, C.A.; Kankar, B. Practical energy management systems for isolated microgrids. *IEEE Trans. Smart Grid* **2019**, *10*, 4762–4775. [CrossRef]

16. Dou, C.X.; An, X.G.; Yue, D. Multi-agent system based energy management strategies for microgrid by using renewable energy source and load forecasting. *Electr. Power Compon. Syst.* **2016**, *44*, 2059–2072. [CrossRef]

17. Wang, Z.; Feng, L.; Low, S.H.; Zhao, C.; Mei, S. Distributed frequency control with operational constraints, Part I: Per-node power balance. *IEEE Trans. Smart Grid* **2019**, *10*, 40–52. [CrossRef]

18. Wang, Z.; Feng, L.; Low, S.H.; Zhao, C.; Mei, S. Distributed frequency control with operational constraints, Part II: Network power balance. *IEEE Trans. Smart Grid* **2019**, *10*, 53–64. [CrossRef]

19. Hamed, M.G.; Mohammad, K. A novel optimal control method for islanded microgrids based on droop control using the ICA-GA algorithm. *Energies* **2017**, *10*, 485.

20. Quynh, T.T.T.; Maria, L.D.S.; Riva Sanseverino, E.; Zizzo, G.; Pham, T. Driven primary regulation for minimum power losses operation in islanded microgrids. *Energies* **2018**, *11*, 2890.

21. Sanseverino, E.R.; Silvestre, M.L.D.; Mineo, L.; Favuzza, S.; Nguyen, N.Q.; Tran, Q.T.T. A multi-agent system reinforcement learning based optimal power flow for islanded microgrids. In Proceedings of the 2016 IEEE 16th International Conference on Environment and Electrical Engineering (EEEIC), Florence, Italy, 7–10 June 2016; pp. 1–6.

22. Li, Q.; Gao, D.W.; Zhang, H.; Wu, Z.; Wang, F.Y. Consensus-based distributed economic dispatch control method in power systems. *IEEE Trans. Smart Grid* **2019**, *10*, 941–954. [CrossRef]

23. Yang, S.; Tan, S.; Xu, J. Consensus based approach for economic dispatch problem in a smart grid. *IEEE Trans. Power Syst.* **2013**, *28*, 4416–4426. [CrossRef]

24. Pourbabak, H.; Luo, J.; Chen, T.; Su, W. A novel consensus-based distributed algorithm for economic dispatch based on local estimation of power mismatch. *IEEE Trans. Smart Grid* **2018**, *9*, 5930–5942. [CrossRef]

25. Tang, Z.; Hill, D.J.; Liu, T. A novel consensus-based economic dispatch for microgrids. *IEEE Trans. Smart Grid* **2018**, *9*, 3920–3922. [CrossRef]

26. Zhang, X.; Yu, T.; Yang, B.; Li, L. Virtual generation tribe based robust collaborative consensus algorithm for dynamic generation command dispatch optimization of smart grid. *Energy* **2016**, *101*, 34–51. [CrossRef]

27. Lu, L.; Chu, C. Consensus-based droop control of isolated micro-grids by ADMM implementations. *IEEE Trans. Smart Grid* **2018**, *9*, 5101–5112. [CrossRef]

28. Liu, Y.; Li, Y.; Xin, H.; Gooi, H.B.; Pan, J. Distributed optimal tie-line power flow control for multiple interconnected AC microgrids. *IEEE Trans. Power Syst.* **2019**, *34*, 1869–1880. [CrossRef]

29. Zhao, B.; Wang, X.; Lin, D.; Calvin, M.M.; Morgan, J.C.; Qin, R.; Wang, C. Energy management of multiple-microgrids based on a system of systems architecture. *IEEE Trans. Power Syst.* **2018**, *33*, 6410–6421. [CrossRef]

30. Bui, V.H.; Hussain, A.; Kim, H.M. A multiagent-based hierarchical energy management strategy for multi-microgrids considering adjustable power and demand response. *IEEE Trans. Smart Grid* **2018**, *9*, 1323–1333. [CrossRef]

31. Asimakopoulou, G.E.; Dimeas, A.L.; Hatziargyriou, N.D. Leader-follower strategies for energy management of multi-microgrids. *IEEE Trans. Smart Grid* **2013**, *4*, 1909–1916. [CrossRef]

32. Xu, Y.; Li, Z. Distributed optimal resource management based on the consensus algorithm in a microgrid. *IEEE Trans. Ind. Electron.* **2015**, *62*, 2584–2592. [CrossRef]

33. Zhang, Z.; Chow, M.Y. Convergence analysis of the incremental cost consensus algorithm under different communication network topologies in a smart grid. *IEEE Trans. Power Syst.* **2012**, *27*, 1761–1768. [CrossRef]

34. Alegria, E.; Brown, T.; Minear, E.; Lasseter, R.H. CERTS microgrid demonstration with large-scale energy storage and renewable generation. *IEEE Trans. Smart Grid* **2014**, *5*, 937–943. [CrossRef]

35. Olivares, D.E.; Canizares, C.A.; Kazerani, M. A centralized energy management system for isolated microgrids. *IEEE Trans. Smart Grid* **2014**, *5*, 1864–1875. [CrossRef]

36. Fang, X.; Yang, Q.; Wang, J.; Yan, W. Coordinated dispatch in multiple cooperative autonomous islanded microgrids. *Appl. Energy* **2016**, *162*, 40–48. [CrossRef]

37. Wang, B.; Li, I. Load balancing task scheduling based on multi-population genetic algorithm in cloud computing. In Proceedings of the 2016 35th IEEE Chinese Control Conference (CCC), Chengdu, China, 27–29 July 2016; pp. 5261–5266.

Article

Energy Model for Long-Term Scenarios in Power Sector under Energy Transition Laws

Gabriela Hernández-Luna [1], Rosenberg J. Romero [1], Antonio Rodríguez-Martínez [1,*],
José María Ponce-Ortega [2], Jesús Cerezo Román [1] and Guadalupe Diocelina Toledo Vázquez [1]

[1] Engineering and Applied Sciences Center Research—CIICAp, Autonomous University of Morelos
State—UAEM, Av. Universidad 1001, Chamilpa, Cuernavaca 62210, Morelos, Mexico;
gabriela.hernandez@uaem.mx (G.H.-L.); rosenberg@uaem.mx (R.J.R.); jesus.cerezo@uaem.mx (J.C.R.);
diocelina.toledo@gmail.com (G.D.T.V.)
[2] Faculty of Chemical Engineering, Universidad Michoacana de San Nicolás de Hidalgo, Av. Francisco J.
Múgica S/N Ciudad Universitaria, Morelia 58030, Michoacán, Mexico; jmponce@umich.mx
* Correspondence: antonio_rodriguez@uaem.mx; Tel.: +52-777-3297-084 (ext. 6262)

Received: 13 August 2019; Accepted: 23 September 2019; Published: 29 September 2019

Abstract: High electricity demand, as well as emissions generated from this activity impact directly to global warming. Mexico is paying attention to this world difficulty and it is convinced that sustainable economic growth is possible. For this reason, it has made actions to face this problem like as launching constitutional reforms in the power sector. This paper presents an energy model to optimize the grid of power plants in the Mexican electricity sector (MES). The energy model considers indicators and parameters from Mexican Energy Reforms. Electricity demand is defined as a function of two population models and three electricity consumption per capita. Prospectives are presented as a function of total annual cost of electricity generation, an optimal number of power plants—fossil and clean—as well as CO_2eq emissions. By mean of the energy model, optimized grid scenarios are identified to meet the governmental goals (energy and environment) to 2050. In addition, this model could be used as a base to identify optimal scenarios which contribute to sustainable economic growth, as well as evaluate the social and environmental impacts of employed technologies.

Keywords: electricity model; power plants prospectives; Mexican prospectives

---

1. Introduction

Power in any region of the world is essential for development and economic growth. The electricity sector has historically developed under uncertainty and constant changes, especially since the end of the 1980s when the damage to the ozone layer was evidenced [1]. The magnitude of the event gave a global social concern, which is reflected in the Montreal Protocol in 1987 [2], the multilateral treaty on the environment has had the most success in all of history. This treaty subsequently gives rise to international agreements such as the Kyoto Protocol in 1998 [3] until reaching the Sustainable Development Goals of the United Nations launched in 2015 [4], also considering the Paris Agreement of 2015 [5] which established keeping the global temperature rise below 2 °C. However, despite the international effort, electric consumption continues, increasing incessantly, as reported by the World Bank in 2017 [6], as can be seen in Figure 1.

This high electricity demand brings an environmental impact, which has reached 30% of the total global emissions of greenhouse gases (GHG) as Figure 2 shows.

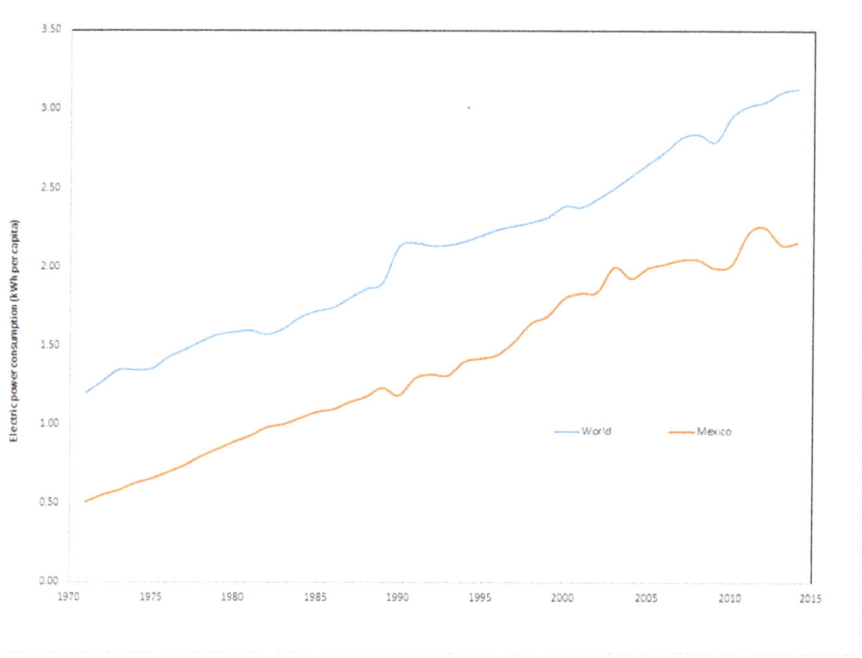

Figure 1. World and Mexico electric power consumptions. Prepared by the authors based on data from cited Reference [6].

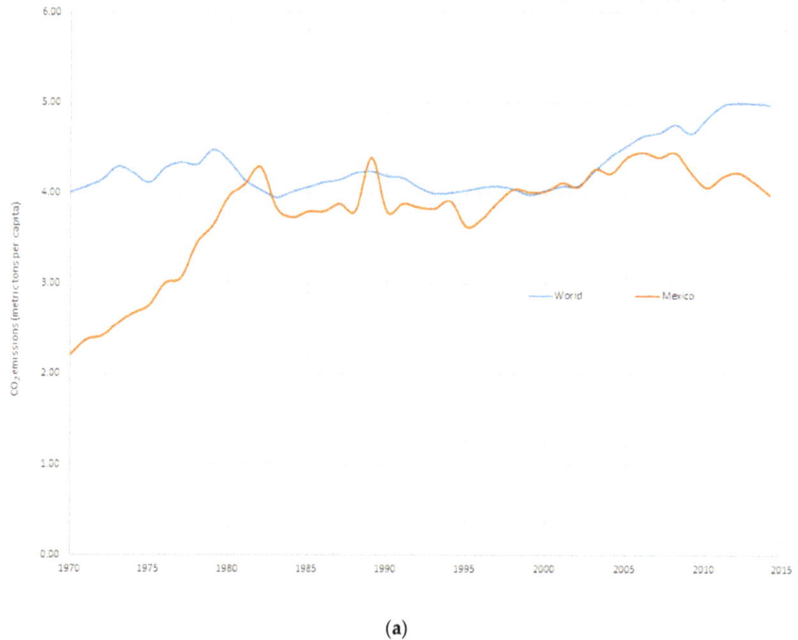

(**a**)

Figure 2. *Cont.*

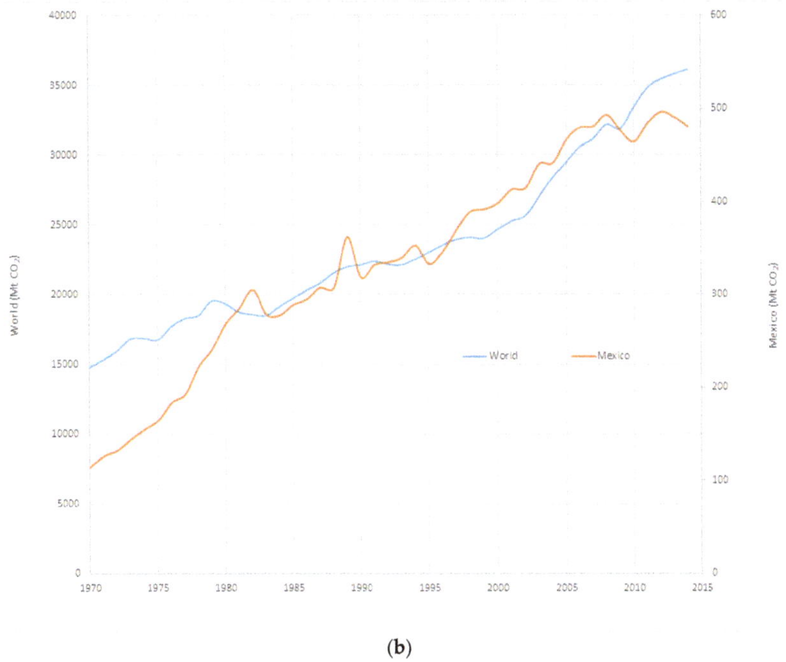

(b)

Figure 2. World and Mexico (**a**) CO_2 emissions per capita and (**b**) CO_2 total emissions. Prepared by the authors based on data from cited Reference [6].

The world faces a challenge between meeting the world's electricity demand and reducing greenhouse gas emissions produced by this sector. Mexico is not only sensitive to this global reality but it is also convinced that economic progress can and should be parallel to environmental protection. Under this premise, the Mexican government has taken actions to face this challenge as issuing reforms related to the energy sector, specifically in the Mexican Electricity System (MES) since this is the second largest emitter of GHG in the country, only after the transport sector. The GHG emissions produced by electricity sector has reached approximately 25% of total national emissions in 2015, as reported by National Institute of Climate Change (INECC) in 2015 [7] (see Figure 3) due to electricity generated in Mexico which is around 80% from fossil energy sources as Energy State Secretary (SENER) indicated in 2017 [8].

The amendments or reforms in energy affairs are also a product of the influence of international organizations and financial institutions as well as of changes in the price of fuels [9]. The main reforms launched in this field are *"General Climate Change Law"* (GCCL) in 2012 [10] and the *"National Climate Change Strategy"* (NCCS) in 2013 [11], whose main objective is minimizing GHG emissions produced by electricity generation and also modernize the Mexican Electricity System. The strategy considers the planning and sustainable growth, low carbon emissions of the MES in long term, reflected in an increase of the so-called "clean technologies" (According to the Energy Transition Law, 2015 [12]) for electricity generation. The goal is to achieve at least 35% of electricity generated from clean sources for the year 2024, in addition in achieving a specific GHG emission reduction of 30% and 50% for the years 2020 and 2050, respectively, with respect to the 2000 year emissions [10,11].

In response to this questioning, the use of renewable technologies is promoted to replace the main conventional technologies and achieve the proposed goals, however, this option can lead to risks such as potential incidences in the planning and construction of plants generating power, in addition to potential situations that would increase the start-up time of the electricity generating plants. This is one of the reasons why the use of simulations and numerical models can be a useful tool that allows strategically planning and visualizing the sector to suggest scenarios where proposed objectives by reforms are achieved.

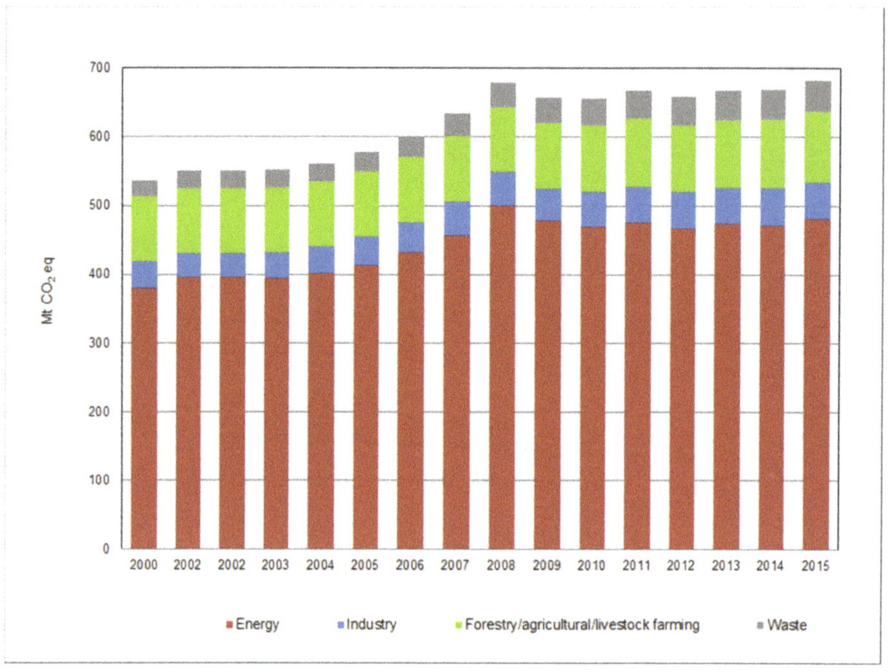

Figure 3. Mexican CO_2 emissions by sector. Prepared by the authors based on data from cited Reference [7].

The energy sector has been the object of study from various points of view; from the economic, technical as well as environmental. This section presents a brief review of the research.

Manzini et al. [13] analyzed the environmental impacts of the use of renewable sources in electricity generation. They basically identified CO_2, CH_4, NOx and SOx emissions for conditions where renewable energy reaches a contribution of 31 and 43% of installed capacity.

The World Bank [14] has also studied this sector and in 2009 they conducted an analysis to identify areas of influence in the reduction of GHG; to identify the most important to the electricity sector. In this study, the authors also identify the minimum costs available to be implemented using the cost-benefit methodology.

This cost-benefit methodology has been used by Lund and Mathiesen [15] for the study of the electrical system in Denmark, analyzing transition scenarios emphasizing areas such as energy efficiency, reduction of emissions of CO_2eq and industrial development, whose objective was is to reach up to 100% of the demand for electrical energy using renewable sources.

In 2014, Santoyo-Castelazo et al. [16] analyzed the environmental implications of decarbonization of the MES, by proposing scenarios with diverse configurations of the energy matrix of technologies used for combined-cycle power plant with life cycle analysis studies. The authors concluded that it is evident the possibility of reducing the environmental impact with the proposals made.

The MES energy transition has been developing also studied by Vidal-Amaro et al. [17]. The authors made a proposal to determine an optimal configuration of the energy matrix consisting of fossil and renewable sources for one moment in time, 2024 with the goal of reaching 35% of electricity from renewable sources, besides the authors managed to identify an optimal configuration as well as quasi-optimal configurations as an alternative.

The review focuses its efforts on analyzing GHG emissions as well as on identifying or evaluating the contributions of renewable energy sources to the MES in a specific scenario; however, no methodologies have been identified that propose identifying the number of power plants generating electricity to meet the demand neither short nor long term. This paper presents a proposal for an alternative electricity generation model to optimize the number of power plants to satisfy demand under Mexico's government policies with at short and at long term environmental impact scenarios.

2. Methods

This paper proposes an energy model to satisfy the short and long term demand for power plants in periods of lustrum from 2020 to 2050 year at several scenarios and it is compared with International Energy Agency (IEA) [18] data. The objective function is to minimize the Total Annual Cost (TAC) of electricity generation as well as determine a matrix made up of a number of electricity generating power plants required, minimizing power plants number whose primary energy source is from fossil resources. The proposed model considers population growth as a function to determine the electricity demand, taken from the World Population Prospects [19] as can be seen in Figure 4.

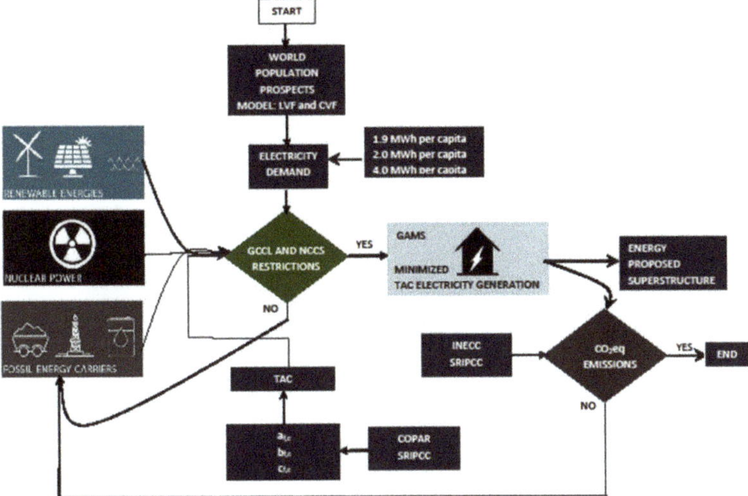

Figure 4. Schematic representation of Mexican Electricity System (MES) optimization.

The proposed matrix (composed by the number of power plants) is based on a superstructure which considers technologies based on fossil and clean energy sources for electricity generation, presented in Figure 4 and the code is in Appendix A. In addition, the software GAMS © identifies the optimal number of power plants generation to satisfy a period demand and CO_2eq emissions. The CO_2eq emissions are compared with NCCS and GCCL, if the value is lower than the constraint then the superstructure is shown as a result.

Electricity demand is determined using two population models from the United Nations (2017) [19], which are detailed next.

2.1. Electricity Demand

The proposed mathematical model assumes that the annual electricity demand is satisfied with electricity generation from two main energy sources: from fossil and clean sources, which is stated as follows:

$$E_d = \sum_i E_i \tag{1}$$

E_d is the annual electricity demand to be met, E_i the total electricity generated for each analyzed period corresponding to the total amount of the electricity coming from fossil sources, E_f, as well as from clean sources, E_c, in other words, $i = f$ for fossil sources, while $i = c$ for clean sources.

Electricity demand, E_d, is defined with two population models used in the realization of World Population Prospects of the Department of Economics and Social Affairs of the United Nations [19]. The population models are the Low Variability of Fertility (LVF) and life expectancy at birth and Constant Variability of Fertility (CVF) and life expectancy at birth. Three demand conditions per capita are assumed in each model: 1.9, 2.0 and 4.0 MWh. The demand of 1.9 MWh per capita is taken from the historical demand of Mexico in the period of 2000 to 2010; the demand of 2.0 MWh per capita considers an increase of 10% in the population demand and finally, the demand of 4.0 MWh per capita is the typical demand of the population consumption from developed countries. Table 1 presents a matrix with the described conditions.

Table 1. Matrix of described conditions.

	Population Model					
	LVF			CVF		
	Electricity Demand [MWh per capita]					
Year	1.9	2.0	4.0	1.9	2.0	4.0
2015	LVF-1.9-2010	LVF-2.0-2010	LVF-4.0-2010	CVF-1.9-2010	LVF-2.0-2010	LVF-4.0-2010
2020	LVF-1.9-2010	LVF-2.0-2010	LVF-4.0-2010	CVF-1.9-2010	LVF-2.0-2010	LVF-4.0-2010
2025	LVF-1.9-2025	LVF-2.0-2025	LVF-4.0-2025	CVF-1.9-2025	LVF-2.0-2025	LVF-4.0-2025
2030	LVF-1.9-2030	LVF-2.0-2030	LVF-4.0-2030	CVF-1.9-2030	LVF-2.0-2030	LVF-4.0-2030
2035	LVF-1.9-2035	LVF-2.0-2035	LVF-4.0-2035	CVF-1.9-2035	LVF-2.0-2035	LVF-4.0-2035
2040	LVF-1.9-2040	LVF-2.0-2040	LVF-4.0-2040	CVF-1.9-2040	LVF-2.0-2040	LVF-4.0-2040
2045	LVF-1.9-2045	LVF-2.0-2045	LVF-4.0-2045	CVF-1.9-2045	LVF-2.0-2045	LVF-4.0-2045
2050	LVF-1.9-2050	LVF-2.0-2050	LVF-4.0-2050	CVF-1.9-2050	LVF-2.0-2050	LVF-4.0-2050

2.2. Model Formulation

Electricity generation from each type of fuel is determined with Equation (2).

$$\sum E_i = \sum_f \sum_c E_{f,c} \leq \sum_f \sum_c \left(E_{f,c_{inst}} + x_{f,c} \cdot E_{f,c_{cap}} \cdot fc \right) \tag{2}$$

The fossil energy sources, $i = f$, consider electricity generation from coal, co; diesel, d; gas g and fuel oil o, while electricity generation from clean energy sources, $i = c$, considers the use of biomass, bm; carbon capture and storage, ccs; eolic, eo; photovoltaic, phv; geothermal, gtr; hydraulic, hdr; nuclear, nc and solar concentration, sc.

The proposed model determines superstructure electricity generation in each analyzed period taking into account electricity contribution generated in a period prior to the analyzed one, $E_{f,c_{ins}}$, the number of power plants generation $x_{f,c}$ and the capacity of each power plant, $E_{f,c_{cap}}$ as well as a coefficient fc to homogenize units.

Total country annual cost of electricity generation from each energy plant as well as the emissions due to this process is determined by Equations (3) and (4), respectively.

$$TAC(E_i) = \sum_f \sum_c \left(E_{f,c} \cdot Cost \left(a_{f,c} + b_{f,c} + c_{f,c} \right) \right) \tag{3}$$

where $a_{f,c}$ represents investment costs, $b_{f,c}$ is the associated cost from fuel type used and finally $c_{f,c}$ represents operation and maintenance cost, considering fixed and variable costs, taken from Generation, Costs and Reference Parameters for the Formulation of Investment Projects for the Mexican Electricity Sector (COPAR) [20] and Special Report of the Intergovernmental Panel of Climate Change (SRIPCC) of 2012 [21].

CO_2eq emissions are determined with the Equation (4).

$$CO_2eqEm(E_i) = \sum_f \sum_c \left(E_{f,c} \cdot CO_2eq_{f,c} \right) \tag{4}$$

$COeq_{f,c}$ factor is particular for each type of fuel, taken from the INECC [7] and SRIPCC [21].

In an integrated manner, for each analyzed electricity demand condition, TAC of optimized electricity generation, matrix formed by the number of power plants required to satisfy the demand, integrated by power plants that use fossil fuel technologies and clean, as well as the emissions generated by each type of technology are calculated.

2.3. Solution Strategy

To determine the set of optimal solutions that satisfy governmental criteria in each electrical demand condition, epsilon constraint method [22] is implemented in the proposed model. The Pareto chart is constructed using that demand condition and optimized TAC. The epsilon constraint method satisfies governmental objectives proposed in the GCCL and NCCS, (2012) and (2013), respectively. These restrictions have implications to be considered in the application of the model, presented in Equations (5) and (6).

$$E_f < \%f \cdot E_d \tag{5}$$

$$E_c < \%c \cdot E_d \tag{6}$$

E_f is the amount of electric energy generated from fossil sources, E_c is the amount of electric energy generated from clean sources, described in Table 2 and in Figure 5.

Table 2. Objectives and goals to electricity generation by source energy defined in the General Climate Change Law (GCCL) Prepared by the authors based on public data from cited Reference [10].

Electricity Generation Percentage by Energy Source, %		
Year	**Fossil, f**	**Clean, c**
2024	65	35
2035	60	40
2050	50	50

The model considers linear tendency and it was formulated in the General Algebraic Modelling System (GAMS) software used for modelling and mathematical optimization. The results obtained for each condition is shown in Table 1 and under the restrictions presented in Table 2 are detailed in the next section.

The restrictions are used in MES planning scenarios throughout the analysis period.

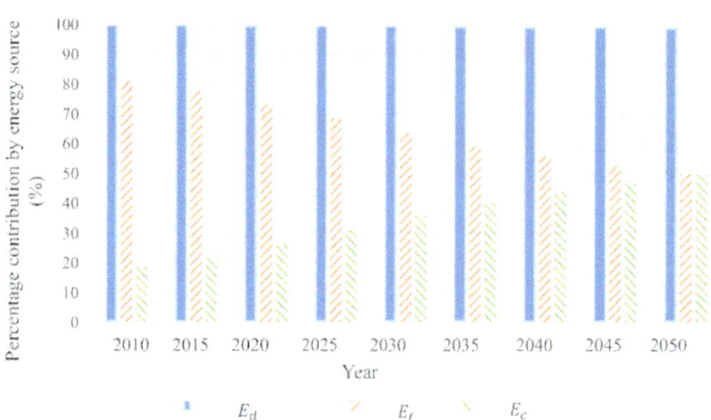

Figure 5. Governmental objectives to satisfy in General Climate Change Law (GCCL). Prepared by the authors based on data from cited reference [10].

3. Results

The modelling scenarios are presented in each electricity demand condition at three different conditions: electricity generation costs, optimized power plants number and generated emissions by this activity.

3.1. Electricity Demand

Electricity demand conditions determined with two population models throughout the analyzed period is presented in Figure 6. Three analyzed demand conditions using LVF presents a growth rate of 13% during all period, meanwhile the corresponding three analyzed demand conditions growth rate using CVF model is 50% are shown.

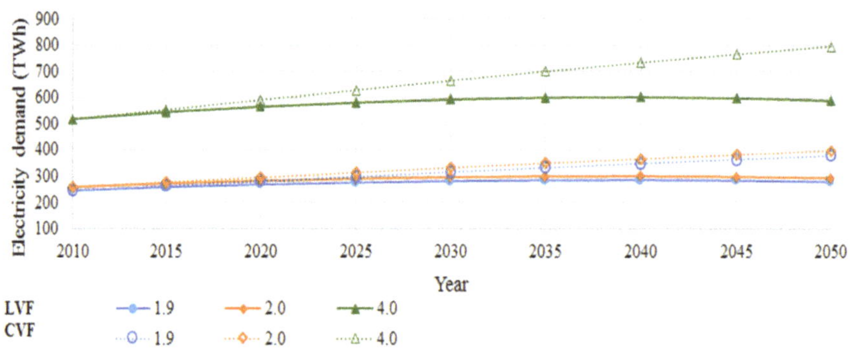

Figure 6. The analyzed electricity demand conditions determined with two population models and three demand conditions per capita. Prepared by the authors based on data from cited reference [19].

Electricity demand for both population models shows similar behavior, at first sight, both vary directly with time, however, it is possible to identify differences. Electricity demand determined with the LVF model presents three different stages throughout the period of analysis. From the beginning of the analysis until the year 2030 a constant growth is observed, followed by a stage without growth until the year 2045; finally, in the last stage of the period, the modelling shows a demand deceleration. The electric demand with the CVF model shows a constant behavior throughout the analysis period, basically an incessant increase.

3.2. TAC of Electricity Power Generation and Optimized Centrals Number

Once the energy demand to be supplied has been defined throughout the analyzed period, the modelling takes the TAC of electric power generation and number of plants for takes two population models (LVF and CVF) and three demand conditions (1.9, 2.0 and 4.0 MWh per capita). In order to determine the number of power plants, priority is given to those that use clean technologies and minimize those of fossil technologies.

Optimized TACs of electricity power generation compute differences between the two population models. The corresponding one from demand determined with LVF population model shows an almost constant growth, however, the demand decreases in the last period of the analysis, specifically in the last two decades. Maximum electricity demand is located in years 2035 and 2045, after this period, electricity demand has a decreasing tendency. This behavior is observed for the three demand conditions.

Optimized TACs present a different behavior using the CVF model—it grows directly proportional to electricity demand during the entire analyzed period. Details for the scenarios using both population models are presented next.

3.2.1. LVF Population Model

Details for the electricity demand determined using LVF population model as a function of TAC and optimized power plants number are analyzed and classified as fossil fuel and clean fuels of energy supply in Figure 7. Total electricity demand is presented by (T), electricity demand from fossil fuels are represented by (F) and electricity demand from clean fuels are represented by (C). Total electricity demand exhibit positive growth behavior meanwhile electricity demand from fossil technologies show an opposite trend as a function of time. Clean technologies have a constant growth meaning the main energy technologies to be employed are from clean energy sources.

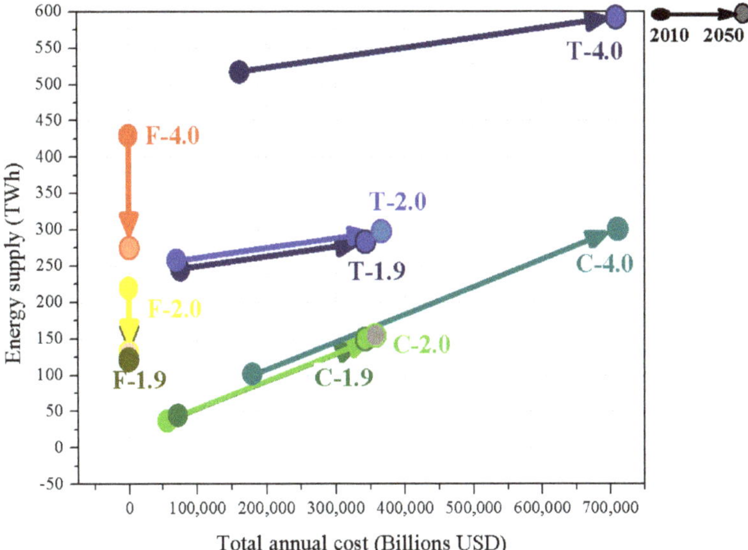

Figure 7. Scheme of the optimized power plants generation centrals number minimizing Total Annual Cost (TAC) of electricity generation for Low Variability of Fertility (LVF) population model [19] and three electricity demand conditions per capita. Prepared by the authors based on data from cited reference.

Clean energy electricity demand shows a growth rate of 204.37%, meanwhile electricity demand from fossil energy sources has a decrease of 30.04% at the end of analyzed period respect year 2010.

If fossil fuel energy (F) does not have new investment, just maintenance, then the value in Figure 7 is relatively low, compared with that of clean energy, so F looks like 0, until the value is 2.1E + 08.

The case of the electricity demand of 1.9 KWh per capita shows total electricity supply from 2010 to 2050 initializing at 2.46 TWh until it rises to 2.80 TWh at the end of the analyzed period. This energy requirement impacts directly in the number of electric power plants necessary to satisfy this demand, presenting a growth rate from 11.27% and 99.51% from fossil and clean energy sources, respectively, at the end of the analyzed period.

Electricity demand of 2.0 KWh per capita, shows that the electricity supply in the analyzed period starts at 2.60 TWh until it raises 2.95 TWh. This energy requirement impacts directly in the number of electric power plants necessary to satisfy this demand, which presents a growth rate from 11.06% and 75.69% from fossil and clean energy sources.

The Electricity demand of 4.0 KWh per capita, shows electricity to supply from from 2010 to 2050 initializing with 5.18 TWh until it raises 5.90 TWh. It is important to highlight that even though this is a necessity, electricity demand presents a growth rate of 9.38% and 74.98%, from fossil and clean energy sources, respectively.

Optimized matrix of number of electricity generating power plants from fossil and clean energy sources to satisfy the demand of 1.9 MWh per capita are presented in Figures 8 and 9.

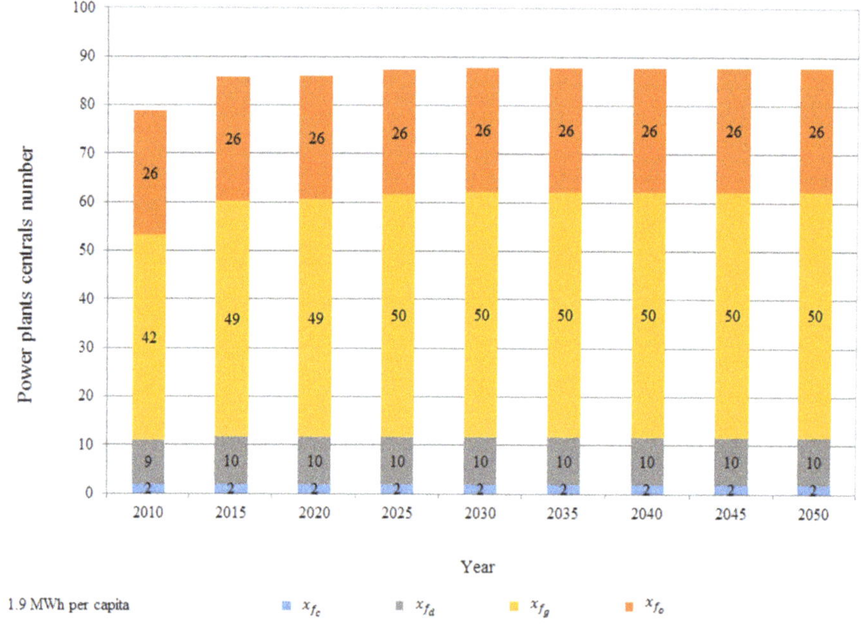

Figure 8. Optimized power plants number from fossil energy resource for 1.9 MWh per capita demand with the LVF model.

As can be appreciated from Figure 8, optimized fossil power plants' number for demand electricity condition of 1.9 MWh per capita, presents a quasi-constant behavior during the analyzed period; only a minimum growth is observed from gas power plants number.

The optimized power plants' number from clean sources presents discrete but constant growth during the whole analyzed period. Power plants which use technology from bioenergy source present higher growth, although their contribution is modest. Hydraulics is a technology source with the greatest contribution within the energy matrix.

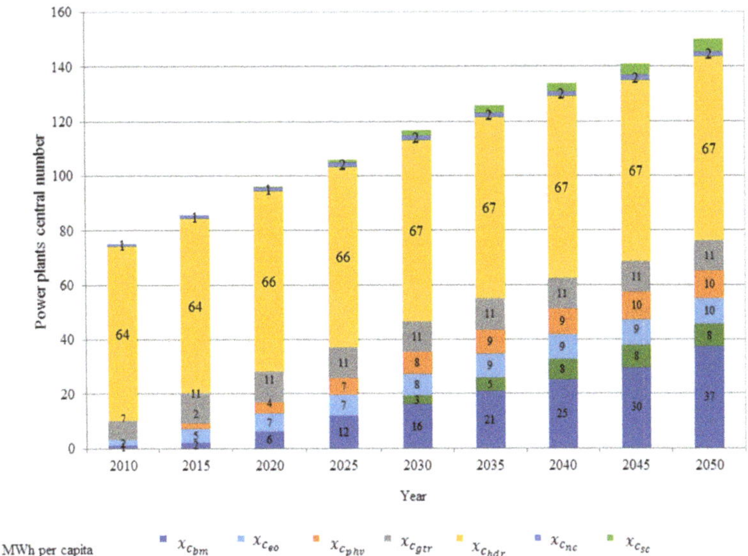

Figure 9. Optimized power plants number from clean energy resource for 1.9 MWh per capita demand with LVF model.

The number of electricity demand per capita of 2.0 MWh of optimized fossil and clean power plants is presented in Figures 10 and 11. The model shows that the optimized fossil power plant's number remains almost constant during the whole analyzed period, meanwhile the optimized clean power plant's number present a discrete growth.

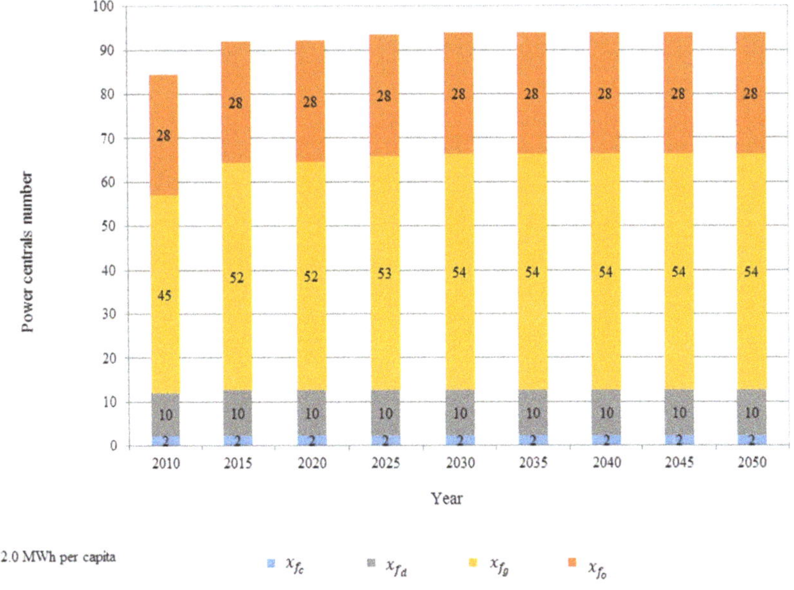

Figure 10. Optimized power plants number from fossil energy resource for 2.0 MWh per capita demand with the LVF model.

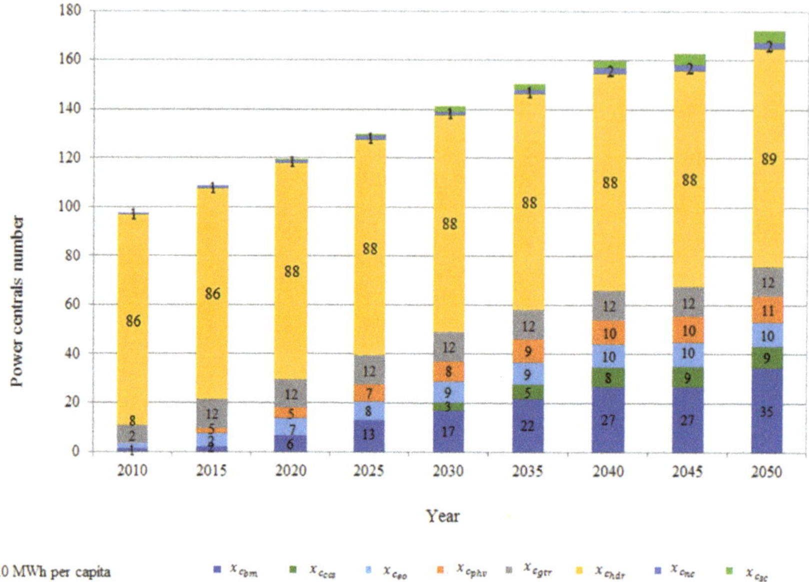

Figure 11. Optimized power plants number from clean energy resource for 2.0 MWh per capita demand with the LVF model.

Optimized fossil and clean power plants number of electricity demand per capita of 4.0 MWh are presented in Figures 12 and 13.

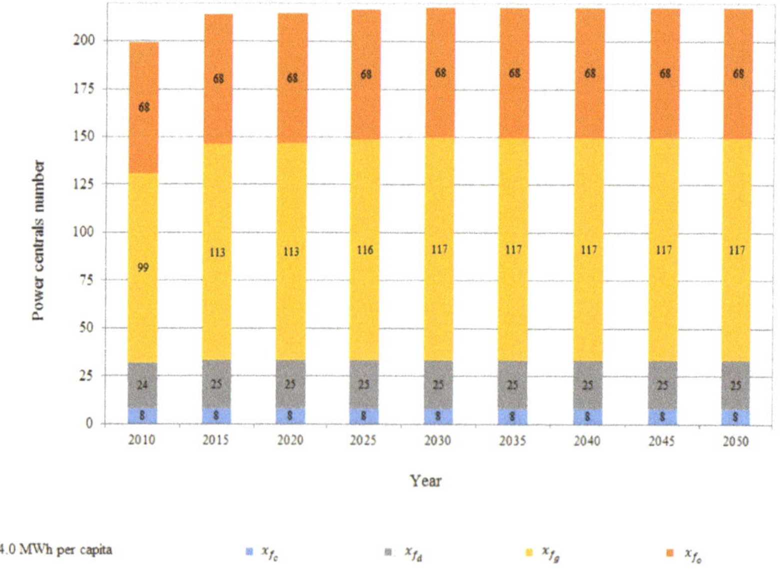

Figure 12. Optimized power plants number from fossil energy resource for 4.0 MWh per capita demand with the LVF model.

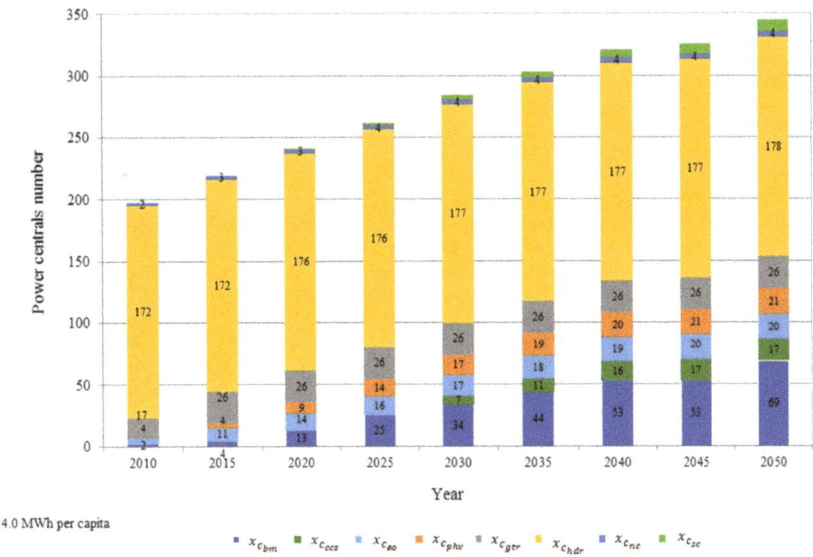

Figure 13. Optimized power plants number from clean energy resource for 4.0 MWh per capita demand with the LVF model.

In this electricity demand case, fossil power plants present an almost constant behavior, contrary to clean energy sources for which growth is constant, discrete but constant. Power plants which use technology from bioenergy and hydraulic sources present higher growth, although their contribution is modest.

3.2.2. CVF Population Model

Electricity demand determined using CVF population model is presented disaggregated between fossil fuel and clean fuels of energy supply the in analyzed period in Figure 14. Total electricity supply is presented by (T), electricity supply from fossil fuels are represented by (F) and electricity supply from clean fuels are represented by (C). Total electricity supply presents positive a growth behavior meanwhile electricity supply from fossil technologies present an opposite trend and at the same time, clean technologies present a constant growth meaning main energy technologies to be employed are from clean energy sources.

Electricity supply for demand from clean energy sources presents a growth rate of 310.24%, meanwhile electricity demand from fossil energy sources presents a decrease of 5.70% at the end of analyzed period respect year 2010. Same aspect for F values close to 0 were explained on Figure 7.

Electricity demand of 2.0 KWh per capita, shows electricity to supply in analyzed period starts with 2.60 TWh until it raises 3.98 TWh impacting directly in the number of electric power plants generation necessary to satisfy this demand, presenting a growth rate from 25.21% and 149.78% from fossil and clean energy sources, respectively, in the end of analyzed period.

The case of electricity demand of 4.0 KWh per capita, shows electricity to supply from year 2010 to year 2050 initializing with 5.18 TWh until it raises 7.95 TWh. This energy increment implies a growth rate of optimized power plants of 21.40% and 148.36%, from fossil and clean energy sources, respectively.

Optimized fossil and clean power plants number of electricity demand per capita of 1.9 MWh are presented in Figures 15 and 16. For this condition, optimized total power plants are calculated to be 313, from which 99 are planned to be from fossil technology and the rest from clean technology.

The optimized power plants' number from fossil fuel sources behaves quite similar to results obtained when the LVF population model is used, that means, power plant number present an almost constant behavior during all analyzed period, only gas power plants slightly increase. Clean power plants number presents a discrete but constant growth being bioenergy power plants the technology which presents higher growth, although their contribution is modest. Hydraulic is also the technology which presents the greatest contribution in the energy matrix.

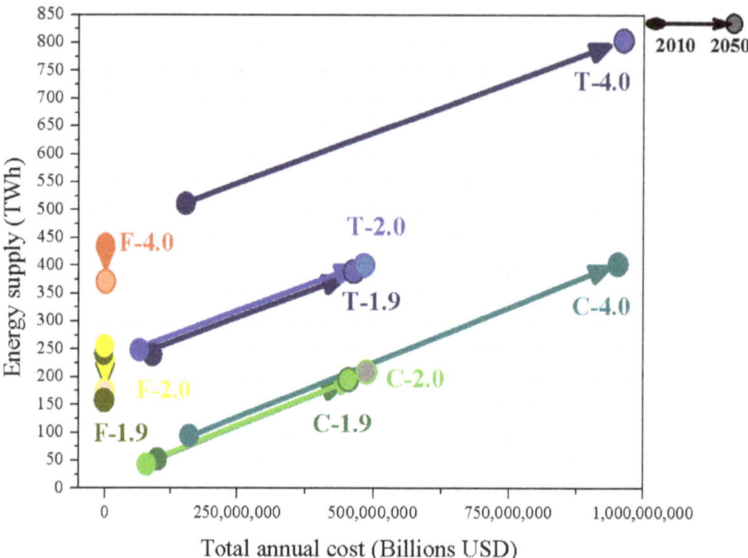

Figure 14. Schematic optimized power plants generation centrals number minimizing total annual cost (TAC) of electricity generation for the CVF population model [19] and three electricity demand conditions per capita. Prepared by the authors based on data from own results.

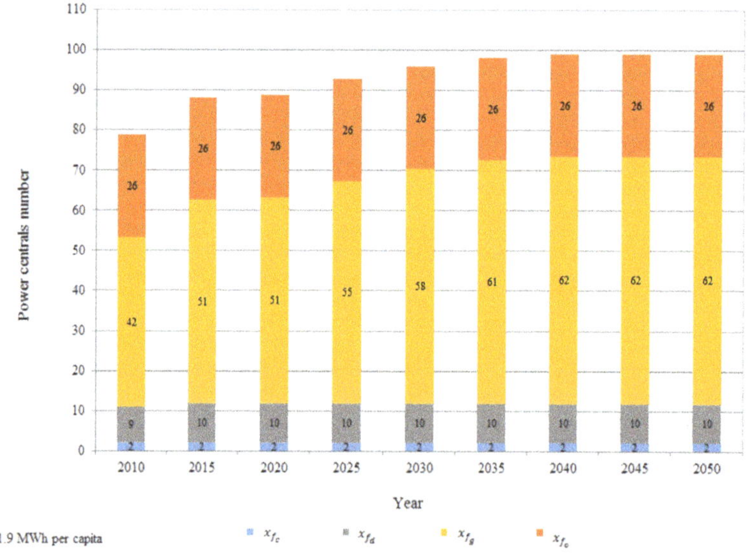

Figure 15. Optimized power plants number from fossil energy resource for 1.9 MWh per capita demand with the CVF model.

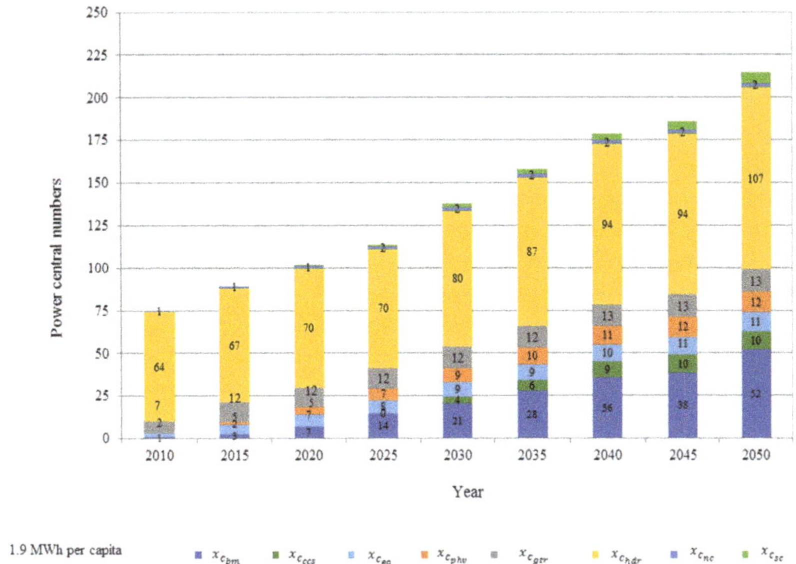

Figure 16. Optimized power plants number from clean energy resource for 1.9 MWh per capita demand with the CVF model.

To satisfy electricity demand of 2.0 MWh per capita, required optimized fossil and clean power plants number are presented in Figures 17 and 18.

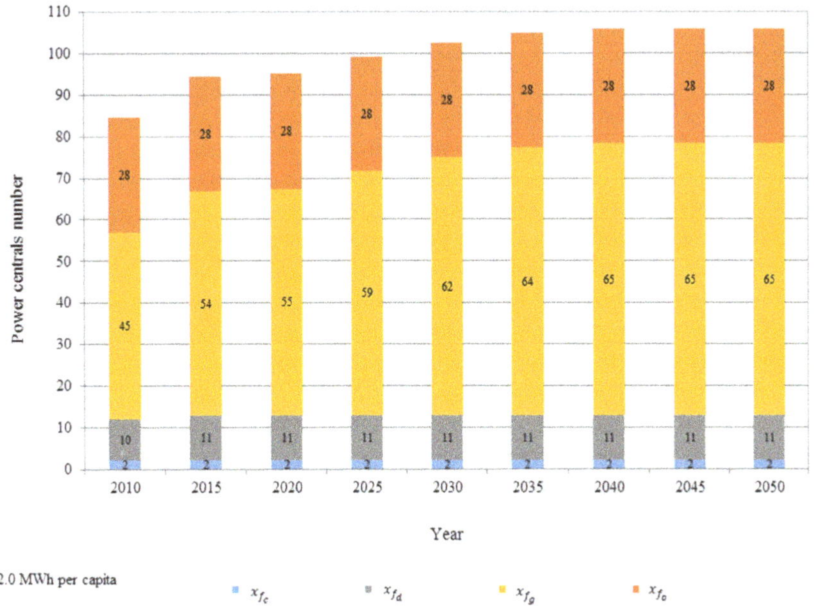

Figure 17. Optimized power plants number from fossil energy resource for 2.0 MWh per capita demand with the CVF model.

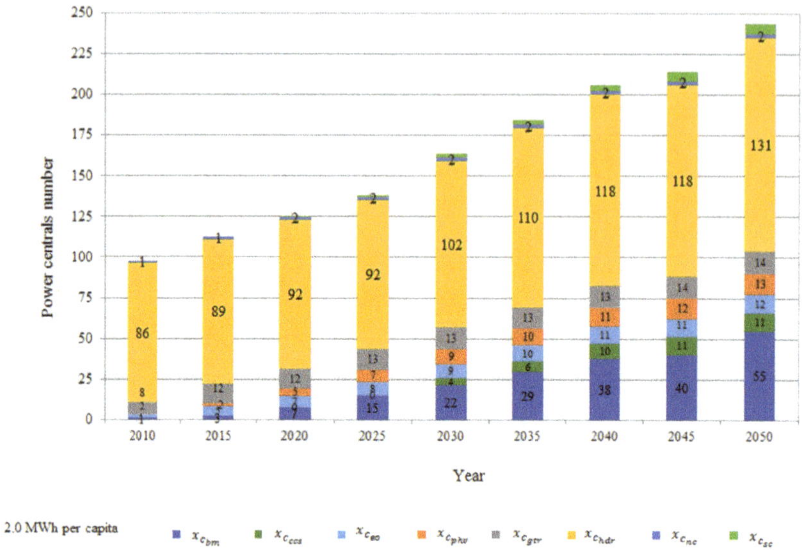

Figure 18. Optimized power plants number from clean energy resource for 2.0 MWh per capita demand with the CVF model.

Electricity demand of 4.0 MWh per capita, required optimized fossil and clean power plants number are presented in Figures 19 and 20.

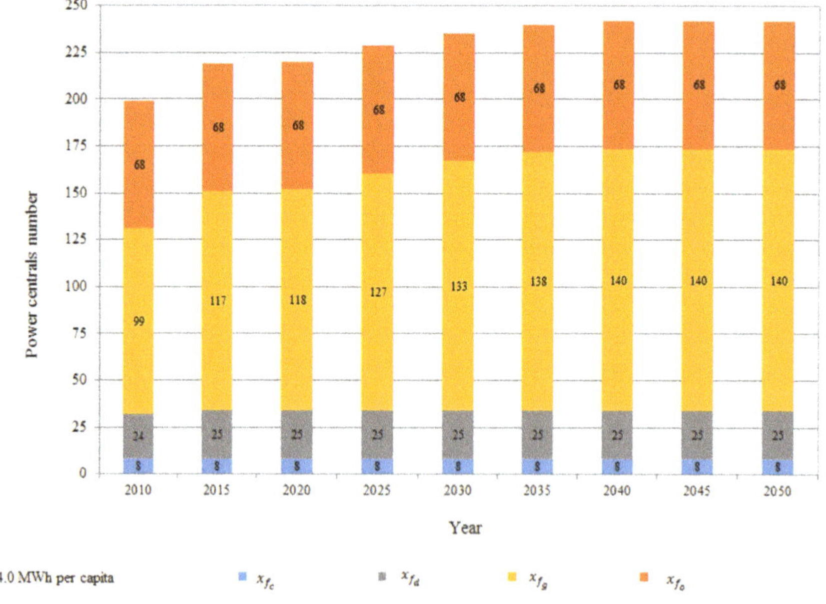

Figure 19. Optimized power plants number from fossil energy resource for 4.0 MWh per capita demand with the CVF model.

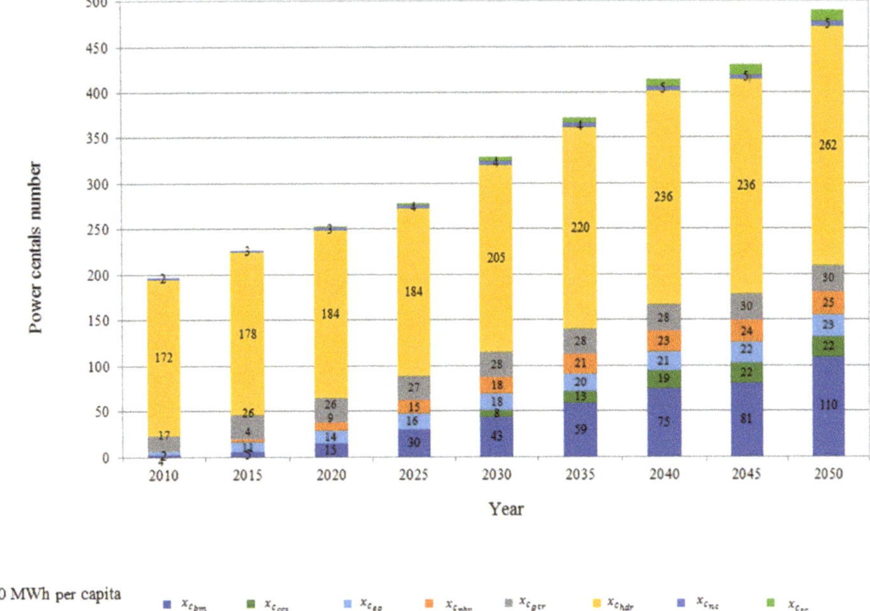

Figure 20. Optimized power plants number from clean energy resource for 4.0 MWh per capita demand with the CVF model.

From Figure 19 a behavior similar to the LVF model is observed which means quasi-constant growth during the analysis period for all technologies except gas. The number of plants that use gas presents an increase almost 50% at the end of the period compared to the initial number. The complement of the matrix of optimized power plants from clean plants is presented in Figure 20. As it is observed, clean power number presents constant growth during analyzed period. Power plants which use bioenergy technology present highest growth, with a modest contribution. The energy source with the greatest contribution within the energy matrix corresponds to the hydraulic resource.

It is important to bear in mind that the determined power plant number corresponds to the required plants in each analyzed period, taking as references the number of plants installed at the beginning of the analysis.

Each optimized TAC of electricity power generation condition, implies the energy matrix definition to satisfy the electricity demand as well as CO_2eq emissions intrinsic to this process. CO_2eq generated due this process are presented below and complement the proposed scenarios in each analyzed period.

3.2.3. CO_2eq Emissions

CO_2eq emissions generated by optimized number of fossil and clean power plants are presented by energy type energy source—fossil and clean.

Produced emissions using fossil sources as primary energy in electrical energy generation for three population demands are presented in Figure 21.

As shown, CO_2eq emissions present a decrease as the analysis period increases, expected behavior given that contribution from this energy type source decreases, satisfying objectives proposed in the General Climate Change Law (GCCL) [10].

The corresponding emissions produced by clean sources as primary energy in electricity generation are presented in Figure 22.

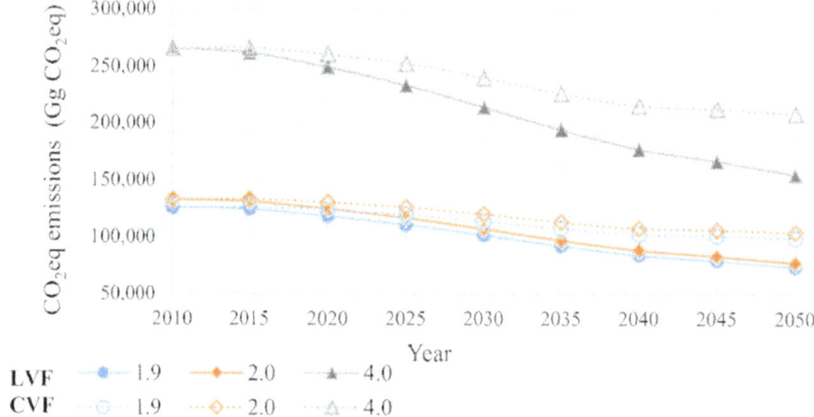

Figure 21. CO_2eq emissions generated by fossil central power plants for three electricity demand per capita and two population models. Prepared by the authors based on own results.

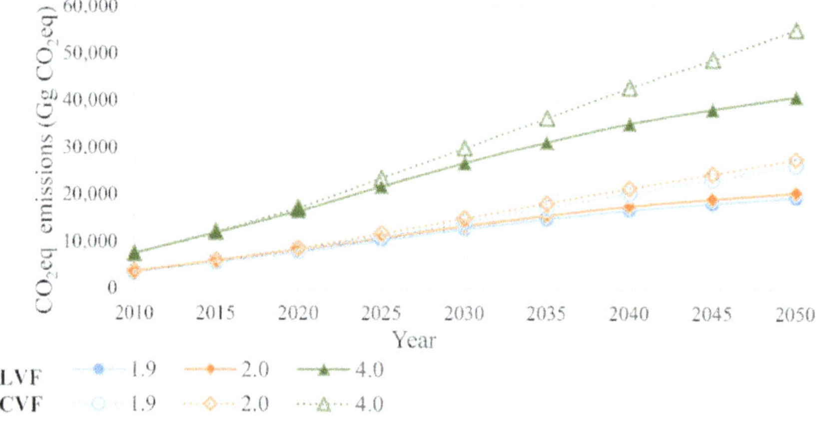

Figure 22. CO_2eq emissions generated by clean central power plants for three electricity demand per capita and two population models. Prepared by the authors based on own results.

CO_2eq emissions generated with clean technologies show an inverse behavior of emissions generated from fossil fuels, that means increasing directly with the period of analysis, however, it is very important to note the amount of emissions generated with this type of energy is, at least, three magnitude orders lower than emissions from fossil sources, satisfying the objective proposed in the General Climate Change Law, GCCL [11].

4. Discussion

Energy planning to explore possible alternative scenarios is a necessary tool for the economic development of the country in the face of global, regional events and international commitments acquired in the face of climate change. This tool serves to guide those responsible for energy policy and regulators in the development of policies to visualize scenarios of energy development that allow to contribute effectively to the sustainable growth of the region as well as to demonstrate the impact of policies and plans established in the short and long term. For two grow population model the Figures 23 and 24 are plotted as follow. The data is obtained from the detailed model in Section 2.2.

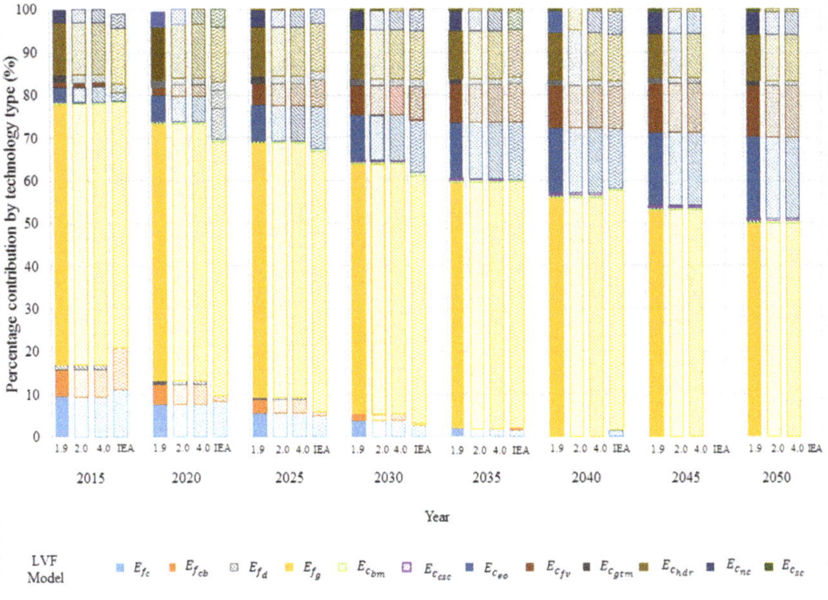

Figure 23. LVF Energy superstructure scenario satisfying governmental objectives from GCCL and National Climate Change Strategy (NCCS) [11,12] compared with IEA scenarios. Prepared by the authors based on own results and public data from cited Reference [18].

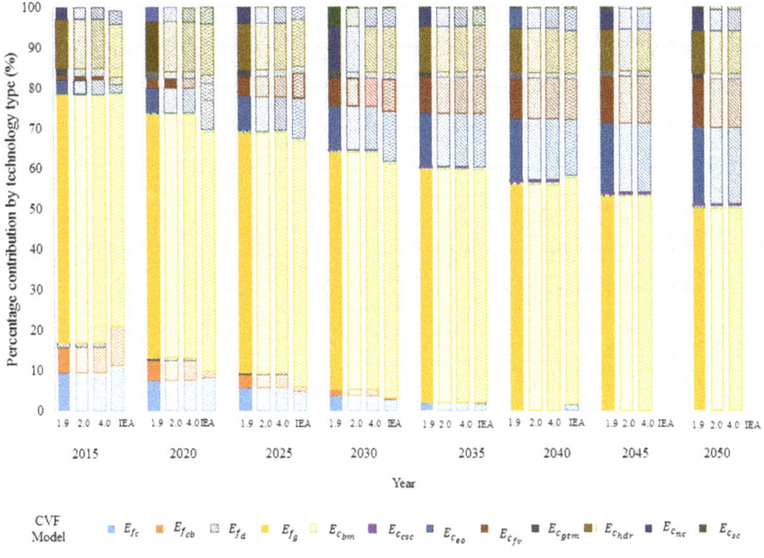

Figure 24. CVF Energy superstructure scenario satisfying governmental objectives from GCCL and NCCS [11,12] compared with IEA scenarios. Prepared by the authors based on own results and public data from cited Reference [18].

In these Figures 23 and 24, for the 2015 year, all scenarios take 10% approximately for carbon transformation in the bottom of the columns. The oil plants are close 5% for 1.9, 2.0 and 4.0% except IEA. For IEA in 2015 about 10% was reported. Diesel plants are very low, close to 1% and for IEA it is not reported. The biggest contribution for this scenarios, including IEA is gas process plant, this model has a 61.2% value and IEA is 57.3%. For clean source the values are 21.9 calculated for this model and 20.7% for IEA data.

The maximum deviations for the gas process plant are from 6.3 to 6.8% based on this modelling and IEA reports in 2015. For that year the clean energy variation is from 5.4 to 5.7% variation.

On 2020 calculation, the fossil energy is 73.5%, 73.5% and 73.8% for 1.9, 2.0 and 4.0 scenarios respectively. For IEA projection, fossil energy is 69.0%. The maximum deviation goes from 6.1 to 6.5%. For clean energy, the values are 26.5, 26.5 and 26.2 for 1.9, 2.0 and 4.0 scenarios respectively. But IEA the predicted value is 31.0%. This mean the largest deviation value for our modelling: 14.5% to 16.9% variation.

On 2025, the fossil energy is computed as 68.7% for any scenario and for IEA is 66.8%. This is a variation close to 2.75%. Clean energy values are 31.3% from this work and 33.2% from IEA. This is a variation from 5.7 to 6.0%.

On 2030 predictions, the values for fossil energy the value is constant at 63.8% and for IEA 61.1%. This variation is just from 4.2 to 4.4% for all scenarios. Clean energy for that year is 36.2% compared with IEA 38.9%. Those values have been 6.9% to 7.4% lower.

For 2035 compared values, fossil energy is 59.5% from our model, compared with 59.6% for IEA prediction. They are almost the same value (0.1% variation). This is similar behavior for clean energy—our modelling reports 40.5 and IEA 40.4 (0.2% variation).

Predictions for 2040 for fossil energy must be 55.9% compared with 57.6% from IEA. These values represent 2.95% variation. Clean energy are 44.1% and 42.4% for this work and IEA, respectively. The variation is close to 3.9%.

The IEA do not report for 2045 and 2050. The values for fossil energy are 52.9% and 49.9% for these years. Then the clean energy 47.1 and 50.1% for same years, respectively. This is the goal for NCCS.

There are considerations in each model described in Section 3.2 and a consequence of these, the resulting data are not the same.

Like other optimized models, there are disadvantages, for example dependency of macroeconomic circumstances such as technology cost trends, as well as demographic dependency population but this dependency is inherent to the modelling process.

5. Conclusions

This paper has proposed a model that satisfactorily defines energy scenarios integrated by the optimized matrix of power plants required to satisfy the electricity demand, by optimized costs of electricity generation as well as CO_2eq emissions produced by this process. In each analysis scenario, governmental objectives with impact on energy security, economy and environmental sustainability are reached satisfactory.

Electricity demand presents a constant growth with both models and three conditions for almost all the period of analysis. This growth is reflected in the gradual increase in the number of plants needed to meet demand. Power plants which use fossil technologies, specifically gas, are those that increase, the rest kept constant. Power plants that use clean source technologies, show a gradual growth, highlighting hydraulics and biomass. The results indicate, in each scenario analyzed, CO_2eq emissions satisfy objectives proposed in the General Climate Change Law [11], thus, despite the increase in electricity demand, CO_2eq emissions of a reverse behavior.

With the application of the proposed model to the MES, scenarios obtained from applying restricted public policies are visualized. In this scenarios each optimal conditions to achieve goals and plans established in the short and long term are visualized. Given the results obtained with this proposed model, bases are laid with possibility of using the model to identify alternative scenarios to the optimal ones as well as to evaluate costs of environmental impacts of the technologies in different stages of modelling.

This modelling would be implemented for another country's conditions with few data to predict good agreement with IEA long-term predictions. For the Mexico case, the energy superstructures percentage contribution for 2025 and forward years shows a 7.4% variation or lower compared with this work and IEA data.

Author Contributions: G.H.-L., J.M.P.-O. and G.D.T.V. were working in GAMS programming for this Project. R.J.R., A.R.-M., J.C.R. analyzed the data from the Mexican electricity system.

Funding: The first author is grateful for the support from the SENER-CONACyT/Energy Sustainability fund for the grant awarded for the postdoctoral stay at the Engineering and Applied Sciences Center Research (CIICAp) of the Autonomous University of Morelos State (UAEM) as well as the support from the Thematic Network on Energy Sustainability, Environment and Society (SUMAS Network), CONACyT project 281101.

Conflicts of Interest: The authors declare no conflicts of interest. All figures were prepared by the authors based on own results and public data from cited References.

Appendix A

The GAMS © code is shown as follow.

VARIABLES EFdemand, TAC; POSITIVE VARIABLES	
Energy contributions (MWh)	
Ffos	FOSSIL
Frenv	CLEAN
Ffoscarb	CARBON
Ffoscomb	FUEL OIL
Ffosdis	DIESEL
Ffosgas	GAS
Frenvbio	BIOMASS
Frenvcscrb	CARBON CAPTURE AND STORAGE
Frenveol	EOLIC
Frenvftv	PHOTOVOLTAIC
Frenvgtm	GEOTHERMIC
Frenvhdr	HYDRAULIC
Frenvnucl	NUCLEAR
Frenvsc	SOLAR CONCENTRATION
COSTS	
TAC	TOTAL ANNUAL COST FROM FOSSIL AND CLEAN
Costfos	FOSSIL
Costrenv	CLEAN
Costfoscarb	CARBON
Costfoscomb	FUEL OIL
Costfosdis	DIESEL
Costfosgas	GAS

Costrenvbio	BIOMASS
Costrenvcscrb	CARBON CAPTURE AND STORAGE
Costrenveol	EOLIC
Costrenvftv	PHOTOVOLTAIC
Costrenvgtm	GEOTHERMIC
Costrenvhdr	HYDRAULIC
Costrenvnucl	NUCLEAR
Costrenvftrm	SOLAR CONCENTRATION
***************************CO_2 Emission (kg/MWh)**	
EFdeman	TOTAL EMISSION FROM FOSSIL AND CLEAN
EFfos	FOSSIL
EFrenv	CLEAN
EFfoscarb	CARBON
EFfoscomb	OIL FUEL
EFfosdis	DIESEL
EFfosgas	GAS
EFrenvbio	BIOMASS
EFrenvcscrb	CARBON CAPTURE AND STORAGE
EFrenveol	EOLIC
EFrenvftv	PHOTOVOLTAIC
EFrenvgtm	GEOTHERMIC
EFrenvhdr	HYDRAULIC
EFrenvnucl	NUCLEAR
EFrenvftrm	SOLAR CONCENTRATION
********************* NUMBER OF POWER PLANTS GENERATION ***********************************	
xtot	TOTAL
xf	FOSSILS
xr	CLEAN
xfoscarb	CARBON
xfoscomb	FUEL OIL
xfosdis	DIESEL
xfosgas	GAS
xrenvbio	BIOMASA
xrenvcscrb	CARBON CAPTURE AND STORAGE
xrenveol	EOLICA
xrenvftv	PHOTOVOLTAIC
xrenvgtm	GEOTHERMIC
xrenvhdr	HYDRAULIC
xrenvnucl	NUCLEAR
xrenvftrm	SOLAR CONCENTRATION

PARAMETERS		
COSTS (USD/MWh neto)		
*************************** INVESTMENT COSTS BY ENERGY FUEL***************************************		
afoscarb	CARBON	/37.39/
afoscomb	OIL FUEL	/41.891/
afosdis	DIESEL	/64.16/
afosgas	GAS	/72.53/
arenvbio	BIOMASS	/150/
arenvcscrb	CARBON CAPTURE AND STORAGE	/47.39/
arenveol	EOLIC	/1.65E6/
arenvftv	PHOTOVOLTAIC	/5.11E6/
arenvgtm	GEOTHERMIC	/3.18E6/
arenvhdr	HYDRAULIC	/2.0E6/
arenvnucl	NUCLEAR	/87.93/
arenvftrm	SOLAR CONCENTRATION	/180/
*************************** ASSOCIATED COST FROM FUEL TYPE ***************************		
bfoscarb	CARBON	/27.57/
bfoscomb	OIL FUEL	/93.89/
bfosdis	DIESEL	/147.06/
bfosgas	GAS	/49.37/
brenvbio	BIOMASS	/7.9/
brenvcscrb	CARBON CAPTURE AND STORAGE	/27.67/
brenveol	EOLICA	/0/
brenvftv	PHOTOVOLTAIC	/0/
brenvgtm	GEOTHERMIC	/50.12/
brenvhdr	HYDRAULIC	/6.38/
brenvftrm	SOLAR CONCENTRATION	/0/
OPERATION AND MAINTENANCE COST, INCLUDING FIXED AND VARIABLE COSTS *		
cfoscarb	CARBON	/8.13/
cfoscomb	OIL FUEL	/9.21/
cfosdis	DIESEL	/12.88/
cfosgas	GAS	/11.33/
crenvbio	BIOMAS	/4.33/
crenvcscrb	CARBON CAPTURE AND STORAGE	/8.13/
crenveol	EOLIC	/1.75/
crenvftv	PHOTOVOLTAIC	/7.72/
crenvgtm	GEOTHERMIC	/19.41/
crenvhdr	HYDRAULICA	/5.71/
crenvftrm	SOLAR CONCENTRATION	/20/

*********************** CO_2eq BY FUEL (Kg CO_2/MWh)***********************		
ECO2eqfoscarb	CARBON	/1089.81/
ECO2eqfoscomb	OIL FUEL	/822/
ECO2eqfosdis	DIESEL	/274.44/
ECO2eqfosgas	GAS	/524/
ECO2eqrenvbio	BIOMAS	/1403.75/
ECO2eqrenvcscrb	CARBON CAPTURE AND STORAGE	/217.2/
ECO2eqrenveol	EOLIC	/210/
ECO2eqrenvftv	PHOTOVOLTAIC	/106/
ECO2eqrenvgtm	GEOTHERMIC	/372/
ECO2eqrenvhdr	HYDRAULIC	/15/
ECO2eqrenvftrm	SOLAR CONCENTRATION	/14/
Fdemand	ANNUAL NATIONAL ELECTRICITY DEMAND MWh	/246381740/

* ELECTRICITY CONTRIBUTION GENERATED IN A PERIOD PRIOR TO THE ANALYZED (MWh)*		
Ffcarbins	CARBON	/18380000/
Ffcombins	OIL FUEL	/6451200/
Ffdisins	DIESEL	/780000/
Ffgasins	GAS	/27166000/
Frbioins	BIOMAS	/0/
Frcscrbins	CARBON CAPTURE AND STORAGE	/0/
Freolins	EOLIC	/5000/
Frftvins	PHOTOVOLTAIC	/00/
Frgtmins	GEOTHERMIC	/729900/
Frhdrins	HYDRAULIC	/26851000/
Frnuclins	NUCLEAR	/1080500/
Frftrmins	SOLAR CONCENTRATION	/0/

*************************** CAPACITY OF POWER PLANT (MW)***************************************		
Fcapfoscarb	CARBON	/2600/
Fcapfoscomb	OIL FUEL	/12711/
Fcapfosdis	DIESEL	/182/
Fcapfosgas	GAS	/7230/
Fcaprenvbio	BIOMASA	/1500/
Fcaprenvcscrb	CARBON CAPTURE AND STORAGE	/1000/
Fcaprenveol	EOLIC	/20000/
Fcaprenvftv	PHOTOVOLTAIC	/1000/
Fcaprenvgtm	GEOTHERMIC	/1000/
Fcaprenvhdr	HYDRAULIC	/10270/
Fcaprenvnucl	NUCLEAR	/1365/
Fcaprenvftrm	SOLAR CONCENTRATION	/960/

	EQUATIONS
R1	ANNUAL ELECTRICY DEMAND
R2	FOSSIL FUEL CONTRUIBUTION TO ELECTRICITY DEMAND
R3	CLEAN FUEL CONTRUIBUTION TO ELECTRICITY DEMAND
R4	ELECTRICITY GENERATION FROM CARBON
R5	ELECTRICITY GENERATION FROM FUEL OIL
R6	ELECTRICITY GENERATION FROM DIESEL
R7	ELECTRICITY GENERATION FROM GAS
R8	ELECTRICITY GENERATION FROM BIOMAS
R9	ELECTRICITY GENERATION FROM CARBON CAPTURE AND STORAGE
R10	ELECTRICITY GENERATION FROM EOLIC
R11	ELECTRICITY GENERATION FROM PHOTOVOLTAIC
R12	ELECTRICITY GENERATION FROM GEOTHERMAL
R13	ELECTRICITY GENERATION FROM HYDRAULIC
R14	ELECTRICITY GENERATION FROM NUCLEAR
R15	ELECTRICITY GENERATION FROM SOLAR CONCENTRATION
	***********************************COSTS***
R16	TOTAL ANNUAL COST OF ELECTRICITY GENERATION
R17	TOTAL ANNUAL COST OF ELECTRICITY GENERATION FROM FOSSIL FUELS
R18	TOTAL ANNUAL COST OF ELECTRICITY GENERATION FROM CLEAN FUELS
R19	TOTAL ANNUAL COST OF ELECTRICITY GENERATION FROM CARBON
R20	TOTAL ANNUAL COST OF ELECTRICITY GENERATION FROM OIL FUEL
R21	TOTAL ANNUAL COST OF ELECTRICITY GENERATION FROM DIESEL
R22	TOTAL ANNUAL COST OF ELECTRICITY GENERATION FROM GAS
R23	TOTAL ANNUAL COST OF ELECTRICITY GENERATION FROM BIOMAS
R24	TOTAL ANNUAL COST OF ELECTRICITY GENERATION FROM CARBON CAPTURE AND STORAGE
R25	TOTAL ANNUAL COST OF ELECTRICITY GENERATION FROM EOLIC
R26	TOTAL ANNUAL COST OF ELECTRICITY GENERATION FROM PHOTOVOLTAIC
R27	TOTAL ANNUAL COST OF ELECTRICITY GENERATION FROM GEOTHERMAL
R28	TOTAL ANNUAL COST OF ELECTRICITY GENERATION FROM HYDRAULIC
R29	TOTAL ANNUAL COST OF ELECTRICITY GENERATION FROM NUCLEAR
R30	TOTAL ANNUAL COST OF ELECTRICITY GENERATION FROM SOLAR CONCENTRATION
	***********************************CO_2eq EMISSIONS***
R31	CO2eq EMISSION FROM TOTAL ELECTRICITY GENERATION
R32	CO2eq EMISSION FROM ELECTRICITY GENERATION FROM FOSSIL FUELS
R33	CO2eq EMISSION FROM ELECTRICITY GENERATION FROM CLEAN FUELS
R34	CO2eq EMISSION FROM ELECTRICITY GENERATION FROM CARBON
R35	CO2eq EMISSION FROM ELECTRICITY GENERATION FROM FUEL OIL
R36	CO2eq EMISSION FROM ELECTRICITY GENERATION FROM DIESEL

R37	CO2eq EMISSION FROM ELECTRICITY GENERATION FROM GAS
R38	CO2eq EMISSION FROM ELECTRICITY GENERATION FROM BIOMAS
R39	CO2eq EMISSION FROM ELECTRICITY GENERATION FROM CARBON CAPTURE AND STORAGE
R40	CO2eq EMISSION FROM ELECTRICITY GENERATION EOLIC
R41	CO2eq EMISSION FROM ELECTRICITY GENERATION PHOTOVOLTAIC
R42	CO2eq EMISSION FROM ELECTRICITY GENERATION GEOTHERMIC
R43	CO2eq EMISSION FROM ELECTRICITY GENERATION HYDRAULIC
R44	CO2eq EMISSION FROM ELECTRICITY GENERATION NUCLEAR
R45	CO2eq EMISSION FROM ELECTRICITY GENERATION SOLAR CONCENTRATION
***************************************RESTRICTIONS***	
R46	ELECTRICITY CONTRIBUTION FROM CARBON
R47	ELECTRICITY CONTRIBUTION FROM FUEL OIL
R48	ELECTRICITY CONTRIBUTION FROM DIESEL
R49	ELECTRICITY CONTRIBUTION FROM GAS
R50	ELECTRICITY CONTRIBUTION FROM BIOMAS
R51	ELECTRICITY CONTRIBUTION FROM CARBON CAPTURE AND STORAGE
R52	ELECTRICITY CONTRIBUTION FROM EOLIC
R53	ELECTRICITY CONTRIBUTION FROM PHOTOVOLTAIC
R54	ELECTRICITY CONTRIBUTION FROM GEOTHERMIC
R55	ELECTRICITY CONTRIBUTION FROM HYDRAULIC
R56	ELECTRICITY CONTRIBUTION FROM NUCLEAR
R57	ELECTRICITY CONTRIBUTION FROM SOLAR CONCENTRATION
R58	ELECTRICITY CONTRIBUTION FROM FOSSIL FUELS
R59	TOTAL NUMBER OF POWER PLANTS GENERATION
R60	FOSSIL NUMBER OF POWER PLANTS GENERATION
R61	CLEAN NUMBER OF POWER PLANTS GENERATION

MODEL	
*************************************ELECTRICIY CALCULUS *************************************	
R1..	Fdemand = E = Ffos + Frenv;
R2..	Ffos = E = Ffoscarb + Ffoscomb + Ffosdis + Ffosgas;
R3..	Frenv = E = Frenvbio + Frenvcscrb + Frenveol + Frenvftv + Frenvftrm + Frenvgtm + Frenvhdr + Frenvnucl;
*********************FÓSILES***	
R4..	Ffoscarb =L= Ffcarbins + (Fcapfoscarb * xfoscarb * 8760 * fafoscarb);
R5..	Ffoscomb =L= Ffcombins + (Fcapfoscomb * xfoscomb * 8760 * fafoscomb);
R6..	Ffosdis =L= Ffdisins + (Fcapfosdis * xfosdis * 8760 * fafosdis);
R7..	Ffosgas =L= Ffgasins + (Fcapfosgas * xfosgas * 8760 * fafosgas);

**************************RENOVABLES***	
R8..	Frenvbio =L= Frbioins + (Fcaprenvbio * xrenvbio * 8760 * farenvbio) ;
R9..	Frenvcscrb =L= Frcscrbins + (Fcaprenvcscrb * xrenvcscrb * 8760 * farenvcscrb) ;
R10..	Frenveol =L= Freolins + (Fcaprenveol * xrenveol * 8760 * farenveol) ;
R11..	Frenvftv =L= Frftvins + (Fcaprenvftv * xrenvftv * 8760 * farenvftv) ;
R12..	Frenvftrm =L= Frftrmins + (Fcaprenvftrm * xrenvftrm * 8760 * farenvftrm) ;
R13..	Frenvgtm =L= Frgtmins + (Fcaprenvgtm * xrenvgtm * 8760 * farenvgtm) ;
R14..	Frenvhdr =L= Frhdrins + (Fcaprenvhdr * xrenvhdr * 8760 * farenvhdr) ;
R15..	Frenvnucl =L= Frnuclins + (Fcaprenvnucl * xrenvnucl * 8760 * farenvnucl) ;
***	
******************************COST CALCULUS**	
R16..	TAC =E= Costfos + Costrenv ;
R17..	Costfos =E= Costfoscarb + Costfoscomb + Costfosdis + Costfosgas ;
R18..	Costrenv =E= Costrenvbio + Costrenvcscrb + Costrenveol + Costrenvftv + Costrenvftrm + Costrenvgtm + Costrenvhdr + Costrenvnucl;
R19..	Costfoscarb =E= Ffoscarb * (afoscarb + bfoscarb + cfoscarb) * 1000 ;
R20..	Costfoscomb =E= Ffoscomb * (afoscomb + bfoscomb + cfoscomb) * 1000 ;
R21..	Costfosdis =E= Ffosdis * (afosdis + bfosdis + cfosdis) * 1000 ;
R22..	Costfosgas =E= Ffosgas * (afosgas + bfosgas + cfosgas) * 1000 ;
R23..	Costrenvbio =E= Frenvbio * (arenvbio + brenvbio + crenvbio) * 1000 ;
R24..	Costrenvcscrb =E= Frenvcscrb * (arenvcscrb + brenvcscrb + crenvcscrb) * 1000 ;
R25..	Costrenveol =E= Frenveol * (arenveol + brenveol + crenveol) * 1000 ;
R26..	Costrenvftv =E= Frenvftv * (arenvftv + brenvftv + crenvftv) * 1000 ;
R27..	Costrenvgtm =E= Frenvgtm * (arenvgtm + brenvgtm + crenvgtm) * 1000 ;
R28..	Costrenvhdr =E= Frenvhdr * (arenvhdr + brenvhdr + crenvhdr) * 1000 ;
R29..	Costrenvnucl =E= Frenvnucl * (arenvnucl + brenvnucl + crenvnucl) * 1000
R30..	Costrenvftrm =E= Frenvftrm * (arenvftrm + brenvftrm + crenvftrm) * 1000 ;
***	
R31..	EFdemand =E= EFfos + EFrenv;
***	
*************************** EMISSIONS CALCULUS**************************************	
R32..	EFfos =E= EFfoscarb + EFfoscomb + EFfosdis + EFfosgas ;
R33..	EFrenv =E= EFrenvbio + EFrenvcscrb + EFrenveol + EFrenvftv + EFrenvftrm + EFrenvgtm + EFrenvhdr;
****************************DE FUENTES FÓSILES**	
R34..	EFfoscarb =E= ECO2eqfoscarb * Ffoscarb ;
R35..	EFfoscomb =E= ECO2eqfoscomb * Ffoscomb ;
R36..	EFfosdis =E= ECO2eqfosdis * Ffosdis ;
R37..	EFfosgas =E= ECO2eqfosgas * Ffosgas ;

****************************DE FUENTES RENOVABLES***	
R38..	EFrenvbio =E= ECO2eqrenvbio * Frenvbio ;
R39..	EFrenvcscrb =E= ECO2eqrenvcscrb * Frenvcscrb ;
R40..	EFrenveol =E= ECO2eqrenveol * Frenveol ;
R41..	EFrenvftv =E= ECO2eqrenvftv * Frenvftv ;
R42..	EFrenvftrm =E= ECO2eqrenvftrm * Frenvftrm ;
R43..	EFrenvgtm =E= ECO2eqrenvgtm * Frenvgtm ;
R44..	EFrenvhdr =E= ECO2eqrenvhdr * Frenvhdr ;
R45..	EFrenvnucl =E= ECO2eqrenvnucl * Frenvnucl ;
**************************PERCENTAGE TECHNOLOGY CONTRIBUTIONS**************************	
R46..	Ffoscarb =L= Ffos ;
R47..	Ffoscomb =L= Ffos ;
R48..	Ffosdis =L= Ffos ;
R49..	Ffosgas =L= Ffos ;
R50..	Frenvbio =L= Fdemand - Ffos ;
R51..	Frenvcscrb =L= Fdemand - Ffos ;
R52..	Frenveol =L= Fdemand - Ffos ;
R53..	Frenvftv =L= Fdemand - Ffos ;
R54..	Frenvftrm =L= Fdemand - Ffos ;
R55..	Frenvgtm =L= Fdemand - Ffos ;
R56..	Frenvhdr =L= Fdemand - Ffos ;
R57..	Frenvnucl =L= Fdemand - Ffos ;
***********FOSSIL PERCENTAGE CONTRIBUTION ***********	
R58..	Ffos =L= Fdemand * 0.8132 ;
****************************CENTTRAL NUMERO DE CENTRALES********************************	
R59..	xtot =E= xf + xr ;
R60..	Xf =E= xfoscarb + xfoscomb + xfosdis + xfosgas ;
R61..	xr =E= xrenvbio + xrenvcscrb + xrenveol + xrenvftv + xrenvftrm + xrenvgtm + xrenvhdr + xrenvnucl;
**********************NUMBER OF POWER PLANTS GENERATION ***	
	Ffoscarb.LO = Fdemand * 0.11 ;
	Ffoscarb.UP = Fdemand * 0.115 ;
	Ffoscomb.LO = Fdemand * 0.075 ;
	Ffoscomb.UP = Fdemand * 0.0781 ;
	Ffosdis.LO = Fdemand * 0.008 ;
	Ffosdis.UP = Fdemand * 0.01 ;
	Ffosgas.LO = Fdemand * 0.45 ;
	Ffosgas.UP = Fdemand * 0.61 ;
	Frenvbio.LO = Fdemand * 0.00 ;

	Frenvbio.UP = Fdemand * 0.003 ;
	Frenvcscrb.LO = Fdemand * 0.00 ;
	Frenvcscrb.UP = Fdemand * 0.00 ;
	Frenveol.LO = Fdemand * 0.190 ;
	Frenveol.UP = Fdemand * 0.013 ;
	Frenvftrm.LO = Fdemand * 0.00 ;
	Frenvftrm.UP = Fdemand * 0.00 ;
	Frenvftv.LO = Fdemand * 0.0055 ;
	Frenvftv.UP = Fdemand * 0.0115 ;
	Frenvgtm.LO = Fdemand * 0.0145 ;
	Frenvgtm.UP = Fdemand * 0.015 ;
	Frenvhdr.LO = Fdemand * 0.11 ;
	Frenvhdr.UP = Fdemand * 0.1194 ;
	Frenvnucl.LO = Fdemand * 0.05 ;
	Frenvnucl.UP = Fdemand * 0.027 ;
MODEL PCL05 /ALL/; SOLVE PCL05 USING MIP MINIMIZING TAC;	

References

1. Farman, J.C.; Gardiner, B.G.; Shanklin, J.D. Large losses of total ozone in Antarctica reveal seasonal ClOx/NOx interaction. *Nature* **1985**, *315*, 207–210. [CrossRef]
2. *Montreal Protocol of Substances that Deplete the Ozone Layer*. United Nations Environment Programme, Ozone Secretariat; Montreal, Canada. 1987. Available online: https://treaties.un.org/doc/Treaties/1989/01/19890101%2003-25%20AM/Ch_XXVII_02_ap.pdf (accessed on 13 August 2019).
3. *Kyoto Protocol to the United Nations Framework Convention on Climate Change*. United Nations Framework Convention on Climate Change; Kyoto, Japan. 1998. Available online: https://unfccc.int/resource/docs/convkp/kpeng.pdf (accessed on 13 August 2019).
4. Sustainable Development Goals. United Nations. Available online: https://www.un.org/sustainabledevelopment/ (accessed on 7 July 2018).
5. Paris Agreement. In Proceedings of the Conference of the Parties of the United Nation Framework Convention on Climate Change, United Nations Framework Convention on Climate Change, Le Bourget, France, 15 December 2015.
6. The World Bank Open Data. World Development Indicators. 2017. Available online: http://datos.bancomundial.org/indicador/EG.USE.ELEC.KH.PC?view=chart (accessed on 26 August 2019).
7. National Ecology and Climate Change Institute (INECC). National Inventory of Emissions of Gases and Compounds of Greenhouse Gases. 2015. Available online: www.gob.mx/inecc/acciones-y-programas/emisiones-80133 (accessed on 26 August 2019).
8. SENER; Mexican Energy State Secretariat; Secretaría de Energía. *Electricity Sector Prospectives 2018–2032*. Mexico City, Mexico. 2018. Available online: https://www.gob.mx/cms/uploads/attachment/file/284345/Prospectiva_del_Sector_El_ctrico_2017.pdf (accessed on 13 August 2019).
9. Ruiz-Mendoza, B.J.; Sheinbaum-Pardo, C. Electricity sector reforms in four Latin-American countries and their impact on carbon dioxide emissions and renewable energy. *Energy Policy* **2010**, *38*, 6755–6766. [CrossRef]
10. SEGOB; Mexican Governing Secretariat; Secretaría de Gobernación. General Climate Change Law. Mexico City, Mexico. 2012. Available online: www.diputados.gob.mx/LeyesBiblio/pdf/LGCC.pdf (accessed on 7 September 2018).

11. SEMARNAT; Mexican Ministry of the Environment and Natural Resources; Undersecretary of Planning and Environmental Policy. *National Climate Change Strategy*; Vision 10-20-40; Secretaría del Medio Ambiente y recusos Naturales, Subsecretaría de Planeación y Política Ambiental; Dirección General de Políticas para el Cambio Climático: Mexico City, Mexico, 2013.

12. SEGOB; Mexican Governing Secretariat. Secretaría de Gobernación. *Energy Transition Law*. Energy Transition Law; Tenth-Five Transitory from the Third Article. Ley de Transición Energética. Transitorio Décimo quinto del Artículo Tercero; Mexico City, Mexico. 2015. Available online: www.diputados.gob.mx/LeyesBiblio/pdf/LTE.pdf (accessed on 5 September 2018).

13. Manzini, F.; Islas, J.; Martínez, M. Reduction of greenhouse gases using renewable energies in México in 2025. *Int. J. Hidrogen Energy* **2001**, *26*, 145–149. [CrossRef]

14. Johnson, T.M.; Alatorre, C.; Romo, Z.; Lui, F. *Low Carbon Development for Mexico*; World Bank: Washington, DC, USA, 2009.

15. Lund, H.; Mathiesen, B.V. Energy system analysis of 100% renewable energy systems—The case of Denmark in years 2030 and 2050. *Energy* **2009**, *34*, 524–531. [CrossRef]

16. Santoyo-Castelazo, E.; Stamford, L.; Azapagic, A. Environmental implications of decarbonizing electricity supply in large economies: The case of Mexico. *Energy Convers. Manag.* **2014**, *85*, 272–291. [CrossRef]

17. Vidal-Amaro, J.J.; Østergar, P.A.; Sheinbaum-Pardo, C. Optimal energy mix for transition from fossil fuel to renewable energy sources—The case of the Mexican electricity system. *Appl. Energy* **2015**, *150*, 80–96. [CrossRef]

18. International Energy Agency (IEA). *CO$_2$ Emissions from Fuel Combustion*, Highlights: Paris, France. November 2017.

19. Department of Economic and Social Affairs; United Nations. World Population Prospects. 2017. Available online: https://esa.un.org/unpd/wpp/Graphs/Probabilistic/POP/TOT/ (accessed on 29 March 2018).

20. Federal Electricity Commission (CFE); COPAR. Costos y Parámetros de Referencia para la Formulación de Proyectos de Inversión en el Sector Eléctrico. *Generation, Costs and Reference Parameters for the Formulation of Investment Projects in the Electricity Sector. Mexico*, Programmation Subdirection; Evaluation Coordination; México, 2012.

21. Intergovernmental Panel of Climate Change; Renewable Energy Sources and Climate Change Mitigation. *Special Report of the Intergovernmental Panel of Climate Change*, 1th ed.; Cambridge University: Cambridge, MA, USA, 2012.

22. Diwekar, U.M. *Introduction to Applied Optimization*, 2nd ed.; Springer: New York, NY, USA, 2008.

 *processes*

Article

Intelligent Energy Management for Plug-in Hybrid Electric Bus with Limited State Space

Hongqiang Guo *, Shangye Du, Fengrui Zhao, Qinghu Cui and Weilong Ren

School of Mechanical & Automotive Engineering, Liaocheng University, Liaocheng 252059, China;
max_becker@163.com (S.D.); isukee@163.com (F.Z.); cui_qinghu@163.com (Q.C.); a1440107906@163.com (W.R.)
* Correspondence: guohongqiang@lcu.edu.cn

Received: 26 August 2019; Accepted: 16 September 2019; Published: 28 September 2019

Abstract: Tabular Q-learning (QL) can be easily implemented into a controller to realize self-learning energy management control of a plug-in hybrid electric bus (PHEB). However, the "curse of dimensionality" problem is difficult to avoid, as the design space is huge. This paper proposes a QL-PMP algorithm (QL and Pontryagin minimum principle (PMP)) to address the problem. The main novelty is that the difference between the feedback SOC (state of charge) and the reference SOC is exclusively designed as state, and then a limited state space with 50 rows and 25 columns is proposed. The off-line training process shows that the limited state space is reasonable and adequate for the self-learning; the Hardware-in-Loop (HIL) simulation results show that the QL-PMP strategy can be implemented into a controller to realize real-time control, and can on average improve the fuel economy by 20.42%, compared to the charge depleting–charge sustaining (CDCS) strategy.

Keywords: plug-in hybrid electric bus; energy management; Q-learning; limited state space; Hardware-in-Loop (HIL) simulation

---

1. Introduction

Plug-in hybrid electric vehicle (PHEV) is a promising approach for energy-saving and lowering emissions, which would help the problems of energy shortage, global warming, and environment pollution [1]. Moreover, the advantage of the PHEV can be maximized by a well-designed energy management strategy (EMS) [2].

Rule-based control strategy is one of the most widely used methods in real-world situation, due to its easy implementation and real-time control performances [3]. Moreover, many investigations have demonstrated that the blended charge depletion (BCD) mode is the most efficient strategy, namely, the SOC can continuously decline to the expected value (such as 0.3) at the destination of route [4,5]. However, known driving conditions are usually indispensable for the BCD strategy, which brings great challenge to practical application.

Since the EMS can be taken as a nonlinearly constrained optimization control problem, the BCD mode can be well realized by optimal control methods [6]. It can be further classified into two categories: The optimization-based and the adaptive strategies [7]. The optimization-based strategies, such as dynamic programming (DP), Pontryagin's Minimum Principle (PMP), and Equivalent Consumption Minimization Strategy (ECMS), can obtain global optimization solutions [8–10]. However, driving conditions need to be known prior to use, which is the reason that they cannot be directly used in real-world situations and are usually only taken as the benchmark for other strategies. In contrast, the adaptive strategy has great potential in practical application [7]. For example, a model predictive control (MPC)-based EMS was proposed by combining Markov chain, DP, and a reference SOC plan method, based on the principle of receding horizon control [11]. An Adaptive PMP (A-PMP)-based-EMS was proposed by combining PI, PMP, and reference SOC plan method, based on the principle of feedback control [12]. An Adaptive ECMS (A-ECMS)-based-EMS was proposed by combining ECMS and particle swarm optimization (PSO) algorithms, where the equivalent factors (EFs) were firstly optimized, and then the actual EF was recognized based on a look-up table constituted by the optimized EFs [13]. A driving pattern recognition-based EMS was proposed based on PSO algorithm, where a series of typical reference cycles were firstly optimized off-line, then the corresponding optimal results were taken as the sampling sets for real-time control [14].

Reinforcement learning (RL) is an intelligent method that can be used in EMS, where Q-learning (QL) is the most popular method. It can be further classified into tabular and deep learning methods. The former can easily solve the self-learning problem, by employing a simplified Q-table. However, the state and the action should be discretized and the "curse of dimensionality" problem may be introduced once the state space is huge. In contrast, the Q-table will be substituted by neural network (NN) in the deep learning-based-EMS, and the control problem with continuous variable can also be solved. For the former, Ref. [15] proposed a Tabular QL-based EMS, where the required power, the velocity, and the SOC were taken as the states, meanwhile, the current of the battery and the shifting instruction of the automated mechanical transmission (AMT) were taken as the actions. Ref. [16] proposed a similar method by employing a Markov chain and a Kullback–Leibler (KL) divergence rate, where the power, the SOC and the state of voltage (SOC) were taken as the states, meanwhile, the current of the battery was taken as the action. Ref. [17] proposed a different Tabular QL-based EMS, where the SOC and the speed of the engine were taken as the states, and the throttle of the engine was taken as the action. For the latter, Ref. [18] proposed a deep QL (DQL)-based EMS based on NN, where the SOC, the engine power, the velocity and the acceleration were taken as the states, and the increment of the engine power was taken as the action. Ref. [19] proposed a similar DQL method, where the required torque and the SOC were taken as the states, and the engine torque was taken as the action. Ref. [20] proposed a different DQL-based EMS with AC (action–critic) framework, where the speed together with the torque of the wheel, the SOC together with the voltage of the battery, and the gear position of the AMT were taken as the states, meanwhile, the power of the motor was taken as the action. Ref. [21] proposed an interesting DQL-based EMS using a neural network, where the required power at the wheels, the SOC together with the distance to destination were taken as the states, and the power of the engine was taken as the action. Nevertheless, the existing tabular QL based-methods usually have huge state space, which is easy to result in the "curse of dimensionality" problem. On the other hand, the control performance of the DQL method may be greatly deteriorated once the fitting precision of NN is low. Moreover, the high computation burden of the deep learning may also restrict its application in currently used controllers.

This paper aims at solving the practical application problem of self-learning energy management for a single-parallel plug-in hybrid electric bus (PHEB). Since the shift instruction (discrete variable) and the throttle of the engine (continuous variable) can simultaneously influence the fuel economy of the vehicle, a mixed control variable with compact format is deployed into the control strategy. The main innovation of this paper is that a QL-PMP-based EMS is proposed, by combining the QL and the PMP algorithms, where the mixed control variables can be indirectly solved by the PMP using a self-learned co-state from the QL. More importantly, since the co-state is mainly dependent on the difference between the feedback SOC and the reference SOC, a limited state space with 50 rows and 23 columns is designed. Because the state space is greatly reduced, the QL-PMP algorithm can be directly implemented into controller.

The remainder of this paper is structured as follows. The configuration, parameters and models of the PHEB are described in Section 2. The QL-PMP algorithm is formulated in Section 3. The training process is discussed in Section 4. The Hardware-in-Loop (HIL) simulation results are detailed in Section 5, and the conclusions are drawn in Section 6.

2. The Configuration, Parameters, and Models of the PHEB

The single-parallel PHEB is shown in Figure 1, which includes an engine, a motor, an AMT, a clutch, and a battery pack.

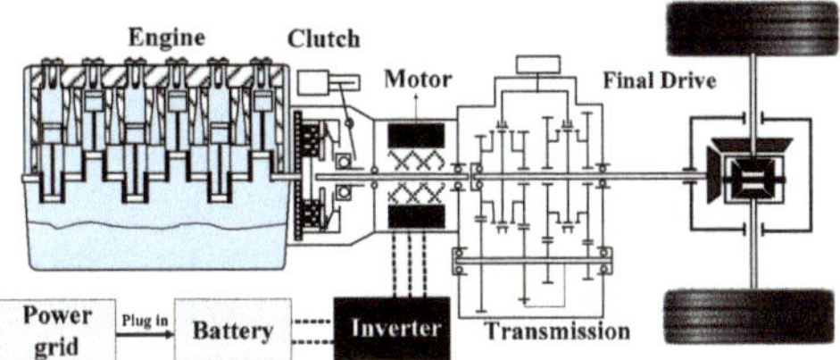

Figure 1. The configuration of the plug-in hybrid electric bus (PHEB).

The engine and the motor will work coordinately to provide the required power of the PHEB, and the battery pack is the energy source for the motor. The clutch can change the driving modes during the vehicle runs. In specific, the regenerative braking and the pure electric driving modes can be realized when the clutch is disengaged, otherwise, the hybrid driving with or without charging mode can be realized. In addition, the detailed parameters of the PHEB are shown in Table 1.

Table 1. The parameters of the PHEB. AMT: automated mechanical transmission.

Item	Description
Vehicle	Curb mass (kg): 8500
Passengers	Maximum number: 60; Passenger's mass (kg): 70
AMT	Speed ratios: 4.09:2.45:1.5:0.81
Final drive	Speed ratio: 5.571
Engine	Max torque (Nm): 639 Max power (kW): 120
Motor	Max torque (Nm): 604 Max power (kW): 94
Battery	Capacity (Ah): 35

2.1. Modeling the Engine

The modeling of the engine can be classified into theoretical and empirical methods. The former is formulated by combustion, fluid mechanics, and dynamic theories. As shown in Figure 2, the instantaneous fuel consumption of the engine can be interpolated by the brake specific fuel consumption (BSFC) map, based on Equation (1).

$$\dot{m}_e = \frac{T_e \cdot \omega_e \cdot b_e \cdot \Delta t}{3,600,000} \tag{1}$$

where $\dot{m}_e$ denotes the instantaneous fuel consumption of the engine; T_e denotes the torque of the engine; ω_e denotes the rotational speed of the engine; b_e denotes the fuel consumption rate of the engine, which can be obtained by the speed and torque of the engine; Δt denotes the sampling time.

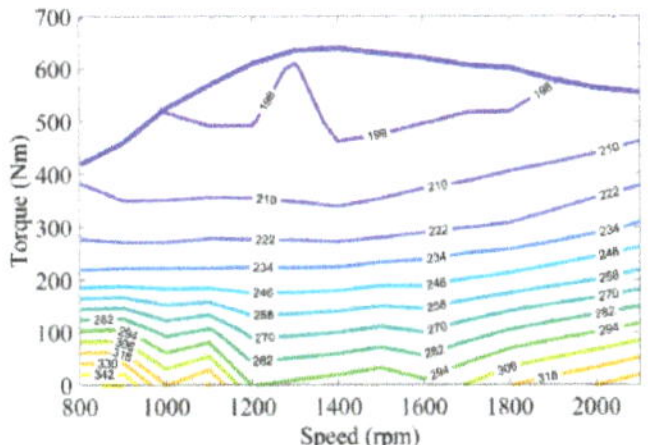

Figure 2. The BSFC map of the engine.

2.2. Modeling the Motor

Since only the working efficiency of the motor is need for the EMS, the empirical modeling method is adequate for the motor. In addition, because the motor can work in driving or regenerative mode, the motor model can be described as

$$P_m = \begin{cases} T_m \cdot \omega_m / \eta_m & T_m > 0 \\ T_m \cdot \omega_m \cdot \eta_g & T_m \leq 0 \end{cases} \tag{2}$$

where P_m denotes the power of the motor, T_m denotes the torque of the motor, ω_m denotes the rotational speed of the motor, and η_m and η_g denote the efficiency of the motor in driving mode and braking mode. When $T_m > 0$, motor works in driving mode, and when $T_m \leq 0$, motor works in braking mode. As shown in Figure 3, the power of the motor can be interpolated by the efficiency map of the motor, based on Equation (2).

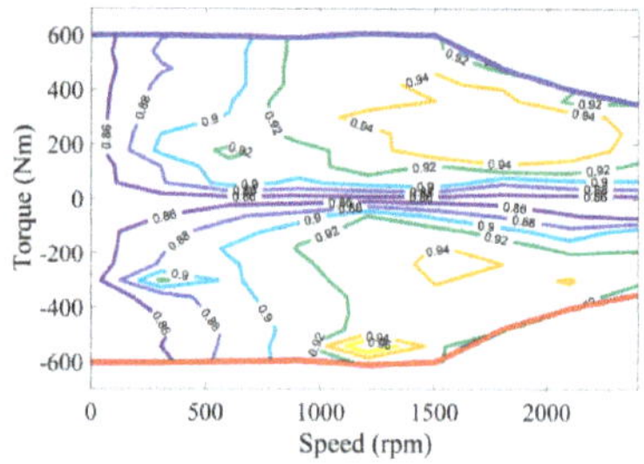

Figure 3. The efficiency map of the motor.

2.3. Modeling the Battery

For the energy management, the key issue of the battery is the estimation of the SOC, based on the voltage, the internal resistance and the current of the battery. Accordingly, a simplified battery model is deployed in Figure 4, and the power of the battery can be described as:

$$P_b = V_{oc}I_b - I_b^2 R_b \tag{3}$$

where P_b denotes the power of the battery, V_{oc} denotes the open-circuit voltage of the battery, I_b denotes the current of the battery and R_b denotes the internal resistance of the battery.

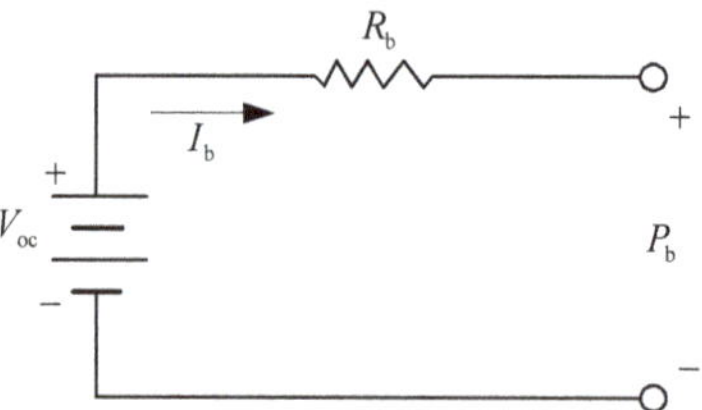

Figure 4. The simplified battery model.

2.4. The Dynamic Model of the Vehicle

Since only the required power is needed for the EMS, the model can be simplified to a lumped mass model, and only the longitudinal dynamic characteristic can be considered. Here, the driving and resistance forces can be described as:

$$F_t - F_R = \delta m \frac{dv}{dt} \tag{4}$$

where F_t denotes the driving force, δ denotes the rotating mass conversion factor, and m denotes the vehicle mass, which is constituted by the curb mass of the vehicle and the stochastic mass of the passengers, F_R denotes the resistance force, which is constituted by:

$$F_R = fmg \cos \alpha_s + mg \sin \alpha_s + \frac{1}{2} C_D A_a \rho_a v^2 \tag{5}$$

where f denotes the coefficient of the rolling resistance, g denotes the gravity acceleration, α_s denotes the road slope, C_D and A_a denote the coefficient of the air resistance and frontal area, respectively, ρ_a denotes the air density, and v denotes the velocity. The required power can be obtained by:

$$P_r = \frac{F_t \cdot v}{3600 \eta_t} \tag{6}$$

where P_r denotes the required power and η_t denotes the efficiency of the transmission system.

3. The Formulation of the QL-PMP Algorithm

Different from the general traffic environment, the route of the PHEB is fixed and repeatable. Therefore, a series of historical driving cycles can be downloaded from remote monitoring system (RMS). In addition, many investigations have demonstrated that the factors of the velocity and the road slope have great effect on the EMS, which essentially points at the importance of the required power of the vehicle [2,7]. In this case, the factor of the stochastic vehicle mass is also a significant factor for EMS based on the Equations (4)–(6). Accordingly, a series of combined driving cycles constituted by the historical driving cycles, the road slope and the stochastic distributions of the vehicle mass are firstly designed. As shown in Figure 5, the training of the QL-PMP requires three steps.

Step 1 A series of co-states with respect to the combined driving cycles are firstly optimized, by an off-line PMP with Hooke–Jeeves algorithm [22]. Then, the average co-state is obtained and the corresponding optimal SOCs are extracted. Finally, a reference SOC model is established by taking the normalized distance as input and the optimal SOCs as output.

Step 2 Based on the known average co-state and the reference SOC model, the training of the QL-PMP is carried out as follows: firstly, the Q-table (denoted by Q_0), that is defined as zeros matrix, will be trained with the combined driving cycle 1; secondly, the trained Q-table (denoted by Q_1) will be taken as the initial Q-table for the second training with the combined driving cycle 2; and then this process will be continued until the Q-table is adequately trained.

Step 3 Verifying the adequately trained QL-PMP algorithm with HIL platform, using different combined driving cycles.

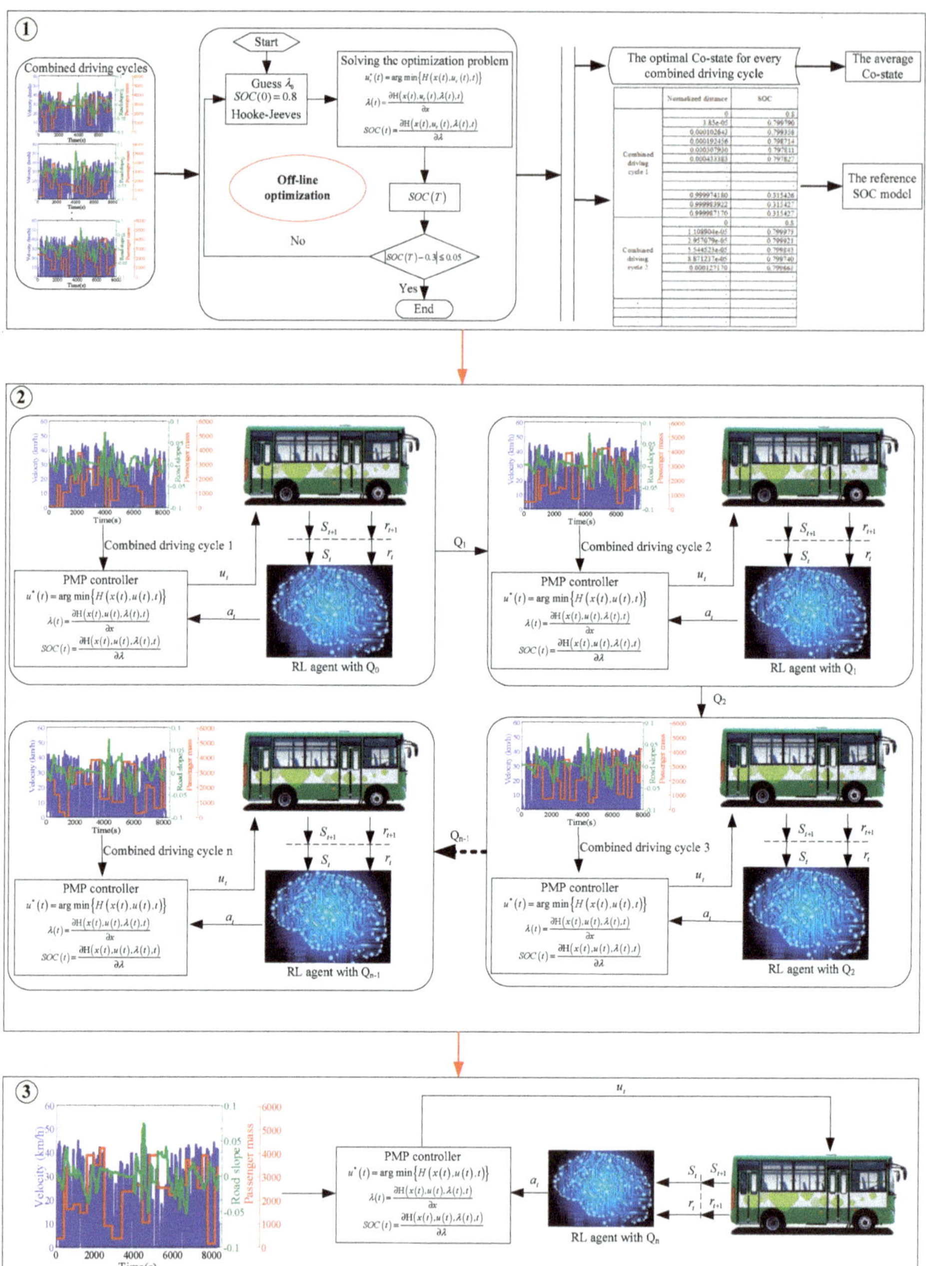

Figure 5. The QL-PMP algorithm (Q-learning and Pontryagin minimum principle (PMP)) algorithm.

3.1. The Combined Driving Cycle

In terms of the combined driving cycle, the historical driving cycle can be downloaded from the RMS, and the road slope can be obtained off-line based on the attitude of the road and the corresponding travelled distance. To simplify the EMS problem, the road slope is implemented into the controller in a prior process by designing a look-up table, through taking the travelled distance as input and the road slop as output. Similar to Ref. [23], the stochastic distributions of the vehicle mass were designed as follows:

Firstly, as shown in Figure 6, 25 road segments are defined based on the number of the neighbored bus stops. Because the distributions of the passenger in different road segments are stochastic, 25 factors with respect to the road segments were defined to describe the stochastic distribution of the vehicle mass over the bus route.

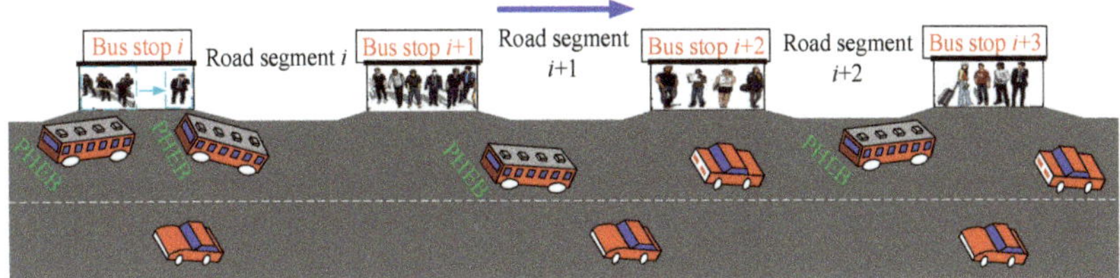

Figure 6. The route of the PHEB.

Secondly, assuming the stochastic distribution of the vehicle mass is reflected by the passenger mass, 60 levels are defined for each factor, based on the maximum passenger number. Moreover, to exhaustively probe the design space constituted by the 25 factors, the Optimal Latin hypercube design (Opt. LHD) is deployed, due to its good spatial filling and equalization performances. Finally, the combined driving cycles are constructed by stochastic matching of the historical driving cycles, the road slope and the stochastic distributions of passenger mass (kg) are generated. As shown in Figure 7, the blue line represents the velocity of the PHEB, the red line represents the total mass of passengers, and the green line represents the slope of the road. Besides, the stochastic vehicle mass is added by the curb vehicle mass and the stochastic distribution of passenger mass.

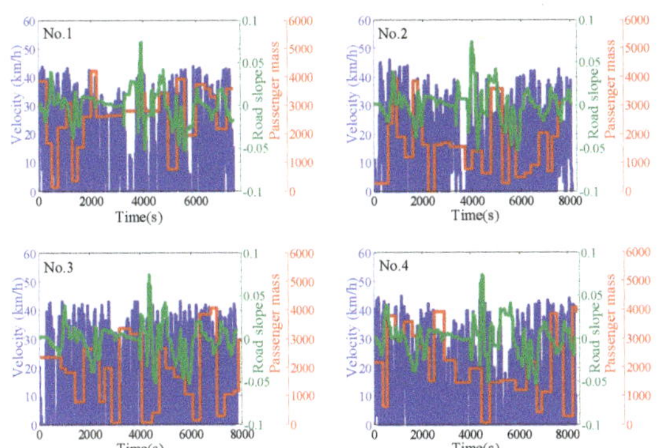

Figure 7. The combined driving cycles.

3.2. The Off-Line PMP Algorithm

Theoretically, EMS can be taken as an optimal control problem, which can make the vehicle reach the optimal state within a time domain through a series of discrete controls. Furthermore, since the bus route is fixed, and the BCD mode can be realized by the optimal control, and the terminal SOC can be easily controlled within the expected zone. This implies that the terminal SOC may be similar, despite of any combined driving cycle. Therefore, the electricity consumptions of the battery are similar for all of the combined driving cycles, and the minimum fuel consumption of the PHEB can be taken as one of the objectives. Moreover, the shift number should also be restricted to improve the control performance. In addition, only the state of the SOC is selected to simplify the control problem, and the control vector is reduced to a one-dimensional format constituted by the shift instruction of the AMT and the throttle of the engine, then the optimal control problem of the energy management can be described as:

$$\min J = \int_{t_0}^{t_f} \dot{m}_e(u(t)) + \beta|s(t)|dt$$

$$\text{S.t.}\begin{cases} \dot{SOC}(t) = -\dfrac{V_{oc} - \sqrt{V_{oc}^2 - 4R_{bat}P_{bat}}}{2R_{bat}Q_{nom}} \\ \omega_{e\_min} \le \omega_e(t) \le \omega_{e\_max} \\ \omega_{m\_min} \le \omega_m(t) \le \omega_{m\_max} \\ p_{e\_min} \le p_e(t) \le p_{e\_max} \\ p_{m\_min} \le p_m(t) \le p_{m\_max} \end{cases} \tag{7}$$

where J denotes the performance index function; $u(t)$ (Equation (8)) denotes the control vector, which is defined as a compacted format and is sampled by Opt. LHD. Here, the shift instruction is denoted by $s(t)$, and the throttle of the engine is denoted by $th(t)$. Moreover, $s(t)$ is defined as −1, 0 and 1, denoting the downshift, hold on and upshift, respectively; $th(t)$ is ranged from 0 to 1.

$$u(t) = \begin{bmatrix} s(t) \\ th(t) \end{bmatrix} \tag{8}$$

where the required power of the vehicle can be defined as P_r, $s(t)$ and $th(t)$ are changed with the state of P_r. If $P_r > 0$, it means that the vehicle runs in the driving state, the AMT can carry out the controls of the downshift, hold on and upshift, and the throttle of the engine can be ranged from 0 to 1; if $P_r = 0$, it means that the vehicle runs in the stopping state, the AMT will be downshifted to the lowest gear and the throttle of the engine is 0; if $P_r < 0$, the vehicle runs in the braking state, the gear of the AMT will be maintained and the throttle of the engine is 0.

In addition, $\dot{SOC}(t)$ denotes the state function, Q_{nom} denotes the battery capacity, $\omega_e(t)$ and $\omega_m(t)$ denote the rotate speeds of the engine and the motor, respectively, meanwhile, ω_{e_min}, ω_{e_max} and ω_{m_min}, ω_{m_max} denote the corresponding rotate speed boundaries of the $\omega_e(t)$ and $\omega_m(t)$, $p_e(t)$ and $p_m(t)$ denote the powers of the engine and the motor, meanwhile, p_{m_min}, p_{e_max} and p_{m_min}, p_{m_max} denote the corresponding power boundaries.

Finally, the Hamilton function can be descried as:

$$\begin{cases} u^*(t) = \text{argmin}\{H(x(t), u(t), t)\} \\ H(x(t), u(t), \lambda(t), t) = \dot{m}(u(t)) + \beta|s(t)| + \lambda(t) \cdot \dot{SOC} \end{cases} \tag{9}$$

where $u^*(t)$ denotes the optimal control vector, $H(x(t), u(t), t)$ denotes the Hamiltonian function and $\lambda(t)$ denotes the co-state.

As shown in Figure 5, the off-line optimization of the PMP can be formulated by

$$\begin{aligned} \min \quad & F_{PMP} = f(\lambda)\tfrac{1}{2} \\ \text{S.t} \quad & 2000 \le \lambda \le 3000 \\ & 0.27 \le SOC_f \le 0.33 \end{aligned} \tag{10}$$

where F_{PMP} denotes the objective function of the Hooke–Jeeves, and SOC_f denotes the terminal SOC, and is defined as a soft constraint to accelerate the convenience of the optimization. Moreover, the offline optimization of the PMP can be categorized into three steps.

Step 1 Taking the co-state as independent value of the Hooke–Jeeves and guessing an initial value of the co-state with a defined initial SOC value (0.8).

Step 2 Solving the dynamic optimal control problem by Equations (7)–(9), based on the given independent value from the Hooke–Jeeves.

Step 3 If the terminal SOC value satisfies the optimization objective, the co-state will be taken as the optimal co-state, otherwise, the iteration will be repeated from Step 1 to Step 3, till the objective is satisfied.

3.3. The Reference SOC Model

The most important innovation of this paper is the employment of a small state space for the Q-table, which is dependent on the difference between the feedback SOC and the reference SOC. Therefore, a reference SOC model should be designed. Since a series of optimal SOCs with respect to different combined driving cycles can be calculated off-line and the city bus route is fixed, the reference SOC model can be constructed by taking the normalized distance as input and the optimal SOC as output, based on partial least squares (PLS) method [23]. Here, the normalized distance can be described as

$$x(t) = \frac{d_{real}(t)}{d_{whole}(t)} \tag{11}$$

where $x(t)$ denotes the normalized distance, $d_{whole}(t)$ denotes the total distance and $d_{real}(t)$ denotes the travelled distance, which can be described as:

$$d_{real}(t) = \sum_{t=1}^{k}\left(v(t)\cdot\Delta t + \frac{1}{2}\cdot a(t)\cdot\Delta t^2\right) \tag{12}$$

The general form of the reference SOC model is described as:

$$SOC_r(t) = \frac{p_1+p_3x(t)+p_5x(t)^2+p_7x(t)^3+p_9x(t)^4+p_{11}x(t)^5+p_{13}x(t)^6+p_{15}x(t)^7+p_{17}x(t)^8+p_{19}x(t)^9}{1+p_2x(t)+p_4x(t)^2+p_6x(t)^3+p_8x(t)^4+p_{10}x(t)^5+p_{12}x(t)^6+p_{14}x(t)^7+p_{16}x(t)^8+p_{18}x(t)^9} \tag{13}$$

where $SOC_r(t)$ denotes the reference SOC; $p_i(i = 1, 2, 3\cdots 19)$ denotes the fitting coefficient. Based on the method in Ref. [23], the parameters of the reference SOC model can be obtained, and are shown in Table 2:

Table 2. The parameters of the reference state of charge (SOC) model.

p_1	p_2	p_3	p_4	p_5	p_6	p_7	p_8	p_9	p_{10}
0.7910	−8.0413	−6.3492	24.939	13.587	−53.294	15.148	143.58	−72.790	−162.76

p_{11}	p_{12}	p_{13}	p_{14}	p_{15}	p_{16}	p_{17}	p_{18}	p_{19}
77.584	−73.520	−46.524	74.766	−23.603	90.256	73.1139	36.364	−8.6263

As shown in Figure 8, a series of combined driving cycles from No.13 to No.20 are also designed to further verify the predictive precision of the reference SOC model.

As shown in Figure 9, the optimal SOC trajectories extracted from the off-line optimization fluctuate around the predicted SOC trajectory. As shown in Figure 10, the relative errors between the predicted SOC and the optimal SOCs are ranged from −0.089 to 0.0257, which implies that the reference SOC trajectory cannot be well predicted. Therefore, it is no requirement for the feedback SOC trajectory to strictly track the reference SOC trajectory. On the contrary, better fuel economy may be realized if it makes the feedback SOC fluctuate around the reference SOC, compared to the strictly tracking control.

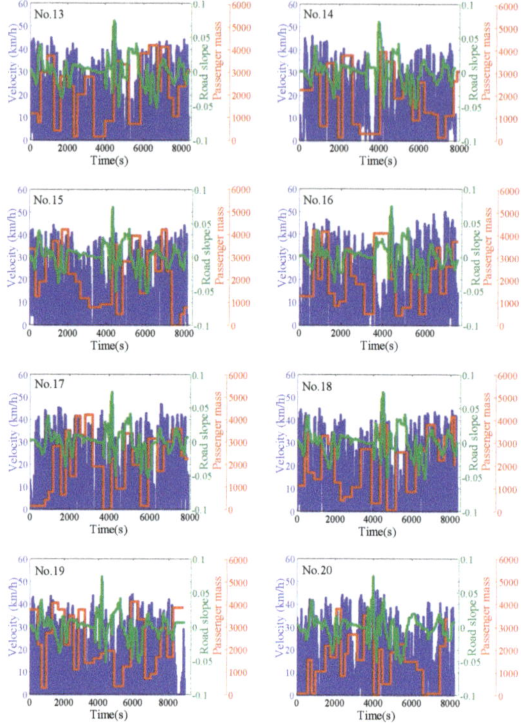

Figure 8. The combined driving cycles from No. 13 to No. 20.

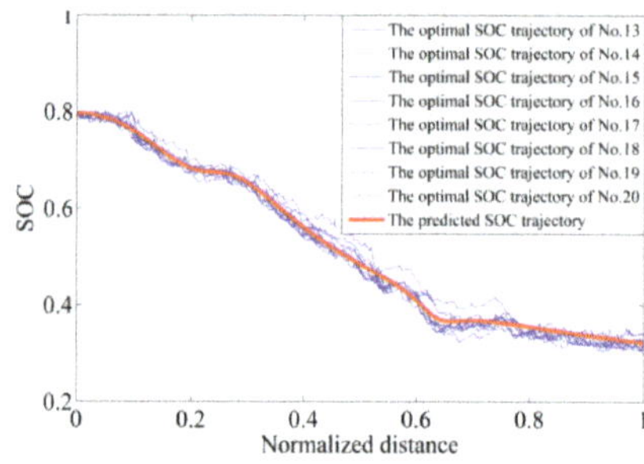

Figure 9. The comparison between the reference SOC and the optimal SOC trajectories.

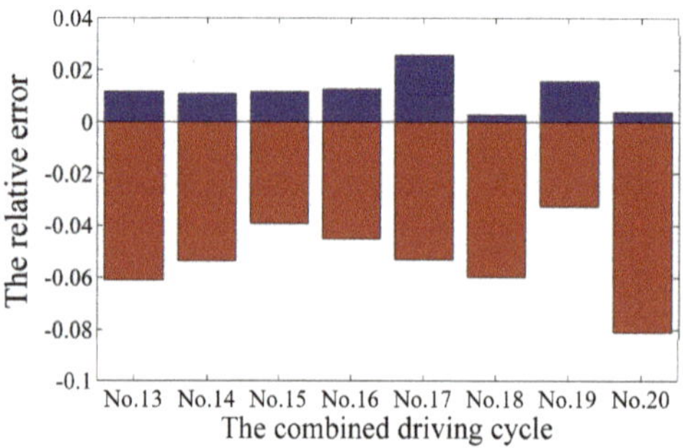

Figure 10. The relative errors between the reference and the optimal SOCs.

3.4. The QL-PMP Algorithm

QL is a value-based algorithm, which is also one of the most important temporal difference (TD) algorithms. It is defined by 4-tuple (S, A, R, γ), where S denotes the state space; A denotes the action space; R denotes the reward function; γ denotes the discount factor, which is defined as 0.8. Tabular QL is one of the most widely used methods, where the state-action function $(Q(s, a))$ is indispensable. The Q-table is the concrete manifestation of the $Q(s, a)$, which is designed to realize the mapping from the state to the action, and can be updated by:

$$Q(s_k, a_k) \leftarrow Q(s_k, a_k) + \alpha_l[r_{k+1} + \gamma Q(s_{k+1}, a_{k+1}) - Q(s_k, a_k)] \tag{14}$$

where $Q(s_k, a_k)$ denotes the old reward; α_l denotes the learning rate, which is defined as 1; $Q(s_{k+1}, a_{k+1})$ denote the maximum reward in future.

In the existing RL-based energy managements, the factors such as the current SOC, the required power, and the velocity are usually designed as the states, and should be discretized intensively to probe good action. This may lead to the "curse of dimensionality" problem, owing to the employment of a large state space in the Q-Table. In this paper, the difference between the reference SOC and the feedback SOC is exclusively designed as the state for the QL-PMP, and there are three reasons for this:

Firstly, since the reference SOC model can provide the corresponding reference SOC at any time step and the purpose of the QL-PMP is to make the feedback SOC track the reference SOC, the state of the difference between the reference SOC and the feedback SOC is a reasonable choice.

Secondly, the adaptive energy management control can be realized by making the feedback SOC fluctuate around the reference SOC, through adjusting the co-state. This implies that the difference of the SOC between the feedback SOC and the reference SOC can be taken as the sole state for the QL-PMP algorithm.

Thirdly, the range of the difference of the SOC is small, which means that the algorithm can find the optimal action with limited state space.

Based on the above discussion, the state of the difference of the SOC is ranged from −0.1 to 0.1, and are uniformly discretized to 50 points. The action is designed as the co-state, and is non-uniformly discretized by 23 points, namely, [−1000, −512, −256, −128, −64, −32, −16, −8, −4, −2, −1, 0, 1, 2, 4, 8, 16, 32, 64, 128, 256, 512, 1000]. The reward function is designed as:

$$r_{ss'}^a = \begin{cases} -\mathrm{abs}(0.8 - SOC(k+1)) & \text{if } SOC(k+1) > 0.8 \\ \frac{1}{\mathrm{abs}(SOC(k+1) - SOC_r(k+1))} & \text{if } SOC(k+1) <= 0.8 \ \& \ SOC(k+1) >= 0.3 \\ -\mathrm{abs}(0.3 - SOC(k+1)) & \text{if } SOC(k+1) < 0.3 \end{cases} \tag{15}$$

where $r_{ss'}^a$ denotes the reward function. It means that if the $SOC(k+1)$ is bigger than 0.8, the punishment will be applied, and the farther from 0.8, the bigger punishment will be; if the $SOC(k+1) <= 0.8 \& SOC(k+1) >= 0.3$, the smaller of the difference of the SOC, the bigger reward will be, otherwise, if the $SOC(k+1) < 0.3$, the punishment will be applied, and the farther from 0.3, the bigger punishment will be.

As shown in Figure 5, at every time step, the RL agent will firstly receive the state (denoted by $\Delta SOC(k)$). Then, an action (denoted by $a(k)$) will be carried out, based on the current state and the greedy algorithm. In specific, if the random number is larger than the threshold value, the action will be selected by the random number, otherwise, it will be selected by the $Q(s_k, a_k)$, by finding the maximum value of the action-state value. Finally, the next state and the $Q(s_k, a_k)$ will be undated based on the selected action. It is worth mentioning that the co-state applied to the PMP algorithm is formulated by

$$\lambda_k = \lambda_{ave} + a_k \tag{16}$$

where λ_{ave} denotes the average co-state. The Algorithm 1 can be described as

---

Algorithm 1. The QL-PMP Algorithm

1: Initialize $Q(s,a) \leftarrow 0$ for all s
2: Initialize $R(s,a) \leftarrow 0$ for all s
3: Initialize $SOC(1) \leftarrow 0.8$
4: Initialize $\lambda \leftarrow \lambda_{ave} + a_1$
5: Repeat
6: Observe $s_k = \{\Delta SOC(k)\}$
7: Generate a random number $x \in [0,1]$
8: **if** $x > \varepsilon$
   select an action randomly $a_k = \{\lambda_k\}$
   else
   choose the optimal action by $a_k = \max(Q(s_k, a_k))$
   end
9: Calculate $\Delta SOC(k+1)$
10: Update the state-value function
   $Q(s_k, a_k) \leftarrow Q(s_k, a_k) + \alpha[r_{k+1} + \gamma Q(s_{k+1}, a_{k+1}) - Q(s_k, a_k)]$
11: $k \leftarrow k+1$
12: Until simulation stop

---

4. The Training Process

4.1. The Average Co-State

As shown in Figure 11, the co-states obtained from off-line optimizations are distributed evenly, and the values range from 1960 to 2280. The average co-state is 2202, which is calculated by the sum of the co-sates divided by the number of the combined driving cycles.

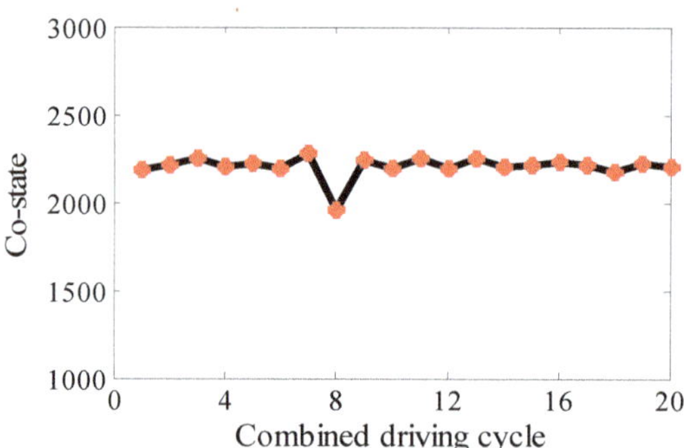

Figure 11. The co-state with different combined driving cycles.

4.2. The Off-Line Training

To train the Q-table of the QL-PMP, the combined driving cycles from No.13 to No.19 in Figure 8 are deployed. The complete training process is shown in Figures 12–17. The left figures denote the training results of the training episodes from 1 to 35, with the greedy factor of 0.3, the middle figures denote the training results of the training episodes from 36 to 45, with the greedy factor of 0.55, and the right figures denote the training results of the training episodes from 46 to 50, with the greedy factor of 0.9. The greedy factor 0.3 means that the QL-PMP tends to explore a new action to maximize the reward. The greedy factor 0.55 means that the QL-PMP tends to balance the chance between the exploration and the best action. The greedy factor 0.9 means that the QL-PMP tends to use the best action to realize the optimal control whilst providing a chance to find a better action.

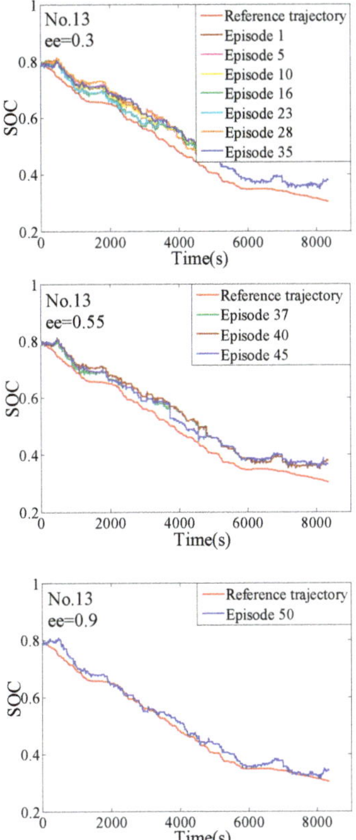

Figure 12. The training process of combined driving cycle 13.

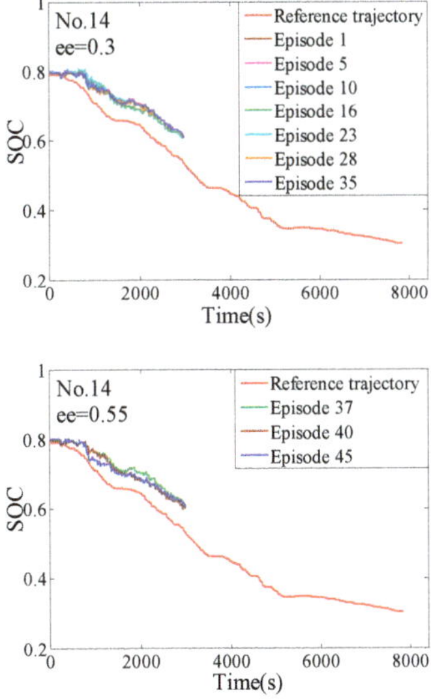

Figure 13. *Cont.*

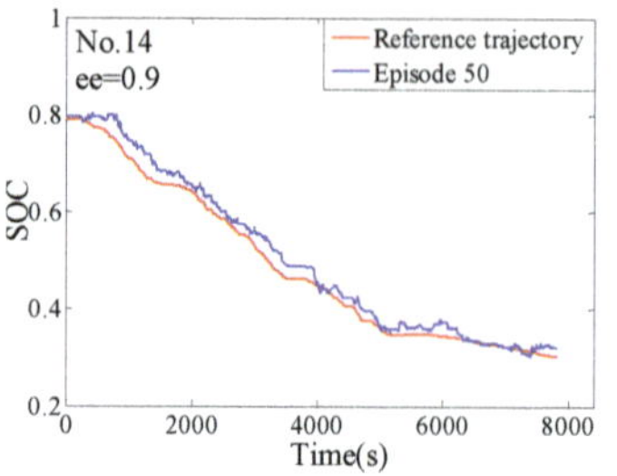

Figure 13. The training process of combined driving cycle 14.

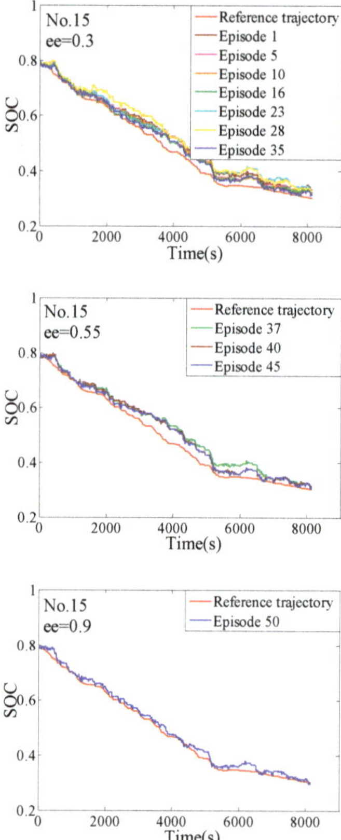

Figure 14. The training process of combined driving cycle 15.

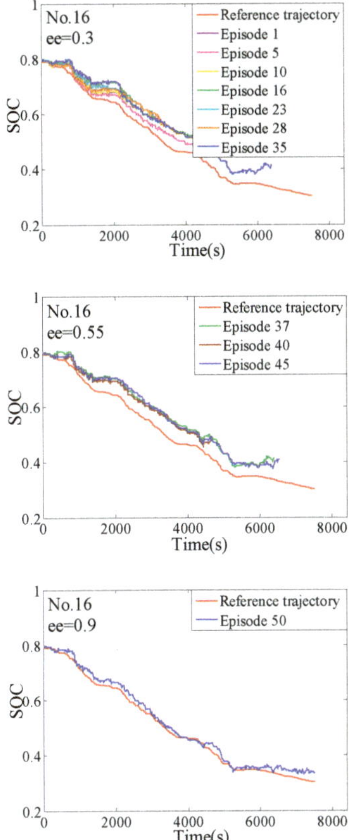

Figure 15. The training process of combined driving cycle 16.

As shown in Figure 12, for the training results of the combined driving cycle 13, the difference between the feedback SOC and the reference SOC is large when the greedy factor is 0.3, and the training fails all the time till the training reaches to episode 35. Moreover, although the feedback SOC trajectory is reasonable at the episode 35, the feedback SOC trajectory severely deviates from the reference trajectory. The QL-PMP becomes better and better with the continuous training, but is far from reliable When the greedy factor is 0.55, most of the feedback SOC trajectories can meet the requirements, but they are also far away from the reference SOC trajectory. When the greedy factor is 0.9, the feedback SOC trajectory of the episode 50 is reasonable and is close to the reference SOC trajectory. This implies that the trained Q-table is reasonable for the combined driving cycle 13.

As shown in Figure 13, for the training results of the combined driving cycle 14, the trainings are always fail when the episode is lower than 45 with the Q_1 (the trained Q-table with the combined driving cycle 13), which means that the Q_1 is not suitable for the combined driving cycle of 14, and the Q-table should be continually trained. Fortunately, the Q-table can be trained well enough through 45 trainings. It is worth noting that the failings of the feedback SOC trajectories do not mean that the Q_1 is not trained well, the failings are also attributed by the smaller greedy factor.

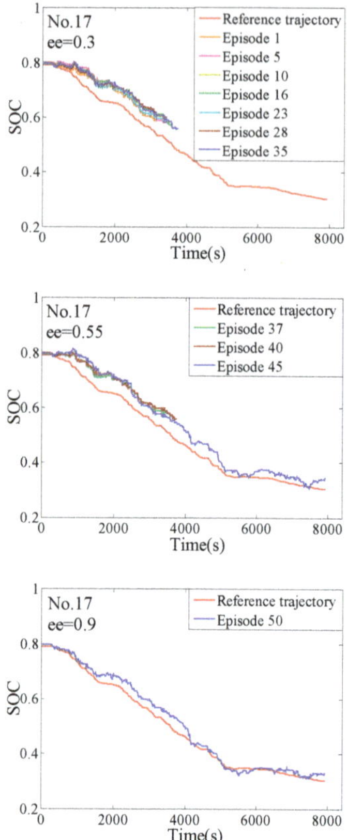

Figure 16. The training process of combined driving cycle 17.

As shown in Figure 14, for the training results of the combined driving cycle 15, the feedback SOC trajectories are always reasonable for any greedy factor and any episode. This implies that the Q_2 (the trained Q-table with the combined driving cycle 14) is suitable for this combined driving cycle. Moreover, the feedback SOC trajectory of the episode 50 with the greedy factor of 0.9 is closer to the reference SOC trajectory than others, which shows that the Q-table becomes more and more reliability through the trainings from episode of 1 to 45.

As shown in Figure 15, for the training results of the combined driving cycle 16, the feedback SOC trajectories always fail when the greedy factor is 0.3. However, the control performance is continuously improved with the increasing of training number. Similarly, the trainings still fail despite the greedy factor being 0.55. This implies that the driving conditions are different from the above combined driving cycles. Therefore, the Q-table should be continuously trained based on Q_3 (the trained Q-Table with the combined driving cycle 15). However, when the episode reaches to 50, the feedback SOC trajectory is reasonable and close to the reference SOC trajectory, which implies that the Q-table has been trained well.

As shown in Figure 16, for the training results of the combined driving cycle 17, although the Q_4 (the, trained Q-Table with the combined driving cycle 16) has been well trained, the training fails before episode 44. This implies that the Q_4 still not be suitable for all of the driving conditions. However, the Q_5 (the trained Q-Table with the combined driving cycle 17) can be suitable for the combined driving cycle 17 through a series of training.

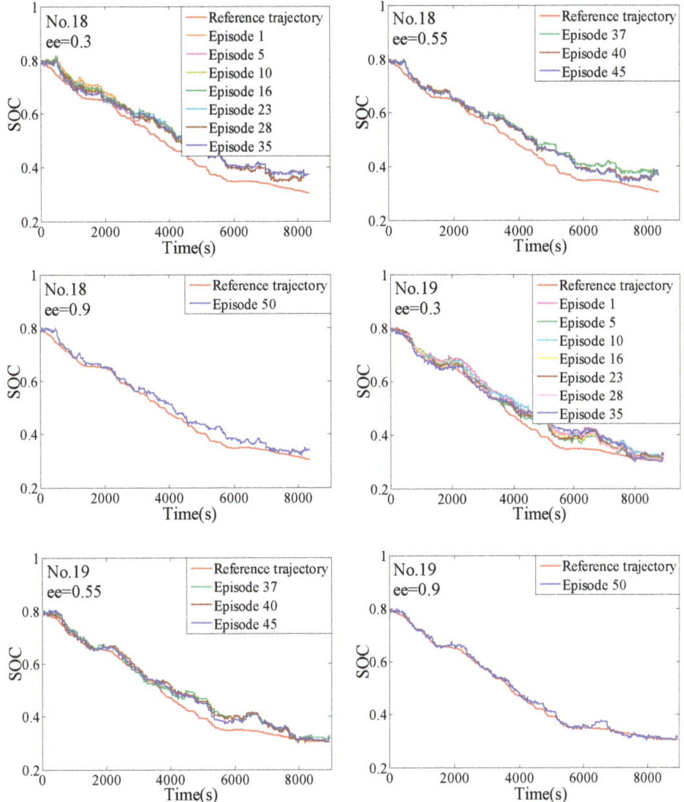

Figure 17. The training process of combined driving cycle 18 and 19.

As shown in Figure 17, the training results of the combined driving cycle 18 and 19 have demonstrated that Q_5 is well trained. The QL-PMP can be directly applied to the on-line control.

5. The Hardware-In-Loop Simulation

5.1. The Introduction of the HIL

To further verify the real-time and robust control performances of the EMS, a hardware-in-loop simulation (HIL) platform was also developed, based on the D2P rapid prototype control system. The D2P development platform mainly includes the MotoHawk development software and the MotoTune debugging software, where the former can bridge the gap between the control code and the MATLAB code; the latter is the debugging software for downloading and viewing the programs.

As shown in Figure 18, The Target Definition is the model selection and configuration module; the Main Power Relay is the main relay control module; the Trigger is the trigger module. In this paper, the controller type is ECM-5554-112-0904-Xd (DEV), the microprocessor is the MPC5554 of Freescale with 32-bit, and the frequency is 80 MHz. In terms of the EMS, the controlled models such as the engine, the motor and the battery are modeled as Simulink models, the algorithm of the QL-PMP will be implemented into the Hybrid Control Unit (HCU). Specifically, the QL-PMP algorithm will be firstly compiled and generated into the product-level code. Then, the code will be implemented into the HCU and the real-time communication between the Simulink models and the HCU will be built through the CAN bus.

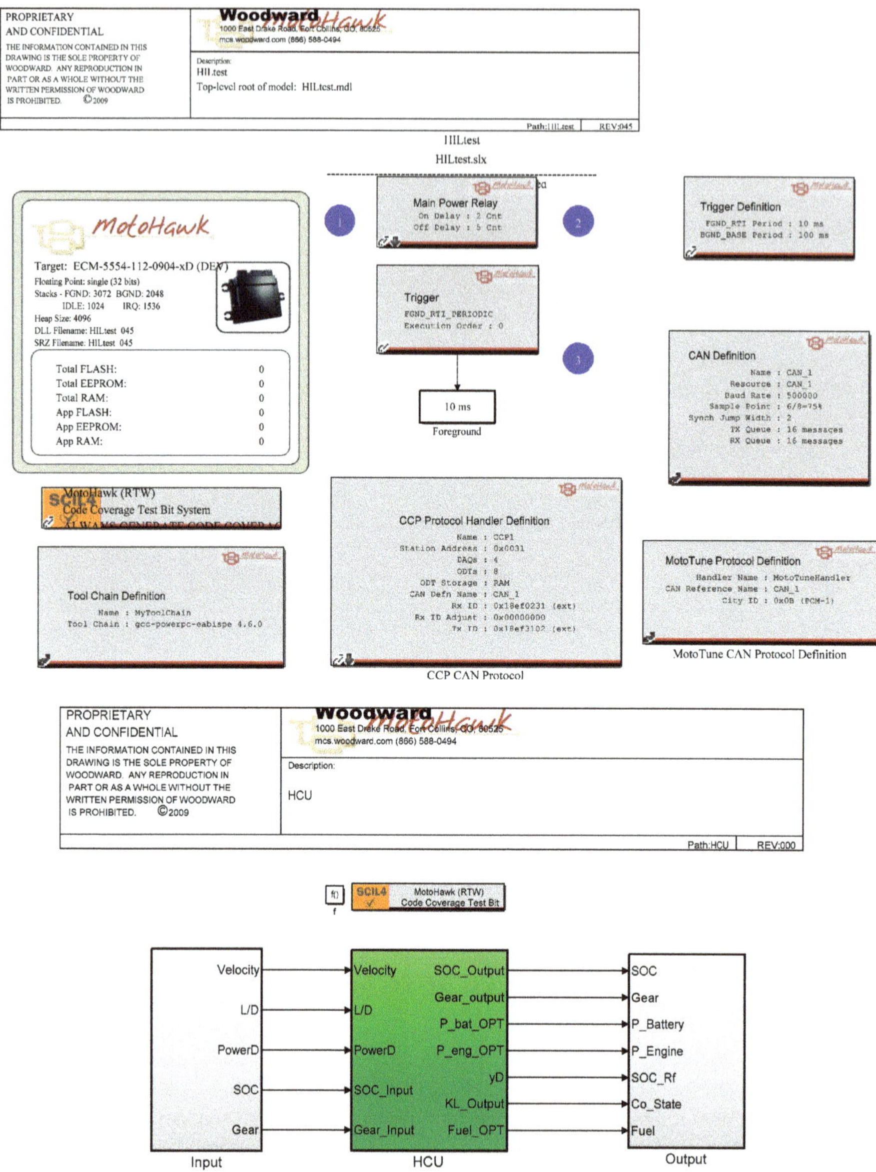

Figure 18. The MotoHawk-project interface.

5.2. The Verification of the HIL

As shown in Figure 19, the computer is the upper computer, which sends the signals of the Simulink models and monitors the signals transmitted from the HCU, through Kvaser. As shown in Figure 20, a series of combined driving cycles from No.21 to No.24 were designed to further verify the reliability of the Q-table and evaluate the fuel economy of the QL-PMP. In addition, a rule-based energy management (named by CDCS control strategy) which is characterized by CD followed by CS control strategy was also deployed to evaluate the fuel economy of the QL-PMP.

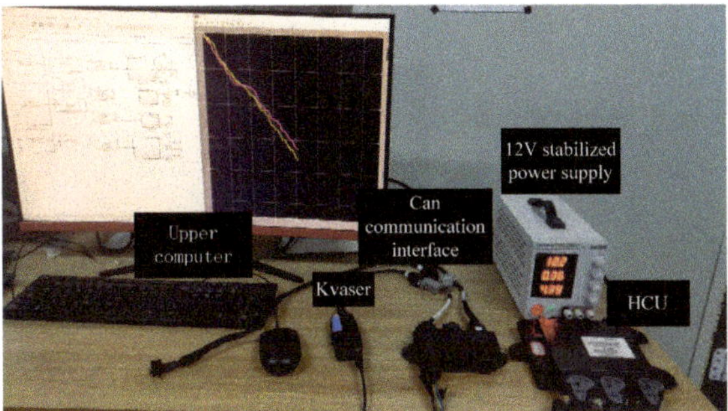

Figure 19. The Hardware-in-Loop (HIL) platform.

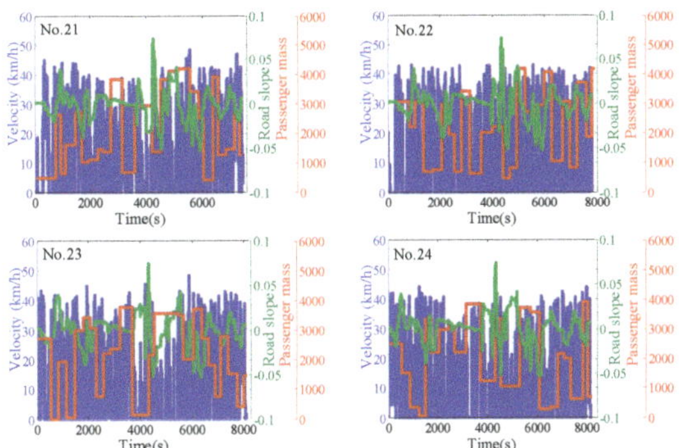

Figure 20. The combined driving cycles from No.21 to No.24.

As shown in Figures 21–24, (a) denotes the gears, (b) denotes the powers of the motor and the engine, (c) denotes the co-state, and (d) denotes the SOC trajectories. From (a), it can be seen that the gears of the AMT tended to maintain high gear to improve the fuel economy of the PHEB. From (b), it can be seen that the engine and the motor work coordinately to provide the required power of the vehicle, meanwhile, the motor works in driving or regenerative braking modes based on the driving conditions. From (c), it can be seen that the co-states are adjusted all the time to make the feedback SOC track the reference SOC trajectory. From (d), it can be seen that the feedback SOC can track the reference SOC trajectory well, despite of any combined driving cycle. Moreover, the SOC trajectory of the CDCS control strategy is far away from the reference SOC trajectory. At the same time, the CDCS control strategy firstly works in the CD mode, and then works in the CS mode.

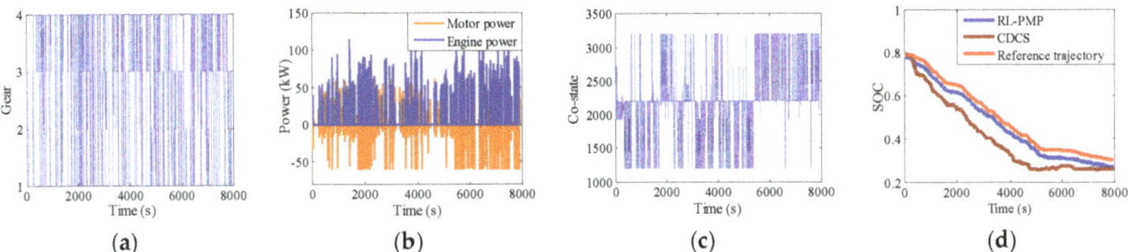

(a) (b) (c) (d)

Figure 21. The HIL results of No.21. (**a**) denotes the gears, (**b**) denotes the powers of the motor and the engine, (**c**) denotes the co-state, and (**d**) denotes the SOC trajectories.

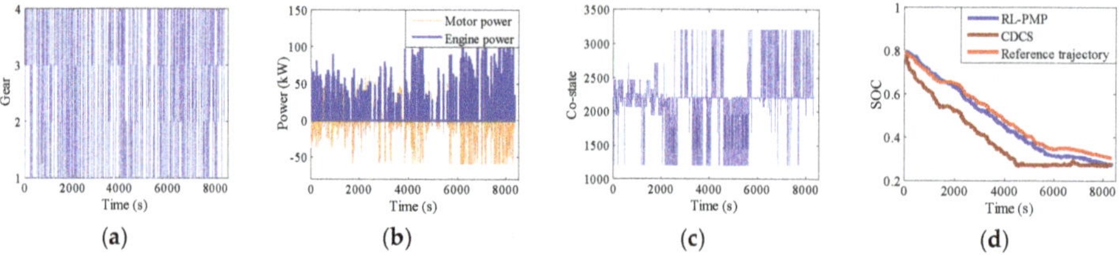

Figure 22. The HIL results of No.22. (**a**) denotes the gears, (**b**) denotes the powers of the motor and the engine, (**c**) denotes the co-state, and (**d**) denotes the SOC trajectories.

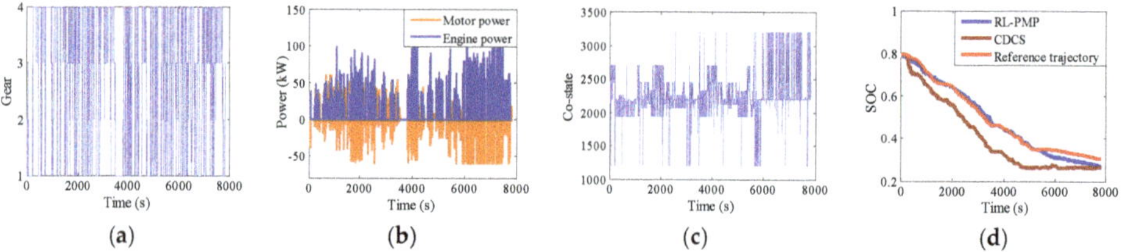

Figure 23. The HIL results of No.23. (**a**) denotes the gears, (**b**) denotes the powers of the motor and the engine, (**c**) denotes the co-state, and (**d**) denotes the SOC trajectories.

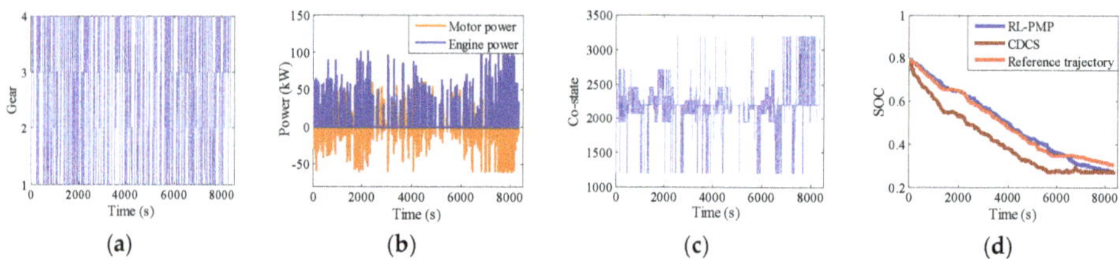

Figure 24. The HIL results of No.24. (**a**) denotes the gears, (**b**) denotes the powers of the motor and the engine, (**c**) denotes the co-state, and (**d**) denotes the SOC trajectories.

As shown in Figure 25, the fuel consumption of QL-PMP is significantly lower than the CDCS control strategy for any combined driving cycle, which further demonstrates the good economic performance of the QL-PMP algorithm. In specific, compared to the CDCS control strategy, the fuel economy can be improved by 23.01%, 23.57%, 14.13%, and 20.96% respectively. In a word, it can be improved by an average of 20.42%.

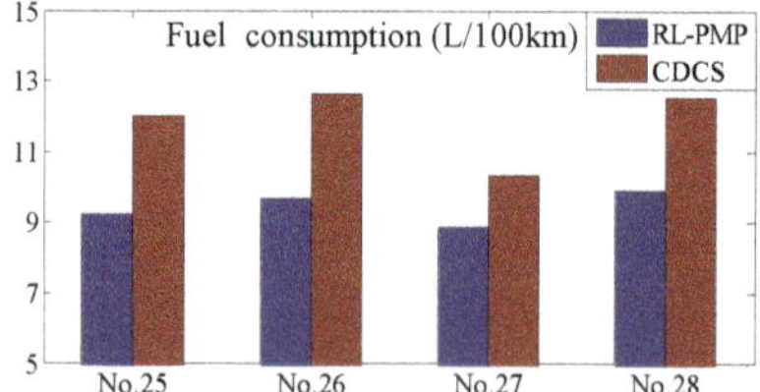

Figure 25. The fuel consumptions of the HIL verification.

6. Conclusions

This paper proposes a QL-PMP based- energy management with a small state space, which can realize real-time control in dynamically stochastic traffic circumstance. The conclusions are summarized as follows:

1. A self-learning energy management constituted by the QL and the PMP algorithms are proposed. The actual control variables of the throttle of the engine and the shift instruction of the AMT are exclusively solved by the PMP and the co-state in the PMP is designed as the action in the QL, which can effectively reduce the dependence of the action on the state and give a chance for the reduced state space design.
2. A limited state space constituted by the difference between the feedback SOC and the reference SOC is proposed. Since the co-state is designed as the action of the QL-PMP and has weak dependence on the state, only a Q-matrix with 50 rows and 25 columns are adequate for the Q-Table. This also gives a chance for the real-time energy management control in real-world.
3. The complete training process of the QL-PMP algorithm and the HIL verification are presented. The training process show that the QL-PMP can realize the self-learning energy management control for any combined driving cycle, where the control performance can be continuously strengthened. The verification of HIL demonstrates that the completely trained QL-PMP has high reliability, and can realize the real-time control, despite of any combined driving cycle. Moreover, the fuel economy can be improved by 20.42%, compared to CDCS control strategy.

Author Contributions: H.G. put forward the idea of intelligent energy management strategy, and analyzed the context of the whole paper. S.D. was mainly responsible for the writing and editing. F.Z., Q.C. and W.R. were mainly responsible for the data collecting, processing and analysis.

Funding: This research was funded by [the Natural Science Foundation of Shandong Province, China] grant number [No. ZR2014EL023].

Conflicts of Interest: The authors declare no conflict of interest.

Nomenclature

$\dot{m}_e$	the instantaneous fuel consumption of the engine
T_e	the torque of the engine
b_e	the fuel consumption rate of the engine
Δt	the sampling time
λ_{ave}	the average co-states
T_m	the torque of the motor
γ	the discount factor
η_m	the efficiency of the motor in driving mode
η_g	denote the efficiency of the motor in braking mode
P_b	the power of the battery
V_{oc}	the open-circuit voltage of the battery
I_b	the current of the battery
R_b	the internal resistance of the battery
F_t	the driving force
δ	the rotating mass conversion factor
m	the vehicle mass
F_R	the resistance force
f	the coefficient of the rolling resistance
g	the gravity acceleration
α_s	the road slope
C_D	the coefficient of the air resistance
A_a	the frontal area
ρ_a	the air density
v	the velocity
P_r	the required power
η_t	the efficiency of the transmission system
H	the Hamiltonian function
$S\dot{O}C$	the state function
Q_{nom}	the battery capacity

ω_e	the rotate speed of the engine
ω_m	the rotate speed of the motor
ω_{m_min}	the lower boundary of the ω_m
ω_{m_max}	the higher boundary of the ω_m
ω_{e_min}	the lower boundary of the ω_e
ω_{e_max}	the higher boundary of the ω_e
p_e	the power of the engine
p_m	the power of the motor
p_{m_min}	the lower boundary of the p_m
p_{m_max}	the higher boundary of the p_m
p_{e_max}	the higher boundary of the p_e
p_{e_min}	the lower boundary of the p_e
u^*	the optimal control vector
λ	the co-state
F_{PMP}	the objective function
SOC_f	the terminal SOC
x	the normalized distance
d_{whole}	the total distance
d_{real}	the travelled distance
SOC_r	the reference SOC
p_i	the fitting coefficient
$Q(s_k, a_k)$	the old reward
α_l	the learning rate
$Q(s_{k+1}, a_{k+1})$	the maximum reward in future
$r^a_{ss'}$	the reward function
s	the shift instruction
S	the state space
A	the action space

References

1. Xie, S.; Hu, X.; Xin, Z.; Brighton, J. Pontryagin's Minimum Principle based model predictive control of energy management for a plug-in hybrid electric bus. *Appl. Energy* **2019**, *236*, 893–905. [CrossRef]
2. Huang, Y.; Wang, H.; Khajepour, A.; Li, B.; Ji, J.; Zhao, K.; Hu, C. A review of power management strategies and component sizing methods for hybrid vehicles. *Renew. Sustain. Energy Rev.* **2018**, *96*, 132–144. [CrossRef]
3. Yan, M.; Li, M.; He, H.; Peng, J.; Sun, C. Rule-based energy management for dual-source electric buses extracted by wavelet transform. *J. Clean. Prod.* **2018**, *189*, 116–127. [CrossRef]
4. Padmarajan, B.V.; McGordon, A.; Jennings, P.A. Blended rule-based energy management for PHEV: System structure and strategy. *IEEE Trans. Veh. Technol.* **2016**, *65*, 8757–8762. [CrossRef]
5. Pi, J.M.; Bak, Y.S.; You, Y.K.; Park, D.H.; Kim, H.S. Development of route information based driving control algorithm for a range-extended electric vehicle. *Int. J. Automot. Technol.* **2016**, *17*, 1101–1111. [CrossRef]
6. Sabri, M.F.M.; Danapalasingam, K.A.; Rahmat, M.F. A review on hybrid electric vehicles architecture and energy management strategies. *Renew. Sustain. Energy Rev.* **2016**, *53*, 1433–1442. [CrossRef]
7. Zhou, Y.; Ravey, A.; Péra, M.C. A survey on driving prediction techniques for predictive energy management of plug-in hybrid electric vehicles. *J. Power Source* **2019**, *412*, 480–495. [CrossRef]
8. Tie, S.F.; Tan, C.W. A review of energy sources and energy management system in electric vehicles. *Renew. Sustain. Energy Rev.* **2013**, *20*, 82–102. [CrossRef]
9. Hou, C.; Ouyang, M.; Xu, L.; Wang, H. Approximate Pontryagin's minimum principle applied to the energy management of plug-in hybrid electric vehicles. *Appl. Energy* **2014**, *115*, 174–189. [CrossRef]
10. Tulpule, P.; Marano, V.; Rizzoni, G. Energy management for plug-in hybrid electric vehicles using equivalent consumption minimization strategy. *Int. J. Electr. Hybrid Veh.* **2010**, *2*, 329–350. [CrossRef]
11. Li, G.; Zhang, J.; He, H. Battery SOC constraint comparison for predictive energy management of plug-in hybrid electric bus. *Appl. Energy* **2017**, *194*, 578–587. [CrossRef]

12. Onori, S.; Tribioli, L. Adaptive Pontryagin's Minimum Principle supervisory controller design for the plug-in hybrid GM Chevrolet Volt. *Appl. Energy* **2015**, *147*, 224–234. [CrossRef]
13. Yang, C.; Du, S.; Li, L.; You, S.; Yang, Y.; Zhao, Y. Adaptive real-time optimal energy management strategy based on equivalent factors optimization for plug-in hybrid electric vehicle. *Appl. Energy* **2017**, *203*, 883–896. [CrossRef]
14. Lei, Z.; Qin, D.; Liu, Y.; Peng, Z.; Lu, L. Dynamic energy management for a novel hybrid electric system based on driving pattern recognition. *Appl. Math. Model.* **2017**, *45*, 940–954. [CrossRef]
15. Lin, X.; Wang, Y.; Bogdan, P.; Chang, N.; Pedram, M. Reinforcement learning based power management for hybrid electric vehicles. In Proceedings of the 2014 IEEE/ACM International Conference on Computer-Aided Design, San Jose, CA, USA, 3–6 November 2014; IEEE Press: Piscataway, NJ, USA, 2014; pp. 32–38.
16. Xiong, R.; Cao, J.; Yu, Q. Reinforcement learning-based real-time power management for hybrid energy storage system in the plug-in hybrid electric vehicle. *Appl. Energy* **2018**, *211*, 538–548. [CrossRef]
17. Liu, T.; Wang, B.; Yang, C. Online Markov Chain-based energy management for a hybrid tracked vehicle with speedy Q-learning. *Energy* **2018**, *160*, 544–555. [CrossRef]
18. Wu, J.; He, H.; Peng, J.; Li, Y.; Li, Z. Continuous reinforcement learning of energy management with deep Q network for a power split hybrid electric bus. *Appl. Energy* **2018**, *222*, 799–811. [CrossRef]
19. Hu, Y.; Li, W.; Xu, K.; Zahid, T.; Qin, F.; Li, C. Energy management strategy for a hybrid electric vehicle based on deep reinforcement learning. *Appl. Sci.* **2018**, *8*, 187. [CrossRef]
20. Liessner, R.; Schroer, C.; Dietermann, A.M.; Bäker, B. Deep Reinforcement Learning for Advanced Energy Management of Hybrid Electric Vehicles. *ICAART* **2018**, *2*, 61–72.
21. Qi, X.; Luo, Y.; Wu, G.; Boriboonsomsin, K.; Barth, M. Deep reinforcement learning enabled self-learning control for energy efficient driving. *Transp. Res. Part C Emerg. Technol.* **2019**, *99*, 67–81. [CrossRef]
22. Moser, I.; Chiong, R. A hooke-jeeves based memetic algorithm for solving dynamic optimisation problems. *Hybrid Artif. Intell. Syst.* **2009**, *5572*, 301–309.
23. Guo, H.; Lu, S.; Hui, H.; Bao, C.; Shangguan, J. Receding horizon control-based energy management for plug-in hybrid electric buses using a predictive model of terminal SOC constraint in consideration of stochastic vehicle mass. *Energy* **2019**, *176*, 292–308. [CrossRef]

 *processes*

Article

Using PSO Algorithm to Compensate Power Loss Due to the Aeroelastic Effect of the Wind Turbine Blade

Ying Zhao [1,2,3]**, Caicai Liao** [1,4,5,]*****, Zhiwen Qin** [1,3,4] **and Ke Yang** [1,3,4]

[1] Institute of Engineering Thermophysics, Chinese Academy of Sciences, Beijing 100190, China; zhaoyingiet@163.com (Y.Z.); qinzhiwen@iet.cn (Z.Q.); yangke@iet.cn (K.Y.)
[2] University of Chinese Academy of Sciences, Beijing 100190, China
[3] Key Laboratory of Wind Energy Utilization, Chinese Academy of Sciences, Beijing 100190, China
[4] National Laboratory of Wind Turbine Blade Research & Development Center, Beijing 100190, China
[5] Dalian National Laboratory for Clean Energy, Chinese Academy of Sciences, Dalian 116023, China
* Correspondence: liaocaicai@iet.cn; Tel.: +86-134-2620-2817

Received: 31 July 2019; Accepted: 12 September 2019; Published: 18 September 2019

Abstract: Power loss due to the aeroelastic effect of the blade is becoming an important problem of large-scale blade design. Prior work has already employed the pretwisting method to deal with this problem and obtained some good results at reference wind speed. The aim of this study was to compensate for the power loss for all of the wind speeds by using the pretwisting method. Therefore, we developed an aeroelastic coupling optimization model, which takes the pretwist angles along the blade as free variables, the maximum AEP (annual energy production) as the optimal object, and the smooth of the twist distribution as one of the constraint conditions. In this optimization model, a PSO (particle swarm optimization) algorithm is used and combined with the BEM-3DFEM (blade element momentum—three-dimensional finite element method) model. Then, the optimization model was compared with an iteration method, which was recently developed by another study and can well compensate the power loss at reference wind speed. By a design test, we found that the power loss can be reduced by pretwisting the origin blade, whether using the optimization model or the iteration method. Moreover, the optimization model has better ability than the iteration method to compensate the power loss with lower thrust coefficient while keeping the twist distribution smooth.

Keywords: aeroelastic effect; pretwisting method; power loss; optimization model; pretwist angle

---

1. Introduction

In modern times, in order to reduce environmental contamination and carbon emission, clean energy such as wind energy are quickly developing. To get more energy from wind energy, many researchers have done lots of works to improve the power output. These works include aerodynamic optimization [1,2], adding flow control devices like vortex generators (VGs) and gurney flaps (GFs) on the blade [3–5], control rules design [6,7] and so on. Besides, some researchers have even applied intelligent algorithms such as reinforcement learning (RL) [6] and evolutionary strategy algorithm [7] to increase aerodynamic performance. These studies are very useful for raising the ratio of wind energy utilization. However, power loss due to the aeroelastic effect is not dedicatedly studied.

Practically, with the increasing unit capacity of wind turbines, blades become longer and softer. Under the action of the different kinds of loads, especially aerodynamic forces, the flexible blade generates non-negligible deflections. These deflections, in turn, change the aerodynamic loads on the blade. Particularly, the induced twist due to the aeroelastic effect has a significant impact on the load and the power performance of the blade. Many researchers have studied the aeroelastic effect and found that an induced twist would not only reduce the blade load, but also decrease the power output [8–12]. This phenomenon is contrary to design objectives, one of which is to capture as much wind energy as possible. To deal with this problem, Lobitz et al. [9] suggested compensating the power loss by pretwisting the blade to adjust for coupling-induced twist. Lee et al. [11] also thought a pretwisting design of blade geometry could provide power performance superior to that of the origin blade after they found that the power loss was more than 13% in some wind speeds. Therefore, understanding how to obtain a suitable pretwist angle distribution along the blade is very important to output the maximum power. Recently, Stäblein et al. [8] presented an iteration procedure to get the pretwist angle distribution at specific wind speed. It can efficiently increase the power at the selected wind speed and make it close to the power without considering the aeroelastic effect. However, the power performances at many other wind speeds are not taken into account. In fact, the induced twist distribution changes with the operational state. This pretwist angle distribution calculated at one wind speed may not suitable for other wind speeds. Hence, it cannot get the maximum annual energy production (AEP), and it even reduces the AEP. Besides, the smooth of the blade profile, which makes the blade feasible in practical, has not been considered yet. Generally, the pretwist angle varies from the blade root to the blade tip. If the pretwist angle distribution is not restrained, after pretwisting the origin blade using these pretwist angles, the twist distribution of the blade will become out of control, and the blade may get an irregular aerodynamic profile.

Therefore, the aim of this study was to compensate for the power loss for all the wind speeds by using the pretwisting method. Firstly, an aeroelastic coupling optimization model, combining the PSO (particle swarm optimization) algorithm with the BEM-3DFEM (blade element momentum—three-dimensional finite element method) model, was built and verified. In this model, we considered the pretwist angle distribution as free variables, taking the maximum AEP (annual energy production) as the optimal object and the smooth of the twist distribution as one of the constraint conditions. Secondly, the optimization model was applied to a 100 kW wind turbine and compared with an iteration method [8], which was recently developed by another study and can well compensate the power loss at reference wind speed. Thirdly, we carried out some analysis and discussion about the results.

2. Aeroelastic Model

In this section, a steady-state aeroelastic coupling model, called the BEM-3DFEM (blade element momentum—three-dimensional finite element method) model, was developed to calculate the power output during the normal operating state.

2.1. Aerodynamic Model and Verification

There are many methods for calculating the aerodynamic performance of a wind turbine, such as the BEM (blade element momentum) method [13,14], the vortex model [14,15], and the CFD (computational fluid dynamics) model [15,16]. Among them, the BEM method is the most widely used, due to its superior comprehensive performance on the computation cost and simulation precision. In this paper, the BEM model, considering the tip loss and hub loss, was used to get the steady aerodynamic performance of the blade. This aerodynamic model was programmed by using the MATLAB language (Matlab R2012a, The MathWorks, Inc., Novi, MI, USA, 2012). To certificate this procedure, it was used to calculate the aerodynamic performance of the NREL (natioanal renewable energy laboratory) phase VI wind turbine (NREL, Golden, CO, USA) [17,18]. The results of this BEM procedure, compared with the experiment results and the results using GH Bladed software [19], are shown in Figure 1. As can be seen, the power coefficients and the powers from the above three methods are close. So, this BEM procedure can be used to evaluate the aerodynamic performance of the wind turbine blade.

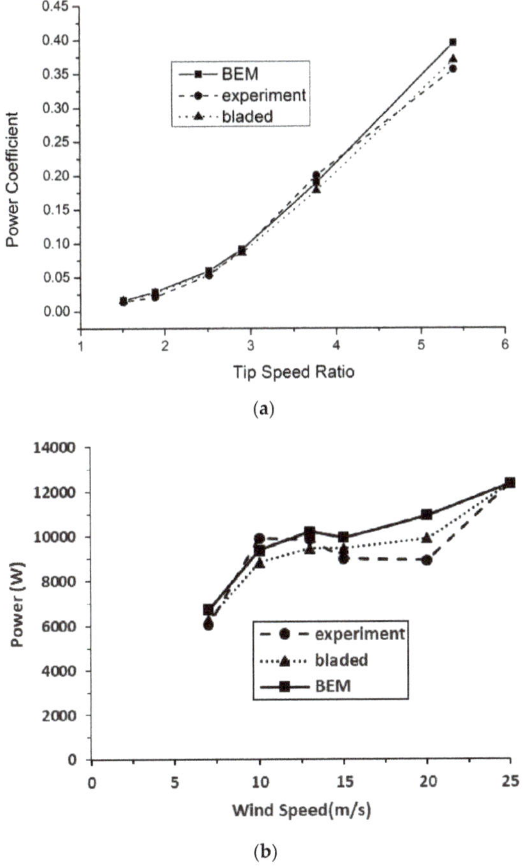

Figure 1. Aerodynamic performance of the NREL phase VI wind turbine from different methods: (**a**) Power coefficient changing with tip speed ratio; (**b**) power changing with wind speed. Note: "BEM" stands for the results obtained from the BEM procedure; "experiment" stands for the measurement results in ref [8]; "bladed" stands for the results by the use of GH Bladed software.

2.2. Structure Model and Verification

There are also lots of structure models for simulating the blade structure properties. These models can be roughly categorized into three groups. They are FEM (finite element method) model, the multi-body model, and the 1D (one-dimensional) equivalent beam model [11,15,20]. Compared with the other two models, the 3D (three-dimensional) FEM model has higher computational precision. This is because it can correctly describe the varieties of the layers in detail. Hence, the 3DFEM model was applied in this article to ensure the accuracy of structure deform analysis. The blade was built by shell elements, which are suitable to characterize composite laminates and sandwich structures (see Figure 2). To help the datum transfer between aerodynamics and the structure model, the structure model was programmed by APDL (ANSYS Parametric Design Language), which includes blade parametric model, deformation calculating, and results processing. To get an accurate result, we can see from Figure 2 that the flap-wise and edge-wise direction loads were distributed on the nodes of spar caps to avoid stress concentration. Because the root stiffnesses of the blade are strong enough to endure the blade loads and moments, the deformations at the root are very small. The six degrees of freedom at the blade root is fixed to simply the analysis.

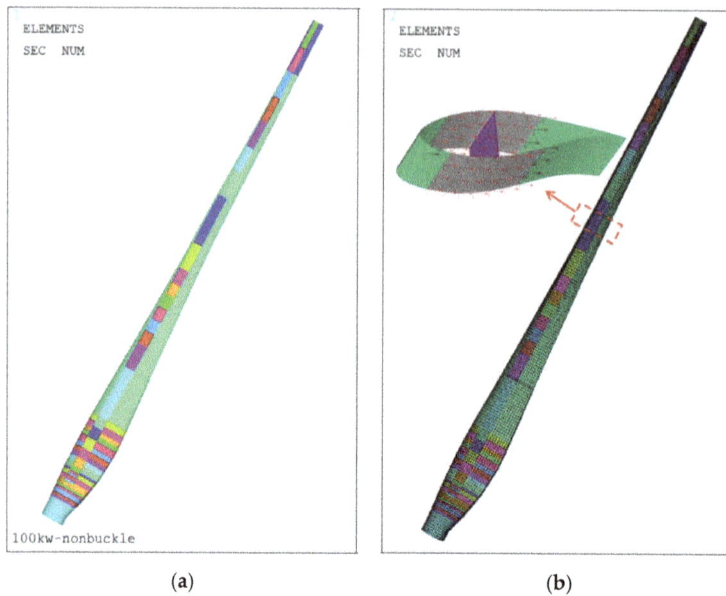

(**a**)   (**b**)

Figure 2. The 3DFEM model of a 100 kW blade: (**a**) Blade laminates in different colors; (**b**) meshes and loading of the blade model.

To verify this FEM procedure, it was also applied to the 100 kW wind turbine blade, which is subjected to a full-scale static test [21]. The numerical simulation results by using this FEM procedure, together with the test results, are shown in Table 1. We can see the tip deflections are very close. It indicates that this 3DFEM model has superior computational accuracy and can be applied to build the aeroelastic model.

Table 1. Tip deflections of a 100 kW blade.

Experiment (m)	FEM (m)	Relative Error (%)
1.369	1.330	2.8

2.3. Building the BEM-3DFEM Model

The aeroelastic process is very complicated. To analyze it, many coupling models have been developed [22–24]. The weak coupling method, due to its advantage of keeping the independence of the aerodynamic model and structure model and its expansibility, has been widely used in fluid-structure coupling analysis cases. In this paper, based on the above BEM method and 3DFEM model, an aeroelastic coupling model, named as BEM-3DFEM, was established by using a weak coupling method. The calculation process of the BEM-3DFEM is shown in Figure 3. During the calculating process, the loads were firstly calculated by the BEM method and loaded on the 3DFEM model of the blade to get the deflections of the blade. Then, the induced twist of every section used in BEM were gotten by interpolating the information of the nodes nearby. After that, the induced twist of every section were used in BEM computation to change the loads on the structure model. So, the deformations of the blade also change by the loads. If the difference of the deformation between two adjacent iterations, expressed as Δ, is less than the tolerance Δ_a, the analysis process ends. At the same time, the final pretwist angle distribution is obtained. At this point, a steady aeroelastic analysis has been done.

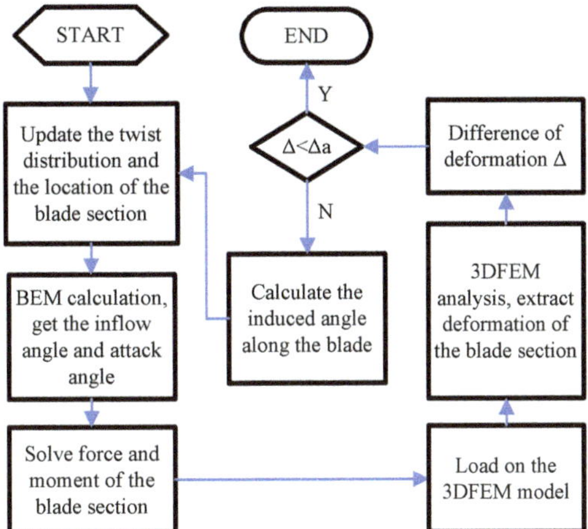

Figure 3. Flow chart of the BEM-3DFEM model.

3. An Aeroelastic Coupling Optimization Model

3.1. Objective Function

The power output is one of the most important properties of the wind turbine, so the maximum *AEP* considering the aeroelastic effect was selected as the objective function. It is shown as follows:

$$\max\{AEP\} = \max\{N_h \int_{v_{in}}^{v_{out}} P(v) \cdot f(v)dv\}$$

where v_{in} and v_{out} are cut-in and cut-out wind speed, respectively; N_h is the number of hours in a year; $P(v)$ is the power of the wind turbine at the wind speed of v, taking the aeroelastic effect into account; and $f(v)$ is the probability of the wind speed of v in a year.

In this article, the Rayleigh distribution with the average wind speed of 8 m/s in a year was used to calculate the $f(v)$.

3.2. Free Variables

The twist plays an important role in the aerodynamic design of a blade. Its change can reflect the aeroelastic effect. Hence, in this optimization model, pretwist angles at different cross-sections of the blade were selected as free variables, which are named as $x_i(i = 1, 2, 3 \cdots n)$, where x_i is the pretwist angle at section i and n is the number of the sections used to compute the power.

After getting the x_i, the origin twist at the section i is pretwisted by x_i.

3.3. Constraints

In this optimization model, to achieve the excellent aerodynamic performance of the blade, some constraints about the x_i are given.

Firstly, on the one hand, x_i should not be too big. This is because a big x_i makes the attack angle of the section i exceed the stall attack angle. On the other hand, x_i should not be too small. This leads to the aerodynamic twist becoming bigger than the allowed value, which is limited by the production and transportation. So, it is expressed as follows:

$$Lb_i - c \leq x_i \leq Ub_i + c$$

where Lb_i is the minimum induced twist from computing the twist change at the different wind speed, considering the aeroelastic effect for the origin blade; Ub_i is the maximum induced twist from computing the twist change at the different wind speed, considering the aeroelastic effect for the origin blade; and c is a constant value, which is used to expand the searching range of x_i. In this article, it was decided by the difference of the Ub_i to the stall attack angle at section i.

Secondly, to keep the smoothness of the aerodynamic shape of the blade, the twist distribution along the blade after pretwisting should meet a curve of nine times polynomial. It is written as:

$$x_{i0} - x_i = a_0 + a_1 r_i + a_2 r_i^2 + a_3 r_i^3 + a_4 r_i^4 + a_5 r_i^5 + a_6 r_i^6 + a_7 r_i^7 + a_8 r_i^8 + a_9 r_i^9$$

where x_{i0} is the twist at the section i of the origin blade; r_i corresponds to the location of the section i span-wise from root to tip; and $a_i (i = 0, 1, \cdots, 9)$ is the coefficient, which is gotten by fitting the twist at different sections after pretwisting.

3.4. Optimization Process

The flow chart of the aeroelastic coupling optimization model to get the maximum AEP, as well as the optimal pretwist angle-distribution, is given in Figure 4.

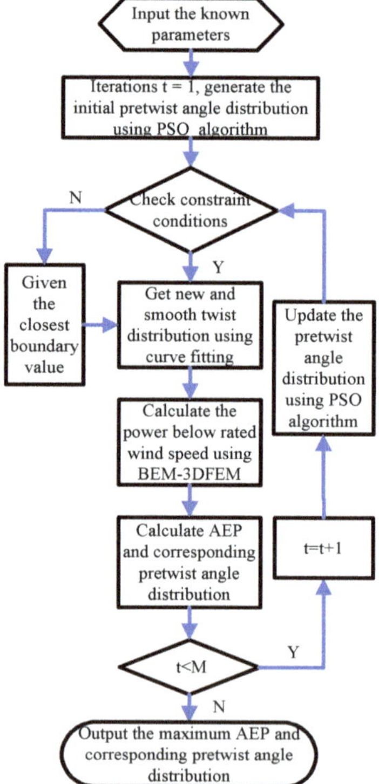

Figure 4. The flow chart of the aeroelastic coupling optimization model.

During the optimization, we generated the initial pretwist angle distribution using the PSO algorithm at first. After checking the constraints about the pretwist angle, we obtained a series of smooth twist distributions for different blades. Then, we used the BEM-3DFEM to evaluate the power performance at different wind speeds by considering the aeroelastic effect. The maximum AEP was gotten at the same probability distribution of annual wind speed. With the increase of iteration number, more suitable pretwist angle distributions were generated using the PSO algorithm [25–28]. Besides, the AEP was calculated again to find better results. While the iteration number was equal to the maximum iteration number M, the optimum AEP, as well as the corresponding pretwist angle distribution, were output.

Therefore, the pretwist angle distribution changed in every iteration by using a linear change inertia weight PSO algorithm. In this way, the attack angle of the section increased in most wind speeds. The power loss due to the aeroelastic influence was made up.

4. Results & Discussion

In this paper, a 100 kW wind turbine blade was analyzed using the above aeroelastic coupling optimization model to compensate the power loss due to the aeroelastic effect. It is a field test wind turbine operated by National Energy Wind Turbine Blade R&D Center and located in Zhangbei County, China. It is also a VSVP (variable speed variable pitch) wind turbine, like the large scale wind turbines. So, the power under the rated wind speed is mainly studied. Besides, the induced twists of the blade are small due to the relatively large stiffness of the blade. A high precision model, especially the structure model, is needed to simulate the deflection of the blade. Fortunately, the BEM-3DFEM model can efficiently reflect the aeroelastic influence of this blade because it simulates the blade layers in detail.

4.1. Pre-Assigned Variables

During the computing process, some of the parameters were fixed. They are as follows:

1. PSO algorithm, such as inertial weight, accelerating factors;
2. Aerodynamic profile, such as chord distribution, twist distribution, relative thickness distribution;
3. Structure layers, such as the number, size, location, materials; and
4. Some other parameters about the wind turbine, such as the hub height, hub radius, cut-in wind speed, cut-out wind speed, rated wind speed.

Some typical parameters among them are given in Table 2. In Table 2, the first six parameters were used in the PSO algorithm. The inertial weights and accelerating factors were selected according to the Ref. [26–28]. The maximum number of iterations and number of individuals were relatively small to reduce the calculating time due to the high computing cost of FEM analysis. The 7th to 10th parameters in this table were used to describe the 100 kW wind turbine. So, their values were fixed. The last parameter was used for evaluating the convergence of the steady aeroelastic analysis. Its value was small enough for 100 kW wind turbine to get the correct blade deflection. However, we should be concerned; this value cannot be too small as this would lead to the non-convergence of the FEM-3DFEM analysis. In addition, this value changes for different wind turbines.

Table 2. Some typical parameters used in this optimization model.

Parameter Names	Values
Maximum inertial weight	1
Minimum inertial weight	0
accelerating factors c_1	1
accelerating factors c_2	1
Maximum number of iterations M	25
Number of individuals	22
Number of the blade sections	16
Radius of the rotor (m)	10.292
Hub height (m)	26.2
Number of blades	3
Δ_a	1.00×10^{-6}

4.2. Results and Analysis

To analyze the benefits of the aeroelastic coupling optimization model, we applied the optimization model and the iteration method in Ref. [8] to the origin blade. Two new blades were gotten, and named as the PSO blade and the iteration blade. The AEP under four different cases was computed. The cases are as follows:

1. The origin blade without considering the aeroelastic coupling effect, which is represented as ucpl.;
2. The origin blade considering the aeroelastic coupling effect, which is represented as cpl.;
3. The PSO blade considering the aeroelastic effect, which is represented as PSO + cpl.; and
4. The iteration blade considering the aeroelastic effect, which is represented as pret. + cpl.

When the AEP was calculated, wind speeds from 4 to 25 m/s with an interval of 1 m/s, as well as the rated wind speed of 10.6 m/s, were used. However, only wind speeds from 4 to 10.6 m/s consider the aeroelastic effect, because the blade was used on a VSVP wind turbine. Besides, for the different reference wind speeds, different pretwisting angle distributions were gotten using the iteration method. This led to different AEP. In this article, lots of AEP using the iteration method were calculated by varying the reference wind speed from 4 to 10.6 m/s. Then, the maximum value of them, with reference wind speed of 10 m/s, was selected as the result of the iteration blade.

The twist distribution along the blade for the origin blade, the PSO blade, and the iteration blade are given in Figure 5. The relative difference in AEP for different cases compared to the ucpl. case is shown in Figure 6. We can see that the AEP of the origin blade decreases due to the aeroelastic effect. This feature is the same as the result of other researches [8–11]. In addition, the AEP of the PSO blade and the iteration blade are higher than that of the origin blade considering the aeroelastic effect. It indicates that the optimization model and the iteration method can both compensate the power loss and make the AEP close to that of the ucpl. case. Based on the varieties of the twist distribution in Figure 5, we can conclude that reducing the twists of the blade, especially that of the outboard part of the blade, leads to the increase of AEP to some extent. By comparing the twist distribution of the PSO blade with that of the iteration blade, it indicates that the AEP does not continuously rise while the twist keeps going down. This is because the ratio of the lift to drag, as well as the lift coefficient, cannot continue to increase while the attack angle gets bigger. So, there exists an optimal twist distribution to get the maximum AEP under the consideration of the aeroelastic influence.

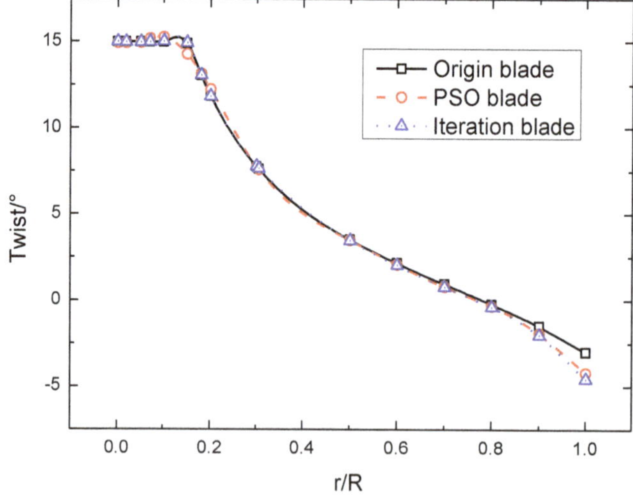

Figure 5. Twist distribution along the blade.

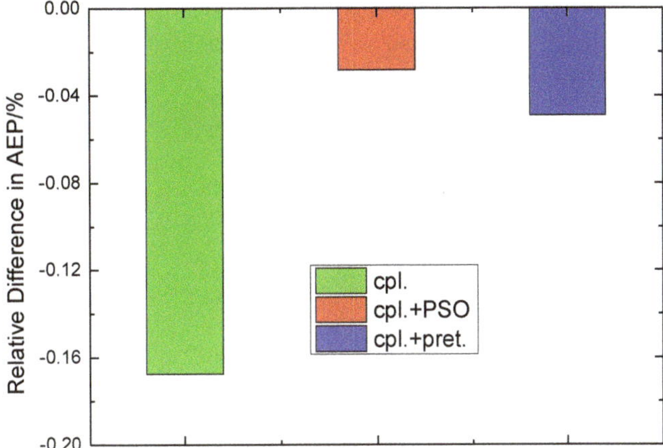

Figure 6. The relative difference in annual energy production (AEP) compared to the ucpl. Case.

From Figures 5 and 6, we can see that by using the optimization model, the PSO blade gets the maximum AEP and its twist distribution is very smooth. It shows that the optimization model is more efficient to compensate for power loss than the iteration method. Therefore, it gives us a good choice to improve the power output of the blade, while the aeroelastic effect cannot be negligible.

Figures 7 and 8 illustrate the relative difference in power and thrust coefficient compared to the cpl. Case, respectively. From Figure 7, we can see, in the case of considering the aeroelastic effect, the powers of the PSO blade and the iteration blade are higher than that of the origin blade at high wind speed. However, the comparison results about the power are converse at low wind speed. It can be concluded that the enhancements of the powers at high wind speed play the leading role in raising the AEP of the PSO blade and the iteration blade. This may be caused by the stronger aeroelastic coupling at high wind speed.

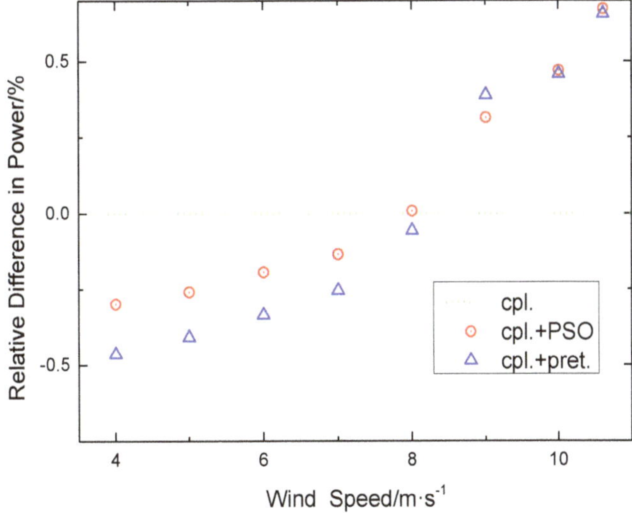

Figure 7. The relative difference in power compared to the cpl. Case.

Figure 8 shows the increase of the thrust coefficients of the PSO blade and the iteration blade at every wind speed. This is mainly due to the increase of the attack angle by pretwisting the origin blade.

In the future design, this should be a constraint condition for getting the maximum AEP. By comparing the PSO blade with the iteration blade, we can find that the thrust coefficients of the PSO blade are less than that of the iteration blade, while the powers of the PSO blade are larger than that of the iteration blade, at most of the wind speeds. This is mainly because the attack angles and the inflow angles of different sections change at the same time after pretwisting. However, the optimization model can help us to find out the more suitable values for them. This is a good advantage of the optimization model.

Certainly, the above result are based on the steady aeroelastic analysis. This is because we just applied the steady BEM model in the BEM-3DFEM. For the evaluation of AEP, this steady model is suitable. However, the wind turbine always operates in unsteady wind conditions. This needs an unsteady aerodynamic model to simulate the aerodynamic force. Hence, to calculate the aeroelastic loads in unsteady conditions, the steady BEM model needs to be corrected by using the dynamic stall model, or even replaced by the vortex method or unsteady CFD model. The BEM-3DFEM model will be improved in the future to compute the effect of the PSO blade under turbulence wind.

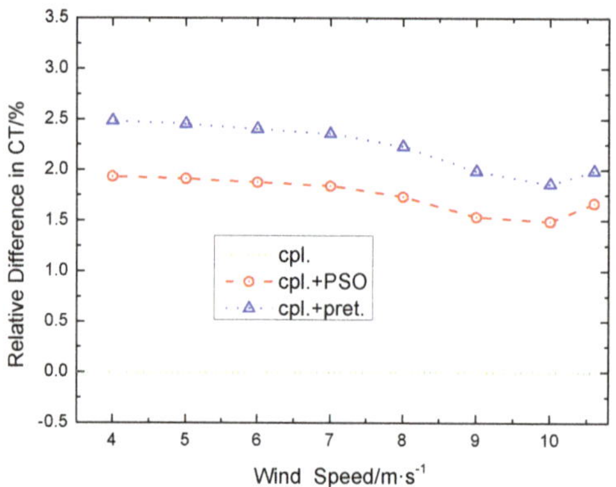

Figure 8. Relative difference in thrust coefficient compared to the cpl. Case.

5. Conclusions

In this article, to get the maximum AEP and the optimal pretwist angle distribution, an aeroelastic coupling optimization model, combining the PSO algorithm with the BEM-3DFEM model, was built to compensate power loss due to the aeroelastic effect.

Then, this aeroelastic coupling optimization model was applied to a 100 kW wind turbine blade to reduce the power loss. The iteration method developed to solve this problem from Ref. [8] was also used for the same blade. Two new blades with different aerodynamic profiles were gotten, respectively. After that, different aerodynamic properties for these two new blades, as well as the origin blade, were calculated and compared.

It is shown that the AEP of the origin blade reduces due to the aeroelastic effect. This feature is the same as the result of other researches [8–11]. In addition, the AEP of the origin blade with the aeroelastic effect can be improved by pretwisting the origin blade, whether using the aeroelastic coupling optimization model or the iteration method. Compared with the iteration method, the optimization model is more efficient to compensate for the power loss, with a lower thrust coefficient while keeping the twist distribution smooth. Therefore, it gives us a good choice to improve the power output of the origin blade, while the aeroelastic effect cannot be negligible.

According to the results, the enhancements of the powers at high wind speed play the leading role in raising the AEP for both the PSO blade and the iteration blade. This may be caused by the stronger aeroelastic process at high wind speed. So, we can conclude that the improvement of the AEP is more obvious when the wind turbine is used in a better wind resource area. However, the thrust coefficients increase after pretwisting the origin blade, since the attack angles enlarge. Hence, to avoid the thrust of the blade exceeding the allowable value, the thrust coefficient should be taken as a constraint condition while getting the maximum AEP in the future.

Besides, the results of compensating for the power loss on the 100 kW wind turbine blade are relatively small. This may be because the aeroelastic effect of the 100 kW wind turbine blade is weak. To further certify this optimization model, it will be used on some large-scale wind turbines such as the NREL 5MW reference wind turbine and the DTU (Danmarks Tekniske Universitet) 10MW reference wind turbine. Certainly, this article focuses on steady aeroelastic analysis. This is because we just applied the steady BEM model in the BEM-3DFEM. To simulate aeroelastic loads in unsteady conditions, the steady BEM model needs to be corrected by using the dynamic stall model, or replaced by the vortex method or even the unsteady CFD model. This BEM-3DFEM model will be improved in the future to compute the effect of the PSO blade under turbulence wind.

Author Contributions: Y.Z. developed most of the procedure, did the numerical computation, and wrote the article; C.L. developed the new idea, did the analysis of the results, and reviewed the article; Z.Q. developed the APDL procedure for FEM analysis and modified the article; K.Y. supervised the work and gave some useful comments for writing.

Funding: This research was funded by the National Natural Science Foundation of China (No. 51606196).

Acknowledgments: The authors would like to thank the China Scholarship Council (CSC) and the International Clean Energy Talent (iCET) Program. This work was also supported by the Strategic Priority Research Program of the Chinese Academy of Science, Grant No. XDA21050303.

Conflicts of Interest: The authors declare no conflict of interest.

References

1. Bavanish, B.; Thyagarajan, K. Optimization of power coefficient on a horizontal axis wind turbine using bem theory. *Renew. Sustain. Energy Rev.* **2013**, *26*, 169–182. [CrossRef]
2. Liu, X.W.; Wang, L.; Tang, X.Z. Optimized linearization of chord and twist angle profiles for fixed-pitch fixed-speed wind turbine blades. *Renew. Energy* **2013**, *57*, 111–119. [CrossRef]
3. Saenz-Aguirre, A.; Fernandez-Gamiz, U.; Zulueta, E.; Ulazia, A.; Martinez-Rico, J. Optimal Wind Turbine Operation By Artificial Neural Network-based Active Gurney Flap Flow Control. *Sustainability* **2019**, *11*, 2809. [CrossRef]
4. Fernandez-Gamiz, U.; Zulueta, E.; Boyano, A.; Ansoategui, I.; Uriarte, I. Five Megawatt Wind Turbine Power Output Improvements By Passive Flow Control Devices. *Energies* **2017**, *10*, 742. [CrossRef]
5. Fernandez-Gamiz, U.; Zulueta, E.; Boyano, A.; Ramos-Hernanz, J.A.; Lopez-Guede, J.M. Microtab Design And Implementation On A 5 Mw Wind Turbine. *Appl. Sci.* **2017**, *7*, 536. [CrossRef]
6. González-González, A.; Etxeberria-Agiriano, I.; Zuluet, E.; Oterino-Echavarri, F.; Lopez-Guede, J.M. Pitch Based Wind Turbine Intelligent Speed Setpoint Adjustment Algorithms. *Energies* **2014**, *7*, 3793–3809. [CrossRef]
7. Kusiak, A.; Zheng, H.Y.; Song, Z. Power optimization of wind turbines with data mining and evolutionary computation. *Renew. Energy* **2010**, *35*, 695–702. [CrossRef]
8. Stäblein, A.R.; Tibaldi, C.; Hansen, M.H. Using Pretwist to Reduce Power Loss of Bend-Twist Coupled Blades. In Proceedings of the 34th Wind Energy Symposium, San Diego, CA, USA, 4–8 January 2016; American Institute of Aeronautics & Astronautics: San Diego, CA, USA, 2016; AIAA 2016-1010.
9. Lobitz, D.W.; Veers, P.S. Load Mitigation with Bending/Twist-coupled Blades on Rotors using Modern Control Strategies. *Wind Energy* **2003**, *6*, 105–117. [CrossRef]
10. Verelst, D.R.; Larsen, T.J. *Load Consequences When Sweeping Blades—A Case Study of a 5 MW Pitch Controlled Wind Turbine*; Risø-R-Report; Risø National Laboratory for Sustainable Energy: Roskilde, Denmark, 2010; Risø-R-1724(EN).
11. Lee, Y.J.; Jhan, Y.T.; Chung, C.H. Fluid–structure interaction of FRP wind turbine blades under aerodynamic effect. *Compos. Part B Eng.* **2012**, *43*, 2180–2191. [CrossRef]
12. Bottasso, C.L.; Campagnolo, F.; Croce, A.; Tibaldi, C. Optimization-Based Study of Bend-Twist Coupled Rotor Blades for Passive and Integrated Passive/Active Load Alleviation. *Wind Energy* **2013**, *16*, 1149–1166. [CrossRef]

13. Hansen, M.O.L. *Aerodynamics of Wind Turbines*; Earthscan Publications Ltd.: London, UK, 2008.
14. Hansen, M.O.L.; Sørensen, J.N.; Voutsinas, S.; Sørensen, N.; Madsen, H.A. State of the art in wind turbine aerodynamics and aeroelasticity. *Prog. Aerosp. Sci.* **2006**, *42*, 285–330. [CrossRef]
15. Wang, L.; Liu, X.W.; Kolios, A. State of the art in the aeroelasticity of wind turbine blades: Aeroelastic modelling. *Renew. Sustain. Energy Rev.* **2016**, *64*, 195–210. [CrossRef]
16. Bai, C.J.; Wang, W.C. Review of computational and experimental approaches to analysis of aerodynamic performance in horizontal-axis wind turbines (HAWTs). *Renew. Sustain. Energy Rev.* **2016**, *63*, 506–519. [CrossRef]
17. Hand, M.M.; Simms, D.A.; Fingersh, L.J.; Jager, D.W.; Cotrell, J.R.; Schreck, S.; Larwood, S.M. *Unsteady Aerodynamics Experiment Phase VI: Wind Tunnel Test Configurations and Available Data Campaigns*; Techinical Report; National Renewable Energy Laboratory: Golden, CO, USA, 2001; NREL/TP-500-29955.
18. Simms, D.; Schreck, S.; Hand, M.; Fingersh, L.J. *NREL Unsteady Aerodynamics Experiment in the NASA-Ames Wind Tunnel: A Comparison of Predictions to Measurements*; Techinical Report; National Renewable Energy Laboratory: Golden, Colorado, USA, 2001; NREL/TP-500-29494.
19. Bossanyi, E.A. *GH Bladed Theory Manual*; Garrad Hassan and Partners Limited: Bristol, UK, 2008.
20. Hayat, K.; Ha, S.K. Load mitigation of wind turbine blade by aeroelastic tailoring via unbalanced laminates composites. *Compos. Struct.* **2015**, *128*, 122–133. [CrossRef]
21. Chen, X.; Qin, Z.W.; Yang, K.; Zhao, X.L.; Xu, J.Z. Numerical Analysis and Experimental Investigation of Wind Turbine Blades with Innovative Features: Structural Response and Characteristics. *Sci. China Tech. Sci.* **2014**, *58*, 1–8. [CrossRef]
22. Li, Y.; Castro, A.M.; Sinokrot, T.; Prescott, W.; Carrica, P.M. Coupled multi-body dynamics and CFD for wind turbine simulation including explicit wind turbulence. *Renew. Energy* **2015**, *76*, 338–361. [CrossRef]
23. Mo, W.W.; Li, D.Y.; Wang, X.N.; Zhong, C.T. Aeroelastic Coupling Analysis of the Flexible Blade of a Wind Turbine. *Energy* **2015**, *89*, 1001–1009. [CrossRef]
24. Lee, J.W.; Lee, J.S.; Han, J.H.; Shin, H.K. Aeroelastic analysis of wind turbine blades based on modified strip theory. *J. Wind Eng. Ind. Aerodyn.* **2012**, *110*, 62–69. [CrossRef]
25. Liao, C.C.; Zhao, X.L.; Xu, J.Z. Blade layers optimization of wind turbines using FAST and improved PSO algorithm. *Renew. Energy* **2012**, *42*, 227–233.
26. Shi, Y.; Eberhart, R.C. Empirical study of particle swarm optimization. Evolutionary Computation. In Proceedings of the 1999 Congress on Evolutionary Computation-CEC99, Washington, DC, USA, 6–9 July 1999; pp. 1945–1950.
27. Shi, Y.; Eberhart, R.C. A modified particle swarm optimizer. Evolutionary Computation. In Proceedings of the IEEE World Congress on Computational Intelligence, Anchorage, AK, USA, 4–9 May 1998; pp. 69–73.
28. Eberhart, R.; Kennedy, J. A New Optimizer Using Particle Swarm Theory. In Proceedings of the Sixth International Symposium on Micro Machine and Human Science, Nagoya, Japan, 4–6 October 1995; pp. 39–43.

Article

The Direct Speed Control of Pmsm Based on Terminal Sliding Mode and Finite Time Observer

Yao Wang *, HaiTao Yu *, Zhiyuan Che, Yuchen Wang and Yulei Liu

School of Electrical Engineering, Southeast University, Nanjing 210096, China; zhiyuanche@foxmail.com (Z.C.); 220172743@seu.edu.cn (Y.W.); wangyhbcd@126.com (Y.L.)
* Correspondence: 230179191@seu.edu.cn (Y.W.); htyu@seu.edu.cn (H.Y.)

Received: 22 August 2019; Accepted: 10 September 2019; Published: 16 September 2019

Abstract: A non-singular terminal sliding mode control based on finite time observer is designed to achieve speed direct control for the permanent magnet synchronous motor (PMSM) drive system. Speed and current are regulated in one loop under the non-cascade structure, taking place of the cascade structure control method in the vector control of PMSM. Based on the second-order speed function of the PMSM, the disturbance and parameters uncertainties are estimated by the designed finite time observer (FTO), and compensate to the drive system. The estimated value of the finite time observer will converge to the actual disturbance value in a finite time. A second-order non-singular terminal sliding mode controller is proposed to realize the speed and current single-loop, which can track the reference speed and reference current in a finite time. Rigorous stability analysis is established. Comparative results verified that the proposed method has faster speed tracking performance and disturbance rejection property.

Keywords: non-singular terminal sliding mode control (NTSMC); finite-time observer (FTO); mismatched/matched disturbance/uncertainties; permanent magnet synchronous motor (PMSM)

1. Introduction

By reason of the high-power density, torque-to-inertia ratio and high efficiency, the permanent magnet synchronous motor (PMSM) are widely used in industrial areas, such as, aerospace, servo control, numerical control machine and robot [1–5]. In these applications, the dynamic response performance and disturbance rejection property of PMSM are very important.

In recent years, with the progress of technology, the control periods between the speed loop and current loop of PMSM gradually decreased, or even vanished [6]; making it possible to realize the speed-current single-loop of PMSM drive system under the non-cascade structure. Generally speaking, in the traditional cascade control method for PMSM, the control period of the speed loop is 5–10 times that of the current loop, reducing the real-time control performance of the speed [7–9]. When the same control algorithm is adopted, different from the cascade control structure, the number of adjustable parameters is reduced and the speed can be directly controlled. These are the virtues of the non-cascade control structure [10,11]. Despite its advantages, there is little research on non-cascade control structures for the PMSM system in recent years. A non-cascade structure control based on model predictive control is proposed in [12], in which the dynamic performance of the system is improved and the computational complexity is reduced, compared with the traditional cascade predictive control method. In [13], under the non-cascade structure, the speed and current are adjusted in one proportion integration differentiation (PID) controller. Rigorous theoretical derivation and experimental analysis verified that the proposed method has better dynamic performance and disturbance rejection ability. Considering the influence of various disturbances on the PMSM system, a new non-cascade structure controller is established in [14], which can directly control the speed of PMSM. PMSM speed and current are adjusted in one loop based on terminal sliding mode and nonlinear disturbance observer under non-cascade structure control in [15]. However, when without the nonlinear disturbance observer, the proposed method has a poor ability to deal with the load sudden change. A direct speed control method based on radial basis function (RBF) is designed in [16], which avoided the control of current, simplified the control structure and improved the control performance. A model predictive direct speed control method based on voltage vector control is proposed in [17]. In this method, the voltage vector does not need to be measured, the computational burden of the system is reduced, and the output current is constrained within a certain range. A model predictive direct speed controller is proposed in [18], which overcomes the shortcoming of cascade linear controller in high-speed control, and the results show that the proposed method has better stability performance. Based on the state-dependent Riccati equation (SDRE) and Convex constrained optimization, a direct speed controller was proposed in [19], which can make the PMSM control system achieve high dynamic and accurate stability performance, and the input voltage and stator current can be constrained.

Due to the nonlinear and strong coupling characteristics of the PMSM drive system, ideal control results can hardly be achieved in traditional PI controller [20,21]. Many nonlinear control methods have been applied in PMSM drive systems, such as sliding mode control, model predictive control, auto-disturbance rejection control, finite time control, etc., [22–26]. Among these methods, it can converge in finite time and has a better disturbance rejection performance. The terminal sliding mode is widely used in control systems. In [27], a new terminal sliding mode controller is designed to adjust the speed of the PMSM servo system, which can make the system reach the reference speed in a finite time, ensuring a fast convergence performance and a better tracking accuracy of the system. In [28], a non-singular terminal sliding mode control based on state observer is investigated to realize the pressure control. In the proposed method, the pressure tracking error can converge to the equilibrium point in finite time and the chattering of the sliding mode is weakened. In [29] according to euler discrete technology, a new discrete time fast terminal sliding mode method is proposed and applied to the control of permanent magnet synchronous linear motor (PMLSM), and the reference position of PMLSM can be quickly tracked. In [30], a fractional-order terminal sliding mode controller based on fractional-order disturbance observer is proposed, under which the speed can converge to the reference speed in a finite time. In [31], a higher speed tracking accuracy can be achieved by a continuous fast terminal sliding mode control, and the robustness of the PMSM system can be improved when the disturbance is feedforward to the system by the extended state observer. In [32], a nonsingular terminal sliding mode based on improved extended state observer is investigated to realize the direct voltage control for the stand-alone doubly-fed induction generator (DFIG) system, which can achieve a balanced stator voltage.

Load change, parameters uncertainty and unmodeled dynamics are considered to be important factors affected the control performance. At present, in order to improve the robustness, the disturbance will be estimated by state observer and feed forward to the system before it affects the system. In [33], to improve the robustness in surface permanent magnet synchronous motor, the lumped disturbance consisted of the external disturbance and mismatched parameters can be estimated by a Luneburg observer, and compensate to the PMSM system. In [34] the parameters uncertainties and disturbances in DC-DC converters are considered as lumped disturbance, estimated by a reduced order generalized proportional integral observer and fed forward to the system, which improves the dynamic performance of the system. In [35], the lumped disturbance in air-breathing hypersonic vehicles is calculated by a disturbance observer, and the accuracy of speed and position control is improved when the disturbance feedforward to the system. In [36], a high-gain generalized proportional integral observer is designed, to estimate the load change and parameters uncertainties in PMSM. In [37], the disturbance is estimated and compensated by a nonlinear disturbance observer to improve the disturbance rejection property of the system. Then, a nonlinear controller is used to control the system, and the semi-global stability of the designed nonlinear controller and nonlinear disturbance observer is proved. In [38], a robust nonlinear observer is proposed for the Lipschitz nonlinear system. On the one hand, the new observer does not need to be added to small Lipschitz constants; on the other hand, the state estimation error of the system can quickly approach zero in the face of large additional disturbances. Disturbance also exists in the PMSM drive system under the non-cascade structure. In order to improve the anti-disturbance ability, it is necessary to estimate and compensate the disturbance to the system.

2. Preliminaries

2.1. The Mathematical Model of Pmsm

The ideal model of a surface mounted PMSM in the d-q frame can be expressed as follows.

$$
\begin{cases}
\frac{di_d}{dt} = \frac{-Ri_d + n_p \omega L i_q}{L} + \frac{1}{L} u_d \\
\frac{di_q}{dt} = \frac{-Ri_q - n_p \omega L i_d - n_p \omega \psi_f}{L} + \frac{1}{L} u_q \\
\frac{d\omega}{dt} = -\frac{B\omega}{J} + \frac{n_p \psi_f}{J} i_q - \frac{T_L}{J}
\end{cases}
\tag{1}
$$

where, i_d, i_q are the d-axis and q-axis stator currents, respectively; u_d, u_q are the d-axis and q-axis stator voltages, respectively; L is the inductor; R is stator resistance; n_p is the number of pole pairs; ω is angular velocity; ψ_f is rotor flux linkage; T_L is load torque; B is viscous frictional coefficient; J is rotor inertia.

2.2. The Mathematic Model of Speed-Current Single-Loop

Let $x_1 = \omega_{ref} - \omega$, and its derivative can be expressed as

$$
\dot{x}_1 = x_2 = \dot{\omega}_{ref} - \dot{\omega} = \dot{\omega}_{ref} + a_1 \omega - a_2 i_q + a_3 T_L.
\tag{2}
$$

where, $a_1 = \frac{B}{J}, a_2 = \frac{n_p \psi_f}{J}, a_3 = \frac{1}{J}$.

When the parameters uncertainties are considered, the following expression can be obtained.

$$
\dot{x}_1 = x_2 + d_1 = \dot{\omega}_{ref} + a_1 \omega - a_2 i_q + a_3 T_L + d_1
\tag{3}
$$

where, $d_1 = \Delta a_1 \omega - \Delta a_2 i_q + \Delta a_3 T_L$ is considered as mismatched uncertainties. And a_1, a_2, a_3 are nominal parameter values, a_{t1}, a_{t2}, a_{t3} are the actual parameter values. $\Delta a_1 = a_{t1} - a_1$, $\Delta a_2 = a_{t2} - a_2$, $\Delta a_3 = a_{t3} - a_3$.

The second order differential equation of speed error can be expressed as follows

$$
\dot{x}_2 = \ddot{\omega}_{ref} - \ddot{\omega} = \ddot{\omega}_{ref} + a_1 \dot{\omega} - a_2 \dot{i}_q + a_3 \dot{T}_L.
\tag{4}
$$

Considering system (1), the (4) can be rewritten as

$$
\dot{x}_2 = \ddot{\omega}_{ref} + a_1 \dot{\omega} + a_2 b_1 i_q + a_2 b_2 \omega i_d + a_2 b_3 \omega - a_2 b_4 u_q + a_3 \dot{T}_L.
\tag{5}
$$

where $b_1 = \frac{R}{L}, b_2 = n_p, b_3 = \frac{n_p \psi_f}{L}$ and $b_4 = \frac{1}{L}$.

Taking the parameters uncertainties and disturbance into consideration, system (5) can be expressed as follows

$$
\dot{x}_2 = -a_1 x_2 - a_2 b_3 x_1 - a_2 b_4 u_q + d_2.
\tag{6}
$$

where,
$$
\begin{aligned}
d_2 = {} & \ddot{\omega}_{ref} + a_1 \dot{\omega}_{ref} + a_2 b_3 \omega_{ref} + a_2 b_1 i_q + a_2 b_2 \omega i_d + a_3 \dot{T}_L - \Delta a_1 x_2 - \Delta a_2 \Delta b_3 x_1 + \\
& \Delta a_1 \dot{\omega}_{ref} + \Delta a_2 \Delta b_3 \omega_{ref} + \Delta a_2 \Delta b_1 i_q + \Delta a_2 \Delta b_2 \omega i_d - \Delta a_2 \Delta b_4 u_q + \Delta a_3 \dot{T}_L
\end{aligned}
$$

b_1, b_2, b_3, b_4 are the nominal parameter values, $b_{t1}, b_{t2}, b_{t3}, b_{t4}$ are the actual parameter values, $\Delta b_1 = b_{t1} - b_1, \Delta b_2 = b_{t2} - b_2, \Delta b_3 = b_{t3} - b_3, \Delta b_4 = b_{t4} - b_4$.

The second-order speed regulation system can be expressed as follows

$$\begin{cases} \dot{x}_1 = x_2 + d_1 \\ \dot{x}_2 = -a_1 x_2 - a_2 b_3 x_1 - a_2 b_4 u_q + d_2 \end{cases} \tag{7}$$

3. Control Design

Based on the method designed in this paper, the PMSM control structure block diagram is shown in Figure 1.

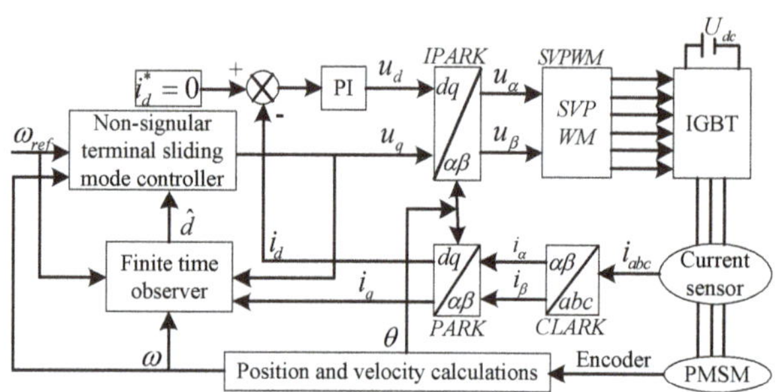

Figure 1. Permanent magnet synchronous motor (PMSM) control system based on the method designed in this paper.

3.1. Finite Time Observer

The disturbance is estimated and feedforward to the system based on the finite time observation method. The stability of the finite time control system is defined as follows.

Lemma 1. *[39] the following system is considered*

$$\dot{x} = f(x), x \in U \subseteq R^n, f(0) = 0 \tag{8}$$

where, $f : U \to R^n$ is a continuous function of x in domain of definition. For the equilibrium solution of the system, $x = 0$ is defined as finite time stability, which requires the system to be both stable and convergent in finite time. Finite time convergence means there are $\forall x_0 \in U_0 \subset R^n$ and a continuous function $T(x)$: $U_0\backslash\{0\} \to (0, +\infty)$, making the solution $x(t, x_0)$ of the system (3) satisfied the following conditions: when $t \in [0, T(x_0))$, $x(t, x_0) \in U_0\backslash\{0\}$ and $\lim\limits_{x \to T(x_0)} x(t, x_0) = 0$ are true; When $t > T(x_0)$, $x(t, x_0) = 0$ is always true. If $U = U_0 = R^n$ existed, the system is considered globally finite time stable.

Notation 1. *For writing convenience, denote $sig^\beta(x) = \text{sgn}(x)|x|^\beta$, where $x, \beta \in \mathbb{R}$, and $\text{sgn}(\cdot)$ is the sign function.*

Lemma 2. *[40] Consider the following system*

$$\dot{x} = f(x) + \hat{f}(x), f(0) = 0, x \in \mathbb{R}^n \tag{9}$$

where, $f(x)$ is a continuous homogeneous vector field, and $f(x)$ has negative homogeneous degree k with respect to expansion vector $(r_1, r_2, \ldots, r_n)$. $\hat{f}(x)$ is the estimated disturbance of the system, which satisfies $\hat{f}(0) = 0$. $x \in \mathbb{R}^n$ refers that x is belonged to the n-dimensional vector. Suppose that the asymptotically stable equilibrium point of system $\dot{x} = f(x)$ is $x = 0$ and satisfies the following conditions, $\forall x \neq 0$

$$\lim\limits_{\varepsilon \to 0} \frac{\hat{f}_i(\varepsilon^{r_1} x_1, \ldots, \varepsilon^{r_n} x_n)}{\varepsilon^{k+r_i}} = 0, i = 1, 2, \ldots, n \tag{10}$$

Then $x = 0$ is a locally finite time equilibrium point of system (9).

Lemma 3. *[41–43] Consider the nonlinear system*

$$\begin{cases} \dot{x}_i = x_{i+1}, i = 1, \ldots, n-1 \\ \dot{x}_n = u \\ y = x_1 \end{cases} \tag{11}$$

where, $x = (x_1, x_2, \ldots, x_n)^T \in R^n$ are the state variables of the system; $u \in R$ and $y \in R$ are the input and output of the system, respectively.

For system (11), the estimated state values $(\hat{x}_1, \hat{x}_2, \ldots, \hat{x}_n)$ can converge to the real states $(x_1, x_2, \ldots, x_n)$ of system (11) in a finite time by the following global finite time observer.

$$\begin{cases} \dot{\hat{x}}_i = \hat{x}_{i+1} + \lambda_i sig^{\beta_i}(x_1 - \hat{x}_1), i = 1, 2, \ldots, n-1 \\ \dot{\hat{x}}_n = u + \lambda_n sig^{\beta_n}(x_1 - \hat{x}_1) \end{cases} \tag{12}$$

where, $\beta_i > 0$, $i = 1, 2, \ldots, n$, it's a Hurwitz polynomial $s^n + \beta_1 s^{n-1} + \cdots + \beta_{n-1} s + \beta_n$.

$$\beta_i = i\beta - (i-1), i = 1, \ldots, n, \beta \in (1 - \frac{1}{n}, 1] \tag{13}$$

Lemma 4. *[44,45] A second-order system can be expressed as follows*

$$\begin{cases} \dot{x}_1 = x_2 - \lambda sig^{\frac{1}{2}}(x_1) \\ \dot{x}_2 = -vsign(x_1) + F(t) \end{cases} \tag{14}$$

If there is a positive real number f^+, $|F(t)| < f^+$ is true, and v, λ satisfies the following description.

$$v > f^+, \lambda > \sqrt{\frac{2}{v - f^+}} \frac{(v + f^+)(1 + \mu)}{(1 - \mu)} \tag{15}$$

where, μ is a constant, $0 < \mu < 1$. Then the state (x_1, x_2) of the system (14) will be converged to the equilibrium point 0 in a finite time, and the system (14) is globally stable in finite time.

Assumption 1. *The disturbance d_1, d_2 in system (7) are second-order and first-order differentiable, respectively.*

Let $\bar{x}_1 = x_1, \bar{x}_2 = x_2 + d_1, d = \dot{d}_1 + d_2 + a_1 d_1$, the following system can be derived from system (7)

$$\begin{cases} \dot{\bar{x}}_1 = \bar{x}_2 \\ \dot{\bar{x}}_2 = -a_1 \bar{x}_2 - a_2 b_3 \bar{x}_1 - a_2 b_4 u_q + d \\ y = \bar{x}_1 \end{cases} \tag{16}$$

Finite time state observers are designed for the state variables of system (7) and system (14) according to Lemma 3.

$$\begin{cases} \dot{\hat{\bar{x}}}_1 = x_2 + \hat{d}_1 - \lambda_1 sig^{\beta_1}(\hat{\bar{x}}_1 - x_1) \\ \dot{\hat{d}}_1 = -\lambda_2 sign(\hat{\bar{x}}_1 - x_1) \end{cases} \tag{17}$$

$$\begin{cases} \dot{\hat{\bar{x}}}_2 = -a_1 \hat{\bar{x}}_{22} - a_2 b_3 \hat{\bar{x}}_{11} - a_2 b_4 u_q + \hat{d}_2 + a_1 \hat{d}_1 - \bar{\lambda}_1 sig^{\beta_2}(\hat{\bar{x}}_2 - \bar{x}_2) \\ \dot{\hat{d}}_2 = -\bar{\lambda}_2 sign(\hat{\bar{x}}_2 - \bar{x}_2) \end{cases} \tag{18}$$

where, $\lambda_1, \lambda_2, \overline{\lambda}_1, \overline{\lambda}_2$ are the observation gain of the finite time observer, β_1, β_2 are the fractional power of the finite time observer, $\hat{\overline{x}}_1, \hat{\overline{x}}_2, \hat{d}_1, \hat{d}_2$ are the estimated values of $\overline{x}_1, \overline{x}_2, d_1, d_2$. The following function can be acquired $\hat{d} = a_1\hat{d}_1 + \hat{d}_2$, where, $\hat{d}$ is the estimated value of d.

Proof: The estimated errors are defined as $\widetilde{\overline{x}}_1 = \hat{\overline{x}}_1 - \overline{x}_1, \widetilde{d}_1 = \hat{d}_1 - d_1$. The error equation obtained by system (7) and system (17) can be expressed as

$$\begin{cases} \dot{\widetilde{\overline{x}}}_1 = \widetilde{d}_1 - \lambda_1 sig^{\beta_1}(\widetilde{\overline{x}}_1) \\ \dot{\widetilde{d}}_1 = -\lambda_2 sig^{\beta_2}(\widetilde{\overline{x}}_1) - \dot{d}_1 \end{cases} \tag{19}$$

From Assumption 1, $-L \le \dot{d}_1 \le L$ can be obtained. When the gain meets (15), the error system can be reached stability within a finite time. Namely, $\hat{d}_1$ can converge to the true value d_1 in finite time. After this moment, $\hat{d}_1 \equiv d_1, \hat{\overline{x}}_1 \equiv \overline{x}_1$ are always true. Then $\overline{x}_2 \equiv x_2 + \hat{d}_1$ is true. The proof of the finite time stability of the error system for the system (17) is the same as above.

In conclusion, the observation state $(\hat{\overline{x}}_1, \hat{\overline{x}}_2, \hat{d}_1, \hat{d}_2)$ estimated by the finite time observer will converge to the actual values $(\overline{x}_1, \overline{x}_2, d_1, d_2)$ of system (7) and system (18) within a finite time.

3.2. Non-Singular Terminal Sliding Mode Control

Consider the following second-order system

$$\begin{cases} \dot{e}_1 = e_2 \\ \dot{e}_2 = f(e) + u + d(t) \end{cases} \tag{20}$$

where, e_1, e_2 are the state variables, $d(t)$ is the disturbance, $|d(t)| \le D$.

The non-singular terminal sliding mode surface is selected as follows

$$s = e_1 + \frac{1}{\eta} e_2^{p/q} \tag{21}$$

where, $\eta > 0$, and $p > q > 0$ are odd.

In order to make the system state converge to the actual value in finite time, the control law can be designed as

$$u = -(D + \varepsilon)sign(s) - f(e) - \eta\frac{q}{p}e_2^{2-p/q} \tag{22}$$

where, ε is the robustness coefficient, $\varepsilon > 0, 1 < p/q < 2$.

In order to prove the stability of the designed system, the Lyapunov function is selected as

$$V = \frac{1}{2}s^2 \tag{23}$$

The derivative of V is as following

$$\begin{aligned} \dot{V} &= s\dot{s} \\ &= s(\dot{e}_1 + \frac{1}{\eta}\frac{p}{q}e_2^{p/q-1}\dot{e}_2) \\ &= s[e_2 + \frac{1}{\eta}\frac{p}{q}e_2^{p/q-1}(f(e) + u + d(t))] \\ &= s[e_2 + \frac{1}{\eta}\frac{p}{q}e_2^{p/q-1}(f(e) + u + d(t))] \\ &= s[e_2 + \frac{1}{\eta}\frac{p}{q}e_2^{p/q-1}(-(D+\varepsilon)sign(s) - \eta\frac{q}{p}e_2^{2-p/q} + d(t))] \\ &= s[\frac{1}{\eta}\frac{p}{q}e_2^{p/q-1}(-(D+\varepsilon)sign(s) + d(t))] \\ &\le \frac{1}{\eta}\frac{p}{q}e_2^{p/q-1}(-\varepsilon)|s| \end{aligned} \tag{24}$$

where p, q are positive odd integers and $1 < p/q < 2$, thus $e_2^{p/q-1} > 0$. Then $\dot{V} < 0$ is always true.

According to the above analysis, the control law (22) designed for system (20) can ensure the system convergence.

Assume that the system state reaches the sliding mode surface at t_r, that is to say $s(t_r) = 0$, then

$$\begin{cases} e_1 + \frac{1}{\eta}e_2^{p/q} = 0 \\ \dot{e}_1 = \eta e_1^{q/p} \end{cases} \tag{25}$$

The time it takes for the system to stabilize to the equilibrium point can be expressed as

$$t_s = \frac{p}{\eta(p-q)}\left|e_1(t_r)\right|^{1-q/p} \tag{26}$$

As can be seen from the time function (26), the larger η is, the smaller t_s is to the stable state; However, if η is too large, the effect of switching item will be strengthened due to the change of s symbol, and the control output will be weakened.

For system (16) (17) (18), the non-singular terminal sliding mode surface function is selected as

$$s = \bar{x}_1 + \frac{1}{\eta}\hat{\bar{x}}_2^{p/q} \tag{27}$$

The control law is designed as

$$u_q = \frac{1}{a_2 b_4}\left[-a_1\hat{\bar{x}}_2 - a_2 b_3\bar{x}_1 + \hat{d} + \eta\frac{q}{p}\hat{\bar{x}}_2^{2-p/q} + (D+\varepsilon)sign(s)\right] \tag{28}$$

Choose the Lyapunov function as

$$V = \frac{1}{2}s^2 \tag{29}$$

Derivation of (29)

$$\begin{aligned} \dot{V} &= s\dot{s} \\ &= s(\dot{\bar{x}}_1 + \frac{1}{\eta}\frac{p}{q}\hat{\bar{x}}_2^{p/q-1}\dot{\hat{\bar{x}}}_2) \\ &= s[\hat{\bar{x}}_2 + \frac{1}{\eta}\frac{p}{q}\hat{\bar{x}}_2^{p/q-1}(-a_1\hat{\bar{x}}_2 - a_2 b_3\bar{x}_1 - a_2 b_4 u_q + d)] \end{aligned} \tag{30}$$

Consider the control law (28)

$$\begin{aligned} \dot{V} &= s[\frac{1}{\eta}\frac{p}{q}\hat{\bar{x}}_2^{p/q-1}(-(D+\varepsilon)sign(s) - \hat{d} + d)] \\ &= s[\frac{1}{\eta}\frac{p}{q}\hat{\bar{x}}_2^{p/q-1}(-\varepsilon sign(s) + d - \hat{d} - Dsign(s))] \\ &\leq \frac{1}{\eta}\frac{p}{q}\hat{\bar{x}}_2^{p/q-1}(-\varepsilon)|s| \\ &\leq 0 \end{aligned} \tag{31}$$

In (31), p, q are positive odd integers and $1 < p/q < 2$, thus $\hat{\bar{x}}_2^{p/q-1} > 0$ is true, and $\eta > 0$, so $\frac{1}{\eta}\frac{p}{q}|s| > 0$ is true, because $\varepsilon > 0$, then $\frac{1}{\eta}\frac{p}{q}\hat{\bar{x}}_2^{p/q-1}(-\varepsilon)|s| < 0$ can be proved.

It can be known from (31) that the second-order PMSM system can reach a stable state in a finite time based on the composite strategy of finite time observer and non-singular fast terminal sliding mode.

4. Simulation and Analysis

In order to verify the effectiveness of the proposed method, comparative simulations are built on the traditional cascade PID, cascade sliding mode, and the proposed method this paper. The simulations are based on Asus notebook FX503VD, Intel(R)Core i7 7700HQ, CPU@2.80GHz, RAM 7.88GB (Hynix

DDR4 2400MHz), SanDiskSD8SN8U128G1002(128GB/solid state disk), Nvidia GeForce GTX 1050 (4GB/Asus), 64-bit operating system, matlab 2017b (ASUSTek Computer Inc., Taiwan, China). In order to ensure the fairness of the comparison, the bus voltage is set to 36 V. The reference speed of PMSM is set at 1000 r/min. The PMSM parameters used for simulation are shown in Table 1. The parameters of the traditional cascade PID, the traditional cascade sliding mode control and NTSMC-FTO proposed in this paper are shown in Tables 2–4, respectively. In cascade SMC controller, SMC and PID are used for speed loop and current loop, respectively. $s = cx_1 + x_2$ is taken as the sliding mode surface of SMC, and $i_q^* = \frac{2J}{3n_p\psi_f}\int_0^t [c(x_2) + Mu \times sign(s) + \kappa s]dt$ as the expression of output.

Table 1. Rated parameters of the permanent magnet synchronous motor (PMSM).

Rated Power	P_N	200	W
line resistance	R	0.33	Ω
line inductance	L	9×10^{-4}	H
magnetic poles	n_p	4	pairs
torque constant	K_t	0.087	N·m/A
rated power	U_N	36	VAC
rated current	I_N	7.5	A
rotor inertia	J	1.89×10^{-5}	kg·m^2
rated speed	n_N	3000	r/min

Table 2. The cascade PID controller.

Description	Parameter	Value
speed loop proportional gain	K_1	0.01
speed loop integral gain	I_1	0.95
speed loop proportional gain	K_2	50
speed loop integral gain	I_2	100,000
current loop Id proportional	K_{p1}	2000
current loop Id integral gain	K_{I1}	100,000

Table 3. The cascade sliding mode control (SMC) controller.

Description	Parameter	Value
error gain of SMC	c	10.8
switch gain of SMC	Mu	100
sliding mode surface gain of SMC	κ	12
speed loop proportional gain	K_2	50
speed loop integral gain	I_2	100,000
current loop Id proportional	K_{p1}	2000
current loop Id integral gain	K_{I1}	10,000

Table 4. The proposed controller this paper.

Description	Parameter	Value
the power of NTSMC	p	37
the power of NTSMC	q	35
proportional gain of NTSMC	η	5100
switch gain of NTSMC	ε	200,000,000,000
the gain of observer1	λ_1	1,000,000
the gain of observer1	λ_2	10
the gain of observer2	$\overline{\lambda}_1$	50,000,000
the gain of observer2	$\overline{\lambda}_2$	500
current loop i_d proportional	K_{p1}	2000
current loop i_d integral gain	K_{I1}	10,000

There are two groups of comparative simulations, one is the response curve at the phase of startup, and the other is the response curve when the load torque suddenly changes at a constant speed stage. It can be found from the comparison results that the NTSMC-FTO proposed in this paper, which regulate the speed and current of PMSM in one loop, has a better dynamic performance and disturbance rejection property than the traditional PID and SMC.

Case I: Phase of start. The reference speed of PMSM is set at 1000 r/min, and the motor starts without load torque. Figure 2a–c are ω, i_q, i_d response curves of startup, respectively. The solid (blue) line is NTSMC-FTO controller, the dotted (pink) line is PID controller, and the dotted (black) line is SMC controller. It can be summarized that when the motor starts without speed overshoot, When the motor starts without speed overshoot, it takes 0.0028 s for NTSMC-FTO to reach the steady state, compared with the cascade SMC and PID 0.045 s is needed. The cost to reach steady state is reduced by 0.0422 s. The d-axis and q-axis currents chattering of NTSMC-FTO are smaller than the cascade SMC and PID controller. The comparative simulation results of startup can be seen in Table 5.

Table 5. The comparative simulation results of start up.

Method	Reference Speed	Time to Reach Steady State
NTSMC-FTO	1000 r/min	0.0028 s
the cascade SMC	1000 r/min	0.045 s
the cascade PID	1000 r/min	0.045 s

Case II: Load torque is changed suddenly. The load torque has a sudden change from $T_L = 0\,\mathrm{N \cdot m}$ to $T_L = 0.1\,\mathrm{N \cdot m}$ at $t = 0.1$ s. Figure 3a–c, are ω, i_q, i_d response curves of load torque sudden change, respectively. When the load torque changed suddenly, the speed of NTSMC-FTO is decreased by 2.5 r/min (0.25%), while SMC and PID are 87 r/min (8.7%) and 74 r/min (7.4%), respectively. The recovery time of NTSMC-FTO, SMC and PID to 1000 r/min are 0.0004s, 0.06s and 0.06 s, respectively. The comparative simulation results of load changed suddenly can be seen in Table 6.

Table 6. The comparative simulation results of load changed suddenly.

Method	Reference Speed	Decreased Value of Speed	The Recover Time of Steady State
NTSMC-FTO	1000 r/min	2.5 r/min	0.0004 s
the cascade SMC	1000 r/min	87 r/min	0.06 s
the cascade PID	1000 r/min	74 r/min	0.06 s

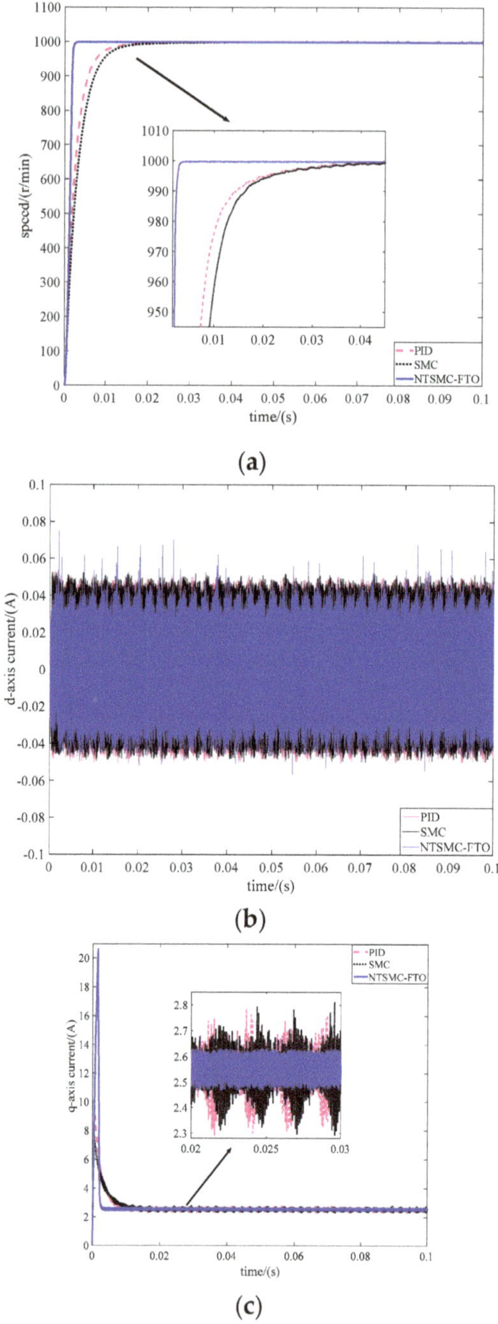

Figure 2. Performance comparisons under the PID, permanent magnet synchronous motor (SMC) and NTSMC-FTO at the phase of startup. (**a**) Speed response curves. (**b**) *d*-axis current curves. (**c**) *q*-axis current curves.

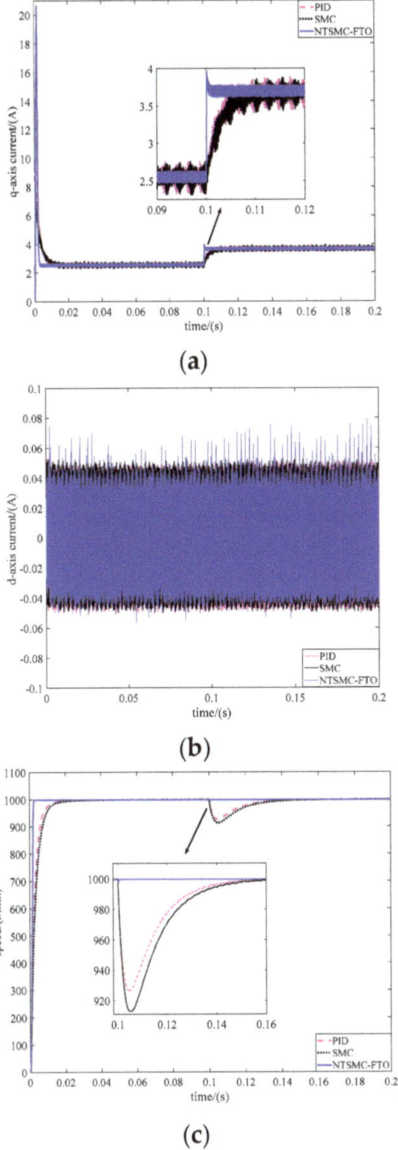

Figure 3. Performance comparisons under the PID, SMC and NTSMC-FTO with sudden load torque change. (**a**) Speed response curves. (**b**) *d*-axis current curves. (**c**) *q*-axis current curves.

Figure 4a,b are disturbance d_1 and disturbance d_2 curves estimated by the finite time observers, respectively.

It can be concluded that compared with the traditional SMC and PID, the NTSMC-FTO proposed in this paper, which put the speed and current in one loop to regulate, has a faster tracking speed and a better disturbance rejection performance, demonstrating that the proposed method in this paper has strong robustness.

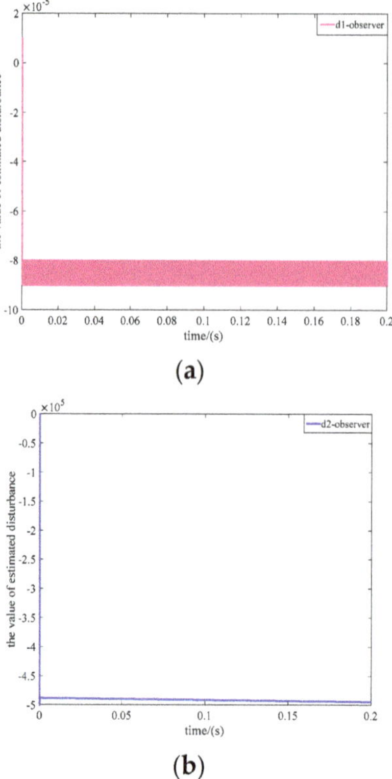

(a)

(b)

Figure 4. The estimated value based on the finite-time observer. (**a**)The estimated value of d_1 based on the finite-time observer. (**b**)The estimated value of d_2 based on the finite-time observer.

5. Conclusions

In this paper, a novel speed-current single-loop controller for the PMSM drive system has been proposed. Simulations have verified that compared with the cascade PID and cascade SMC method, the proposed method has a faster start up response and a better disturbance rejection performance. The disturbance can be accurately estimated and compensated by the proposed disturbance observer. Future research can be carried out from the state constraint of the proposed method to reduce the q-axis transient current and improve the safety of the system.

Author Contributions: This is a joint work and the authors were in charge of their expertise and capability: Y.W. (Yao Wang) for investigation and analysis; H.Y. for funding support; Z.C., Y.W. (Yuchen Wang), and Y.L. for manuscript revision.

Funding: This research was supported by National Natural Science Foundation of China (NSFC) under Grant 41576096.

Acknowledgments: The authors would like to express their gratitude to all those who helped them during the writing of this paper. And the authors would like to thank the reviewers for their valuable comments and suggestions.

Conflicts of Interest: The authors declare no conflict of interest.

References

1. Lara, J.; Xu, J.; Chandra, A. Effects of Rotor Position Error in the Performance of Field Oriented Controlled PMSM Drives for Electric Vehicle Traction Applications. *IEEE Trans. Ind. Electron.* **2016**, *63*, 1. [CrossRef]
2. Kommuri, S.K.; Defoort, M.; Karimi, H.R.; Veluvolu, K.C. A Robust Observer-Based Sensor Fault-Tolerant Control for PMSM in Electric Vehicles. *IEEE Trans. Ind. Electron.* **2016**, *63*, 7671–7681. [CrossRef]
3. Liu, H.X.; Li, S.H. Speed control for PMSM servo system using predictive functional control and extended state observer. *IEEE Trans. Ind. Electron.* **2012**, *59*, 1171–1183. [CrossRef]
4. Chaoui, H.; Khayamy, M.; Aljarboua, A.A. Adaptive Interval Type-2 Fuzzy Logic Control for PMSM Drives with a Modified Reference Frame. *IEEE Trans. Ind. Electron.* **2017**, *64*, 3786–3797. [CrossRef]
5. Valente, G.; Formentini, A.; Papini, L.; Gerada, C.; Zanchetta, P. Performance improvement of bearingless multisector PMSM with optimal robust position control. *IEEE Trans. Power Electron.* **2019**, *34*, 3575–3585. [CrossRef]
6. Sun, Z.X.; Li, S.H.; Wang, J.G.; Zhang, X.H.; Mo, X.H. Adaptive composite control method of permanent magnet synchronous motor systems. *Trans. Inst. Meas. Control* **2018**, *11*, 3345–3357. [CrossRef]
7. Yu, J.; Shi, P.; Zhao, L. Finite-time command filtered backstepping control for a class of nonlinear systems. *Automatica* **2018**, *92*, 173–180. [CrossRef]
8. Liang, W.; Fei, W.; Luk, P.C.-K. An Improved Sideband Current Harmonic Model of Interior PMSM Drive by Considering Magnetic Saturation and Cross-Coupling Effects. *IEEE Trans. Ind. Electron.* **2016**, *63*, 4097–4104. [CrossRef]
9. Ren, H.P.; Liu, D. Nonlinear feedback control of chaos in permanent magnet synchronous motor. *IEEE Trans. Circuits Syst. II Express Briefs* **2016**, *53*, 45–50.
10. Formentini, A.; Trentin, A.; Marchesoni, M.; Zanchetta, P.; Wheeler, P. Speed Finite Control Set Model Predictive Control of a PMSM Fed by Matrix Converter. *IEEE Trans. Ind. Electron.* **2015**, *62*, 6786–6796. [CrossRef]
11. Cheema, M.A.M.; Fletcher, J.E.; Farshadnia, M.; Xiao, D.; Rahman, M.F. Combined Speed and Direct Thrust Force Control of Linear Permanent-Magnet Synchronous Motors with Sensorless Speed Estimation Using a Sliding-Mode Control with Integral Action. *IEEE Trans. Ind. Electron.* **2017**, *64*, 3489–3501. [CrossRef]
12. Lang, X.Y.; Yang, M.; Xu, H.D.; Long, J.; Xu, D.G. A non-cascade predictive speed and current controller with PWM modulation for PMSM. In Proceedings of the IECON 2016-42nd Annual Conference of the IEEE Industrial Electronics Society, Florence, Italy, 23–26 October 2016.
13. Guo, T.L.; Sun, Z.X.; Wang, X.Y.; Li, S.H.; Zhang, K.J. A simple current-constrained controller for permanent-magnet synchronous motor. *IEEE Trans. Ind. Inform.* **2019**, *15*, 1486–1495. [CrossRef]
14. Yan, Y.; Yang, J.; Sun, Z.; Zhang, C.; Li, S.; Yu, H. Robust Speed Regulation for PMSM Servo System with Multiple Sources of Disturbances via an Augmented Disturbance Observer. *IEEE/ASME Trans. Mechatron.* **2018**, *23*, 769–780. [CrossRef]
15. Liu, X.; Yu, H.; Yu, J.; Zhao, L. Combined Speed and Current Terminal Sliding Mode Control with Nonlinear Disturbance Observer for PMSM Drive. *IEEE Access* **2018**, *6*, 29594–29601. [CrossRef]
16. Chaoui, H.; Khayamy, M.; Okoye, O. Adaptive RBF network based direct voltage control for interior PMSM based vehicles. *IEEE Trans. Veh. Technol.* **2018**, *67*, 5740–5749. [CrossRef]
17. Zhang, X.G.; He, Y.K. Direct voltage-selection based model predictive direct speed control for PMSM drives without weighting factor. *IEEE Trans. Power Electron.* **2019**, *34*, 7838–7851. [CrossRef]
18. Preindl, M.; Bolognani, S. Model predictive direct speed control with finite control set of PMSM drive systems. *IEEE Trans. Power Electron.* **2013**, *28*, 1007–1015. [CrossRef]
19. Smidl, V.; Janous, S.; Adam, L.; Peroutka, Z. Direct Speed Control of a PMSM Drive Using SDRE and Convex Constrained Optimization. *IEEE Trans. Ind. Electron.* **2018**, *65*, 532–542. [CrossRef]
20. Liu, B.; Zhou, B.; Ni, T.H. Principle and stability analysis of an improved self-sensing control strategy for surface-mounted PMSM drives using second-order generalized integrators. *IEEE Trans. Energy Convers.* **2018**, *33*, 126–136. [CrossRef]
21. Mynar, Z.; Veselý, L.; Vaclavek, P. PMSM Model Predictive Control with Field-Weakening Implementation. *IEEE Trans. Ind. Electron.* **2016**, *63*, 5156–5166. [CrossRef]
22. Liu, J.; Li, H.W.; Deng, Y.T. Torque ripple minimization of PMSM based on robust ILC via adaptive sliding mode control. *IEEE Trans. Power Electron.* **2018**, *33*, 3655–3671. [CrossRef]

23. Tarczewski, T.; Grzesiak, L.M. Constrained state feedback speed control of PMSM based on model predictive approach. *IEEE Trans. Ind. Electron.* **2016**, *63*, 3867–3875. [CrossRef]

24. Wang, W.-C.; Liu, T.-H.; Syaifudin, Y. Model Predictive Controller for a Micro-PMSM-Based Five-Finger Control System. *IEEE Trans. Ind. Electron.* **2016**, *63*, 3666–3676. [CrossRef]

25. Xia, C.; Li, S.; Shi, Y.; Zhang, X.; Sun, Z.; Yin, W. A Non-Smooth Composite Control Approach for Direct Torque Control of Permanent Magnet Synchronous Machines. *IEEE Access* **2019**, *7*, 45313–45321. [CrossRef]

26. Zhang, G.Q.; Wang, G.L.; Yuan, B.H. Active disturbance rejection control strategy for signal injection-based sensorless IPMSM drives. *IEEE Trans. Transp. Electrif.* **2018**, *1*, 330–339. [CrossRef]

27. Li, S.H.; Zhou, M.M.; Yu, X.H. Design and Implementation of terminal sliding mode control method for PMSM speed regulation system. *IEEE Trans. Ind. Inform.* **2013**, *9*, 1879–1891. [CrossRef]

28. Li, S.H.; Wu, C.; Sun, Z.X. Design and implementation of clutch control for automotive transmissions using terminal-sliding-mode control and uncertainty observer. *IEEE Trans. Veh. Technol.* **2016**, *65*, 1890–1898. [CrossRef]

29. Du, H.B.; Chen, X.P.; Wen, G.H.; Yu, X.H.; Lü, J.H. Discrete-time fast terminal sliding mode control for permanent magnet linear motor. *IEEE Trans. Ind. Electron.* **2018**, *65*, 9916–9927. [CrossRef]

30. Wu, F.; Li, P.; Wang, J. FO improved fast terminal sliding mode control method for permanent-magnet synchronous motor with FO disturbance observer. *IET Control. Theory Appl.* **2019**, *13*, 1425–1434. [CrossRef]

31. Xu, W.; Junejo, A.K.; Liu, Y.; Islam, M.R. Improved Continuous Fast Terminal Sliding Mode Control with Extended State Observer for Speed Regulation of PMSM Drive System. *IEEE Trans. Veh. Technol.* **2019**, in press. [CrossRef]

32. Guo, L.; Wang, D.; Diao, L.; Peng, Z. Direct voltage control of stand-alone DFIG under asymmetric loads based on non-singular terminal sliding mode control and improved extended state observer. *IET Electr. Power Appl.* **2019**, *13*, 958–968. [CrossRef]

33. He, L.; Wang, F.; Wang, J.; Rodriguez, J. Zynq Implemented Lunenberger Disturbance Observer Based Predictive Control Scheme for PMSM Drives. *IEEE Trans. Power Electron.* **2019**, in press. [CrossRef]

34. Yang, J.; Cui, H.Y.; Li, S.H.; Zolotas, A. Optimized active disturbance rejection control for DC-DC buck converters with uncertainties using a reduced-order GPI observer. *IEEE Trans. Circuits Syst. I Regul. Pap.* **2018**, *65*, 832–841. [CrossRef]

35. An, H.; Liu, J.X.; Wang, C.H.; Wu, L.G. Disturbance observer-based anti-windup control for air-breathing hypersonic vehicles. *IEEE Trans. Ind. Electron.* **2016**, *63*, 3038–3049. [CrossRef]

36. Hebertt, S.R.; Jesús, L.F.; Carlos, G.R.; Marco, A.C.O. On the control of the permanent magnet synchronous motor: An active disturbance rejection control approach. *IEEE Trans. Control Syst. Technol.* **2014**, *22*, 2056–2063.

37. Chen, W.-H. Disturbance Observer Based Control for Nonlinear Systems. *IEEE/ASME Trans. Mechatron.* **2004**, *9*, 706–710. [CrossRef]

38. Chen, M.-S.; Chen, C.-C. Robust Nonlinear Observer for Lipschitz Nonlinear Systems Subject to Disturbances. *IEEE Trans. Autom. Control.* **2007**, *52*, 2365–2369. [CrossRef]

39. Khalil, H. *Nonlinear Systems*, 2nd ed.; Prentice-Hall: Upper Saddle River, NJ, USA, 1996.

40. Hong, Y.; Huang, J.; Xu, Y. On an output feedback finite-time stabilization problem. *IEEE Trans. Autom. Control* **2001**, *46*, 305–309. [CrossRef]

41. Shen, Y.J.; Huang, Y.H. Uniformly observable and globally Lipschitzian nonlinear systems admit global finite-time observers. *IEEE Trans. Autom. Control* **2009**, *54*, 2621–2625. [CrossRef]

42. Perruquetti, W.; Floquet, T.; Moulay, E. Finite-Time Observers: Application to Secure Communication. *IEEE Trans. Autom. Control.* **2008**, *53*, 356–360. [CrossRef]

43. Du, H.B.; Qian, C.J.; Yang, S.Z.; Li, S.H. Recursive design of finite time convergent observers for a class of time varying nonlinear systems. *Automatica* **2013**, *49*, 601–609. [CrossRef]

44. Davila, J.; Fridman, L.; Levant, A. Second-order sliding-mode observer for mechanical systems. *IEEE Trans. Autom. Control.* **2005**, *50*, 1785–1789. [CrossRef]

45. Lin, C.-K. Nonsingular Terminal Sliding Mode Control of Robot Manipulators Using Fuzzy Wavelet Networks. *IEEE Trans. Fuzzy Syst.* **2006**, *14*, 849–859. [CrossRef]

Article

Extended State Observer-Based Predictive Speed Control for Permanent Magnet Linear Synchronous Motor

Yao Wang, Haitao Yu *, Zhiyuan Che, Yuchen Wang and Cheng Zeng

School of Electrical Engineering, Southeast University, Nanjing 210096, China
* Correspondence: htyu@seu.edu.cn

Received: 19 August 2019; Accepted: 6 September 2019; Published: 11 September 2019

Abstract: Combining the feedback of predictive function control and the feedforward of extended state observer, a composite control strategy is proposed for the permanent magnet linear synchronous motor (PMLSM). The mathematical model of the PMLSM vector control system is established based on the basic structure and operation mechanism of PMLSM. Then, a speed regulator based on predictive function control (PFC) is designed to improve the speed tracking performance of the PMLSM drive system. The state and disturbance of the PMLSM system estimated by the extended state observer (ESO) transferred to the PMLSM drive system, and the robustness of the drive system will be improved. Comparative simulation and experiment results show that the proposed method has better speed tracking performance and disturbance rejection property.

Keywords: permanent magnet linear synchronous motor (PMLSM); extended state observer (ESO); predictive function control (PFC); composite control; robustness

1. Introduction

Permanent magnet linear synchronous motors (PMLSMs) have been widely used in industrial applications due to the high speed, high acceleration, high accuracy, and high-power density [1]. However, the PMLSM is a nonlinear system characterized by time-varying, strong coupling, and external disturbances [2], thus the control of the PMLSM servo system is still challenging. For the sake of meeting the requirements of system static and dynamic performances performances, it is necessary to study novel control strategies and improve the robustness of the vector control system [3].

A considerable number of research literatures have been reported in the motor control community [3–10]. Generally, there are two aspects that can be considered to achieve a good dynamic performance for a PMLSM. One type of efficient method is to optimize the structure design of the PMLSM, for example, an optimal design of the auxiliary-teeth model was carried out to decrease the circulating current of PMLSM, and the effect of the unbalanced phase based on the auxiliary-teeth installation was considered [4]. However, for the PMLSM herein, another method is to design an appropriate control strategy to improve the robustness of PMLSM, which is of great significance. For the wafer stage of the lithography systems, Song et al. [3] presented an iterative learning control (ILC)-based feedforward compensation scheme. In the presence of parameter uncertainties and external disturbances, a reduced-order proportional-integral (PI)-based robust cascade control was presented to enhance the robust performance of current and speed regulations [5]. The sliding mode control (SMC) has been extensively investigated for a few decades because of its particular robustness to suppress the disturbances [6], and an extended state observer (ESO)-based SMC control for the permanent magnet synchronous motor (PMSM) servo system was presented in the work by the authors of [7]. In the framework of the active disturbance rejection control (ADRC) technique [8], Sira-Ramirez et al. [9] designed an active disturbance rejection control scheme for the angular velocity trajectory tracking.

In the work by the authors of [10], a comprehensive overview on disturbance/uncertainty estimation and attenuation techniques in PMSMs were given, where the various disturbances and uncertainties were discussed, and the relevant control methods were summarized.

In recent years, the model predictive control (MPC) technology has been well developed due to its conceptual simplicity, high dynamic response, and easy implementation etc. [11]. As mentioned in the work by the authors of [12], the MPC is one of the most popular control strategy in industrial applications, where satisfactory system performances can be guaranteed. One can regard the MPC as an optimal control method [13], which predicts the output of the state variable by employing the dynamic model of the controlled plant [14] and generates the future control input according to the optimization of the performance index function or cost function during each sampling period [15]. In the work by the authors of [16], three predictive current control schemes were compared by theoretical analyses, simulation results, and experiments. By replacing the conventional cascade structure, the speed and current controllers were combined together, and the MPC algorithm was applied to electrical drives [17]. Zhang et al. [18] presented a composite observer-based robust fault-tolerant predictive current control algorithm for PMSMs, which can eliminate the influence of motor parameter perturbation and permanent magnet demagnetization. Based on the field-oriented control (FOC) and vector control approaches, Liu et al. [19] proposed a composite speed regulation comprised by ESO feedforward compensation and predictive functional control (PFC) feedback control. In addition, an ESO can timely estimate the "lumped disturbances", including both internal unmodelled dynamics and external disturbances in the system [20], and thus the ESO-based control approaches (usually called ADRC schemes) can handle large uncertainties as well as achieve the desired performance [21]. It should be emphasized that most of the existing results are concentrated on PMSM servo control. However, the PMLSM is widely used in many applications. Thus, the investigation about designing an ESO-based MPC/PFC law to compensate the adverse effects of uncertainties and disturbances on a PMLSM is of significant importance, which remains as an open area.

The rest of this paper is organized as follows. In Section 2, the mathematical description of a PMLSM is analyzed and its simplified model is established. The main results are given in Section 3. The PFC controller is designed, and an ESO feedforward compensation is incorporated in the PFC feedback item, thus a composite speed regulate scheme is developed. Simulation results and comparative experiments are given to confirm the advantages and disturbance rejection performance of the proposed method in Section 4. Section 5 concludes the paper.

2. Mathematical Model of the Pmlsm

The objective of the servo system is tracking the reference signals as closely as possible, while guaranteeing control system and anti-disturbance performances. In this section, the mathematical model of the PMLSM is established in Laplace domain.

2.1. Mathematical Description of a Pmlsm

Generally, the basic equations of the PMLSM servo control system include the voltage equation, flux linkage equation, electromagnetic force equation, and the mechanical motion equation [22]. With the assumptions that the three-phase primary windings of the PMLSM are completely symmetrical and the traveling wave magnetic field is sinusoidally distributed [23], one can describe the PMLSM on the two-phase orthogonal synchronous rotating coordinate (d-q) as follows,

$$\begin{cases} L_d \frac{di_d}{dt} = u_d - Ri_d + \frac{\pi}{\tau}v\psi_q \\ L_q \frac{di_q}{dt} = u_q - Ri_q - \frac{\pi}{\tau}v\psi_d \\ F_e = \frac{3\pi}{2\tau}\left[\psi_f i_q + \left(L_d - L_q\right)i_d i_q\right] \\ M\frac{dv}{dt} = F_e - F_L - B_f v \end{cases} \tag{1}$$

where i_d, u_d, L_d, ψ_d, i_q, u_q, L_q, and ψ_q are the d-axis and q-axis currents, voltages, inductances, and flux linkages, respectively; ψ_f is the permanent magnet flux linkage; R is the winding resistance; τ is the pole pitch; F_e is the electromagnetic force; F_L is the load torque; B_f is the viscous friction coefficient; v and M are the velocity and the mass of linear motor rotor, respectively.

Considering a surface-mounted PMLSM, and thus $L_d = L_q = L$. We can assume that there no hysteresis, no eddy currents, no cogging, no bearing components, no magnetic saturation, no temperature rise leading to a higher resistance R and lower magnet flux, perfect magnetic alignment with an ideal symmetric 3-phase motor amplifier are exhibited [24]. In addition, we define the force constant as $K_f = 3\pi/2\tau \cdot \psi_f$. As a result, the Equation (1) can be rewritten as

$$\begin{cases} u_d = L\frac{di_d}{dt} + Ri_d - \frac{\pi}{\tau}vLi_q \\ u_q = L\frac{di_q}{dt} + Ri_q + \frac{\pi}{\tau}v\left(Li_d + \psi_f\right) \\ M\frac{dv}{dt} = K_fi_q - F_L - B_fv \end{cases} \tag{2}$$

where $\psi_q = Li_q$ and $\psi_d = Li_d + \psi_f$, respectively.

2.2. A Simplified Model of the Pmlsm

By the framework of FOC and vector control technique [25], a cascade structure comprising the speed regulation and current controllers is usually adopted in the PMLSM servo system [26]. Because the excitation flux linkage is established by permanent magnets, the current i_d is controlled to be zero in the inner loop, while the electromagnetic force is proportional to the current i_q, whose command current is determined by the outer speed loop. It should be emphasized that the dynamics of current loop is usually much faster than speed loop response, that is, the current loop has a higher bandwidth that of speed loop. As a result, the current i_q nearly equals to its expected value i_q^*, and thus the transfer function of inner current loop can be regarded as 1, i.e., $G_i(s) \approx 1$ [27]. To this end, the simplified control diagram of the PMLSM servo system is shown in Figure 1. In Figure 1, e denotes the velocity tracking error satisfying $e = v^* - v$, and v^* is the given signal.

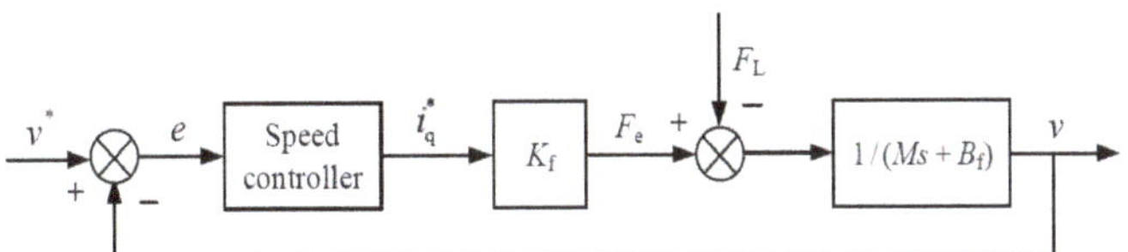

Figure 1. The simplified control diagram of the PMLSM servo system.

According to the control scheme of PMLSM given in Figure 1, the first-order system model in the Laplace domain can be expressed as

$$\Omega(s) = \frac{K_fI_q^*(s) - F_L(s)}{Ms + B_f} \tag{3}$$

3. ESO-Based PFC Design

In this section, a PFC law is firstly designed based on the mathematical model obtained in Section 2. Then, an ESO is constructed to estimate the disturbances in the PMLSM servo system, which are incorporated in the PFC feedback controller. Finally, the whole block diagram for the PMLSM control system is derived.

3.1. PFC Design

Among all the MPC approaches, PFC is characterized by the low online computation, while maintaining the advantages of the MPC method. In general, the PFC design methodology mainly comprises the following five components.

- (1) Basic functions

$$i_q^*(k+i) = \sum_{j=1}^{N} \mu_j u_{bj}(i), \qquad i = 1, 2, \ldots, P \tag{4}$$

where i_q^* is the future control variables, N is the number of base functions, μ_j is the coefficients related to the performance optimization index, $u_{bj}(i)$ is the base functions determined by the reference signals, and P is the optimization horizon.

In this paper, we select the base function as the form of step response, i.e., $u_{bj}(i) = 1$, and set $N = 1$. As a result, the following equation can be derived.

$$i_q^*(k) = \mu_1 \tag{5}$$

- (2) Prediction model

The prediction model of the plant output can be given in its discrete time form, the following formulation is easily obtained by applying the Euler discretization method to Equation (3).

$$v_m(k+1|k) = \alpha_m v_m(k) + K_m(1 - \alpha_m)i_q^*(k) \tag{6}$$

where $v_m(k+1|k)$ is the predicted velocity at time $t = (k+1)T_s$ and T_s is the sample time; K_m and α_m are coefficients of the prediction model (differential equation form).

For the $(k+2)$th sampling time, we have

$$v_m(k+2|k) = \alpha_m v_m(k+1) + K_m(1 - \alpha_m)i_q^*(k+1) \tag{7}$$

According to the well-known mean-level control strategy [28], the control variables are regarded as a constant during the prediction, that is,

$$i_q^*(k) = i_q^*(k+1) = \cdots = i_q^*(k+P-1) \tag{8}$$

By replacing $v_m(k+1|k)$ with $v_m(k+1)$ and combining Equations (6)–(8) we get

$$v_m(k+2|k) = \alpha_m^2 v_m(k) + K_m(1 - \alpha_m^2)i_q^*(k) \tag{9}$$

Obeying the same principle, the prediction output generated at the $(k+2)$th sampling time is as follows,

$$v_m(k+i|k) = \alpha_m^i v_m(k) + K_m(1 - \alpha_m^i)i_q^*(k) \tag{10}$$

Correspondingly, we can described the prediction output as the matrix form based on the Equations (5), (6), (9), and (10), which is as follows,

$$V_m(k) = V_o(k) + V_b(k)\mu_1 \tag{11}$$

where

$$
\begin{aligned}
V_m(k) &= \begin{bmatrix} v_m(k+1|k) & v_m(k+2|k) & \cdots & v_m(k+P|k) \end{bmatrix}^{\mathrm{T}} \\
V_o(k) &= \begin{bmatrix} \alpha_m & \alpha_2^P & \cdots & \alpha_m^P \end{bmatrix}^{\mathrm{T}} v_m(k) \\
V_b(k) &= K_m \begin{bmatrix} (1 - \alpha_m) & (1 - \alpha_2^P) & \cdots & (1 - \alpha_m^P) \end{bmatrix}^{\mathrm{T}}
\end{aligned}
\tag{12}
$$

- **(3) Error correction**

Because the presence of the model mismatch, unknown disturbance, parameter variation, and noise, the prediction error between prediction model output and actual output is existed. When PFC is applied to control systems with a small sampling period, it is generally believed that the error in this processing period remains constant [19]. Since the sampling period of the PMLSM control system in this paper is 10^{-4} s, it is assumed that all prediction errors are equal only within this interruption period. The error can be expressed as follows,

$$e(k) = e(k+1) = e(k+2) = \cdots = e(k+P-1) = v(k) - v_{\mathrm{m}}(k) \tag{13}$$

where $v(k)$ is the measured velocity of the practical control system.

- **(4) Reference trajectory**

A first-order reference trajectory is provided in the following form.

$$v_{\mathrm{r}}(k+i) = v^*(k+i) - \alpha_{\mathrm{r}}^i \left[v^*(k) - v(k) \right] \qquad i = 1, 2, \ldots, P \tag{14}$$

where v_{r} is the velocity reference trajectory, v^* is the set point value, $\alpha_{\mathrm{r}} = e^{-T_{\mathrm{s}}/T_{\mathrm{r}}}$ is the time constant, and T_{r} is the desired response time.

- **(5) Evaluation mechanism**

Considering the tracking error between the velocity reference trajectory v_{r} and the predicted output v_{m}, and minimizing the sum of squared error e with a penalization on the control input i_{q}^*, yield the following quadratic performance index.

$$\min J = \sum_{i=1}^{P} q_i^2 [v_{\mathrm{r}}(k+i) - v_{\mathrm{m}}(k+i|k) - e(k+i)]^2 + r^2 i_{\mathrm{q}}^{*2}(k) \tag{15}$$

where q_i are the parameters imposed on each of the tracking errors and e; r is a weighting parameter of the control input.

Define the following matrices.

$$
\begin{aligned}
V_{\mathrm{r}}(k) &= \left[\begin{array}{cccc} v_{\mathrm{r}}(k+1) & v_{\mathrm{r}}(k+2) & \cdots & v_{\mathrm{r}}(k+P) \end{array} \right]^{\mathrm{T}} \\
E(k) &= \left[\begin{array}{cccc} e(k+1) & e(k+2) & \cdots & e(k+P) \end{array} \right]^{\mathrm{T}} \\
Q &= \begin{bmatrix} q_1^2 & 0 & \cdots & 0 \\ 0 & q_i^2 & \cdots & 0 \\ \vdots & \vdots & \ddots & \vdots \\ 0 & 0 & \cdots & q_i^2 \end{bmatrix}, \qquad R = r^2
\end{aligned} \tag{16}
$$

Substituting Equations (12) and (16) into (15) leads to the following cost function.

$$J = Q\|V_{\mathrm{r}}(k) - V_{\mathrm{m}}(k) - E(k)\|^2 + Ri_{\mathrm{q}}^{*2}(k) \tag{17}$$

where $\|\cdot\|$ denotes the Euclidean norm.

According to the optimization theory [29], we can obtained the following controller by calculating $\partial J / \partial i_{\mathrm{q}}^{*2} = 0$.

$$i_{\mathrm{q}}^* = (V_{\mathrm{b}}^{\mathrm{T}} Q V_{\mathrm{b}} + R)^{-1} V_{\mathrm{b}}^{\mathrm{T}} Q \left[V_{\mathrm{r}}(k) - V_{\mathrm{o}}(k) - E(k) \right] \tag{18}$$

In a summary, the PFC law is derived as Equation (18).

3.2. ESO Design

As mentioned in the work by the authors of [30], the observer-based feedforward compensation method can improve the disturbance rejection performance. In this note, we introduce an ESO to estimate the disturbances, whose estimates will be taken the above designed PFC into account.

Extracting the mechanical motion equation from Equation (2), and rewriting it as the following differential equation form.

$$\dot{v} = \frac{K_f}{M}i_q - \frac{F_L}{M} - \frac{B_f}{M}v = \frac{K_f}{M}i_q^* + d(t) \tag{19}$$

where the lumped disturbance $d(t)$ (containing tracking error of q-axis current loop, external load torque and friction) is as follows:

$$d(t) = \frac{K_f}{M}\left(i_q - i_q^*\right) - \frac{F_L}{M} - \frac{B_f}{M}v \tag{20}$$

whose derivative is assumed to be bounded, namely, satisfying $\dot{d}(t) \leq D$.

The function $d(t)$ is generally unknown and the ESO can generate the disturbance estimate only depending on the input and output information of the control plant [20], which is free of the specific mathematical models. Here, we define the following state variables.

$$\begin{cases} x_1 = v \\ x_2 = d(t) \end{cases} \tag{21}$$

Substituting Equation (21) into (19) yields

$$\begin{cases} \dot{x}_1 = x_2 + bi_q^* \\ \dot{x}_2 = D \end{cases} \tag{22}$$

where b is a coefficient determined by the system parameters, i.e., $b = K_f/M$.

In this paper, we construct the ESO as follows,

$$\begin{cases} \varepsilon = z_1 - x_1 \\ \dot{z}_1 = z_2 - c_1\varepsilon + b_o i_q^* \\ \dot{z}_2 = -c_2\varepsilon \end{cases} \tag{23}$$

where z_1 and z_2 are the estimates of the state variables x_1 and x_2; c_1 and c_2 are the gains of designed ESO to be determined; b_o is the estimate of parameter b.

Based on the stability analysis in the work by the authors of [8], if the gains c_1 and c_2 are appropriately selected, the estimate errors between z_1, z_2 and x_1, x_2 can be guaranteed to be zero, that is,

$$\begin{cases} z_1 \rightarrow x_1 \rightarrow v \\ z_2 \rightarrow x_2 \rightarrow d(t) \end{cases} \tag{24}$$

After the total disturbance $f(t)$ is accurately estimated by the above ESO, one can be incorporated in aforementioned Equation (18) as a feedforward compensation to improve the anti-disturbance performance [31], and thus the composite speed regulator can be obtained as follows.

$$i_q^* = (V_b^T Q V_b + R)^{-1} V_b^T Q\left[V_r(k) - V_o(k) - E(k)\right] - z_2/b_0 \tag{25}$$

As a result, the whole block diagram of the proposed composite control for the PMLSM servo system is given in Figure 2.

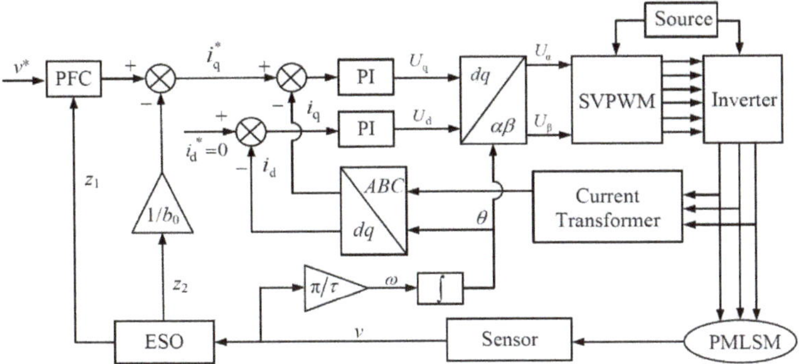

Figure 2. The ESO-based PFC control for the PMLSM.

4. Simulation and Experimental Results

In this section, some simulation results and comparative experiments will be conducted to demonstrate the effectiveness and advantages of the proposed ESO-based PFC composite control strategy. The parameters of the PMLSM are exhibited in Table 1.

Table 1. Parameters of the permanent magnet linear synchronous motor (PMLSM).

Parameter	Value	Unit
Rated voltage (U)	24	V
Pole pitch (τ)	32	mm
Winding resistance (R)	1.25	Ω
Inductance (L)	5.25	mH
Viscous friction coefficient (B_f)	2.12	N·m/s
Flux linkage (ψ_f)	0.0385	Wb
Mass (M)	14	kg

4.1. Simulation Study

In order to illustrate the advantages of the proposed composite control scheme over the conventional PID control method, the simulation comparisons are performed in this subsection. The simulation results are shown in Figures 3 and 4.

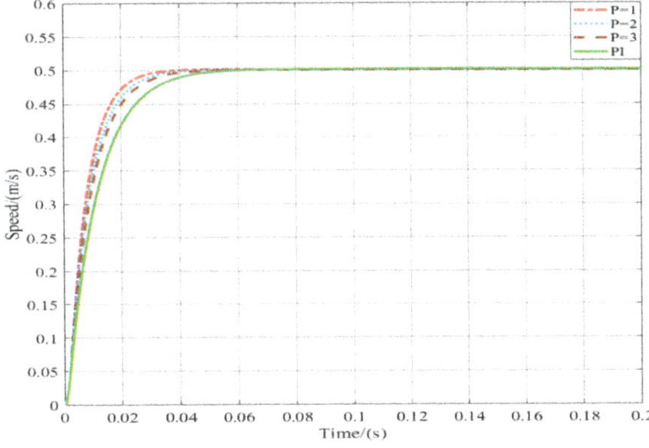

Figure 3. Speed tracking for proportional-integral (PI) and composite controllers (without load torque).

For the case of no external load torque, Figure 3 shows the velocity evolutions based on PI controller and that of ESO-based PFC controller. According to the simulation results, the speed response of PFC controller is significantly faster than that of traditional PI controller, which reflects the superiority of the predictive control algorithm. In PFC controller, the value of prediction depth (also called optimization horizon) P can be arbitrarily set. The simulation curves show that when the optimal prediction depth is set as $P = 1$, the response speed is the fastest, and with the P increasing, the response speed will slow down. Figure 4 shows the system performance of the PI and PFC + ESO when an external disturbances is suddenly loaded. It can be concluded that PFC + ESO control scheme has superior anti-disturbance performance.

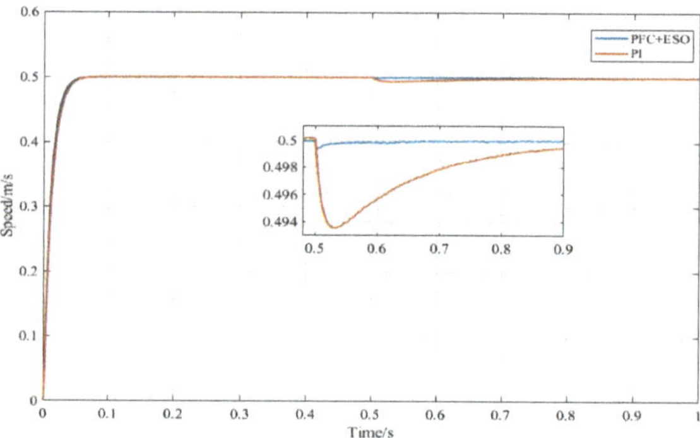

Figure 4. Performance comparison between PI and predictive function control (PFC) + extended state observer (ESO).

4.2. Experiment Comparisons

To evaluate the performance of the proposed method, the experimental setup for a PMLSM test system has been developed, which is shown in Figure 5. The overall speed regulate algorithm including the space vector pulse width modulation (SVPWM) implemented on the STM32 control board, intelligent power module (IPM) integrated driven circuit, external AD7606 conversion circuit, current and position sampling circuits (Hall-effect devices), oscilloscope, computer, Emulator, PMLSM, DC 15V and DC 24V power source, etc.

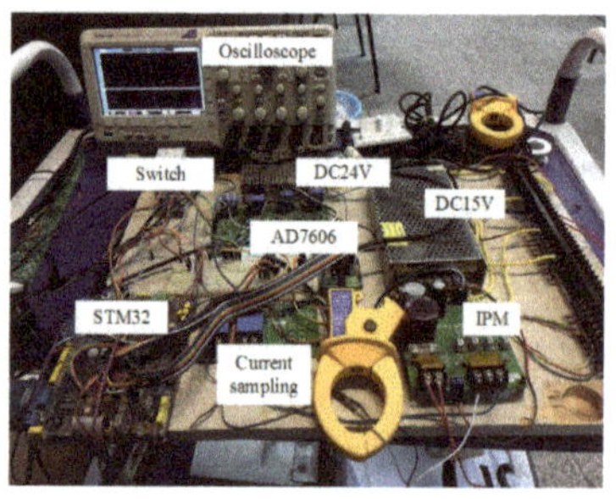

(**a**) Configuration of the test system

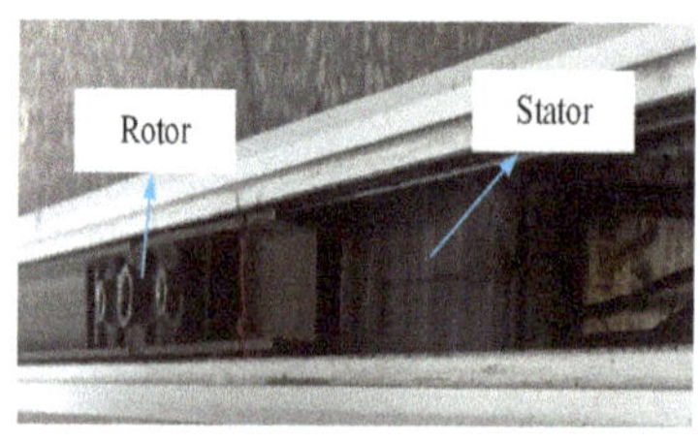

(**b**) Permanent magnet synchronous linear motor

Figure 5. Experiment setup.

In order to verify the advantages and effectiveness of the composite control scheme, comparative simulations and experiments are established on optimized PID controller and the proposed method, respectively; with no speed overshoot as the constraint and the parameters tracking the reference speed in the shortest time are taken as the optimal parameters of the PID. The parameters of the proposed method are determined by the same method.

4.2.1. No-Load Experiments

Experiment results based on PI controller and ESO-based PFC controller (the predicted depth is set as $P = 1$) are performed without external load torque, which are shown in Figure 6. In Figure 6, the reference speed is given as $v_{ref} = 0.5$ m/s, it can be seen from the experiment results that the speed response of PFC controller is faster than that of PI controller, and the system has superior performance.

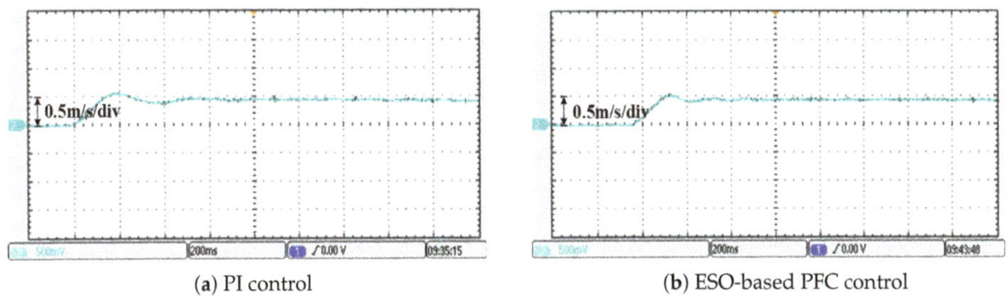

(**a**) PI control   (**b**) ESO-based PFC control

Figure 6. Speed response curves (no load).

4.2.2. Load Torque Experiments

During the stable operation of the PMLSM, a load disturbance is suddenly loaded in a step form, the speed response curves based on PI and PFC + ESO controllers are shown in Figure 7. By comparing the results, it can be concluded that the proposed control scheme has faster response speed, and stronger robustness to external disturbance. In addition, the proposed ESO algorithm can well estimate the system state and disturbance.

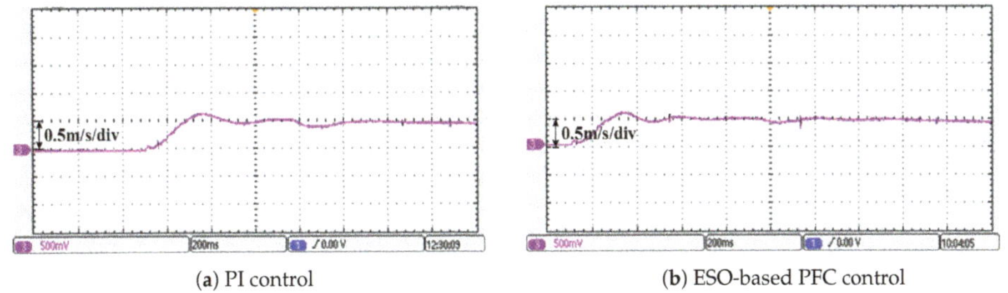

(**a**) PI control   (**b**) ESO-based PFC control

Figure 7. Speed response curves.

5. Conclusions

The speed regulation problem for a PMLSM servo system has been investigated in this paper. The mathematical model of the PMLSM is established in the Laplace domain. A composite control scheme based on the PFC feedback and the ESO-based feedforward compensation has been presented. The simulation and experiment results confirm that the proposed composite control method has a superior system performance compared with the traditional PID control strategy, such as faster response speed and stronger robustness. Our future work will extend the proposed method to deal with the special force ripple and end effect problems in the PMLSM servo system.

Author Contributions: This is a joint work and the authors were in charge of their expertise and capability: Y.W. (Yao Wang) for investigation and analysis; H.Y. for validation and revision; Z.C. for methodology and data analysis; Y.W. (Yuchen Wang) for writing and revision; C.Z. for manuscript revision.

Funding: This work was supported by the National Natural Science Foundationof China (NSFC) under Grant 41576096.

Acknowledgments: The authors would like to express their gratitude to all those who helped them during the writing of this paper. Also, the authors would like to thank the reviewers for their valuable comments and suggestions.

Conflicts of Interest: The authors declare no conflicts of interest.

References

1. Yang, C.Y.; Ma, T.T., Che, Z.Y.; Zhou, L.N. An adaptive-gain sliding mode observer for sensorless control of permanent magnet linear dynchronous motors. *IEEE Access* **2018**, *6*, 3469–3478. [CrossRef]
2. Wang, M.Y.; Yang, R.; Zhang, C.M.; Cao, J.W.; Li, L.Y. Inner loop design for PMLSM drives with thrust ripple compensation and high-performance current control. *IEEE Trans. Ind. Electron.* **2018**, *65*, 9905–9915. [CrossRef]
3. Song, F.Z.; Liu, Y.; Xu, J.X.; Yang, X.F.; Zhu, Q. Data-driven iterative feedforward tuning for a wafer stage: A high-order approach based on instrumental variables. *IEEE Trans. Ind. Electron.* **2019**, *66*, 3106–3116. [CrossRef]
4. Jang, K.B.; Kim, J.H.; An, H.J.; Kim, G.T. Optimal design of auxiliary teeth to minimize unbalanced phase due to end effect of PMLSM. *IEEE Trans. Magn.* **2011**, *47*, 1010–1013. [CrossRef]
5. Son, Y.I.; Kim, I.H.; Choi, D.S.; Shim, H. Robust cascade control of electric motor drives using dual reduced-order PI observer. *IEEE Trans. Ind. Electron.* **2015**, *62*, 3672–3682. [CrossRef]
6. Yang, C.Y.; Che, Z.Y.; Zhou, L.N. Integral sliding mode control for singularly perturbed systems with mismatched disturbances. *Circuits Syst. Signal Process.* **2019**, *38*, 1561–1582. [CrossRef]
7. Li, S.H.; Zong, K.; Liu, H.X. A composite speed controller based on a second-order model of permanent magnet synchronous motor system. *Trans. Inst. Meas. Control* **2011**, *33*, 522–541.
8. Huang, Y.; Xue, W.C. Active disturbance rejection control: Methodology and theoretical analysis. *ISA Trans.* **2014**, *53*, 963–976. [CrossRef]
9. Sira-Ramirez, H.; Linares-Flores, J.; Garcia-Rodriguez, C.; Contreras-Ordaz, M.A. On the control of the permanent magnet synchronous motor: An active disturbance rejection control approach. *IEEE Trans. Control Syst. Technol.* **2014**, *22*, 2056–2063. [CrossRef]
10. Yang, J.; Chen, W.H.; Li, S.H.; Guo, L.; Yan, Y.D. Disturbance/uncertainty estimation and attenuation techniques in PMSM drives—A survey. *IEEE Trans. Ind. Electron.* **2017**, *64*, 3273–3285. [CrossRef]
11. Ganesh, H.S.; Edgar, T.F.; Baldea, M. Model predictive control of the exit part temperature for an austenitization furnace. *Processes* **2016**, *4*, 53. [CrossRef]
12. Gong, Z.; Wu, X.J.; Dai, P.; Zhu, R.W. Modulated model predictive control for MMC–based active front–end rectifiers under unbalanced grid conditions. *IEEE Trans. Ind. Electron.* **2019**, *66*, 2398–2409. [CrossRef]
13. Yan, Y.D.; Zhang, C.L.; Narayan, A.; Yang, J.; Li, S.H.; Yu, H.Y. Generalized dynamic predictive control for nonparametric uncertain systems with application to series elastic actuators. *IIEEE Trans. Ind. Inform.* **2018**, *14*, 4829–4840. [CrossRef]
14. Yang, J.; Wu, H.; Hu, L.; Li, S.H. Robust predictive speed regulation of converter-driven DC motors via a discrete-time reduced-order GPIO. *IEEE Trans. Ind. Electron.* **2019**, *66*, 7893–7903. [CrossRef]
15. Vaccari, M.; Pannocchia, G. A modifier-adaptation strategy towards offset-free economic MPC. *Processes* **2017**, *5*, 2. [CrossRef]
16. Morel, F.; Lin-Shi, X.F.; Retif, J.M.; Allard, B.; Buttay, C. A comparative study of predictive current control schemes for a permanent-magnet synchronous machine drive. *IEEE Trans. Ind. Electron.* **2009**, *56*, 2715–2728. [CrossRef]
17. Bolognani, S.; Bolognani, S.; Peretti, L.; Zigliotto, M. Design and implementation of model predictive control for electrical motor drives. *IEEE Trans. Ind. Electron.* **2009**, *56*, 1925–1936. [CrossRef]

18. Zhang, C.F.; Wu, G.P.; Rong, F.; Feng, J.H.; Jia, L.; He, J.; Huang, S.D. Robust fault-tolerant predictive current control for permanent magnet synchronous motors considering demagnetization fault. *IEEE Trans. Ind. Electron.* **2018**, *65*, 5324–5334. [CrossRef]

19. Liu, H.X.; Li, S.H. Speed control for PMSM servo system using predictive function control and extended state observer. *IEEE Trans. Ind. Electron.* **2012**, *59*, 1171–1183. [CrossRef]

20. Xue, W.C.; Bai, W.Y.; Yang, S.; Song, K.; Huang, Y.; Xie, H. ADRC with adaptive extended state observer and its application to air-fuel ratio control in gasoline engines. *IEEE Trans. Ind. Electron.* **2015**, *62*, 5847–5857. [CrossRef]

21. Wang, J.X.; Li, S.H.; Yang, J.; Wu, B.; Li, Q. Extended state observer-based sliding mode control for PWM-based DC-DC buck power converter systems with mismatched disturbances. *IET Control Theory Appl.* **2015**, *9*, 579–586. [CrossRef]

22. Chen, S.Y.; Chiang, H.H.; Liu, T.S.; Chang, C.H. Precision motion control of permanent magnet linear synchronous motors using adaptive fuzzy fractional-order sliding-mode control. *IEEE/ASME Trans. Mechatron.* **2019**, *24*, 741–752. [CrossRef]

23. Jin, H.Y.; Zhao, X.M. Complementary sliding mode control via elman neural network for permanent magnet linear servo system. *IEEE Access* **2019**, *7*, 82183–82193. [CrossRef]

24. Hu, J.G.; Liu, J.B.; Xu, L.Y. Eddy current effects on rotor position estimation and magnetic pole identification of PMSM at zero and low speeds. *IEEE Trans. Power Electron.* **2008**, *23*, 2565–2575. [CrossRef]

25. Yang, C.Y.; Liu, J.H.; Li, H.; Zhou, L.N. Energy modeling and parameter identification of dual-motor-driven belt conveyors without speed sensors. *Energies* **2018**, *11*, 3313. [CrossRef]

26. Yang, R.; Wang, M.Y.; Li, L.Y.; Zenggu, Y.M.; Jiang, J.L. Integrated uncertainty/disturbance compensation with second-order sliding-mode observer for PMLSM-driven motion stage. *IEEE Trans. Power Electron.* **2019**, *34*, 2597–2607. [CrossRef]

27. Lu, S.W.; Tang, X.Q.; Song, B.; Zheng, S.Q.; Zhou, F.X. Identification and compensation of force ripple in PMSLM using a JITL technique. *Asian J. Control* **2015**, *17*, 1559–1568. [CrossRef]

28. Mendez, J.A.; Kouvaritakis, B.; Rossiter, J.A. State-space approach to interpolation in MPC. *Int. J. Robust Nonlinear Control* **2000**, *10*, 27–38. [CrossRef]

29. Yang, C.Y.; Zhang, Q.L.; Sun, J.; Chai, T.Y. Lur'e Lyapunov function and absolute stability criterion for Lur'e singularly perturbed systems. *IEEE Trans. Autom. Control* **2011**, *56*, 2666–2671. [CrossRef]

30. Liu, M.; Zhang, L.X.; Shi, P.; Zhao, Y.X. Fault estimation sliding mode observer with digital communication constraints. *IEEE Trans. Autom. Control* **2018**, *63*, 3434–3441. [CrossRef]

31. Zhou, L.N.; Che, Z.Y.; Yang, C.Y. Disturbance observer-based integral sliding mode control for singularly perturbed systems with mismatched disturbances. *IEEE Access* **2018**, *6*, 9854–9861. [CrossRef]

Article

Economic Dispatch of Multi-Energy System Considering Load Replaceability

Tao Zheng [1,2,3,*], Zemei Dai [1,2,3], Jiahao Yao [4], Yufeng Yang [1,2,3] and Jing Cao [1,2,3]

[1] NARI Group Corporation (State Grid Electric Power Research Institute), Nanjing 211106, China
[2] NARI Technology Development Co. Ltd., Nanjing 211106, China
[3] State Key Laboratory of Smart Grid Protection and Control, Nanjing 211106, China
[4] Key Laboratory of Measurement and Control of Complex Systems of Engineering, Ministry of Education, School of Automation, Southeast University, Nanjing 210096, China
* Correspondence: zhengtao2@sgepri.sgcc.com.cn

Received: 28 July 2019; Accepted: 23 August 2019; Published: 28 August 2019

Abstract: By integrating gas, electricity, and cooling and heat networks, multi-energy system (MES) breaks the bondage of isolated planning and operation of independent energy systems. Appropriate scheduling of MES is critical to the operational economy, and it is essential to design scheduling strategies to achieve maximum economic benefits. In addition to the emergence of energy conversion systems, the other main novelty of MES is the multivariate of load, which offers a great optimization potential by changing load replaceability (flexibly adjusting the composition of loads). In this paper, by designing load replaceability index (LRI) of composite load in MES, its interaction mechanism with scheduling optimum is systematically analyzed. Through case studies, it is proven that the optimum can be improved by elevating load replaceability.

Keywords: multi-energy system; economic dispatch; load replaceability; multi-energy conversion

---

1. Introduction

Energy flexibility is crucial to the construction of more environmental-friendly and efficient power generation (consumption) patterns. Unlike conventional single-form energy-based energy system, which is usually detached from other types of energy systems from both the planning and operational perspectives, multi-energy system (MES) comprises various forms of subsystems including electricity, heat, cooling and gas networks, contributing to the interactions and corresponding energy flexibility through energy converters. Therefore, MES technologies gradually become irreplaceably key components in building the modern energy systems.

Compared with single-from energy-based system, the most significant difference of MES is the energy conversion function achieved by energy conversion equipments, which serve as in-between interfaces and assist in the formation of coupling relations. Energy converters are foundations of transformation, fusion and decomposition of energy flows among different subsystems. Because of the possibility of convertibility of different energy forms, MES outperforms systems of pure energy form by complementing the advantages of each subsystem.

An important concept in MES is the energy hub (EH), which gives an matrix model of production, conversion, consumption and storage of different energy carriers. Each of the matrix elements represent the abstract connection and conversion coefficients of internal components, which can be embodied by various types of multi-generators, among which combined heat and power (CHP) or combined heat cooling and power (CHCP) plants are the most widely investigated cogeneration or trigeneration plants.

Various research studies have recently begun to show intensive concern for the design and operation of MES, either from the perspective of microscopic (CHP and CHCP plants) or macroscopic (EHs). The main idea of optimal operation of CHP (CHCP) plants is similar to that of power systems, both starts from the optimization model building and ends with model solving by specific optimization algorithms like mathematical or evolutionary programming methods. Like economic electrical power dispatch, the first priority of multi-generation is economy. Some researchers leverage the strong computing capabilities of meta-heuristics to solve economic dispatch of CHP plants [1,2]. Apart from deterministic optimization, uncertainties stemming from volatile renewable energy are gradually introduced by stochastic programming. For example, a novel chance constrained programming model is used to formulate economic dispatch with CHP plants and wind power simultaneously [3]. Li groundbreakingly investigates how to use the temperature dynamics of heating networks to enhance the utilization of wind power, and a CHP dispatch model considering operational economy and wind power integration is built and solved by iterative methods [4]. As for economic dispatch of CHCP, a nonlinear economic optimal operation model is developed and solved by lingo solver [5]. Backward dynamic programming technique is used to obtain the optimal operational set-points of combined heat, cooling and power systems [6].

Besides designing operation schedules solely in pursuit of the economic potential elevation, optimization of (CHP) CCHP plants considering multiple operational output performance begins to capture researchers' interests. Under most circumstances, economy, energy utilization efficiency and environmental preservation are the three main optimization goals. By employing Weighting method and fuzzy optimum selection theory, the integrated performances of operation of CCHP plants considering these objectives are evaluated [7]. Dynamic object method-particle swarm optimization method is used to achieve the goal of carbon emission reduction, energy efficiency elevation and system cost minimization [8]. Additionally, some new evaluation indices evaluating the average useful output and total heat transfer area are studied, correspondingly, a multi-objective optimization-based schedule is designed [9]. Modified Bacterial Foraging Optimization method is adopted to minimize the operational cost and pollutant emission [10].

Optimal system-level operation of MES is usually implemented under EH framework by designing the synergies among generation, consumption and storage components. An optimal operation strategy of MES containing CHP plants, boiler, battery and water tank is studied [11], and particle swarm optimization (PSO) algorithm is used to obtain optimal scheduling of each device. An optimal expansion planning of EH containing CHP plants and natural gas furnaces is designed to minimize investment cost [12]. Under some circumstances MES have some uncertainties involving demand fluctuation, cost dynamics and converter efficiency change. In this situation deterministic framework no longer adapts to uncertain environments, hence robust optimization approach is used to increase the robustness towards uncertain bounded parameters [13]. In [14], uncertainties of demand, market price and renewable energy are considered by stochastic programming, and optimal operation set-points are obtained to minimize operation costs. Instead of treating EH as an individual system, studies of multi-EH systems take into consideration the mutual influence and restriction. By analyzing competitive and cooperative relation between multi-EHs, game theoretic optimal scheduling is established through quantum particle swarm optimization method [15]. Moreover, resilience of MES is researched in last several years, and coordination between different subsystems in post-disaster repair is investigated to reduce load shedding and repair duration [16].

Most of the existing MES research focuses on optimal planning or operation by considering multi-energy generation or conversion on the generation side. Recently, some research studies begin to consider the characteristics of multi-energy loads, which play essential roles in MES planning & operation as well as other applications like load prediction and demand response programs. Some researchers investigate the influence of multi-energy loads upon MES planning, and the capacity of gas turbine under different thermoelectric ratios is studied [17]. In respect to load prediction, load prediction for integrated cooling and heat system is implemented by considering the intrinsic coupling among multi-energy loads using multivariate phase space reconstruction technique [18]. By considering the operating attributes of loads, a quick cooling and heating load prediction model is constructed [19]. Deep-learning technique is adopted to predict various types of load in regional energy system integration [20]. Some researchers also study how to adjust the consumption pattern to better participate in the demand response program [21,22].

Conventionally, the load composition from each energy network is assumed to be fixed without replaceability, which is characterized by the feasibility of substitution of one form of load with another one. Load considering replaceability can be regarded as a composite load, it characterizes the multi-energy consumption on the load side and is essential to flexible and efficient energy consumption. Composite load can adjust the proportion of heterogeneous energy consumption based on the external excitation signal such as market price, thus achieving simultaneous operation cost reduction and the load and generation balance.

In this paper, an optimal scheduling approach for MES consisting of gas, electricity, cooling and heat networks considering load replaceability is designed to maximize the economic potential of operation. For the convenience of analysis of the influence of load replaceability on the optimal scheduling, load replaceability index (LRI), which measures the overall load replaceability of MES, is analyzed in respect to the optimal scheduling results. The intrinsic relations between load replacement capability and optimum under various operating conditions are investigated to improve the scheduling performance by observing specific rules of load replaceability adjustment.

The remaining of the paper is organized as follows: Section 2 addresses load replaceability and the definition of LRI in MES; with the aid of load replaceability, Section 3 presents the economic scheduling model for an integrated electricity, gas, heat and cooling networks; Section 4 studies the relation between optimal scheduling results and load replacement capability; concluding remarks are given in Section 5.

2. Load Replaceability in Multi-Energy System

In this section, load replaceability and its relation with optimal scheduling of MES are systematically analyzed. The definition and calculation method of LRI are presented.

2.1. Load Replaceability and Composite Load

Load replaceability is used to characterize the selection margin of multi-energy consumption. If one specific load can choose two or more than two types of energy flow to meet its own load demand, and then this load is called replaceable load. From the perspective of consumers, the greater the number of energy flow types is and the greater the capacity of multi-generators is, the more the selection margin of multi-energy consumption is.

Generally speaking, loads in MES can be categorized as cooling, heat, power and gas load, corresponding to their respective type of energy network. Usually, due to the natural barrier between heterogenous energy systems in the isolated operating mode, the load can only be supplied by homogenous energy flow. Nevertheless, with the coupling and integration of heterogenous systems assisted by energy conversion systems, it should be emphasized that the load can be satisfied by heterogenous energy flow as well.

In this section, heat load is taken as an example to illustrate the replaceability phenomenon. In independent heat network, heat load can only be supported by the heat flow through steam pipelines. While heat load in MES can be supplied by various forms of energy. For example, on the one hand heat energy can still be transmitted through pipelines just like in independent heat network; on the other hand electric furnaces or gas heaters can transform electric or gas into heat energy. In this case, in addition to the thermal attribute, heat load has electric attribute (electric heater) and gaseous attribute (gas furnace). Electricity and gas are just intermediate transition states. Seen from the exterior, the final state is still heat load since heat flow is the only energy form in the consumption phase.

Though the final load state belongs to heat load in either segregated independent system or integrated dependent system, the introduction of intermediate transition states makes the load property change from monotype load in independent system into composite load in dependent system. The main specialty of composite load is flexibility, especially the market response flexibility. The operator can adjust the proportion of each component in the composite load, based on the market price of each form of energy flow at different operating interval, thus reducing operation cost. The emergence of composite load enhances the optimization space in the consumption phase and further increases economic benefits. In order to vividly illustrate how load replaceability (composite load) influences the operation economy, an exemplary integrated gas power and cooling system in Figure 1 is briefly studied.

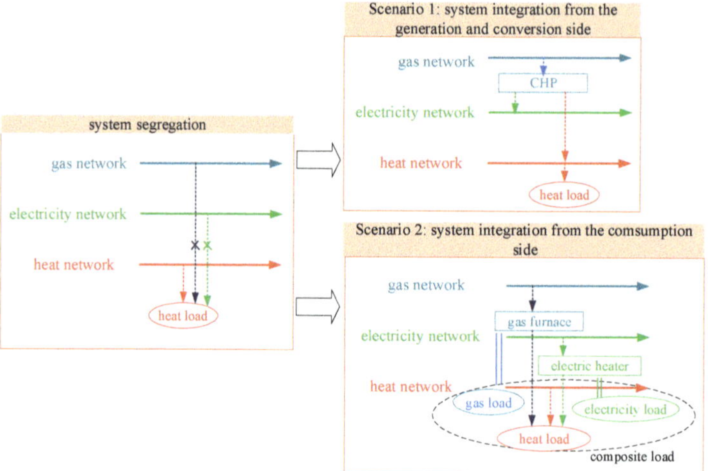

Figure 1. Schematic diagram of two-area classic system.

Suppose the current market prices of gas electricity and heat are 6 \$/kWh, 1 \$/kWh and 15 \$/kWh, respectively. And the amount of heat load demand (100 kWh) is much more than that of electricity or gas load demand, which is set by zero for simplicity. The maximal transmission capacity of each network is 50 kW. The following two scenarios are considered for operation cost calculation.

- **Scenario 1:** There exists no replaceable heat load in the system; the multi-generation plant is the CHP plant. The overall energy conversion ratio is 0.8; the energy proportion of electricity generated by the CHP plant is 0.6; the energy proportion of heat generated by the CHP plant is 0.4. And the maximal capacity of the CHP plant is supposedly to be much bigger than the maximal transmission capacity of the network.
- **Scenario 2:** There exists replaceable heat load in the system, and the load is composite load which contains electric heater and gas furnace. The maximal capacity of both devices is supposedly to be much bigger than the maximal transmission capacity of power and gas network. The energy efficiency ratio for both devices is 0.5.

In Scenario 1, the operation cost is

$$6 \times 50 \times 0.8 \times 0.4 + 15 \times (100 - 50 \times 0.8 \times 0.4) = 1356. \tag{1}$$

In Scenario 2, the operation cost is

$$1 \times 50 \times 0.5 + 6 \times 50 \times 0.5 + 15 \times (100 - 50) = 925. \tag{2}$$

As can be seen, due to the load replaceability, consumers can satisfy the demand by consuming much cheaper transformed energy. In this case it is the transformed heat flow from electric heaters and gas furnaces, thus reducing the overall operation cost. Nevertheless, the multi-generation plant (the CHP plant) can only supplement heat flow to a very limited degree. Heat flow is just a byproduct during the intermediate generation section and the multi-generation plant is not ultimately consumption-oriented, which cannot sufficiently use the price difference between different energy forms to reduce operation cost. As is shown by the results in (1) and (2), operation cost (925$) considering load replaceability is obviously smaller than that is only with multi-generation plants.

2.2. Load Replaceability Index

In Section 2.1, basic information of load replaceability and composite load is addressed. From the deduction of a simple 2-scenario integrated energy system, it can be learned that load replaceability enhances the optimization space from the consumption side. In this section, instead of focusing on substitutes of one specific type of load, we give some quantitative indices to evaluate the broad system-level load replaceability. The LRI can reflect to what extent one type of load (the final load state) can be substituted by the other (the intermediate load state).

LRI can be defined from two perspectives: (1) the potential load replaceability and (2) the actual load replaceability. The former evaluates the potential of load substitution of the system before it is ever put into operation. The latter evaluates the actual load substitution ability during the operation.

2.2.1. Load Replaceability Index Reflecting Potential Load Substitution Ability

As for an integrated gas, power, heat and cooling system, there exist four possible types of load in the system. Denote the maximal convertible load demand of the ith composite load by L_i^{max}, and denote the load component for the ith load from each network by X_i^{load}, P_i^{load} and G_i^{load}. X_i^{load} can either be C_i^{load} or H_i^{load} based on the final energy form. LRI of the ith load can be calculated by

$$\alpha_i = \frac{\left(x_i X_i^{load} + u_{i1} p_i P_i^{load} + u_{i2} g_i G_i^{load} - L_i^{max} \right)}{L_i^{max}}, \tag{3}$$

where x_i, p_i, and g_i represents the energy efficiency ratio of respective intermediate load. Binary variables $u_{i1}, u_{i2} \in \{0,1\}$ represent ith load is ($u_{i1}, u_{i2} = 1$) or is not ($u_{i1}, u_{i2} = 0$) supplied by specific intermediate loads (P_i^{load} or G_i^{load}). Suppose there exist m loads in the integrated energy system, the overall LRI can be expressed as

$$\alpha = \frac{\sum\limits_{i=1}^{m} \alpha_i L_i^{max}}{\sum\limits_{i=1}^{m} L_i^{max}}. \tag{4}$$

LRI in (4) actually uses the proportion of remaining alternative load reserve in the maximal load demand to quantify the substitution potential.

2.2.2. Load Replaceability Index Reflecting Actual Load Substitution Ability

As can be seen from (4), all the parameters are actually the upper bounds of corresponding replaceable loads. That is to say, (4) reflects the remaining load conversion reserve at critical states (the maximum values). Nevertheless, replaceable loads in the actual scheduling would not certainly be equal to the maximum; values of replaceable loads should depend on the scheduling, which will be presented in detail in the next section. Therefore, a modified online scheduling-based LRI of the ith load is expressed by

$$\beta_i = \frac{u_{i1} p_i P_i^{load} + u_{i2} g_i G_i^{load}}{L_i}, \tag{5}$$

where P_i^{load} and G_i^{load} represent the actual intermediate load amount in the scheduling; L_i represents the actual composite load demand. Then the system-level LRI can be expressed by

$$\beta = \frac{\sum\limits_{i=1}^{m} \beta_i L_i}{\sum\limits_{i=1}^{m} L_i}.$$

(6)

Notice that the replaceable load term $x_i X_i^{load}$ disappears in (5); it is because that $x_i X_i^{load} + u_{i1} p_i P_i^{load} + u_{i2} g_i G_i^{load} = L_i$ in the scheduling. If (3) is still used, LRI is always zero. In this case only the proportion of energy converted by intermediate loads in the composite load is used to quantify the degree of flexibility.

3. Optimal Economic Scheduling of Multi-Energy System Considering Load Replaceability

It can be learned from Section 2 that load replaceability offers redundant energy flow path and the optimization space of multi-energy complementary coordination. In response to the market price of different energy flow, the consumption side can participate in economic dispatch by adaptively adjusting the proportion of the intermediate load of various energy form in the composite load. Unlike conventional demand response program in power systems, because of the load replaceability capability through conversion systems, this multi-energy demand response in MES does not cause the change in the ultimate total energy consumption. The consumers only choose the intermediate energy source from which they buy to obtain the ultimate energy rather than change the consumption pattern like they do in electric utility demand response programs. That is to say, operation economy along with customer satisfaction on energy supply is guaranteed.

In this section, by considering load replaceability the economic dispatch model is established for MES in Figure 2 consisting of gas, power, and cooling and heat networks.

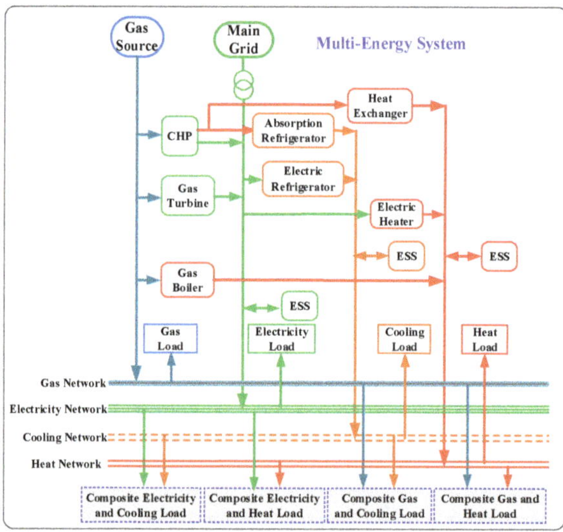

Figure 2. Schematic diagram of a multi-energy system (MES) consisting of gas, power, and cooling and heat networks. CHP = combined heat and power; ESS = energy storage system.

3.1. Objective Function of Optimal Economic Scheduling of MES

From the perspective of economy, operation cost mainly comes from three aspects including the cost of purchasing gas from external gas networks, the cost of purchasing electricity from external electric grids and depreciation expense of the multi-energy storage systems.

The cost of gas purchase O_G is written by the accumulative product of gas price and purchase amount (which is procured from the external gas source shown in Figure 2) during the whole dispatching intervals

$$O_G = \sum_{t=1}^{T} r_{gas}(t) \frac{G_{ext}(t)}{R_{gas}} t, \tag{7}$$

where t represents the dispatch interval; T represents the total number of dispatch intervals; $r_{gas}(t)$ represents gas price at time t; $G_{ext}(t)$ represents gas purchase amount at time t; $R_{gas} = 9.77 \left(kw \cdot h/m^3\right)$ represents the calorific value.

The supplementary electricity is procured from the external main grid as shown in Figure 2. Similarly, we can compute the cost of electricity purchase O_P as

$$O_P = \sum_{t=1}^{T} r_{ele}(t) P_{ext}(t) t, \tag{8}$$

where t and T have the same meaning as in (7); $r_{ele}(t)$ represents electricity price at time t; and $P_{ext}(t)$ represents electricity purchase amount at time t.

The depreciation cost steams from ageing of storage system during the repeated energy charge/discharge processes. In this paper, it is assumed that gas is either directly supplied to gas load or converted to other energy flow, and there is no gas storage system. For simplicity, it is expressed as the linear combination of storage amount of different energy flow:

$$O_S = \sum_{t=1}^{T} \left[r_{PS} P_{PS}^S(t) + r_{HS} H_{HS}^S(t) + r_{CS} C_{CS}^S(t) \right], \tag{9}$$

where r_{PS}, r_{HS}, and r_{CS} represent the depreciation rate per unit quantity of power, and heat and cooling energy; $P_{PS}^S(t)$, $H_{HS}^S(t)$ and $C_{CS}^S(t)$ represent the electricity, and heat and cooling energy stored at time t.

3.2. Constraints of Optimal Economic Scheduling of MES

3.2.1. Energy Supply and Demand Balance

Similar to power balance equation in electric grids, the energy supply and demand balance should be guaranteed at the first place. Nevertheless, due to the coupling enabled by conversion systems, the energy generation side might contain transformed energy flow (the output of conversion systems), the energy consumption side might contain energy flow to be transformed (the input of conversion systems), which is exactly the intermediate load previously addressed in Section 2. Therefore, under the circumstances of MES, the load for each network can be categorized by convertible load and inconvertible load, which are equivalent to intermediate and ultimate load in Section 2.

The energy supply and demand balance equation for the cooling system can be written by

$$C_{AR}(t) + C_{ER}(t) + C_{CS}^R(t) - C_{CS}^S(t) = C_{load}^D(t) + C_{load}^{PC}(t) + C_{load}^{GC}(t), \tag{10}$$

where t represents the dispatch interval; $C_{AR}(t)$ represents cooling energy generated by the absorption refrigerator; $C_{ER}(t)$ represents cooling energy generated by the electric refrigerator; $C_{CS}^R(t)$ represents cooling energy released by cooling energy storage system; $C_{CS}^S(t)$ represents cooling energy stored by cooling energy storage system; $C_{load}^{D}(t)$ represents inconvertible cooling load; $C_{load}^{PC}(t)$ represents convertible electricity-based cooling load; and $C_{load}^{GC}(t)$ represents convertible gas-based cooling load.

Similarly, the energy supply and demand balance equation for the heat system can be written by

$$H_{HE}(t) + H_{GB}(t) + H_{EH}(t) + H_{HS}^{R}(t) - H_{HS}^{S}(t) = H_{load}^{D}(t) + H_{load}^{PH}(t) + H_{load}^{GH}(t), \qquad (11)$$

where $H_{HE}(t)$ represents heat energy generated by heat exchanger; $H_{GB}(t)$ represents heat energy generated by gas boiler; $H_{EH}(t)$ represents heat energy generated by electric heater; $H_{HS}^{R}(t)$ represents heat energy released by heat energy storage system; $H_{HS}^{S}(t)$ represents heat energy stored by heat energy storage system; $H_{load}^{D}(t)$ represents inconvertible heat load; $H_{load}^{PH}(t)$ represents convertible power-based heat load; and $H_{load}^{GH}(t)$ represents convertible gas-based heat load.

The energy supply and demand balance equation for the gas system can be written by

$$G_{ext}(t) - G_{CHP}(t) - G_{GT}(t) - G_{GB}(t) = G_{load}^{D}(t) + G_{load}^{GC}(t) + G_{load}^{GH}(t), \qquad (12)$$

where $G_{CHP}(t)$ represents gas energy consumed by the CHP plant; $G_{GT}(t)$ represents gas energy consumed by the gas turbine; $G_{GB}(t)$ represents gas energy consumed by the gas boiler; $G_{load}^{D}(t)$ represents inconvertible gas load; $G_{load}^{GC}(t)$ represents convertible cooling-oriented gas load; and $G_{load}^{GH}(t)$ represents convertible heat-oriented gas load.

The energy supply and demand balance equation for the power system can be written by

$$P_{ext}(t) + P_{CHP}(t) + P_{GT}(t) + P_{PS}^{R}(t) - P_{PS}^{S}(t) - P_{EH}(t) - P_{ER}(t) = P_{load}^{D}(t) + P_{load}^{PC}(t) + P_{load}^{PH}(t), \qquad (13)$$

where $P_{CHP}(t)$ represents electricity energy generated by the CHP plant; $P_{GT}(t)$ represents electricity energy generated by the gas turbine; $P_{PS}^{R}(t)$ represents electricity energy released by the electricity energy storage system; $P_{PS}^{S}(t)$ represents electricity energy stored by the electricity energy storage system; $P_{EH}(t)$ represents electricity energy consumed by the electric heater; $P_{ER}(t)$ represents electricity energy consumed by the electric refrigerator; $P_{load}^{D}(t)$ represents inconvertible power load; $P_{load}^{PC}(t)$ represents convertible cooling-oriented power load; and $P_{load}^{PH}(t)$ represents convertible heat-oriented power load.

As can be seen from (10)–(13), there exist inconvertible load $C_{load}^{D}(t)$, $H_{load}^{D}(t)$, $G_{load}^{D}(t)$, and $P_{load}^{D}(t)$; meanwhile, the convertible load during the intermediate transition state are $C_{load}^{PC}(t)$, $C_{load}^{GC}(t)$, $H_{load}^{PH}(t)$, $H_{load}^{GH}(t)$, $G_{load}^{GC}(t)$, $G_{load}^{GH}(t)$, $P_{load}^{PC}(t)$, and $P_{load}^{PH}(t)$. Moreover, the first four $C_{load}^{PC}(t)$, $C_{load}^{GC}(t)$, $H_{load}^{PH}(t)$, and $H_{load}^{GH}(t)$ are actually the final state of the convertible load, while the last four $G_{load}^{GC}(t)$, $G_{load}^{GH}(t)$, $P_{load}^{PC}(t)$, and $P_{load}^{PH}(t)$ are actually the initial state of the convertible load.

The total amount of energy of composite load in the final energy form (cooling or heat) can thus be expressed by

$$\begin{cases} p_{load}^{PC} P_{load}^{PC}(t) + C_{load}^{PC}(t) = L_{load}^{PC}(t) \\ g_{load}^{GC} G_{load}^{GC}(t) + C_{load}^{GC}(t) = L_{load}^{GC}(t) \\ p_{load}^{PH} P_{load}^{PH}(t) + H_{load}^{PH}(t) = L_{load}^{PH}(t) \\ g_{load}^{GH} G_{load}^{GH}(t) + H_{load}^{GH}(t) = L_{load}^{GH}(t) \end{cases}, \qquad (14)$$

where p_{load}^{PC} represents the energy efficiency ratio of convertible cooling-oriented power load; $L_{load}^{PC}(t)$ represents the composite power and cooling load; g_{load}^{GC} represents the energy efficiency ratio of convertible cooling-oriented gas load; $L_{load}^{GC}(t)$ represents the composite power and gas load; p_{load}^{PH} represents the energy efficiency ratio of convertible heat-oriented power load; $L_{load}^{PH}(t)$ represents the composite power and heat load; g_{load}^{GH} represents the energy efficiency ratio of convertible heat-oriented gas load; and $L_{load}^{GH}(t)$ represents the composite gas and heat load.

3.2.2. Energy Balance of Conversion Systems

As can be seen from Figure 2, the conversion systems comprise the CHP plant, gas turbine, gas boiler, electric heater and electric furnace. In this section, the input and output energy balance of respective conversion system is presented.

The CHP plant is the use of natural gas to generate electricity and useful heat simultaneously. The relation between input gas power and output electricity or heat power of the back pressure CHP plant is

$$P_{CHP}(t) = p_{CHP}G_{CHP}(t) \tag{15}$$

$$Q_{CHP}(t) = q_{QP}P_{CHP}(t), \tag{16}$$

where p_{CHP} represents the power generation efficiency of the CHP plant; q_{QP} represents the ratio of electricity-to-heat; G_{CHP} represents the input gas power of the CHP plant; and Q_{CHP} represents the surplus heat, which is transformed to cooling and heat energy through the absorption refrigerator and heat exchanger, respectively.

$$C_{AR}(t) = \eta_{AR}(1 - \gamma_{HC})Q_{CHP}(t), \tag{17}$$

$$H_{HE}(t) = \eta_{HE}\gamma_{HC}Q_{CHP}(t). \tag{18}$$

where η_{AR} and η_{HE} represent the energy efficiency ratio of the absorption refrigerator and heat exchanger; γ_{HC} represents the proportion ratio.

The gas turbine is used to transform gas energy into electricity energy

$$P_{GT}(t) = \eta_{GT}G_{GT}(t), \tag{19}$$

where η_{GT} represents the energy efficiency ratio of the gas turbine.

The gas boiler is used to transform gas energy into heat energy

$$H_{GB}(t) = \eta_{GB}G_{GB}(t), \tag{20}$$

where η_{GB} represents the energy efficiency ratio of the gas boiler.

The electric refrigerator is used to transform electricity energy into cooling energy

$$C_{ER}(t) = \eta_{ER}P_{ER}(t), \tag{21}$$

where η_{ER} represents the energy efficiency ratio of the the electric refrigerator.

The electric heater is used to transform electricity energy into heat energy

$$H_{EH}(t) = \eta_{EH}P_{EH}(t), \tag{22}$$

where η_{EH} represents the energy efficiency ratio of the electric heater.

3.2.3. Energy Storage Systems

Operating constraints of the energy storage system mainly comprise the storage state and some output constraints no matter it is the electricity, heat or cooling storage system. Take the electricity energy storage system as an example, the typical operating model can be written as

$$\begin{cases} S_{PS}(t) = (1 - \eta_{PS}) S_{PS}(t-1) + \eta_{PS}^S P_{PS}^S(t) - P_{PS}^R(t)/\eta_{PS}^R \\ S_{PS}^{\min} \leq S_{PS}(t) \leq S_{PS}^{\max} \\ S_{PS}(t_1) = S_{PS}(t_{end}) \\ U_{PS}^S(t) P_{PS}^{\min} \leq P_{PS}^S(t) \leq U_{PS}^S(t) P_{PS}^{\max} \\ U_{PS}^R(t) P_{PS}^{\min} \leq P_{PS}^R(t) \leq U_{PS}^R(t) P_{PS}^{\max} \\ U_{PS}^R(t) + U_{PS}^S(t) \leq 1 \\ U_{PS}^R(t), U_{PS}^S(t) \in (0,1) \end{cases} \tag{23}$$

where $S_{PS}(t)$ represents the remaining power in the electricity energy storage system; η_{PS} represents the attrition rate during the charge/discharge process; η_{PS}^S represents the charge efficiency; η_{PS}^R represents the discharge efficiency; $U_{PS}^S(t)$ represents the on/off charge state; $U_{PS}^R(t)$ represents the on/off discharge state; $S_{PS}^{\min}$ and $S_{PS}^{\max}$ represents the capacity limit; $S_{PS}(t_1)$ represents the initial power in the beginning of one cycle of operation; and $S_{PS}(t_{end})$ represents the final power in the end of one cycle of operation. When considering the periodicity of the operation of ESS (energy storage system), the final power of the previous cycle should be equal to the initial power of the next cycle; hence, $S_{PS}(t_{end}) = S_{PS}(t_1)$ should be satisfied.

3.2.4. Other Constraints

In addition to constraints characterizing energy supply and demand balance, input and output energy relation of conversion systems, other mentionable operating constraints are those that describe the output limit of specific components in MES.

Firstly, the convertible load in each energy network has its lower/upper limit:

$$\begin{cases} \underline{P}_{load}^{PC} \leq P_{load}^{PC}(t) \leq \overline{P}_{load}^{PC} \\ \underline{C}_{load}^{PC} \leq C_{load}^{PC}(t) \leq \overline{C}_{load}^{PC} \\ \underline{P}_{load}^{PH} \leq P_{load}^{PH}(t) \leq \overline{P}_{load}^{PH} \\ \underline{H}_{load}^{PH} \leq H_{load}^{PH}(t) \leq \overline{H}_{load}^{PH} \\ \underline{G}_{load}^{GC} \leq G_{load}^{GC} \leq \overline{G}_{load}^{GC} \\ \underline{C}_{load}^{GC} \leq C_{load}^{GC}(t) \leq \overline{C}_{load}^{GC} \\ \underline{G}_{load}^{GH} \leq G_{load}^{GH} \leq \overline{G}_{load}^{PH} \\ \underline{H}_{load}^{GH} \leq H_{load}^{GH}(t) \leq \overline{H}_{load}^{GH} \end{cases} \tag{24}$$

Secondly, the output of conversion system has its lower/upper limit:

$$\begin{cases} \underline{P}_{CHP} \leq P_{CHP}(t) \leq \overline{P}_{CHP} \\ \underline{P}_{GT} \leq P_{GT}(t) \leq \overline{P}_{GT} \\ \underline{H}_{GB} \leq H_{GB}(t) \leq \overline{H}_{GB} \\ \underline{C}_{ER} \leq C_{ER}(t) \leq \overline{C}_{ER} \\ \underline{H}_{EH} \leq H_{EH}(t) \leq \overline{H}_{EH} \end{cases} \tag{25}$$

Thirdly, the purchase amount of electricity $P_{ext}(t)$ and gas $G_{ext}(t)$ cannot be unlimited; it is assumed that they cannot surpass the allowable maximum.

$$\begin{cases} 0 \leq P_{ext}(t) \leq P_{ext}^{\max} \\ 0 \leq G_{ext}(t) \leq G_{ext}^{\max} \end{cases} \tag{26}$$

Based on (26) it can be learned that the minimal purchase amount from external electricity or gas network is nonnegative. Particularly with respect to gas purchase, since gas is the only crude source which is used as the primitive input energy of all conversion systems and multi-generation plants, it cannot be less than zero. Otherwise, it means MES can autonomously inject gas toward external network out of nothing, which is against the law of conservation of energy. On the other hand, the electricity purchased takes the role of auxiliary, which can act as supplementary energy provider in due time. Because of the existence of gas load, electricity purchased cannot single-handedly solve the energy supply problem. Nevertheless, the energy supplier for heat and cooling load could be electricity purchased.

The model (7)–(26) belongs to mixed integer linear programming (MILP) and is solved by commercial optimization software package CPLEX under general algebraic modelling system (GAMS).

4. Case Study

In this section, MES in Figure 2 is used for numerical analyses. Specifically, the intrinsic relation between load replaceability (as is quantified by (4) and (6)) and optimal scheduling results is studied, and thus giving a priori knowledge of improving economic scheduling performance by reformulating the composite load.

4.1. Simulation Settings

Parameters of main system components are given in Table 1.

Table 1. Parameters and corresponding values of MES.

Parameter	Value
maximum allowable electricity procurement P_{ext}^{max}	12,000 kW
maximum allowable gas procurement G_{ext}^{max}	60,000 kW
maximum output of electric heater $\overline{H}_{EH}$	2000 kW
maximum output of electric refrigerator $\overline{C}_{ER}$	2000 kW
maximum output electricity of CHP plant $\overline{P}_{CHP}$	8000 kW
maximum output of heat exchanger $\overline{H}_{HE}$	4000 kW
maximum output of absorption refrigerator $\overline{C}_{AR}$	6000 kW
maximum output of gas boiler $\overline{H}_{GB}$	2000 kW
maximum output of gas turbine $\overline{P}_{GT}$	8000 kW
power generation efficiency of CHP plant p_{CHP}	0.3
electricity-to-heat ratio of CHP plant q_{QP}	2/3
energy efficiency ratio of heat exchanger η_{HE}	0.9
energy efficiency ratio of absorption refrigerator η_{AR}	0.9
energy efficiency ratio of electric heater η_{EH}	0.95
energy generation proportion ratio γ_{HC}	0.6
energy efficiency ratio of electric refrigerator η_{ER}	3.5
energy efficiency ratio of gas boiler η_{GB}	0.9
maximum charge/discharge energy of electricity ESS P_{PS}^{max}	2000 kW
maximum charge/discharge energy of heat ESS H_{HS}^{max}	1000 kW
maximum charge/discharge energy of cooling ESS P_{PS}^{max}	2000 kW
maximum remaining energy of electricity ESS S_{PS}^{max}	8000 kWh
maximum remaining energy of heat ESS S_{HS}^{max}	4000 kWh
maximum remaining energy of cooling ESS S_{CS}^{max}	8000 kWh
attrition rate of electricity ESS η_{PS}	0.2%
attrition rate of heat ESS η_{HS}	0.3%
attrition rate of cooling ESS η_{CS}	0.2%
operating cost of electricity ESS r_{PS}	0.32 \$/kWh
operating cost of heat ESS r_{HS}	0.25 \$/kWh
operating cost of cooling ESS r_{CS}	0.2 \$/kWh

The load status described by loading rate at every scheduling interval is shown in Figure 3.

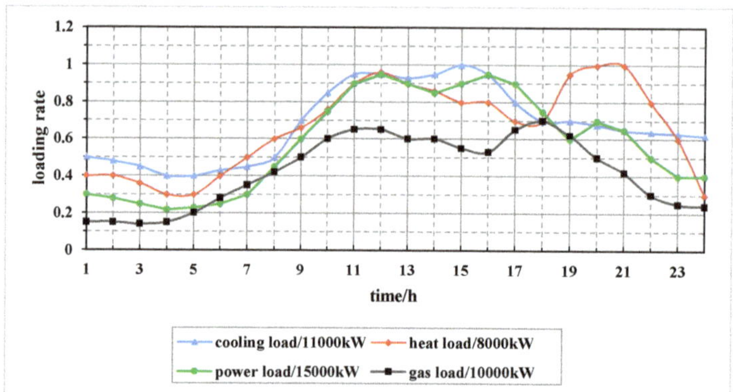

Figure 3. Hourly load status of gas, power, and heat and cooling load of MES.

The rated power of cooling, heat, and power and gas load is 11,000 kW, 8000 kW, 15,000 kW, and 10,000 kW, respectively. The day-ahead price information from the electricity and gas market is given in Table 2.

Table 2. Time-of-Use price of gas and electricity in a day.

Periods	r_{gas}	Periods	r_{ele}
00:00–07:00	2.7	00:00–08:00	0.35
07:00–12:00	3.3	08:00–12:00	1.65
12:00–16:00	3.0	12:00–17:00	0.95
16:00–20:00	3.3	17:00–21:00	1.65
20:00–24:00	3.0	21:00–24:00	0.95

4.2. Simulation Results and Discussions

After presenting the basic simulation settings, optimal scheduling and its relation with load replaceability is studied by calculating the scheduling model in (7)–(9) in Section 3 and LRI in (3)–(6) in Section 2.2. As previously addressed in Section 2.2, there exist two ways of describing the load replacement capability of MES; therefore, by special setting of composite load in (14), we change the values of LRI both in (4) and (6), and analyze the resulting influence upon optimal scheduling results through a sensitivity analysis tests. Specifically, the following 2 scenarios are considered.

4.2.1. Scenario 1

It is supposed that the composite load is of electricity-cooling and electricity-heat types. As for composite electricity-cooling load, the rated capacity of alternative electricity load is fixed to 2000 kW, while the capacity of alternative cooling load is increased from 600 kW to 6000 kW at the step size of 600 kW. Similarly, the rated capacity of alternative electricity load is fixed to 3750 kW in composite electricity-heat load, while the the capacity of alternative heat load is increased from 300 kW to 3000 kW at the step size of 300 kW. Consequently, 10 LRIs can be calculated based on (4) and (6), and the results are shown in Figure 4.

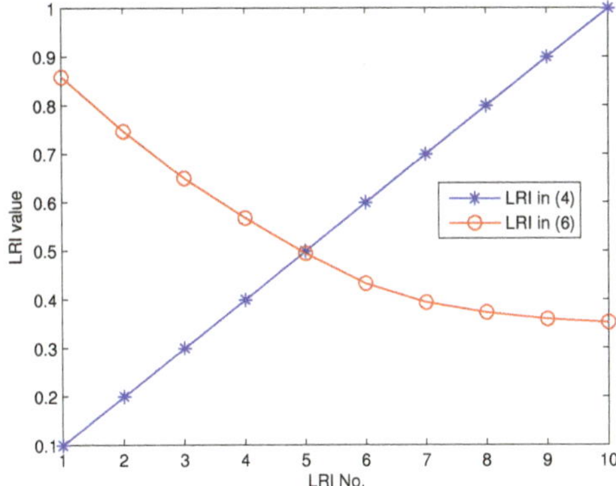

Figure 4. load replaceability index (LRI) with and without optimal scheduling in Scenario 1.

As can be seen from the blue curve in Figure 4, with the increase of alternative cooling and heat load, the remaining alternative load reserve is elevated from 900 kW to 9000 kW, thus enhancing LRI from 0.1 to 1. Meanwhile, the red curve shows that the proportion of energy converted by intermediate loads in the actual scheduling drops as the rated capacity of alternative heat (cooling) load increases, which means the operator chooses to obtain heat or cooling energy through redundant conversion paths. Take the electricity-cooling load as an example, the redundant conversion path is shown in dashed black line in Figure 5.

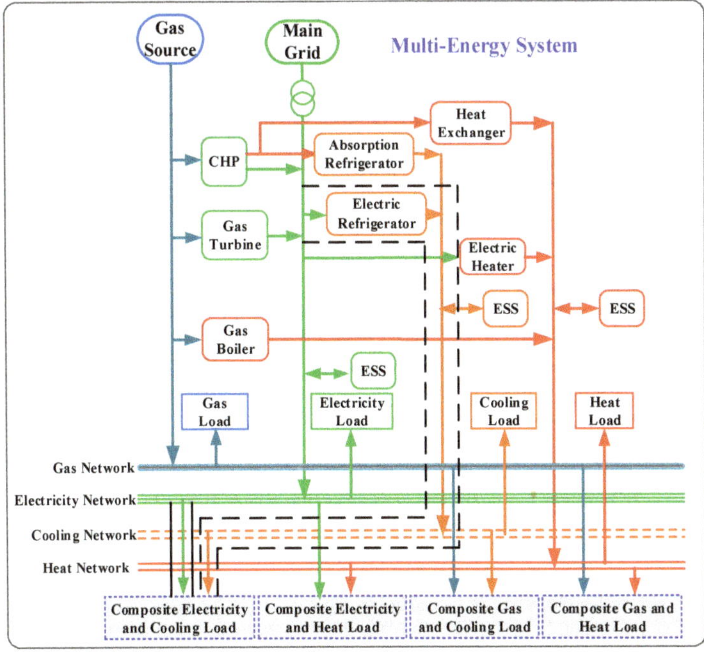

Figure 5. Redundant conversion path of electricity-cooling load in Scenario 1.

Due to the fact that energy efficiency ratio of electric refrigerator is much higher than that of alternative electricity load in the composite electricity and cooling load. The operator chooses redundant path such that more cooling energy can be harvested with equal amount of electricity energy. Similar conclusions can be made in respect to composite electricity and heat load. Therefore, the proportion of energy converted by intermediate electricity load in the composite load is decreased with the increase of transmission capacity of redundant paths, which is equivalent to the drop of LRI (from 0.86 to 0.35). Furthermore, the relation between LRI and optimal scheduling results in Scenario 1 is shown in Figures 6 and 7.

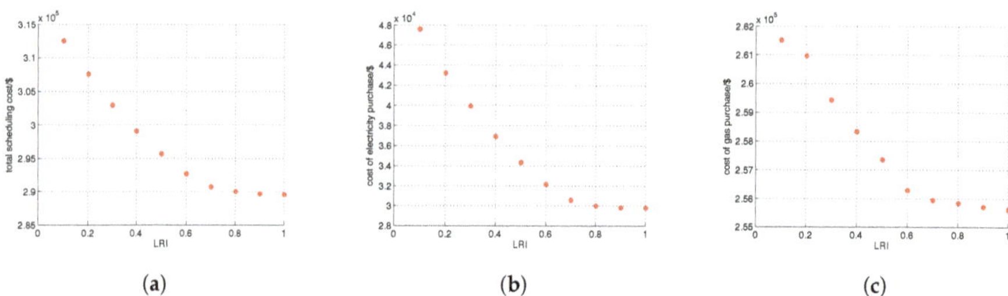

Figure 6. Relation between offline potential-based LRI and optimal scheduling results in Scenario 1. (a) The relation between LRI and optimal energy cost. (b) The relation between LRI and optimal cost of electricity purchase. (c) The relation between LRI and optimal cost of gas purchase.

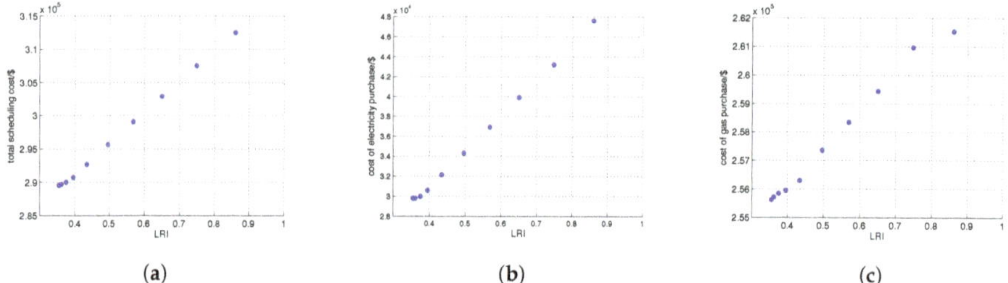

Figure 7. Relation between online scheduling-based LRI and optimal scheduling results in Scenario 1. (a) The relation between LRI and optimal energy cost. (b) The relation between LRI and optimal cost of electricity purchase. (c) The relation between LRI and optimal cost of gas purchase.

As can be seen from Figure 6, the cost including cost of gas and electricity purchase decreases with the increase of offline potential-based LRI. Contrarily, there exists a positive correlation between energy cost and online scheduling-based LRI. It verifies that the optimization space can be enhanced by elevating the capacity of alternative load, thus enabling the decrease of minimal scheduling costs.

4.2.2. Scenario 2

It is supposed that the composite load is of gas-cooling and gas-heat types. As for composite gas-cooling load, the rated capacity of alternative gas load is fixed to 5000 kW, while the capacity of alternative cooling load is increased from 600 kW to 6000 kW at the step size of 600 kW. Similarly, the rated capacity of alternative gas load is fixed to 6000 kW in composite gas-heat load, while the the capacity of alternative heat load is increased from 300 kW to 3000 kW at the step size of 300 kW. Consequently, 10 LRIs can be calculated based on (4) and (6), and the results are shown in Figure 8.

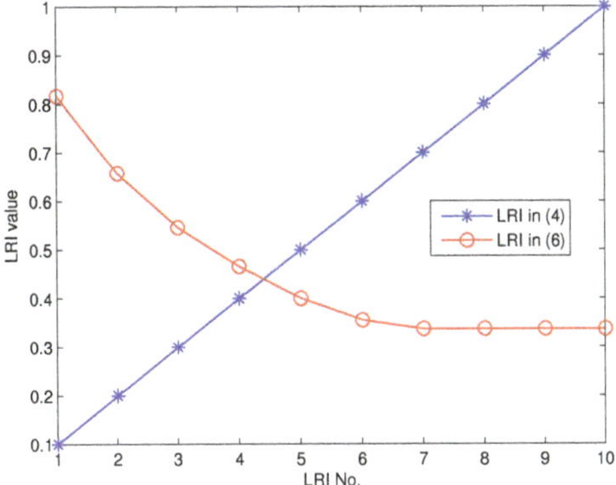

Figure 8. LRI with and without optimal scheduling in Scenario 2.

The relation between LRI and optimal scheduling results in Scenario 2 is shown in Figures 9 and 10.
Similar to Scenario 1, it can be learned that the optimal energy cost can be decreased by elevating
the offline potential based LRI, which quantifies the maximum alternative load capacity of the system.
In addition, it can be found that the optimal cost does not decrease when LRI reaches the critical state.
It means that there is no unrestricted expansion of optimization space by single-handedly elevating
alternative load capacity, which could improve the expectancy of optimum under the joint force of
other factors such as system configuration and market price signals.

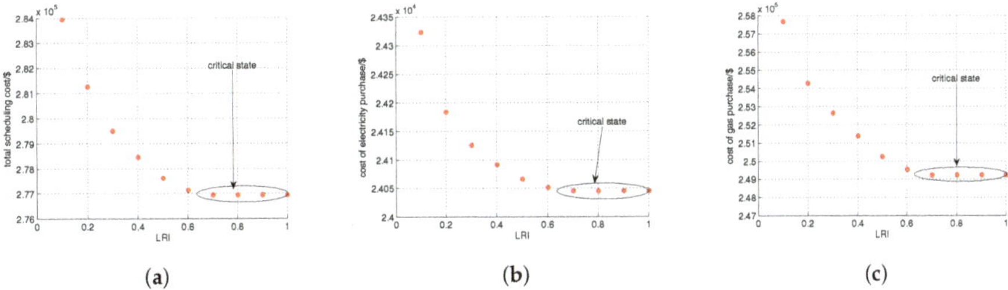

Figure 9. Relation between offline potential-based LRI and optimal scheduling results in Scenario 2.
(**a**) The relation between LRI and optimal energy cost. (**b**) The relation between LRI and optimal cost of
electricity purchase. (**c**) The relation between LRI and optimal cost of gas purchase.

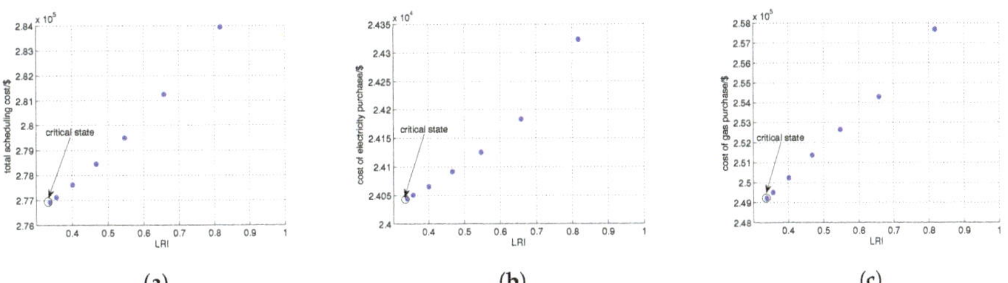

Figure 10. Relation between online scheduling-based LRI and optimal scheduling results in Scenario 2.
(**a**) The relation between LRI and optimal energy cost. (**b**) The relation between LRI and optimal cost of
electricity purchase. (**c**) The relation between LRI and optimal cost of gas purchase.

5. Conclusions

In this paper, optimal economic scheduling of multi-energy systems considering load
replaceability is presented. Two types of LRI reflecting the offline potential and online scheduling

performance are designed, and the relation between LRI and optimal scheduling results is analyzed. Through numerical analysis, it can be found that optimal scheduling performance can be improved by increasing the alternative load capacity (offline potential-based LRI). And this improvement is not unrestricted and could only maintain by changing load replaceability in a limited range, after which the optimum reaches the saturation phase.

Author Contributions: T.Z. and Z.D. conceived and designed the experiments; J.Y. performed the experiment; T.Z. wrote the paper; Y.Y. and J.C. reviewed the paper.

Funding: This work was supported by the National Key Research and Development Program of China (2018YFB0905000) and the project of 'Study and application of multi-energy micro-grid source-network-load-storage coordinated control system' of NARI Group Corporation (State Grid Electric Power Research Institute).

Conflicts of Interest: The authors declare no conflict of interest.

References

1. Vasebi, A.; Fesanghary, M.; Bathaee, S. Combined heat and power economic dispatch by harmony search algorithm. *Int. J. Electr. Power Energy Syst.* **2007**, *29*, 713–719. [CrossRef]
2. Vögelin, P.; Koch, B.; Georges, G.; Boulouchos, K. Heuristic approach for the economic optimisation of combined heat and power (CHP) plants: Operating strategy, heat storage and power. *Energy* **2017**, *121*, 66–77. [CrossRef]
3. Wu, C.; Jiang, P.; Sun, Y.; Zhang, C.; Gu, W. Economic dispatch with chp and wind power using probabilistic sequence theory and hybrid heuristic algorithm. *J. Renew. Sustain. Energy* **2017**, *9*, 013303. [CrossRef]
4. Li, Z.; Wu, W.; Shahidehpour, M.; Wang, J.; Zhang, B. Combined heat and power dispatch considering pipeline energy storage of district heating network. *IEEE Trans. Sustain. Energy* **2015**, *7*, 12–22. [CrossRef]
5. Hashemi, R. A developed offline model for optimal operation of combined heating and cooling and power systems. *IEEE Trans. Energy Convers.* **2009**, *24*, 222–229. [CrossRef]
6. Facci, A.L.; Andreassi, L.; Ubertini, S. Optimization of CHCP (combined heat power and cooling) systems operation strategy using dynamic programming. *Energy* **2014**, *66*, 387–400. [CrossRef]
7. Li, M.; Mu, H.; Li, N. Optimal design and operation strategy for integrated evaluation of CCHP (combined cooling heating and power) system. *Energy* **2016**, *99*, 202–220. [CrossRef]
8. Tan, Z.; Guo, H.; Lin, H.; Tan, Q.; Yang, S.; Gejirifu, D.; Ju, L.; Song, X. Robust scheduling optimization model for multi-energy interdependent system based on energy storage technology and ground-source heat pump. *Processes* **2019**, *7*, 27. [CrossRef]
9. Wang, M.; Wang, J.; Zhao, P.; Dai, Y. Multi-objective optimization of a combined cooling, heating and power system driven by solar energy. *Energy Convers. Manag.* **2015**, *89*, 289–297. [CrossRef]
10. Motevasel, M.; Seifi, A.R.; Niknam, T. Multi-objective energy management of CHP (combined heat and power)-based micro-grid. *Energy* **2013**, *51*, 123–136. [CrossRef]
11. Li, K.; Yan, H.; He, G.; Zhu, C.; Liu, K.; Liu, Y. Seasonal Operation Strategy Optimization for Integrated Energy Systems with Considering System Cooling Loads Independently. *Processes* **2018**, *6*, 202. [CrossRef]
12. Zhang, X.; Shahidehpour, M.; Alabdulwahab, A.; Abusorrah, A. Optimal expansion planning of energy hub with multiple energy infrastructures. *IEEE Trans. Smart Grid* **2015**, *6*, 2302–2311. [CrossRef]
13. Parisio, A.; Del Vecchio, C.; Vaccaro, A. A robust optimization approach to energy hub management. *Int. J. Electr. Power Energy Syst.* **2012**, *42*, 98–104. [CrossRef]
14. Vahid-Pakdel, M.; Nojavan, S.; Mohammadi-Ivatloo, B.; Zare, K. Stochastic optimization of energy hub operation with consideration of thermal energy market and demand response. *Energy Convers. Manag.* **2017**, *145*, 117–128. [CrossRef]
15. Huang, Y.; Zhang, W.; Yang, K.; Hou, W.; Huang, Y. An optimal scheduling method for multi-energy hub systems using game theory. *Energies* **2019**, *12*, 2270. [CrossRef]
16. Lin, Y.; Chen, B.; Wang, J.; Bie, Z. A combined repair crew dispatch problem for resilient electric and natural gas system considering reconfiguration and DG islanding. *IEEE Trans. Power Syst.* **2019**, *34*, 2755–2767. [CrossRef]

17. Liu, Z.; Yang, P.; Xu, Z. Capacity allocation of integrated energy system considering typical day economic operation. *Electr. Power Constr.* **2017**, *38*, 51–59.

18. Zhao, F.; Sun, B.; Zhang, C. Cooling, heating and electrical load forecasting method for CCHP system based on multivariate phase space reconstruction and Kalman filter. *Proc. CSEE* **2016**, *36*, 399–406.

19. Chen, F.; Xu, J.; Wang, C. Research on building cooling and heating load prediction model on user's side in energy internet system. *Proc. CSEE* **2015**, *35*, 3678–3684.

20. Shi, J.; Tan, T.; Guo, J. Multi-task learning based on deep architecture for various types of load forecasting in regional energy system integration. *Power Syst. Technol.* **2018**, *42*, 698–706.

21. Sheikhi, A.; Rayati, M.; Bahrami, S. Integrated demand side management game in smart energy hubs. *IEEE Trans. Smart Grid* **2015**, *6*, 675–683. [CrossRef]

22. Bahrami, S.; Sheikhi, A. From demand response in smart grid toward integrated demand response in smart energy hub. *IEEE Trans. Smart Grid* **2016**, *7*, 650–658. [CrossRef]

Article

Mathematical Modeling and Simulation on the Stimulation Interactions in Coalbed Methane Thermal Recovery

Teng Teng [1,2,*], Yingheng Wang [3], Xiang He [1] and Pengfei Chen [1]

[1] School of Energy and Mining Engineering, China University of Mining and Technology, Beijing 100083, China
[2] State Key Laboratory of Coal Resources and Safe Mining, China University of Mining and Technology, Beijing 100083, China
[3] School of Electronics and Information Engineering, Tongji University, Shanghai 201804, China
* Correspondence: T.Teng@cumtb.edu.cn

Received: 26 June 2019; Accepted: 6 August 2019; Published: 8 August 2019

Abstract: Heat stimulation of coalbed methane (CBM) reservoirs has remarkable promotion to gas desorption that enhances gas recovery. However, coalbed deformation, methane delivery and heat transport interplay each other during the stimulation process. This paper experimentally validated the evolutions of gas sorption and coal permeability under variable temperature. Then, a completely coupled heat-gas-coal model was theoretically developed and applied to a computational simulation of CBM thermal recovery based on a finite element approach of COMSOL with MATLAB. Modeling and simulation results show that: Although different heat-gas-coal interactions have different effects on CBM recovery, thermal stimulation of coalbed can promote methane production effectively. However, CBM thermal recovery needs a forerunner heating time before the apparent enhancement of production. The modeling and simulation results may improve the current cognitions of CBM thermal recovery.

Keywords: coalbed methane thermal recovery; thermal stimulation interaction; heat-gas-coal model; modeling and simulation

---

1. Introduction

Coalbed methane (CBM) is a cleaner and cheaper resource among the fossil fuels [1–3]. It is reported that the accumulated reserve of Chinese CBM is about 10 billion cubic meters, and the recoverable resource represents about 47% [4]. In earlier years, CBM is treated as a nerve-wracking hazard in coal mining. However, people's attitudes are changing that CBM becomes an efficiently and environmentally friendly fuel now. However, the production of CBM meets a great challenge of lower reservoir permeability and higher methane sorption capacity in China. A survey research shows that the permeability of Chinese coal ranges from 1×10^{-4} mD to 1×10^{-3} mD while it ranges from 0.1 mD to 1 mD of American or Australian coal [5]. As a result, the traditional direct-recovery method of CBM cannot satisfy the demands of effective production. Therefore, researchers and engineers in this field tried some unconventional methods by using a series of manual treatments to the methane reservoirs, such as the fracturing by water and gas, the displacement of adsorption by carbon dioxide and the stimulation of increasing temperature.

Thermal stimulation to deep CBM reservoirs is an effective method to promote gas recovery. Researchers have explored many different basic theories and mining technologies [6–10]. Before the production of methane, water with high temperature of 80 °C is assumed to be injected into a hypothetical CBM reservoir to causes helpful coal-gas interactions [11]. The simulation results indicate that the reservoir temperature increases 30 °C in twelve years to finally expand the methane production by 58% compared to the conventional method of direct recovery. Li et al. [12] established a mathematical model before demonstrating the complicated couplings among coalbed, methane and temperature. The model was then applied to a CBM thermal recovery of microwave heating. Research results show that thermal stimulation with microwave expands methane recovery more than 40% due to the deformation induced by gas sorption. Khoshnevis et al. [13] investigated the synergy type of gas production by injecting geothermal water. They established a three-dimension model to discuss the production potential of gas field and the influences of injection rate, reservoir permeability, saturation condition on the ultimate methane production. Lu [14] proposed a numerical simulation of heat injection into a three-dimensional temperature field by ANSYS to evidence the increasing production of CBM with temperature due to the easier desorption with higher temperature. Shahtalebi et al. [15] point out that the costs in coalbed methane thermal production limit the economic effectiveness. Only when the price of natural gas is comparatively higher and the demand for clean energy is stronger, the method of thermal production can become an economically attractive option.

CBM thermal recovery benefits a lot from the enhanced gas adsorption behavior [16–18]. Sakurovs et al. [19] presented three sets of experimental sorption evolutions at variable temperature. The results show that the methane adsorption capacity has great dependency with coal temperature. The sorption of coal to methane at different temperature ranges may indicate different behaviors, and a traditional Langmuir equation cannot describe the different trends. For example, Guan et al. [20,21] estimated the adsorption isotherms at temperatures from 283 K to 343 K. They find that the adsorption capacity of CBM decreases linearly with increasing temperature at the range of 283 K to 323 K, and keeps constant at the range of 323 K to 343 K. Similarly, Crosdale et al. [22] also observed the influence of temperature on gas desorption. Besides, at temperature lower that 30 °C, the adsorption capacity is averagely elevated by 10%, where it has no significant influences at the temperatures that higher than 60 °C. After a series of detailed theoretical and experimental research, Zhang et al. [23] concluded that the hysteresis degree was to be influenced by the microstructures of coal, especially the surface area. Further, Liu et al. [24] find that the influence of temperature on methane adsorption is more remarkable in coals that have smaller pores rather than larger pores. The adsorption capacity of methane decreases by approximately 19%, 32% and 45% for coals at micro-pore with sizes of 0.7–0.9 nm, 1.0–1.3 nm and size that larger than 1.4 nm respectively. Although many researchers have pointed out the effects of temperature on gas sorption, a theoretical presentation based on experiments that can concisely and usefully describe the effects is still lack.

Change of reservoir temperature causes a succession of interactions to CBM seepage behavior and the evolution of coal permeability [25–27]. Wang et al. [28] found that the evolution of coalbed permeability with increasing temperature may be divided into two stages: coal permeability decreases due to the internal swelling of coal matrix at the first stage and increases at the domination of rising gas pressure. Li et al. [12] performed a set of thermo-hydro-mechanical experiments to observe coal permeability by a servo-controlled equipment. The results indicate that evolutions of coal permeability with rising temperature are closely related with coal deformation. Yin et al. [29] goes further to observe the changing of coal permeability at different deformation stages. Coal permeability decreases with the rising temperature remarkably before the peak deformation, and it keeps almost constant after the peak deformation. That is because the growth of fractures in coal has more influences than the temperature on coal permeability. Focusing on the fractured bituminous coal, Perera et al. [30] investigated the permeability evolution under five different temperatures that ranges from 25 °C to 70 °C. The testing result shows that the CO_2 permeability increases linearly with temperature when coal temperature is higher than 90 °C due to the temperature-sensitive sorption behavior of CO_2. Besides, temperature change causes thermal expansion and fracturing [31]. Teng et al. explored the mechanism of these coal-gas interactions and tried to establish a permeability model for methane flow under thermo-hydro-mechanical situations. To conclude, one can find that, the above experimental results are mainly qualitative, it is necessary to propose a theoretically permeability model that can better connect the experimental rules with CBM thermal production.

CBM thermal recovery donates a multi-physical issue of coal, gas and heat [32,33]. Abed et al. [34] described a thermo-hydro-mechanical (THM) framework that suitable for modeling the behavior of unsaturated soils and rock. Due to the CBM thermal recovery method of microwave heating, Gao et al. [35] illustrated the interactions among temperature field, coalbed compaction, and methane transfer. According to their thermo-hydro-mechanical model developed in a finite element environment, higher stimulation temperature results in larger recovery radius. Fan et al. [36] discussed the competitive sorption of carbon dioxide with methane and its interactions with water under a thermo-hydro-mechanical-chemical condition. A two-phase-flow model was developed for CO_2 enhanced CBM recovery. Xia et al. [37] established a coupled hydro-thermo-mechanical model for the spontaneous combustion of underground coal and quantitatively predicted the spontaneous combustion locations of Dongtan coal mine. These precursor works have enlightening meanings for the study of multi-physical interactions in CBM thermal recovery. However, a fully coupled heat-gas-coal model is still necessary.

Following on our previous establishment of coal permeability model, this paper developed a completely coupled heat-gas-coal model for deep CBM thermal recovery by considering the interactions among coalbed deformation, methane delivery and heat transport. To evaluate these interactions among three physical fields and the production efficiency of CBM recovery, a numerical simulation using the finite element approach was validated. Finally, a series of analysis work based on the modeling and simulation are carried out.

2. Experimental Observation of Methane Sorption Under Variable Temperature

2.1. Experimental Program

In this section, the sorption characteristics of coal to pure CH_4 under variable temperature is observed in a self-developed gas adsorption apparatus, see Figure 1. Coal samples that acquired from the exploratory borehole of Jinjia lignite mine China are powdered into size of 150 μm. The experimental temperature varies from 25 °C to 85 °C, concretely 25 °C, 45 °C, 65 °C and 85 °C as representatives. At each temperature, the adsorption content was tested in a pressure vessel at the pressure that increases from the atmospheric pressure to a maximum of 10 MPa.

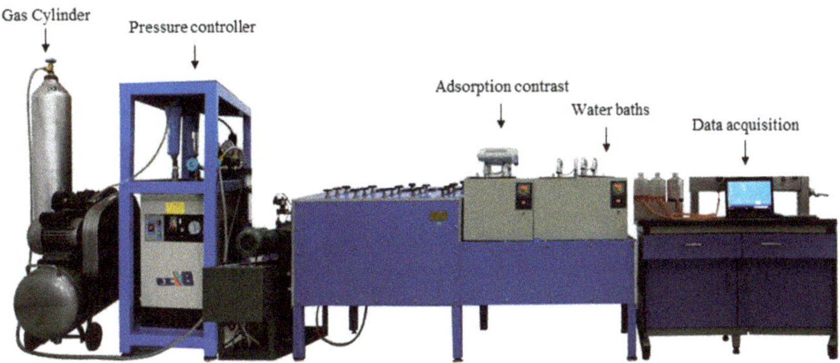

Figure 1. A self-developed isothermal gas adsorption apparatus.

2.2. Effects of Temperature on Methane Adsorption Content

The directly observed methane adsorption content under variable temperature is represented by the colorful spots in Figure 2. One can obtain that temperature has a significant influence on the adsorption capacity of Jinjia lignite. For example, when the temperature increases from 25 °C to

45 °C, 65 °C and 85 °C, the adsorption content at pressure of 10 MPa decreases 14.1%, 22.5%, 30.3%, respectively. It is because that the surface free energy of coal decreases when the temperature increases, as a result more CH_4 is released from the micro-pore.

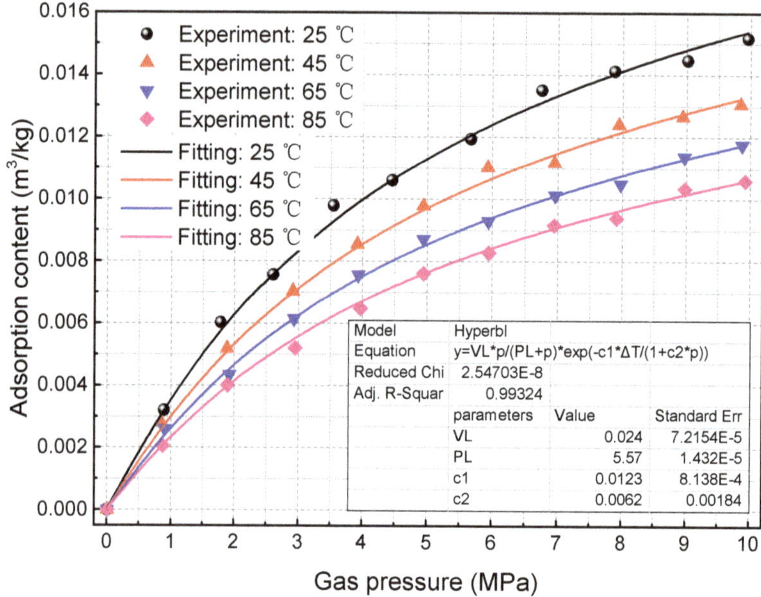

Figure 2. Validation of temperature modified Langmuir equation.

A Langmuir equation is often used to describe the methane adsorption content under constant temperature [38–40] as:

$$V_{sg} = V_L \frac{bp}{1 + bp} \tag{1}$$

where, V_{sg} is the methane adsorption content, m³/kg, V_L and b donate the Langmuir volume and pressure constants.

However, Equation (1) must be modified if the temperature changes. Here, an additional exponential term is introduced to revise the equation as:

$$V_{sg} = \frac{V_L p}{P_L + p} \exp\left[-\frac{c_2\left(T - T_{ref}\right)}{1 + c_1 p} \right] \tag{2}$$

where, P_L donates the Langmuir pressure constant, Pa. T_{ref} is the reference temperature for methane sorption, K. c_1 and c_2 are the coefficients for pressure and temperature, Pa^{-1} and K^{-1}, respectively. Figure 2 also shows the validation results of Equation (2) by the obtained experimental data. The well match of the fitting curves with the experimental results indicates that the modified Langmuir Equation (2) can effectively describe the evolution of methane sorption under variable temperature. This domino effect of gas sorption with changeable temperature indicates useful implication for CBM thermal recovery.

2.3. Volumetric Stain Induced by Methane Desorption

Coal matrix expands when it adsorbs gas. According to [41], the gas sorption-induced volumetric strain has a linear relationship with the gas adsorption content as:

$$\varepsilon_s = \alpha_{sg} V_{sg} \tag{3}$$

where α_{sg} is the expansion coefficient.

Thus, the volume deformation of coal caused by methane adsorption at different temperature can be expressed as:

$$\varepsilon_s = \frac{\alpha_{sg} V_L p}{P_L + p} \exp\left(-\frac{c_2 \Delta T}{1 + c_1 p}\right) - \frac{\alpha_{sg} V_L p_0}{P_L + p_0} \tag{4}$$

where $\Delta T = T - T_0$ shows the change of temperature.

3. Mathematical Model

Thermal recovery of CBM is a coupled process of coalbed deformation, methane delivery and heat transport. Among these three fields, one affects another. For example, the compaction of coalbed block causes the reduction of coal porosity and permeability for gas flow while the decrease of methane pressure changes the effective stress to promote coalbed deformation. Moreover, thermal expansion that caused by temperature change affects both coal deformation and gas flow. Following section is to establish a completely coupled heat-gas-coal model for CBM thermal recovery by considering the interactions among coalbed deformation, methane flow and heat transport. Before any derivation, coalbed is assumed as one kind of homogeneous, isotropic continuum while methane is treated as ideal gas. Methane transport in porous coalbed is treated as a Darcy's flow.

3.1. Coalbed Deformation Equation

Based on rock elasticity theory, the deformation of coal reservoir can be governed by a Navier-type equilibrium equation. In this paper, the equilibrium equation is optimized by the variable temperature [42] as:

$$Gu_{i,kk} + \frac{G}{1 - 2v} u_{k,ki} - \alpha \cdot p_{,i} - K\varepsilon_{s,i} - K\alpha_T T_{,i} + F_i = 0 \tag{5}$$

in which F_i and p represent the body force of coal and the pressure of methane, MPa. $G = 0.5E/(1+v)$ and $K = E/(3(1-2v))$ are the shear and bulk modulus, respectively. E and v represent the elasticity modulus and Poisson's ratio, respectively. $\alpha = 1 - K/K_s$, $(\alpha < 1)$ means the Biot's coefficients, where K_s donates the modulus of coal grains.

The volumetric stain that induced by temperature change is defined [43,44] as:

$$\varepsilon_T = \alpha_T \Delta T \tag{6}$$

where, α_T is the thermal expansion coefficient, K^{-1}.

3.2. Methane Flow Equation

The CBM flow in reservoirs obeys a mass conservation equation [37]:

$$\frac{\partial m}{\partial t} + \nabla \cdot (\rho_g v_g) = Q_s \tag{7}$$

in which Q_{sk} represents the methane source. v_g is the velocity vector of flow, m/s; m is the content of methane in coalbed that can be expressed as:

$$m = \rho_g \phi + \rho_c \rho_{ga} \frac{V_L p}{P_L + p} \exp\left[-\frac{c_2(T - T_{ref})}{1 + c_1 p}\right]$$

(8)

where, the first and second terms represent the free and adsorbed components, respectively. ϕ is the porosity of coal, ρ_c represents coal density while ρ_{ga} represents the methane density at standard conditions. The real gas density can be expressed as:

$$\rho_g = \frac{M_g p}{RT}$$

(9)

in which, R donates the universal gas constant and M_g means the molar mass.

Darcy's velocity vector v_g in Equation (7) is proportional to the pressure gradient ∇p and permeability k:

$$v_g = \frac{-k}{\mu} \cdot \nabla p$$

(10)

where μ represents the viscosity coefficient of methane.

Following on our previous work [31], the permeability of coalbed can be expressed as:

$$\frac{k}{k_0} = \left(\frac{\phi}{\phi_0}\right)^3 = \left\{1 - \frac{3K}{\phi_0 K + 3K_f}\left[(\varepsilon_s + \varepsilon_T - \varepsilon_v) - (\varepsilon_{s0} + \varepsilon_{T0})\right]\right\}^3$$

(11)

in which, K_f donates the equivalent modulus of coal fracture. Volume strain ε_v that induced by effective stress is expressed as:

$$\varepsilon_v = \frac{1}{K}(\bar{\sigma} + \alpha p) + \varepsilon_s + \varepsilon_T$$

(12)

in which, $\bar{\sigma}$ is the mean stress of coal.

Substituting Equations (4)–(6) into Equation (12), the permeability model reads as

$$\frac{k}{k_0} = \left(\frac{\phi}{\phi_0}\right)^3 = \left\{1 - \frac{3K}{\phi_0 K + 3K_f}\left[\left(\frac{\alpha_{sg} V_L p}{P_L + p} e^{-\frac{c_2 \Delta T}{1 + c_1 p}} + \alpha_T \Delta T - \varepsilon_v\right) - \frac{\alpha_{sg} V_L p_0}{P_L + p_0}\right]\right\}^3$$

(13)

In the experimental testing of coal permeability, Equation (13) can be modified as:

$$\frac{k}{k_0} = \left(\frac{\phi}{\phi_0}\right)^3 = \left\{A - Be^{-C\Delta T} - D\Delta T\right\}^3$$

(14)

where,

$$A = 1 - \frac{3K\varepsilon_v}{\phi_0 K + 3K_f} - \frac{\alpha_{sg} V_L p_0}{P_L + p_0}$$
$$B = \frac{3K}{\phi_0 K + 3K_f}\frac{\alpha_{sg} V_L p}{P_L + p}$$
$$C = \frac{c_2}{1 + c_1 p}$$
$$D = \frac{3K\alpha_T}{\phi_0 K + 3K_f}$$

(15)

Figure 3 shows the fitting results of the experimental data [30] by the proposed permeability model. Table 1 lists the fitting parameters of Equation (14). The matching result shows that the proposed coal permeability model can be well used to describe the evolution of coal permeability under variable temperature.

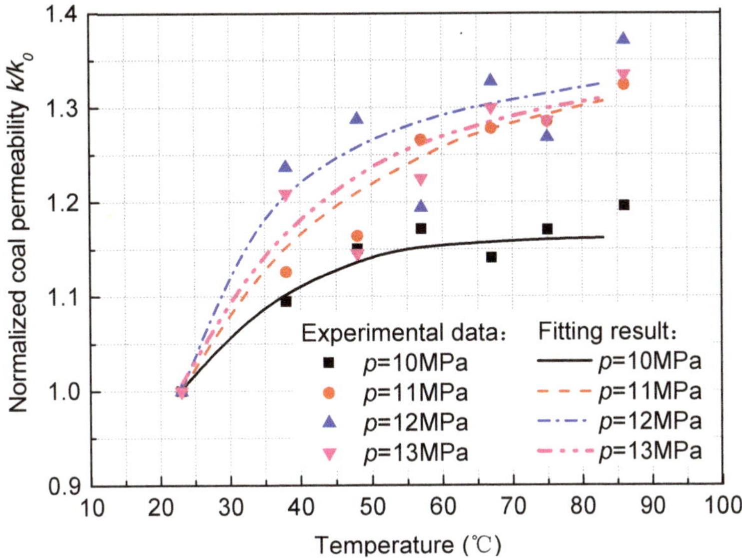

Figure 3. Validation of coal permeability model with experimental data.

Table 1. Fitting parameters for coal permeability evolution.

Coal Samples		Fitting Parameters				
		A	**B**	**C**	**D**	**R^2**
Gas pressure (MPa)	10	5.00×10^{-2}	0.09	6.14×10^{-4}	1.05×10^{-5}	0.915
	11	4.92×10^{-2}	0.11	8.23×10^{-4}	1.03×10^{-5}	0.854
	12	4.83×10^{-2}	0.08	7.01×10^{-4}	1.16×10^{-5}	0.975
	13	4.76×10^{-2}	0.10	1.50×10^{-3}	1.50×10^{-5}	0.962

Substituting Equations (8)–(13) into Equation (7), we can obtain the methane flow equation:

$$\left\{ \frac{\phi}{T} - \frac{N\varepsilon_L p}{T} \cdot \exp\left(-\frac{c_2(T-T_{ref})}{1+c_1 p}\right) \cdot \left[\frac{P_L}{(P_L+p)^2} + \frac{p}{P_L+p} \cdot \frac{c_1 c_2(T-T_{ref})}{(1+c_1 p)^2}\right]\right\} \frac{\partial p}{\partial t}$$
$$+ \left\{\frac{N\varepsilon_L p}{T} \cdot \exp\left(-\frac{c_2(T-T_{ref})}{1+c_1 p}\right) \cdot \left[\frac{p}{P_L+p} \cdot \frac{c_2}{(1+c_1 p)^2}\right] - \frac{\phi p}{T^2} - \frac{Np}{T}\left[\alpha_T + \frac{\vartheta \phi_0}{(1-\vartheta(T-T_0))^2}\right]\right\} \frac{\partial T}{\partial t} \quad (16)$$
$$+ \nabla \cdot \left(-\frac{kp}{\mu T}\nabla p\right) + \nabla \cdot \left[-D\nabla\left(\frac{p}{T}\phi\right)\right] + \frac{Np}{T}\frac{\partial \varepsilon_v}{\partial t} = 0$$

in which, $N = 3\phi_0 K / (\phi_0 K + 3K_f)$.

3.3. Heat Transport Equation

The conservation of energy for CBM thermal recovery obeys an equilibrium [45] as:

$$\frac{\partial(C_{eq}T)}{\partial t} + \nabla\left(-K_{eq}\nabla T\right) + p\nabla \cdot v_g + K\alpha_T T \frac{\partial \varepsilon_v}{\partial t} = Q_T \quad (17)$$

in which, the specific heat capacity of coalbed reads:

$$C_{eq} = \phi\rho_g C_g + (1-\phi)\rho_c C_c \quad (18)$$

K_{eq} is the effective thermal conductivity. C_g and C_c represent the specific heat coefficients for methane and coal. Q_T is the heat source.

Combining Equation (18) with Equation (17), one gets the modified conservation equation of energy as:

$$C_{eq}\frac{\partial T}{\partial t} + p\nabla \cdot \left(-\frac{k}{\mu}\nabla p\right) + K\alpha_T T\frac{\partial \varepsilon_v}{\partial t} = K_{eq}\nabla^2 T + \rho_g C_g \frac{k}{\mu}\nabla p\nabla T \tag{19}$$

The Equations (5), (16) and (19) make up a completely coupled heat-gas-coal model. The mathematical model is used in simulating CBM thermal production.

4. Modeling on CBM thermal Recovery

Figure 4 chooses a modeling and simulation domain to represent a half area of CBM thermal recovery from the research of Shahtalebi et al. [15]. Based on a partial-differential-equation solver of COMSOL with MATLAB that runs on Windows 10 environment with Intel Core i7-6500U and RAM 16.0 GB hardware, the simulator is implemented in 1049 s. In Figure 4, the simulation domain consists of 4208 elements. The computation time is 10^9 s that are derived into 100 steps.

The length and width of the simulation domain are one hundred meters and forty meters, respectively. The depth of methane recovery well into coalbed is seventy meters. A thermal stimulation well in size of thirty meters locates in the center. For the CBM reservoir, the initial methane pressure and temperature are 3.5 MPa and 298 K, respectively. For coal deformation, the boundary AB and BC are confined to normal force of 15 and 8 MPa, while the boundary AD and CD are constrained by normal displacement. For gas flow, the boundary AB, BC, CD and DE are symmetric boundary with no flow, while AE is common pressure boundary of 0.1 MPa. For heat transfer, the boundary AB, BC, CD and DA are thermal insulation, while FG is common temperature boundary of 373 K. To contrast simulation results, comparison sites, no. 1, 2, 3 and 4, are pre-set. Other simulation parameters are listed in Table 2.

Table 2. Parameters used in simulation of coalbed methane thermal recovery.

Variable	Parameter	Value Used	Sources
E	Young's modulus of coal, (MPa)	2713	[45]
E_s	Young's modulus of coal grains, (MPa)	4070	[45]
v	Poisson's ratio of coal	0.339	[45]
ρ_c	Density of coal, (kg/m^3)	1.25×10^3	[41]
ρ_{ga}	Density of gas at standard condition, (kg/m^3)	0.717	-
ε_L	Sorption coefficient for volumetric strain, (kg/m^3)	0.0156	[41]
k_0	Initial permeability of coal, (mD)	0.001	[31]
ϕ_0	Initial porosity of coal	0.02	Given
μ	Dynamic viscosity coefficient of gas, (Pa·s)	1.84×10^{-5}	[41]
V_L	Langmuir volume constant, (m^3/kg)	0.048	[46]
P_L	Langmuir pressure constant, (MPa)	1.57	[46]
c_1	Pressure coefficient, (MPa^{-1})	0.07	[44]
c_2	Temperature coefficient, (K^{-1})	0.02	[44]
T_a	Temperature at standard condition, (K)	273	-
T_0	Temperature of coal seam, (K)	298	Given
p_0	Initial value of gas pressure, (MPa)	3.5	Given
p_a	Pressure at standard condition, (MPa)	0.103	-
α_T	Thermal expansion coefficient, (K^{-1})	2.4×10^{-5}	[45]
C_c	Specific heat capacity of coal, (kJ/(kg·K))	1.25	[44]
C_g	Specific heat capacity of gas, (kJ/(kg·K))	1.62	[44]
K_{eq}	Effective thermal conductivity of coal, (J/(m·s·K))	0.2	[44]

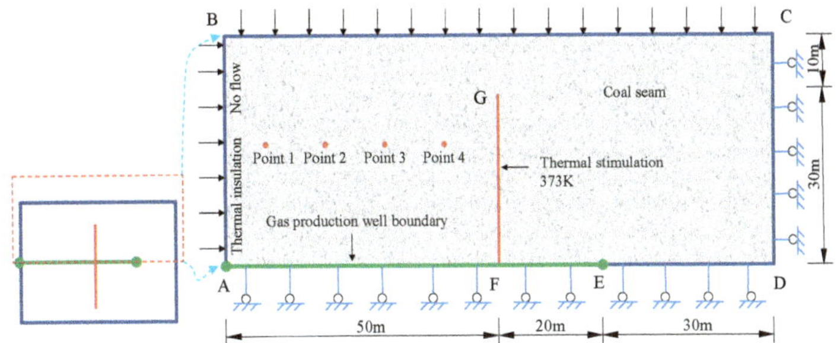

Figure 4. Numerical simulation model.

5. Modeling Results

5.1. Distribution of Coalbed Temperature

Figures 5 and 6 show the contour distribution of temperature in reservoirs during CBM thermal recovery, especially the temperature at four observation points.

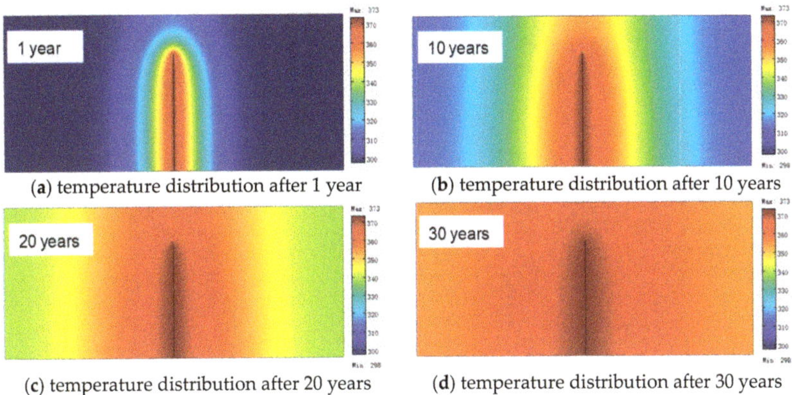

(**a**) temperature distribution after 1 year

(**b**) temperature distribution after 10 years

(**c**) temperature distribution after 20 years

(**d**) temperature distribution after 30 years

Figure 5. Distribution of coal temperature (K) after different production years.

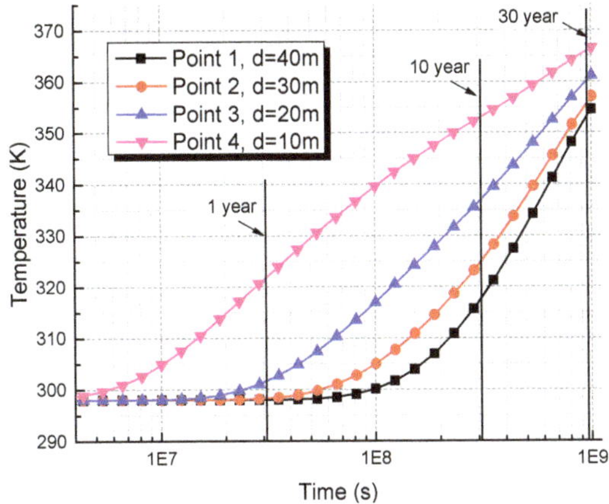

Figure 6. Evolution of reservoir temperature (K) at different observation sites.

From the figures, one can obviously find that the larger distances from thermal stimulation well correspond to lower increment of reservoir temperature. For earlier years, temperature change is only confined to the vicinity of thermal well. It indicates that CBM thermal recovery needs a long period of heating time to get beneficial output. For example, after one year (about 3×10^7 s), the reservoir temperature keeps almost it initial temperature, the temperature at point 4 is 320 K while the temperatures at point 3, 2 and 1 are lower than 300 K. However, when the heating time is longer than 10 years, obvious increment of temperature is observed in each observation point. After a heating time of 30 years, the reservoir temperature is higher than 350 K.

5.2. Evolution of Methane Pressure

The distribution of methane pressure in reservoirs changes a lot during CBM thermal recovery. It can be seen from the contour distribution of methane pressure after 1, 10, 20 and 30 years in Figure 7 that the methane pressure decreases in the first 10 years. From the figure, we can easily find that the mean gas pressure is about 2.1 MPa after 10 years recovery, whereas 0.65, 0.33 and 0.24 MPa after 20, 30 and 40 years.

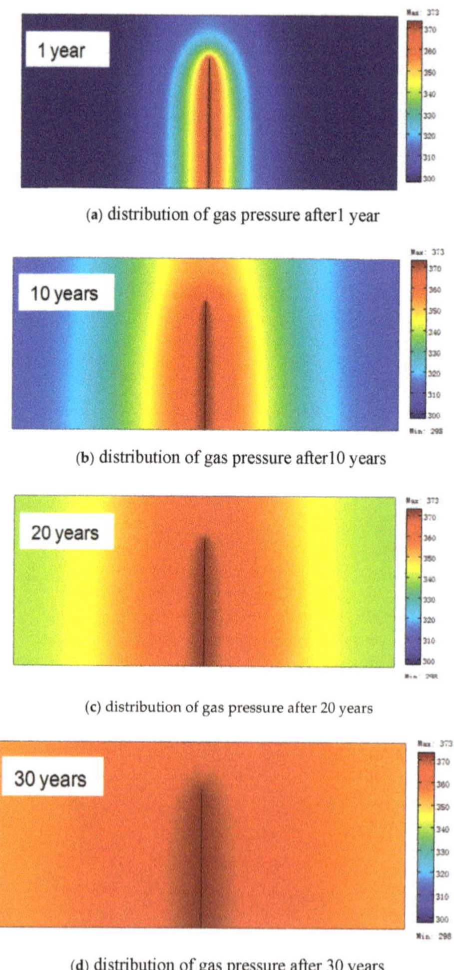

(a) distribution of gas pressure after1 year

(b) distribution of gas pressure after10 years

(c) distribution of gas pressure after 20 years

(d) distribution of gas pressure after 30 years

Figure 7. Distribution of gas pressure (MPa) after different production times.

According to the state equation of ideal gas, higher temperature leads to a higher gas pressure in a confined space. Figure 7 shows that higher coalbed temperature doesn't have prominent domination to methane pressure in reservoirs, even in the area near the thermal stimulation source. This is because the CBM is not confined in coal blocks seriously, but strongly transported out from the narrow fracture network.

5.3. Evolution of Coalbed Permeability

Figure 8 shows the evolution of coalbed permeability at different comparison sites. From the figure, one can draw two conclusions. The first is that coal permeability enlargers firstly and then decreases a little tiny bit. The second is that the permeability ratio increases with the decreasing distance from thermal stimulation well. Further, the normalized permeability ratio enlargers to the maximal value of about 1.53 firstly and then decreases. Although the evolution trends of normalized permeability ratio at 4 different observation points are similar, the corresponding production time node for the maximum permeability ratio delays with the increasing distance from the heat injection well, circumstantiate 8×10^8, 7×10^8, 5×10^8 and 2×10^8 s, respectively. From Equation (8), and the numerical results of Figures 6 and 8, one can explain the permeability evolution trend as that the accelerative effect of thermal desorption due to increased temperature plays a dominant role in enlarging coal permeability in the increasing stage, however the effective stress induced coal compaction gradually takes over the dominant role to reduce coal permeability in the decreasing stage where the reservoir temperature is stabilized.

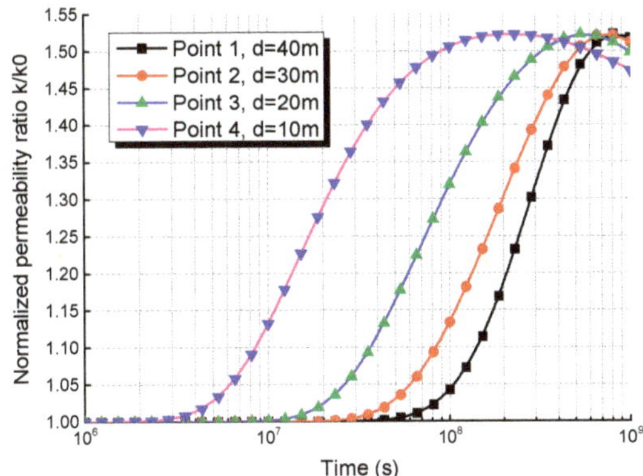

Figure 8. Evolution of coal permeability at different observation sites.

5.4. Methane Production with Different Thermal Stimulation Temperature

Figure 9a,b show the cumulative methane production and recovery efficiency in CBM thermal recovery cases with different stimulation temperatures. From the cumulative production, one can conclude that the thermal recovery with higher stimulation temperature has greater promotion to the final production of methane. However, the equal increment of coalbed temperature generates a larger promoting efficiency at lower temperature level due to the great enhancement of temperature to methane desorption. Figure 9b shows the efficiency of the increasing yield of coalbed methane thermal production. From the figure, one can see that the maximum production efficiencies of methane with stimulation temperatures of 50 °C, 75 °C and 100 °C are enlarged by 20%, 30% and 33% respectively compared with the conversional production method. However, Figure 9b also indicates that thermal recovery of coalbed methane needs a forerunner heating time to obtain apparent enhancement of production.

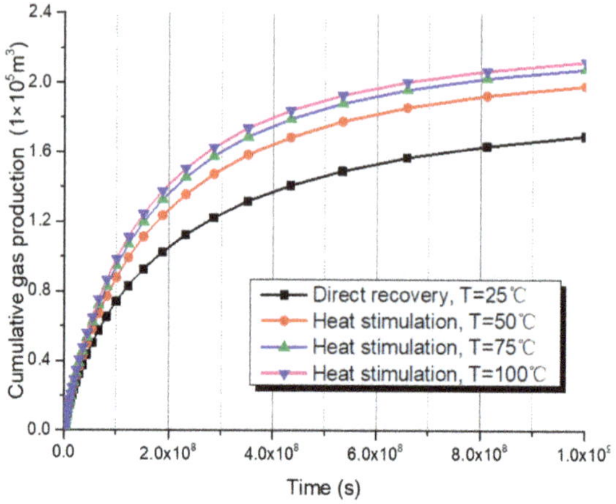

(a) Cumulative production.

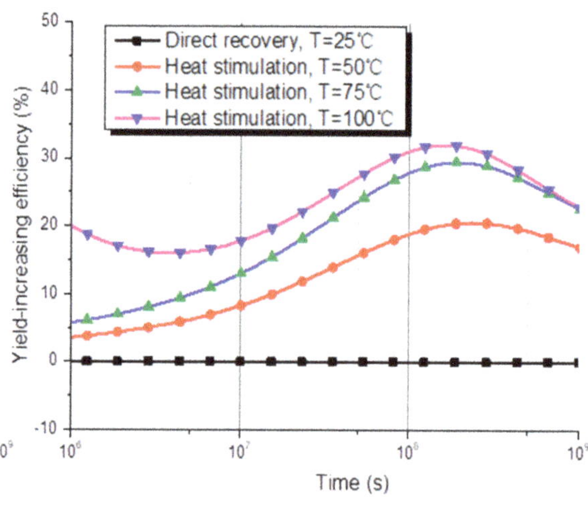

(b) production efficiency.

Figure 9. Gas production with different stimulation temperatures.

5.5. Methane Production with Different Initial Permeability

To evaluate the effects of initial reservoirs permeability on CBM thermal recovery, five numerical cases with different permeability and consistent stimulation temperature of 100 °C are carried out. Evolutions of total methane production are shown in Figure 10a. We can conclude that the methane production enlargers exponentially with production time. To measure the production efficiency, we define an index of output proportion as a percentage of cumulative production to total content of reservoir. Figure 10b shows the evolution of production time with the initial permeability when the output proportion percentage takes value of 25%, 50%, 75% and 90%, respectively. From Figure 10, one can obtain that larger value of initial permeability represents higher methane production in earlier time. For example, it takes 17 and 0.5 years respectively to recover 75% content of the total reserve when the reservoir permeability is 5×10^{-19} m^2 and 5×10^{-17} m^2, respectively.

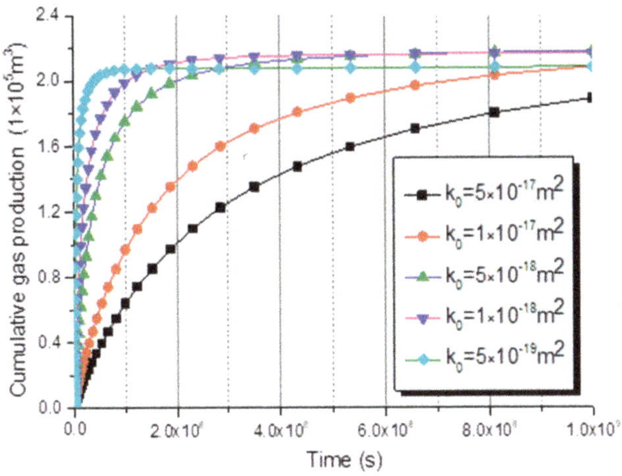

(**a**) Cumulative production.

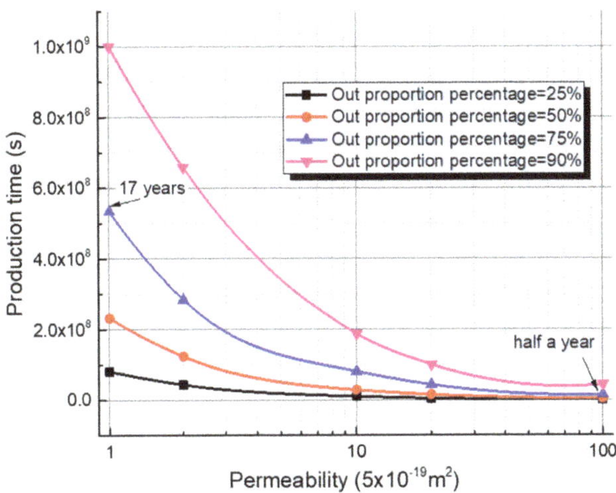

(**b**) production time for different output proportion

Figure 10. Gas production with different initial reservoir permeability.

6. Conclusions

Methane adsorption and coal permeability decrease with increasing temperature. This paper experimentally validated the evolutions of gas sorption and coal permeability under variable temperature. It also established a completely coupled heat-gas-coal model including multi-physics of coalbed deformation, methane delivery and heat transport. The mathematical model was applied to a computational simulation of CBM thermal recovery and solved by COMSOL with MATLAB in a finite element approach environment. To evaluate the thermal recovery process and the production efficiency, a series of analysis work were carried out. The results show that: (1) Thermal stimulation with higher temperature contributes to greater promotion to CBM recovery production. However, it needs a forerunner heating time before an apparent enhancement of production. (2) The normalized coal permeability ratio increases to the maximal value of about 1.53 firstly, and then decreases slightly.

The corresponding production time node for the maximum permeability ratio delays with the distance from thermal stimulation well. This completely coupled heat-gas-coal model can improve the current understandings of coalbed methane thermal recovery.

Author Contributions: T.T. established the model and performed the numerical simulation; X.H. wrote the original draft preparation; Y.W. and P.C. revised the paper.

Funding: This research was funded by the financial support from the National Natural Science Foundations of Jiangsu (Grant No. BK20170457) and the National key research and development plan (2016YFC0600705).

Conflicts of Interest: The authors declare no conflict of interest.

References

1. Al-Jubori, A.; Johnston, S.; Boyer, C.; Lambert, S.W.; Bustos, O.A.; Pashin, J.C.; Wray, A. Coalbed methane: Clean energy for the world. *Oilfield Rev.* **2009**, *21*, 4–13.
2. Yuan, L. Theory and practice of integrated coal production and gas extraction. *Int. J. Coal Sci. Technol.* **2015**, *2*, 3–11. [CrossRef]
3. Towler, B.; Firouzi, M.; Underschultz, J.; Rifkin, W.; Garnett, A.; Schultz, H.; Witt, K. An overview of the coal seam gas developments in Queensland. *J. Nat. Gas Sci. Eng.* **2016**, *31*, 249–271. [CrossRef]
4. Zhou, F.; Xia, T.; Wang, X.; Zhang, Y.; Sun, Y.; Liu, J. Recent developments of coal mine methane extraction and utilization in china: A review. *J. Nat. Gas Sci. Eng.* **2016**, *31*, 437–458. [CrossRef]
5. Noack, K. Control of gas emissions in underground coal mines. *Int. J. Coal Geolo.* **1998**, *35*, 57–82. [CrossRef]
6. Slastunov, S.V.; Prezent, G.M.; Kolikov, K.S. Experimental work on methane recovery at the Lenin Mine of the Karaganda Basin. *J. Min. Sci.* **1999**, *35*, 531–535. [CrossRef]
7. Mutyala, S.; Fairbridge, C.; Paré, J.J.; Bélanger, J.M.; Ng, S.; Hawkins, R. Microwave applications to oil sands and petroleum: A review. *Fuel Process. Technol.* **2010**, *91*, 127–135. [CrossRef]
8. Singh, A.K.; Baumann, G.; Henninges, J.; Görke, U.J.; Kolditz, O. Numerical analysis of thermal effects during carbon dioxide injection with enhanced gas recovery: A theoretical case study for the Altmark gas field. *Environ. Earth Sci.* **2012**, *67*, 497–509. [CrossRef]
9. Wang, H.; Ajao, O.; Economides, M.J. Conceptual study of thermal stimulation in shale gas formations. *J. Nat. Gas Sci. Eng.* **2014**, *21*, 874–885. [CrossRef]
10. Xue, L.; Dai, C.; Wang, L.; Chen, X. Analysis of Thermal Stimulation to Enhance Shale Gas Recovery through a Novel Conceptual Model. *Geofluids* **2019**, *2019*. [CrossRef]
11. Salmachi, A.; Haghighi, M. Feasibility study of thermally enhanced gas recovery of coal seam gas reservoirs using geothermal resources. *Energy Fuels* **2012**, *26*, 5048–5059. [CrossRef]
12. Li, B.; Yang, K.; Xu, P.; Xu, J.; Yuan, M.; Zhang, M. An experimental study on permeability characteristics of coal with slippage and temperature effects. *J. Pet. Sci. Eng.* **2019**, *175*, 294–302. [CrossRef]
13. Khoshnevis, N.; Khosrokhavar, R.; Nick, H.; Bruhn, D.F.; Bruining, H. Injection of disposal water from a geothermal reservoir into a gas reservoir. *J. Pet. Sci. Eng.* **2019**, *178*, 616–628. [CrossRef]
14. Lu, B.X. Numerical simulations of temperature field of coalbed methane with heat injection based on ANSYS. In Proceedings of the 2012 International Conference on Computer Science and Electronics Engineering, Hangzhou, China, 23–25 March 2012; Volume 2.
15. Shahtalebi, A.; Khan, C.; Dmyterko, A.; Shukla, P.; Rudolph, V. Investigation of thermal stimulation of coal seam gas fields for accelerated gas recovery. *Fuel* **2016**, *180*, 301–313. [CrossRef]
16. Li, H.; Kang, J.; Zhou, F.; Qiang, Z.; Li, G. Adsorption heat features of coalbed methane based on microcalorimeter. *J. Loss Prev. Process Ind.* **2018**, *55*, 437–449. [CrossRef]
17. Miklová, B.; Staf, M.; Kyselová, V. Influence of ash composition on high temperature CO2 sorption. *J. Environ. Chem. Eng.* **2019**, *7*, 103017. [CrossRef]
18. Okolo, G.N.; Everson, R.C.; Neomagus, H.W.; Sakurovs, R.; Grigore, M.; Bunt, J.R. The carbon dioxide, methane and nitrogen high-pressure sorption properties of South African bituminous coals. *Int. J. Coal Geol.* **2019**, *209*, 40–53. [CrossRef]
19. Sakurovs, R.; Day, S.; Weir, S.; Duffy, G. Temperature dependence of sorption of gases by coals and charcoals. *Int. J. Coal Geol.* **2008**, *73*, 250–258. [CrossRef]

20. Charrière, D.; Pokryszka, Z.; Behra, P. Effect of pressure and temperature on diffusion of CO2 and CH4 into coal from the Lorraine basin (France). *Int. J. Coal Geol.* **2010**, *81*, 373–380. [CrossRef]

21. Guan, C.; Liu, S.; Li, C.; Wang, Y.; Zhao, Y. The temperature effect on the methane and CO2 adsorption capacities of Illinois coal. *Fuel* **2018**, *211*, 241–250. [CrossRef]

22. Crosdale, P.J.; Moore, T.A.; Mares, T.E. Influence of moisture content and temperature on methane adsorption isotherm analysis for coals from a low-rank, biogenically-sourced gas reservoir. *Int. J. Coal Geol.* **2008**, *76*, 166–174. [CrossRef]

23. Zhang, L.; Aziz, N.; Ren, T.X.; Wang, Z. Influence of temperature on coal sorption characteristics and the theory of coal surface free energy. *Procedia Eng.* **2011**, *26*, 1430–1439. [CrossRef]

24. Liu, Y.; Zhu, Y.; Liu, S.; Li, W.; Tang, X. Temperature effect on gas adsorption capacity in different sized pores of coal: Experiment and numerical modeling. *J. Pet. Sci. Eng.* **2018**, *165*, 821–830. [CrossRef]

25. Akbarzadeh, H.; Chalaturnyk, R.J. Structural changes in coal at elevated temperature pertinent to underground coal gasification: A review. *Int. J. Coal Geol.* **2014**, *131*, 126–146. [CrossRef]

26. Niu, S.; Zhao, Y.; Hu, Y. Experimental ivestigation of the temperature and pore pressure effect on permeability of lignite under the in situ condition. *Trans. Porous Media* **2014**, *101*, 137–148. [CrossRef]

27. Zhang, X.G.; Ranjith, P.G.; Perera, M.S.A.; Ranathunga, A.S.; Haque, A. Gas transportation and enhanced coalbed methane recovery processes in deep coal seams: A review. *Energy Fuels* **2016**, *30*, 8832–8849. [CrossRef]

28. Wang, K.; Du, F.; Wang, G. Investigation of gas pressure and temperature effects on the permeability and steady-state time of Chinese anthracite coal: An experimental study. *J. Nat. Gas Sci. Eng.* **2017**, *40*, 179–188. [CrossRef]

29. Yin, G.; Jiang, C.; Wang, J.G.; Xu, J. Combined effect of stress, pore pressure and temperature on methane permeability in anthracite coal: An experimental study. *Trans. Porous Media* **2013**, *100*, 1–16. [CrossRef]

30. Perera, M.S.A.; Ranjith, P.G.; Choi, S.K.; Airey, D. Investigation of temperature effect on permeability of naturally fractured black coal for carbon dioxide movement: An experimental and numerical study. *Fuel* **2012**, *94*, 596–605. [CrossRef]

31. Teng, T.; Wang, J.G.; Gao, F.; Ju, Y.; Jiang, C. A thermally sensitive permeability model for coal-gas interactions including thermal fracturing and volatilization. *J. Nat. Gas Sci. Eng.* **2016**, *32*, 319–333. [CrossRef]

32. Pandey, S.N.; Chaudhuri, A.; Kelkar, S. A coupled thermo-hydro-mechanical modeling of fracture aperture alteration and reservoir deformation during heat extraction from a geothermal reservoir. *Geothermics* **2017**, *65*, 17–31. [CrossRef]

33. Tong, F.; Jing, L.; Zimmerman, R.W. A fully coupled thermo-hydro-mechanical model for simulating multiphase flow, deformation and heat transfer in buffer material and rock masses. *Int. J. Rock Mech. Min. Sci.* **2010**, *47*, 205–217. [CrossRef]

34. Abed, A.A.; Sołowski, W.T. A study on how to couple thermo-hydro-mechanical behaviour of unsaturated soils: Physical equations, numerical implementation and examples. *Comput. Geotech.* **2017**, *92*, 132–155. [CrossRef]

35. Gao, F.; Xue, Y.; Gao, Y.; Zhang, Z.; Teng, T.; Liang, X. Fully coupled thermo-hydro-mechanical model for extraction of coal seam gas with slotted boreholes. *J. Nat. Gas Sci. Eng.* **2016**, *31*, 226–235. [CrossRef]

36. Fan, C.; Elsworth, D.; Li, S.; Zhou, L.; Yang, Z.; Song, Y. Thermo-hydro-mechanical-chemical couplings controlling CH4 production and CO2 sequestration in enhanced coalbed methane recovery. *Energy* **2019**, *173*, 1054–1077. [CrossRef]

37. Xia, T.; Zhou, F.; Liu, J.; Kang, J.; Gao, F. A fully coupled hydro-thermo-mechanical model for the spontaneous combustion of underground coal seams. *Fuel* **2014**, *125*, 106–115. [CrossRef]

38. Zhang, R.; Liu, S. Experimental and theoretical characterization of methane and CO2 sorption hysteresis in coals based on Langmuir desorption. *Int. J. Coal Geol.* **2017**, *171*, 49–60. [CrossRef]

39. Rani, S.; Padmanabhan, E.; Prusty, B.K. Review of gas adsorption in shales for enhanced methane recovery and CO2 storage. *J. Pet. Sci. Eng.* **2018**, *175*, 634–643. [CrossRef]

40. Kim, J.; Jang, Y.; Seomoon, H.; Lee, H.; Sung, W. A new sorption-corrected deconvolution method for production data analysis in a shale gas reservoir containing adsorbed gas. *Int. J. Oil Gas Coal Technol.* **2019**, *20*, 55–68. [CrossRef]

41. Zhang, H.; Liu, J.; Elsworth, D. How sorption-induced matrix deformation affects gas flow in coal seams: A new FE model. *Int. J. Rock Mech. Min. Sci.* **2008**, *45*, 1226–1236. [CrossRef]

42. Wu, Y.; Liu, J.; Chen, Z.; Elsworth, D.; Pone, D. A dual poroelastic model for CO_2-enhanced coalbed methane recovery. *Int. J. Coal Geol.* **2011**, *86*, 177–189. [CrossRef]

43. Liang, B. Study on temperature effects on the gas absorption performance. *J. Heilongjiang Min. Inst.* **2000**, *10*, 20–22.

44. Zhu, W.C.; Wei, C.H.; Liu, J.; Qu, H.Y.; Elsworth, D. A model of coal-gas interaction under variable temperatures. *Int. J. Coal Geol.* **2011**, *86*, 213–221. [CrossRef]

45. Qu, H.; Liu, J.; Chen, Z.; Wang, J.; Pan, Z.; Connell, L.; Elsworth, D. Complex evolution of coal permeability during CO2 injection under variable temperatures. *Int. J. Greenh. Gas Control* **2012**, *9*, 281–293. [CrossRef]

46. Ibrahim, A.F.; Nasr-El-Din, H.A. A comprehensive model to history match and predict gas/water production from coal seams. *Int. J. Coal Geol.* **2015**, *146*, 79–90. [CrossRef]

 *processes*

Article

A Modular Framework for Optimal Load Scheduling under Price-Based Demand Response Scheme in Smart Grid

Ghulam Hafeez [1,2,*], Noor Islam [3], Ammar Ali [1], Salman Ahmad [1,4], Muhammad Usman [2] and Khurram Saleem Alimgeer [1]

[1] Department of Electrical and Computer Engineering, COMSATS University Islamabad, Islamabad 44000, Pakistan
[2] Department of Electrical Engineering, University of Engineering and Technology, Mardan 23200, Pakistan
[3] Department of Electrical Engineering, CECOS University of IT & Emerging Sciences, Peshawar 25124, Pakistan
[4] Department of Electrical Engineering, Wah Engineering College, University of Wah, Wah Cantt 47070, Pakistan
*   Correspondence: ghulamhafeez393@gmail.com; Tel.: +92-300-5003574

Received: 1 July 2019; Accepted: 29 July 2019; Published: 1 August 2019

Abstract: With the emergence of the smart grid (SG), real-time interaction is favorable for both residents and power companies in optimal load scheduling to alleviate electricity cost and peaks in demand. In this paper, a modular framework is introduced for efficient load scheduling. The proposed framework is comprised of four modules: power company module, forecaster module, home energy management controller (HEMC) module, and resident module. The forecaster module receives a demand response (DR), information (real-time pricing scheme (RTPS) and critical peak pricing scheme (CPPS)), and load from the power company module to forecast pricing signals and load. The HEMC module is based on our proposed hybrid gray wolf-modified enhanced differential evolutionary (HGWmEDE) algorithm using the output of the forecaster module to schedule the household load. Each appliance of the resident module receives the schedule from the HEMC module. In a smart home, all the appliances operate according to the schedule to reduce electricity cost and peaks in demand with the affordable waiting time. The simulation results validated that the proposed framework handled the uncertainties in load and supply and provided optimal load scheduling, which facilitates both residents and power companies.

Keywords: smart grid; demand response; load scheduling; home energy management; enhanced differential evolution; hybrid gray wolf-modified enhanced differential evolutionary algorithm

---

1. Introduction

With the emergence of information and communication technology (ICT), smart grid (SG) can make a robust and reliable system for the energy management of residential homes. ICT and sensors have moved the world towards automation. Thus, excessive use of electricity for every activity has increased demand-side energy consumption. The high demand for electricity and limited fossils fuels lead to increased penetration of renewable energy resources (RERs) [1]. Electricity production from RERs is not a part of this discussion. However, through scheduling and coordination of appliances, this high energy consumption can be managed. In [2], the authors reported that 38% increase in electricity consumption of power sector and 16% increase in electricity consumption of both residential and commercial sectors are expected by the year 2020.

Considering this repaid energy consumption growth, there is a need for a system to manage the resident demand according to generation in such a manner to alleviate the gap between demand

and supply [3]. In this regard, the traditional grid is renovated by SG with the integration of ICT. Advanced metering infrastructure (AMI) is responsible for bi-directional communication between the power company and the resident [4]. In a SG, the power company organizes the consumer demand using particular set of programs. These programs are known as demand response (DR) programs [5]. Various DR incentives schemes are introduced by the power companies for the encouragement of the residents to efficiently use available resources as explained in [6]. Price-based DR schemes such as real-time pricing scheme (RTPS), time of use pricing scheme (TOUPS), critical peak pricing scheme (CPPS), flat-rate pricing scheme (FRPS), a day-ahead pricing scheme (DAPS), and inclined block rate scheme (IBRPS) are widely used for load scheduling. The HEM controller (HEMC) receives the pricing signal from the electric power company and electric load profile from the resident to schedule the household load. The HEMC schedule the household load using a pricing signal and the load of the residents. The home appliances are synchronized with the schedule through infrared, ZigBee, Z-Wave, and Wi-Fi [7].

The main focus of research and development (R&D) is on load shifting from ON-peak timeslots to OFF-peak timeslots using demand-side management (DSM) strategies such as peak clipping, strategic conservation, peak shifting, and valley filling. Load shifting helps in two ways: minimize electricity cost by shifting the load to low-price timeslots and minimize peaks in demand by building load in OFF-peak timeslots [8]. However, load shifting reduces electricity cost at the expense of increase user frustration in terms of waiting time. To reduce electricity cost and peaks in demand with affordable waiting time, heuristic techniques are mostly adopted because they are fast converging and simple.

To overcome this rapidly increasing electricity demand of the residential sector, a hybrid gray wolf-modified differential evolution (HGWmEDE) algorithm is proposed to resolve this problem and enhance the sustainability of the electric grid. The proposed algorithm under the price-based DR encourages resident to take part in DSM via load scheduling. In this work, the main focus is on optimal load scheduling based on HGWmEDE under price-incentive-based DR schemes in smart homes. The main contribution and distinguish features of this paper are as follows:

- A modular framework is introduced for optimal load scheduling, which has four modules: power company module, restricted Boltzmann machine (RBM)-based forecaster module, HGWmEDE-based HEMC module, and resident module. Furthermore, smart home appliances are classified into three categories based on power rating and behavior: schedulable, non-schedulable, and controllable. Moreover, each appliance has a different length of operational time (LOT) and each home has different operation time interval (OTI). Four parameters, i.e., energy consumption, electricity cost, peaks in demand, and waiting time are taken into account.
- A deep neural network technique, i.e., RBM is adopted to forecast the pricing signals of price-based DR scheme for optimal load scheduling.
- Finally, the HGWmEDE algorithm is proposed, which is a hybrid of gray wolf optimization and enhanced differential evolutionary algorithms. The proposed algorithm has global powerful search capability and generalization. The proposed algorithm optimizes the performance by fine-tuning the control parameters.

To analyze the proposed scheme in terms of electricity expense, peaks in demand, and discomfort, simulations are conducted in MATLAB 2016. Moreover, the convergence rate and performance trade-off are also evaluated.

The organization of the paper is as follows: In Section 2, recent and relevant work is demonstrated. The proposed modular framework is presented in Section 3. Section 4 describes the proposed scheme. In Section 5, simulations results and discussion are described. The paper is concluded along with future research directions in Section 6.

2. Recent and Relevant Work

In the last few years, a lot of research has been conducted in the area of HEM based on optimization algorithms in the SG to economically use electrical energy. Some recent and relevant research work is presented, in this section.

A heuristic algorithm (genetic algorithm (GA) and bacterial foraging algorithm (BFA))-based HEM model is proposed in [9]. The performance evaluation of these algorithms is conducted using three price-based DR schemes, i.e., RTPS, TOUS, and CPPS. The focus of the authors is to shift the load from ON-peak timeslots to OFF-peak timeslots to minimize electricity cost and smooth out the demand curve.

In [10], three heuristic algorithms, i.e., differential evolution algorithm (DEA), GA, and binary-particle swarm optimization algorithm (BPSOA) were implemented for load scheduling to minimize electricity cost and PAR. On the other hand, carbon emission was alleviated using RERs. DAPS was chosen as a DR scheme. The primary aim was not load scheduling, but also to prioritize the operation of appliances according to resident demand. The grid sustainability was maintained by keeping a balance between the demand and supply side. However, the balance is maintained at the expense of user discomfort.

In [11], authors presented DR program under the corporate sector. The purpose is to perform HEM and maintain using two different pricing schemes, i.e., DAPS and RTPS. In [12], a GA-based optimization model was proposed for electric load scheduling for 24 h time horizon. The energy consumption and load pattern are calculated using power rating and status of appliances for overall time horizon. However, the user-comfort is compromised, and the convergence rate is reduced.

An intelligent decision support system (IDSS) was used for resolving certain HEM problems [13]. Moreover, IDSS was integrated to AMI for bi-directional communication between the power company and residents. Wind-driven optimization (WDO) algorithm with knapsack (K-WDO) was implemented for electricity cost minimization and user-comfort maximization in [14]. The minimum-maximum constraints of K-WDO were defined. The smart home appliances were classified based on consumer behavior and name-plat power rating. TOUPS was used to shift load from ON-peak hours to OFF-peak hours according to user preference and priority; however, peaks in demand emerged at the expense of increased system complexity.

An ant colony optimization algorithm (ACOA)-based model was proposed by [15] for optimal power flow (OPF). The OPF objective was to determine the load to satisfy the end user by providing a continuous energy supply. In the literature, some statistical methods, i.e., newton method, linear programming (LP), non-LP (NLP), and the interior point method were used to solve such problems.

In [16], load balancing via load scheduling was the main focus of the authors. Thus, a multi-agent system was proposed for load balancing, in this system each consumer act as an independent agent and the consumer electric load was divided into time frames for each agent. Power was supplied at a particular time frame for each agent. In this paper, three sectors of demand-side were considered, i.e., residential, commercial, and industrial. In [17,18], the basic concepts of DEA and enhanced version of DEA (EDEA) with five trial vectors were discussed. The mutant vector and trial vectors were created to update population. Moreover, DE-based scheduling model for electricity cost reduction was also presented.

The multi-objective optimization problem was discussed in [19]. The optimization problem was tested on pareto sets (PS) using DEA. The proposed model was also named as multi-objective evolutionary algorithm (MOEA) and was capable in complex PS shapes mapping. A hybrid evolutionary approach-based forecasting model was proposed to cater varying electricity prices by [20]. The model forecasts the day-ahead and week-ahead price profiles. The hybrid evolutionary approach was a combination of PSO and a neuro-fuzzy logic network. This hybrid approach was used to handle uncertainty in the pricing rates of the electricity market. In [21], price and load correlation were developed to modify the energy consumption pattern of ON-peak timeslots and OFF-peak timeslots. Both generalized mutual information (GMI) and wavelet packet transform (WPT) was adopted to formulate the multiple inputs and multiple output model. Electricity price was forecasted to analyze variation in their pattern. The ACOA was applied for optimization purposes.

Teaching and learning-based optimization algorithm (TLBOA) and shuffled-frog leaping (SFL)-based energy management was presented in [22]. The proposed framework was validated using different tariff schemes such as TOUPS, RTPS, CPPS, and without pricing scheme. The household load scheduling was conducted for varying time interval and pricing schemes. Power storage is incorporated in the system model to ensure continuous operation of the sensitive load [23]. A day-ahead of schedule is generated by virtual power play for load and energy consumption. The increased energy demand encourage electricity market participator generation from distributed generation. The intensive demand of residents was catered using a vehicle to grid station (V2GS) strategy. Load balancing among multiple distribution units was performed using BPSO along with MILP. A complex mathematical model was formulated for day-ahead electricity price forecasting. For experimental evaluation 1000, electric vehicle stations and 180 distributed units were used.

Energy consumption is a crucial parameter in electricity bill and peaks in demand reduction. Taking into account this fact, the resident load was scheduled using a hybrid of GA and artificial neural network (ANN-GA) scheme [24]. The load was scheduled on a weekly basis for a single home with four bedrooms. Obtained results show 25%, 40%, and 10% reduction in grid electricity consumption. However, dynamic and different OTI were not taken into account while all homes have not same OTI.

In [25], the authors focused on residential sector DSM. Multiple homes considered were smart homes and have bi-directional communication between the power company and residents. The GA, BPSO, WDO, and BFOA-based HEMC was installed for home load scheduling. The proposed model was evaluated in terms of electricity bill, user-comfort, and peaks in demand. However, the trade-off effect of conflicting parameters was ignored.

A distributed algorithm was used in [26] for energy management of 2560 households. The purpose was to reduce electricity bill with reasonable appliances waiting time. The authors in [27] proposed a harmony search algorithm (HSA)-based model for load scheduling. However, in [28], authors focused on electricity bill reduction. Game theory-based framework in [29,30] was proposed to reduce PAR by load scheduling and DR program.

In the aforementioned recent and relevant literature, the authors and R&D did not completely use the key features of SG. Some authors minimized peaks in demand, electricity cost, and waiting time. On the other hand, some authors focused on user-comfort and user discomfort in terms of waiting time. However, the conflicting parameters were not catered simultaneously in R&D by any of the authors. Furthermore, dynamic and different OTI were not catered while all homes in a city have not same OTI and energy consumption. In this work, the electricity cost reduction and peaks in demand reduction with affordable discomfort are catered simultaneously. The objective is to cope with the increasing demand of residents with the generation of the power company and reduce the burden on both parties. The comprehension of recent and relevant work is listed in Table 1.

Table 1. Summary of recent and relevant work.

Methodologies	Features	Targets Achieved	Limitations and Remarks
MILP	Optimal domestic load scheduling [5]	Electricity cost reduction	The cost was reduced at the expense of user discomfort
Greedy algorithm	Heuristic optimization base generic model [13]	Both electricity bill and user frustration are reduced	The PAR was compromised and complexity of the system is increased
Multi-agent system	DSM via load shifting and DR programs in SG [16]	Electricity cost reduction via load scheduling	The user-comfort is compromised due to the trade-off between electricity cost and user-comfort

Table 1. *Cont.*

Methodologies	Features	Targets Achieved	Limitations and Remarks
BPSO and neuro-fuzzy logic	Hybrid evolutionary-adaptive-based price forecasting model [21]	Improved forecast accuracy	The convergence rate and complexity increased while improving forecast accuracy
MIMO	Future price and load forecasting in SG [22]	Improved load and price profile forecast accuracy	The convergence and complexity were ignored which have a direct influence on the forecast accuracy
TLBOA and SFL	DR programs optimization in SG [23]	Cost reduction	The cost can further be reduced with RERs integration
PSO and MILP	A multi-objective model was used for resource and load scheduling [24]	Scheduling of virtual power play	Requirements to improve reliable power grids was ignored
MIMO	A hybrid optimization algorithm was used for both price and load forecasting [25]	Accuracy improvement	Computational time and execution time was impractical
GWO	Economic dispatch optimization under the GWO technique [31]	Optimization of dispatch problems	The problem arborized of handling the constraints
ANN-GA	Smart energy management using ANN-GA [32]	Efficiency improvement	The model was limited for a small number of appliances
Game theory algorithm (GTA)	Game theory-based household load scheduling under DR [33]	Electricity cost reduction	RERs integration was ignored
MILP and heuristic algorithms	Household load scheduling [34]	Household load balancing	Electricity cost reduction is ignored
MINLP	Efficient household appliances scheduling under DR [35]	Electricity cost reduction	PAR was compromised
GWO and ILP	GWO-based economic load dispatch [36]	Electric load dispatching in low-price timeslots	The electric load is economically dispatched

3. Proposed Modular Framework

The main objective of home energy management in this work is to minimize electricity cost and peaks in demand under price-based DR scheme by scheduling the smart home appliances. The overall proposed modular framework is demonstrated in Figure 1. The proposed framework has four modules: power company module, forecaster module, HEMC module, and resident module. Different electricity pricing signals (RTPS, TOUS, CPPS, DAPS, IBR, and variable time pricing) are defined by the power company for residents to take part in the price-based DR. Timeslots in which consumer demand reaches to the maximum value is known as peak timeslots. Electricity tariffs are usually high in these peak timeslots. However, in this paper, the power company module provides price-based DR information (RTPS and CPPS) and load pattern to the forecaster module. The forecaster module is based on RBM. The primary goal of this module is to devise a framework which is enabled through learning to forecast future load and pricing signals (RTPS and CPPS). The data for training RBM is collected from [37,38]. The predicted pricing signals must be accurate to obtain optimal load scheduling. The forecasted profile of load and pricing signals (RTPS and CPPS) is illustrated in Figure 2a–c. It is obvious that RBM-based forecast closely follows the real curve. This observation in terms of the numerical value is 0.4% for load, 0.5% for RTPS, and 0.2% for CPPS, respectively. The reason for this accurate performance is the adaption of deep learning technique, i.e., RBM. The forecaster module provides forecasted load pattern and pricing signals to the HEMC module, which is based on the proposed HGWmEDE algorithm. The HEMC based on HGWmEDE schedule the household load under the pricing signals provided by the forecaster module. The schedule developed by HEMC module is forwarded to the resident module. The resident module comprised of a smart home with 17 appliances [39]. Each appliance has its own power rating and behavior. These appliances are scheduled to according to HEMC schedule to achieve the objective function.

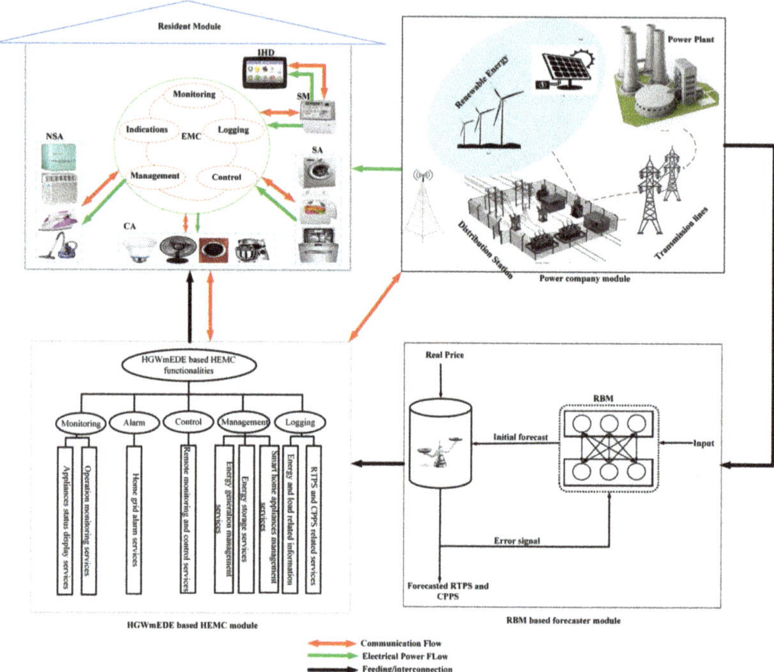

Figure 1. Proposed modular framework.

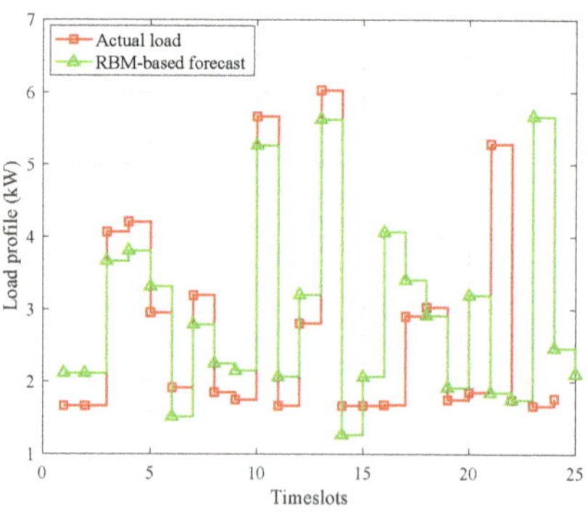

(**a**) RBM-based load forecast

Figure 2. *Cont.*

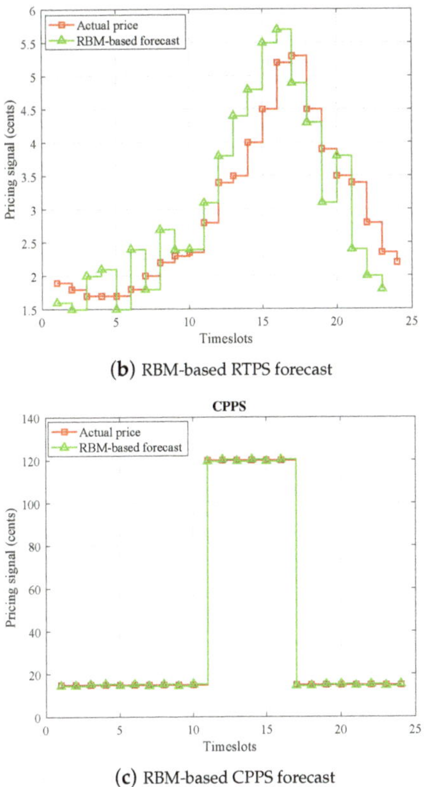

(**b**) RBM-based RTPS forecast

(**c**) RBM-based CPPS forecast

Figure 2. Price-based DR schemes.

In the proposed modular framework bi-directional communication exists between the power company and residents via AMI. Power company sent price-based DR information to the forecaster module, the forecaster module forecast load and pricing signals of RTPS and CPPS. The HEMC module based on HGWmEDE algorithm receives forecasted results to schedule the household load on the basis of pricing signals such that high power-rating appliances cannot be switched on in peak timeslots. The HEMC module dispatches the load schedule to the power company, and ultimately, the power company sent demanded power of residents. The HEMC can perform the functionalities of logging, management, control, monitoring, and alarm. The purpose is to optimally schedule household load and reduce the frustration of both power company and residents. The profiles of forecasted RTPS and CPPS are illustrated in Figure 2a,b. These two pricing schemes are generated using the data of [37,38].

3.1. Smart Home Appliances Categorization

Smart home appliances are classified into three categories based on operating behavior and energy consumption pattern. The smart home total number of appliances is denoted by a set A_n^t which includes shiftable appliances, non-shiftable appliances and controllable appliances and is defined by Equation (1).

$$A_n^t = \{A_s^s,\ A_n^s,\ A_c^s\},\tag{1}$$

where A_s^s denotes shiftable appliances, A_n^s represents non-shiftable appliances, and A_c^s denotes controllable appliances. The detailed description of the classification is as follows.

3.1.1. Shiftable Appliances

In shiftable appliances, shifting to any timeslot is allowed, but interruption during operational time is prohibited. Such type of appliances are also known as deferrable appliances. Once operation of such type appliances is started, it cannot be stopped/interrupted until to finish the assigned task [17]. These appliances are the subset of total appliances and are defined by Equation (2).

$$A_s^s = \{WM,\ DW,\ HS,\ HD,\ MW,\ TP,\ CP,\ OV, CK,\ IR,\ TR,\ EK,\ PR\},\tag{2}$$

where WM denotes washing machine, DM represents dishwasher, HS indicates hair straightener, HD denotes hair dryer, MW represents microwave, TP denotes telephone, CP represents computer, OV represents oven, CK denotes cooker, IR represents iron, TR represents toaster, EK, represents electric kettle, and PR represents printer. Each appliance power rating and status at a particular timeslot t is denoted by p_r^a and X_A^S, respectively. Equation (3) indicates the ON and OFF status of an appliance.

$$X_A^S = \begin{cases} 1 & ON \\ 0 & \text{otherwise} \end{cases}\tag{3}$$

3.1.2. Controllable Appliances

The controllable appliances have constant operational time and cannot be changed; for example, heating system, lightning, and air conditioning. Such type of appliances are also known as interruptible appliances. The controllable appliances are given by Equation (4).

$$A_c^s = \{AC,\ LT,\ HT\},\tag{4}$$

where AC represents air conditioner, LT denotes lighting, and HT represents heater.

3.1.3. Non-Shiftable Appliances

Non-shiftable appliances are also known as base appliances. Such type of appliances are uncontrollable, and their operational behavior and energy consumption cannot be altered. Televisions and refrigerators are kept in this category due to resembles. The set of non-shiftable appliances is defined by Equation (5).

$$A_n^s = \{TV,\ RG\},\tag{5}$$

where TV represents television and RG denotes refrigerator. The overall categorization of smart home appliances is presented in Table 2.

Table 2. Smart home appliances classification and parameters.

Categories	Type	Power Rating (kW)	Start Time (hours)	End Time (hours)	LOT (hours)
	Washing machine	1.4	6	10	1–3
	Dish washer	1.32	15	20	1–3
	Hair straightener	0.055	18	8	1–2
	Hair dryer	1.8	18	8	1–2
	Microwave	1.2	18	8	3–5
	Telephone	0.005	9	17	1–24
Shiftable appliances	Computer	0.15	18	24	6–12
	Oven	2.4	6	10	1–3
	Cooker	0.225	18	24	2–4
	Iron	2.4	18	24	3–5
	Toaster	0.8	9	17	1–2
	Electric Kettle	2	18	8	1–2
	Printer	0.011	18	24	1–2
Non-shiftable appliances	TV	0.095	8	16	6–14
	Refrigerator	1.75	0	23	0–23
Controllable appliances	Air conditioner	1.14	16	23	6–8
	Lightning	0.1	0	23	12–20

4. Problem Description and Formulation

In DSM, optimal household load scheduling and alignment of residents' random demand under the generation of the power company is a challenging task. In the literature, various models have been developed to address energy management via the resident load scheduling. For example, the scheduling algorithm is proposed for the energy management of smart homes to reduce the electricity bill, reverse power flow, and peak load shaving [40]. A GWDO algorithm is proposed to schedule the household load under RTPS + IBRS to increase the revenue of residents and reduce the peak to average ratio [41]. Load scheduling is performed using GA under RTPS to alleviate cost and PAR. However, peaks in demand may emerge in OFF-peak hours and user-comfort may compromise while reducing electricity bill because these parameters are conflicting parameters. Thus, a modular framework is proposed, which is based on our proposed algorithm HGWmEDE, for a load scheduling of smart homes with three types of appliances: shiftable appliances, non-shiftable appliances, and controllable appliances, in order to alleviate electricity cost and peaks in demand with affordable appliances waiting time. The purpose is to facilitate both residents and power companies by reducing burden (electricity bill and generation, respectively) on both parties. The peaks in demand reduction are favorable for both residents and power companies because it alleviates the need for peak power plants, which power companies operate when peaks in demand emerged and charged more cost from the residents. To perform effective load scheduling dynamic OTI, RTPS, and CPPS are used. The formulation for energy consumption, cost, peaks in demand, and load scheduling are demonstrated as follows.

4.1. Smart Home Appliances Energy Consumption

The HEMC schedule the smart home appliances under forecasted RTPS and CPPS over a 24-hours time horizon. These smart home appliances when operating according to the schedule consume electrical energy, which can be defined as the electrical energy used by an appliance in unit time and can be measured in the kWh unit. The electrical energy consumed by an appliance can be calculated by Equation (10);

$$E_c^a(t) = p_r^a \times X_s^a,\qquad(6)$$

where $E_c^a(t)$ is the energy consumption of an appliance a at timeslot t, p_r^a and X_s^a is the power rating and status of an appliance, respectively. The aggregated electrical energy consumption of the smart home is calculated from the following formula:

$$E_T^a = \sum_{t=1}^{T}\left(\sum_{a=1}^{n} E_c^a(t)\right)\qquad(7)$$

where E_T^a represents the aggregated energy consumption of the smart home appliances.

4.2. Smart Home Appliances Electricity Cost

The electricity cost is defined as the bill deposited by the residents to the power company for the used energy per unit time and per unit price. It is measured in the units of cents. For electricity cost determination, the power company provides various pricing schemes such as RTPS, CPPS, TOUPS, DAPS, and FPS; however, we adopted RTPS and CPPS for the proposed scheme. The RTPS and CPPS are the Midwest independent system operator (MISO) daily electricity pricing schemes taken from federal energy regulatory commission (FERC) [41,42]. The cost for the energy used by the residents under forecasted RTPS and CPPS is as follows:

- electricity cost of resident energy consumption is determined using forecasted RTPS as:

$$C_R^a = \sum_{t=1}^{T}\left(\sum_{a=1}^{n} E_c^a(t) \times p_s^r(t)\right),\qquad(8)$$

where C_R^a is the total electricity cost by the resident to power company under RTPS denoted by $p_s^r(t)$.

- electricity cost paid by the resident to power company under forecasted CPPS is determined as:

$$C_p^a = \sum_{t=1}^{T} \left(\sum_{a=1}^{n} E_c^a(t) \times p_s^c(t) \right), \tag{9}$$

where C_p^a is the total electricity cost paid by the resident to power company under CPPS denoted by $p_s^c(t)$.

4.3. Peaks in Demand

Peaks in demand is defined as the highest demand emerged over a specified horizon of time such as daily, weekly, monthly, annually, and seasonally, or the highest points of resident electricity consumption. The electric power company cost high from the resident during the peak demand periods because they supplied continues power to load by bringing peak power plants online. It is measure in the units of power (Watts). The proposed framework tries to smooth out the demand curve by reducing the peaks in demand to avoid blackout situation. The peaks in demand can be determined as:

$$D_d^p(t) = \max\left(E_c^a(t)\right), \tag{10}$$

where $D_d^p(t)$ represents the highest possible peaks in the demand over a specified horizon of time.

4.4. Smart Home Load Scheduling Formulation

The smart home load scheduling problem is formulated as a minimization problem because the main objectives of this work are to alleviate peaks in demand and electricity cost with affordable appliances waiting time.

$$\min\left(\sum_{t=1}^{T} \left(\sum_{a=1}^{n} E_c^a(t) \times p_s^r(t) \right) \right) + \min\left(D_d^p(t) \right) \tag{11}$$

subjected to:

$$E_T^a \leqslant capacity \tag{12a}$$

$$\sum_{a \in A_n^t} E_T^{a,\,unsch} = \sum_{a \in A_n^t} E_T^{a,\,sch} \tag{12b}$$

$$\sum_{a \in A_n^t} T_0^{a,\,unsch} = \sum_{a \in A_n^t} T_0^{a,\,sch} \tag{12c}$$

$$X_a^{s,\,unsch} \neq X_a^{s,\,sch} \tag{12d}$$

The constraint (12a) ensures that total energy consumption of residents must be under the capacity of power company. The constraint (12b) and (12c) ensure that the total energy consumption of residents before and after scheduling must be equal subjected to fair comparison. The scheduling of smart appliance is conformed from constraint (12d).

5. Description of Adapted and Proposed Algorithms

In this section, the adapted and proposed heuristic algorithms for smart home load scheduling are discussed. Electricity is consumed in the three demand-side sectors, i.e., residential, commercial, and industrial sectors. However, the main focus is to perform DSM via optimal residential load scheduling. For this purpose, HEMC based on GWO, mEDE, and the proposed HGWmEDE algorithms schedule the smart home appliances to reduce peaks in demand and electricity cost with affordable user discomfort. In the literature, various optimization schemes have been proposed for load scheduling.

Some of these techniques outperform in cost reduction and others well perform either in PAR or user-discomfort minimization. In this regard, an optimization technique is proposed, i.e., HGWmEDE, which is a hybrid of mEDE and GWO algorithm. The proposed scheme simultaneously caters the minimization objectives of electricity cost, peaks in demand, and waiting time. The existing and proposed techniques are implemented in MATLAB and their detail description is as follows.

5.1. mEDE

The mEDE is a modified enhanced version of DE. The DE at very first time proposed by Storn in 1995 and enhanced and modified by [43]. It is a meta-heuristic population-based algorithm, which includes four main steps, i.e., population creation, crossover, mutation, and selection [44]. Initially, random population is generated by Equation (13) as follows:

$$P_{k,n} = B_n^l + (rand \times (B_n^u - B_j^l)). \tag{13}$$

To form a mutant vector, a random function is generated to create three vectors, i.e., v_{r1}, v_{r2}, and v_{r3}. First vector is the target vector and mutant vectors are generated using Equation (14) as given below:

$$m_{k,G+1} = v_{r1,G} + S(v_{r2,G} - v_{r3,G}), \tag{14}$$

where S is a scaling factor. Mutant vector is generated, then, first three trial vectors are generated by Equations (15)–(17). Then, the best trail vector is selected by comparing with the target vector to update the population with the best trial vectors.

$$B_{n,k,G+1}^u = \begin{cases} m_{n,k,G+1} & if \ randb(n) \le 0.30 \\ v_{n,k,G} & Otherwise \end{cases} \tag{15}$$

$$B_{n,k,G+1}^u = \begin{cases} m_{n,k,G+1} & if \ randb(n) \le 0.60 \\ v_{n,k,G} & Otherwise \end{cases} \tag{16}$$

$$B_{n,k,G+1}^u = \begin{cases} m_{n,k,G+1} & if \ randb(n) \le 0.90 \\ v_{n,k,G} & Otherwise \end{cases} \tag{17}$$

Then, 4th and 5th trial vectors are generated by Equations (18) and (19), respectively, as given below:

$$B_{n,k,G+1}^u = randb(n) \cdot v_{n,k,G} \tag{18}$$

$$B_{n,k,G+1}^u = randb(n) \cdot m_{n,k,G} + (1 - randb(n)) \cdot v_{n,k,G} \tag{19}$$

The mEDE pseudocode is depicted in Algorithm 1. The maximum iterations are denoted by $Mx.itr$; the total population is represented by $POPLAT$, it shows the number of possible solutions. The crossover ratio is denoted by R_c, which is taken as 0.30, 0.60, and 0.90. The mutant, trial, and target vectors are represented by m, μ, and v, respectively.

Algorithm 1: mEDE

Parameters initialization $Mx.itr, R_c, POPLAT, and hour$;

Initially, population is randomly generated by Equation (13) ;

for $a = 1{:}T$ **do**

    Compute mutant vector by Equation (14);

    for $itr = 1{:}Mx.itr$ **do**

        Compute 1st trial vector with crossover rate of 0.30;

        if $rand(.) \leq 0.30$ **then**

            $\mu_n = m_n$

            else

            $\mu_n = v_n$

        end

        Compute 2nd trial vector with crossover rate of 0.60;

        if $rand(.) \leq 0.60$ **then**

            $\mu_n = m_n$

            else

            $\mu_n = v_n$

        end

        Compute 3rd trial vector with crossover rate 0.90;

        if $rand(.) \leq 0.90$ **then**

            $\mu_n = m_n$

            else

            $\mu_n = v_n$

        end

        Create 4th and 5th trial vector using Equations (18) and (19);

        Find out trial vector which is best ;

        $X_{new} \leftarrow$ best of μ_n ;

        Compare trial vector and target vector;

        if $(P_{new}) < (P_n)$ **then**

            $P_n = P_{new}$

        end

    end

end

5.2. GWO

It is a heuristic technique, motivated by the wolves hunting and leadership nature [44]. For leadership 4 levels are defined: α, β, δ, and γ. The α is the most intuitive leader among the group, which provides guidance on hunting strategies to other wolves. The β and δ come after α in the chronological order, and γ is the feebler member among the group. Thus, γ has a lack of leadership qualities and cannot be considered. In HEMC, α is taken as the fittest member to schedule the smart home load to reduce cost and peaks in demand. Initially, the population is randomly generated by Equation (20):

$$P(k,n) = rand(POPLAT, A_n^t),\tag{20}$$

where $POPLAT$ represents the population of gray wolves and A_n^t is the overall appliances in the smart home, which is used in the proposed framework. The objective function of each search agent can be evaluated using co-efficient D and E.

5.2.1. Encircling Prey

Before hunting the gray wolves encircle a prey. The encircling behavior of gray wolves is mathematically modeled using Equations (21) and (22). These Equations taken from [45].

$$P(t+1) = P_p(t) - D \times A_n^t, \tag{21}$$

$$A_n^t = |E \times P_{p(t)} - P(t)|, \tag{22}$$

where the position of prey is represented by P_p, while P is gray wolf position at t_{th} epoch, which is calculated using Equation (21). The co-efficient vectors D and E are determined according to Equation (23) and Equation (24), respectively:

$$\vec{D} = 2\vec{b} \times \vec{r}_1 - \vec{b} \tag{23}$$

$$\vec{E} = 2 \times \vec{r_2} \tag{24}$$

where $\vec{r_1}$ and $\vec{r_2}$ are vectors with random values between 0 and 1. Value of D after multiple epochs is reduced from 2 to 0 while the value of E is randomly taken between 0 and 2. This value of E defines the weight of attractiveness for prey.

5.2.2. Hunting

The α provides guidance for hunting, while the β, and δ are secondary participants. The secondary participants follow α, due to best knowledge about the prey position. The 3 best solutions are achieved and the other participants such as γ update its position according to the best solution. The wolves' position is updated using Equation (25).

$$\vec{P_{t+1}} = \frac{\vec{v_1} + \vec{v_2} + \vec{v_3}}{3} \tag{25}$$

where $\vec{v_1}, \vec{v_2}$ and $\vec{v_3}$ are determined by Equations (26)–(28).

$$\vec{v_1} = \vec{v_\alpha} - \vec{D_1} \times (\vec{d_\alpha}) \tag{26}$$

$$\vec{v_2} = \vec{v_\beta} - \vec{D_2} \times (\vec{d_\beta}) \tag{27}$$

$$\vec{v_3} = \vec{v_\delta} - \vec{D_3} \times (\vec{d_\delta}) \tag{28}$$

where $\vec{v_\alpha}, \vec{v_\beta}$ and $\vec{v_\delta}$ are the best solutions obtained at the t_{th} iteration; $\vec{D_1}, \vec{D_2}, \vec{D_3}$ are determined using Equation (23), while $\vec{P_\alpha}, \vec{P_\beta}, \vec{P_\delta}$ are determined using Equations (29)–(31):

$$\vec{P_\alpha} = \vec{E_1} \times \vec{v_\alpha} - \vec{v} \tag{29}$$

$$\vec{P_\beta} = \vec{E_2} \times \vec{v_\beta} - \vec{v} \tag{30}$$

$$\vec{P_\delta} = |\vec{E_3} \times \vec{v_\delta} - \vec{v}, \tag{31}$$

where E_1, E_2, and E_3 are calculated using Equation (24). The gradation of variable g is conducted in the last step; exploration and exploitation trade-off is controlled by considering value between 0 to 2 in each epoch as depicted in Equation (32).

$$g = 2 - t\frac{2}{Mx.itr} \tag{32}$$

Mathematical modeling of objective function is shown in Equation (33), which use the power rating and status of an appliance.

$$FitnessF = p_r^a \times X_a^S(t) \tag{33}$$

From Algorithm 2, the maximum iterations are represented by $Mx.itr$, the total population is denoted by $POPLAT$, the total number of smart home appliances is A_n^t and $FitnessF$ is the fitness function. α is best among the group participants, which provides a primary optimal solution in the hunting behavior, while β and δ come after α in the group and provide secondary optimal solutions.

Algorithm 2: GWO

Parameters initialization $Mx.itr, POPLAT, A_n^t, \alpha, \beta,$ and δ;
Initially, gray wolves population is generated $Pk(k = 1, 2, 3, ..., Pn)$;
$P(k, n) = rand(POPLAT, A_n^t)$;
while $itr < Mx.itr$ **do**
    for $k = 1$:POPLAT **do**
        Compute fitness by Equation (33);
        if *fitness* $< \alpha_{score}$ **then**
            α_{score} = fitness;
            $\alpha_{Pos} = P(k, :)$;
        end
        if *fitness* $> \alpha_{score}$ *and fitness*$< \beta_{score}$ **then**
            β_{score} = fitness;
            $\beta_{Pos} = P(k, :)$;
        end
        if *fitness* $> \alpha_{score}$ *and fitness*$> \beta_{score}$ *and fitness*$< \delta_{score}$ **then**
            δ_{score} = fitness;
            $\delta_{Pos} = P(k, :)$;
        end
    end
    for $k = 1$:POPLAT **do**
        for $n = 1$:Appliances **do**
            Randomly generate r_1 and r_2 by rand command;
            Compute fitness coefficients D and E using Equations (23) and (24);
            Update 3 vectors (α, β, δ) by Equations (29)–(31);
        end
    end
end

To evaluate the best hunting leader fitness function, compare the fitness of α, β, and δ. The positions are updated according to Equations (29)–(31).

5.3. HGWmEDE

In this section, the proposed hybrid algorithm is demonstrated in detail. Initial population in mEDE is generated by four phases, i.e., initialization phase, mutation phase, crossover phase, and selection phase and the population are updated by comparatively analyzing trial vector with the target vector. The procedure of trail vector selection is effective in choosing the best trial vector from available vectors. The GWO comprised of three steps, i.e., encircling prey, hunting, and wolves position update within the pack. All search agents' positions are updated according to the leader α within the pack. In GWO, unlike the mEDE agents, α with β, and δ are not compared. However, the β

and δ are closer to the prey as compared to α. To conduct a comparison of all search agents a crossover phase of mEDE is adopted. Thus, the best search agent is selected according to the crossover phase of mEDE and search agents' position is updated according to GWO. The HGWmEDE is proposed by combining the taking key characteristics of both mEDE and GWO.

In Algorithm 3, the detailed stepwise procedure of HGWmEDE is presented. The main stages of the proposed HGWmEDE algorithm are initialization stage, encircling prey stage, best search agent selection stage, and wolves position update stage. Initially, the population of wolves is generated randomly using Equation (20). The best search agent is selected by following the steps presented in Algorithm. The mutant vector m is generated using Equation (14). The fitness of m, α, β, and δ is computed by Equation (33). Crossover phase is conducted to select the best search agent using the following Equations:

$$\alpha_{new} = \begin{cases} m_n & if \quad \text{fitness of} m_n \leq \alpha \\ \alpha & \quad Otherwise \end{cases} \tag{34}$$

$$\beta_{new} = \begin{cases} m_n & if \quad \text{fitness of} m_n \leq \beta \\ \beta & \quad Otherwise \end{cases} \tag{35}$$

$$\delta_{new} = \begin{cases} m_n & if \quad \text{fitness of} m_n \leq \delta \\ \delta & \quad Otherwise \end{cases} \tag{36}$$

Algorithm 3: HGWmEDE

Parameters initialization $Mx.itr, POPLAT, A_n^t, \alpha, \beta, \delta$;
Initial gray wolves population generation $Pk(k = 1, 2, 3, ..., Pn)$;
$P(k, n) = rand(POPLAT, A_n^t)$;
while $itr < Mx.itr$ **do**
 for k = 1:POPLAT **do**
  Create a mutant vector by Equation (14) from mEDE;
  Compute mutant vector fitness as $cost \times v_n$;
  Generate randomly α, β, and δ;
  Compute α, β, and δ fitness by Equation (33);
  if *fitness of* $m(n) < \alpha_{score}$ **then**
   $\alpha_{position} = m_n$;
  end
  if *fitness* $m_n > \alpha_{score}$ *and* $m_n < \beta_{score}$ **then**
   $\beta_{position} = m(j)$;
  end
  if *fitness* $v_n > \alpha_{score}$ *and fitness* $v_n > \beta_{score}$ *and fitness* $v_n < \delta_{score}$ **then**
   $\delta_{position} = m_n$);
  end
 end
 for $k = 1:POPLAT$ **do**
  for $n = 1:Appliances$ **do**
   Randomly generate $r1$ and $r2$ by rand command;
   Compute coefficients D and E fitness by Equations (23) and (24);
   Update three vectors (α, β, δ) by Equations (29)–(31);
  end
 end
end

When the best search agents are selected then search agents' position is updated according to GWO. The position is updated using Equation (25).

The detail description of Algorithm 3 for each step is as follows. In 1st step, parameters are initialized. In 2nd step, randomly population is generated, and the counter is adjusted to maximum epochs. Crossover phase of mEDE is conducted to compare the fitness of the mutant vector with α, β, and δ. The search agent status is updated using GWO. The procedure is repeated for several epochs until the termination criteria are reached.

6. Simulation Results and Discussion

Simulation results and discussions of the proposed modular framework are demonstrated, in this section. The pricing signals used for load scheduling is forecasted using RBM. The aim is to evaluate the performance (peaks in demand and electricity cost reduction) of proposed and existing schemes under RBM-based forecasted RTPS and CPPS for different OTI. The HEMC is responsible for scheduling the appliances using the forecasted RTPS and CPPS pricing signals. The performance is evaluated for different OTI such as 15, 30, and 60 timeslots. The home appliances and their parameters such as OTI, LOT, starting time, ending time, and power rating are adopted from [39]. Simulation results and discussion of the proposed and existing algorithms are demonstrated in the succeeding sections in terms of peaks in the demand and electricity cost reduction with affordable appliances waiting time. The detail description is as follows.

6.1. Electricity Cost Evaluation under Price-Based DR

The power company provides various pricing schemes such as RTPS, CPPS, TOUPS, DAPS, and FPS for electricity cost calculation; however, we adopted RTPS and CPPS for the proposed framework. The RTPS and CPPS are the Midwest independent system operator (MISO) daily electricity pricing signals taken from the federal energy regulatory commission (FERC). The electricity cost using RTPS and CPPS is individually discussed in the succeeding sections.

6.2. Electricity Cost Evaluation Using RTPS

To evaluate the cost parameters of the proposed scheme simulations are conducted using different OTI, i.e., 15, 30, and 60 min. The proposed HGWmEDE algorithm reduced electricity cost as compared to GWO and mEDE by scheduling smart home appliances using forecasted RTPS. The scheduled appliances sustain coordination among pricing scheme and the consumption pattern in a particular timeslot of a day to alleviate the electricity cost. The proposed algorithm shifts smart home appliances from ON-peak timeslots to OFF-peak timeslots in an optimal manner to alleviate electricity cost and peaks in demand.

The energy consumption pattern in terms of electricity cost of the proposed and existing algorithms with 15 min OTI is illustrated in Figure 3a. In this Figure 3a, both scheduled and unscheduled scenarios are observed. The peaks in demand are high in case of unscheduled load, which reveals that prices are high in these particular hours. Thus, the use of appliances in these hours results in high electricity cost. However, scheduling smart home appliances using the proposed and existing algorithms eliminate these peaks in demand and reduce the electricity cost. Thus, the proposed HGWmEDE-based framework outperforms both GWO and mEDE in terms of peaks in demand and electricity cost reduction. The electricity cost pattern of the proposed and existing algorithms with 30 min OTI under RTPS is illustrated in Figure 3b. Both proposed and existing algorithms-based HEMC can schedule smart home appliances. However, the electricity cost of GWO is high at the starting timeslots, while HGWmEDE has minimum electricity cost throughout the 24 h. Likewise, in Figure 3c, the electricity cost pattern of the proposed and existing algorithms for 60 min OTI under RTPS is depicted. The proposed HGWmEDE algorithm has reduced the electricity cost by optimally scheduling smart home appliances of resident's, which is one of our main objectives.

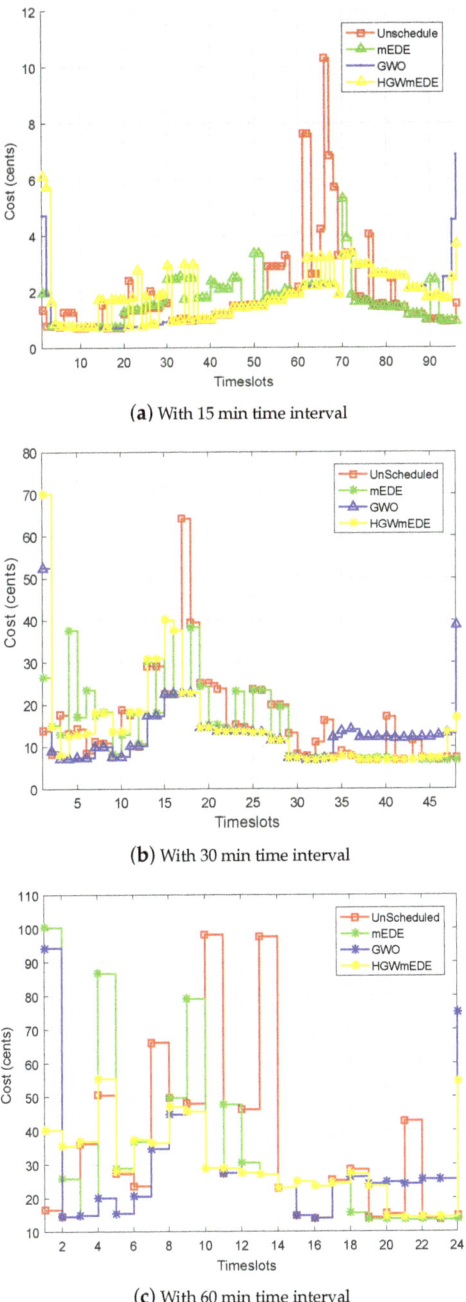

(**a**) With 15 min time interval

(**b**) With 30 min time interval

(**c**) With 60 min time interval

Figure 3. Electricity cost per timeslot evaluation for different OTI under RTPS.

The Figure 4 illustrate that the proposed framework optimally scheduled the smart home appliances as compared to mEDE and GWO under forecasted RTPS and CPPS and reduced the overall aggregated electricity cost of the residents.

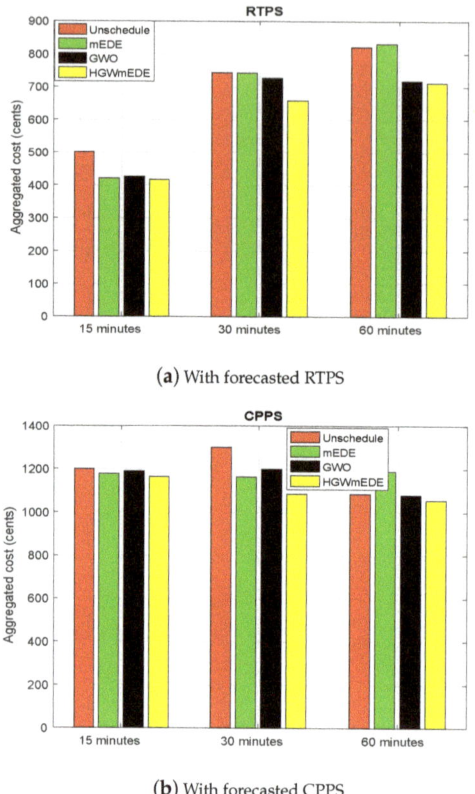

(a) With forecasted RTPS

(b) With forecasted CPPS

Figure 4. Aggregated electricity cost evaluation under forecasted RTPS and CPPS.

Figure 4a presents the overall electricity bill of 15, 30, and 60 OTI under forecasted RTPS. The electricity cost of unscheduled load for 15 min OTI is measured as 500.4827 cents. However, with scheduling smart home appliances using mEDE and GWO reduced the overall electricity cost to 420.5381 cents and 426.0508 cents, respectively. The proposed HGWmEDE scheme reduced electricity cost up to 416.7468 cents, which is the maximum reduction as compared to mEDE and GWO. In a similar fashion, electricity cost reduction behavior of the proposed and existing schemes can be observed for both 30 and 60 min OTI.

Electricity Cost Evaluation under CPPS for Different OTI

The load scheduling is favorable for both residents and power company because it reduces electricity cost, which is favorable for residents; and peaks in the demand, which is favorable for the power company. Cost reduction facilitates residents to deposit less electricity bill and peaks reduction in the demand facilitate power company in the optimal management of supply with demand. In this subsection, electricity cost evaluation is conducted under the forecasted CPPS profile. The electricity bill reduction evaluation of the proposed and existing algorithms are performed using 15, 30, and 60 min OTI under CPPS and is illustrated in Figure 5. The electricity cost profile of 15 min OTI is depicted in

Figure 5a. Generally, the forecasted CPPS remains constant except during critical peak hours where the electricity price reaches to its maximum value [46]. The timeslots from 40 to 65 are critical periods, where the electricity price is high. In unscheduled load scenario, the maximum peak is at 181.55 cents. The scheduled load scenario, where smart home appliances are scheduled, and the peak is reduced to 83.07 cents. The electricity cost for 30 min OTI is illustrated in Figure 5b. The electricity cost is varying for 48 timeslots and the remaining electricity cost profile is as the same as observed in Figure 5a. In our proposed HGWmEDE algorithm-based scenario, no peaks in demand are emerged except at the starting time of day, which is 56 cents. The electricity cost for 60 min OTI is depicted in Figure 5c. The unscheduled appliance electricity cost reaches to 766.8 cents, which is reduced to 203.46 cents when these smart home appliances are scheduled using our proposed HGWmEDE algorithm. This means that the proposed HGWmEDE algorithm optimally scheduled the smart home appliances. The overall cost for the proposed and existing optimization schemes is illustrated in Figure 4b. The overall unscheduled cost is 1300.891 cents, which is reduced to 1085.91 cents when smart home appliances are scheduled using the proposed HGWmEDE algorithm. The proposed HGWmEDE algorithm outperforms both mEDE and GWO algorithms in terms of electricity cost reduction. The overall electricity cost reduction for 30 and 60 min OTI is depicted in Figure 4b. A brief comparison of electricity cost under forecasted RTPS and CPPS is listed in Table 3 for 15, 30, and 60 min OTI.

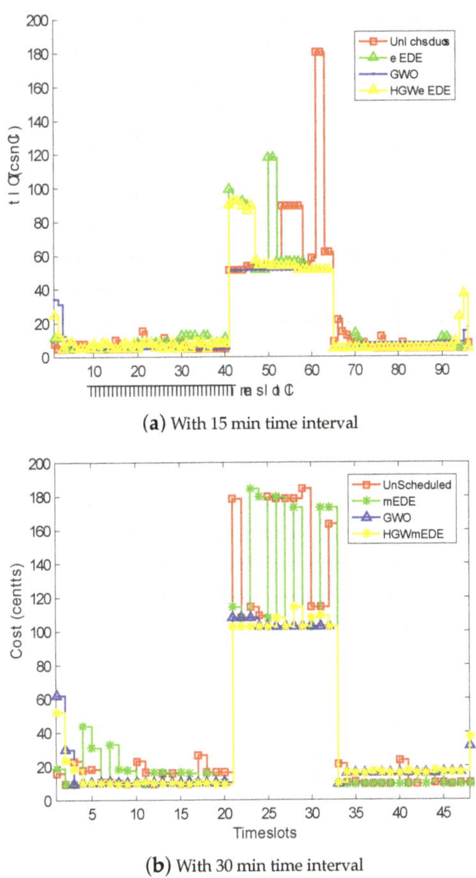

(**a**) With 15 min time interval

(**b**) With 30 min time interval

Figure 5. *Cont.*

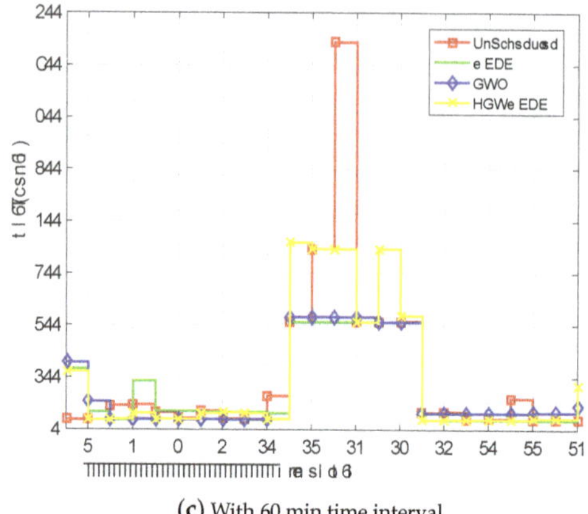

(**c**) With 60 min time interval

Figure 5. Electricity cost per timeslot evaluation under forecasted CPPS for different OTI.

Table 3. Overall electricity cost comparative evaluation for 24 h time horizon under forecasted RTPS and CPPS.

Scenarios	Electricity Cost (Cents) under RTPS			Electricity Cost (Cents) under CPPS		
	15 min	30 min	60 min	15 min	30 min	60 min
Without scheduling	500.4821	743.4871	822.1561	1200.1561	1300.8910	1085.6481
mEDE	420.5381	743.1951	831.2132	1178.0461	1164.4901	1190.6901
GWO	426.0507	727.1431	717.9402	1190.5122	1200.9612	1080.4091
HGWmEDE	416.7468	658.6502	712.7292	1164.4901	1085.9022	1056.7891

6.3. Smart Home Energy Consumption

The smart home appliances energy consumption for both RTPS and CPPS are discussed in detail in the following subsection:

6.3.1. Smart Home Energy Consumption Using RTPS

Energy consumed by smart home appliances under RTPS in each timeslot for 15, 30, and 60 min OTIs is illustrated in Figure 6. In Figure 6a, the smart home energy consumption profile for 15 min OTI is depicted. From the figure it is obvious that at the start and end timeslots of the day have low per unit electricity price; Thus, HEMC based on our proposed HGWmEWDE shifted the load to these low pricing timeslots. In this manner, the proposed scheme optimally curtailed peak load on the power company.

The smart home energy consumption profile for 30 min time interval is illustrated in Figure 6b. The HMEC based on our proposed HGWDE algorithm results in optimal energy consumption profile as compared to GWO and mEDE-based HEMC. The HGWmEDE eliminated load peaks, which curtailed burden on both power company in terms of peak power generation and on residents in terms reduce electricity bill deposit. The mEDE and GWO also reduced peaks in demand and reduced the burden on both power company and residents as compared to without scheduling scenario. The burden reduction of the proposed scheme is more as compared to the existing schemes (GWO and mEDE); thus, the proposed scheme outperforms the existing schemes.

In Figure 6c, energy consumption pattern for 60 min OTI is depicted. The electricity prices are low during starting timeslots and ending timeslots in forecasted RTPS profile. Thus, the proposed and existing schemes shifted most of the load to these low-price timeslots. The prices are maximum from 3 to 7 p.m., so, our proposed scheme not scheduled appliance in these timeslots because the operation of appliances during these timeslots results in high electricity cost. Load shifting from ON-peak timeslots to OFF-peak timeslots results in user discomfort as the residents must stay to switch on a particular smart home appliance because of the trade-off between electricity cost and user-comfort.

The smart home overall energy consumption for 15 min OTI is 56.3108 kWh, which remains the same before and after scheduling subjected to a fair comparison. The scheduled and unscheduled energy consumption for 30 min OTI is same and recorded as 57.7656 kWh. Likewise, for 60 min OTI, smart home overall energy consumption for both scheduled and unscheduled scenarios is the same and recorded as 64.5661 kWh.

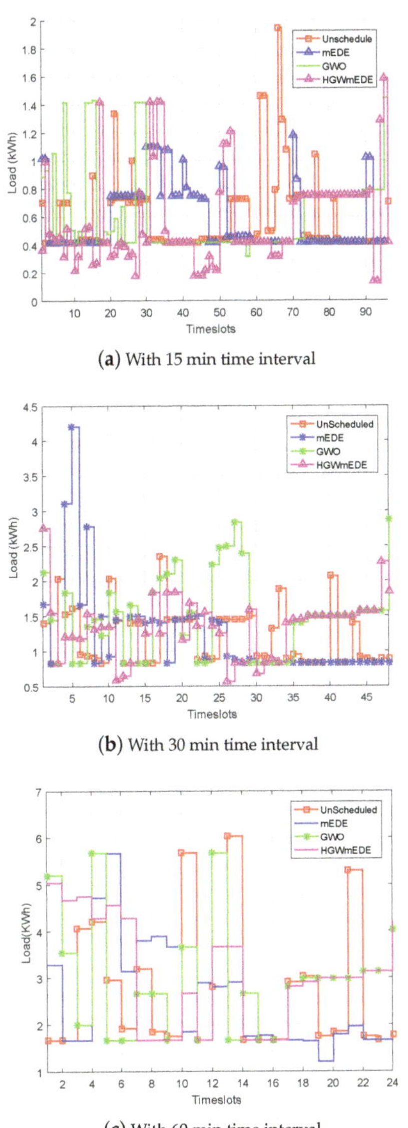

(**a**) With 15 min time interval

(**b**) With 30 min time interval

(**c**) With 60 min time interval

Figure 6. Electricity energy consumption evaluation per timeslot under forecasted RTPS.

6.3.2. Energy Consumption Using CPPS

In this subsection, the proposed HGWmEDE scheme is comparatively evaluated under forecasted CPPS for 15, 30, and 60 min OTI. The comparison is illustrated in Figure 7. The smart home energy consumption profile for 96 timeslots is depicted in Figure 7a. The proposed scheme shifted load to the timeslots where electricity price is low to reduce cost and peaks in demand. In CPPS, at starting and ending timeslots electricity price is constant, while during 40 to 65 timeslots price is maximum as illustrated in Figure 2b. The profile GWO and HGWmEDE is almost similar, except some peaks of GWO are much higher than HGWmEDE at starting timeslots. The proposed HGWmEDE eliminates the peaks in demand for 30 min OTI case; however, the existing schemes have high peaks as compared to the proposed HGWmEDE scheme.

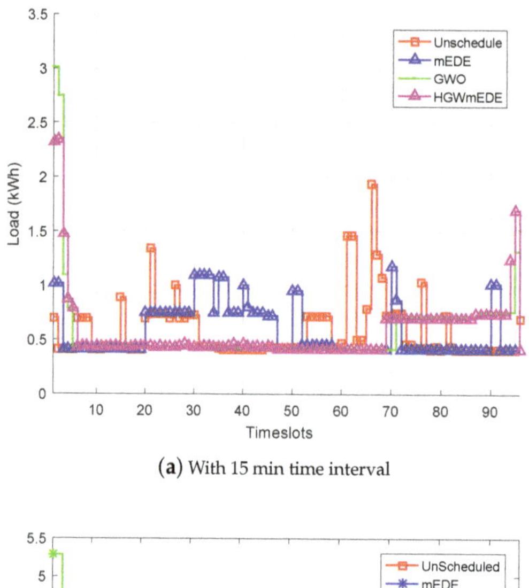

(**a**) With 15 min time interval

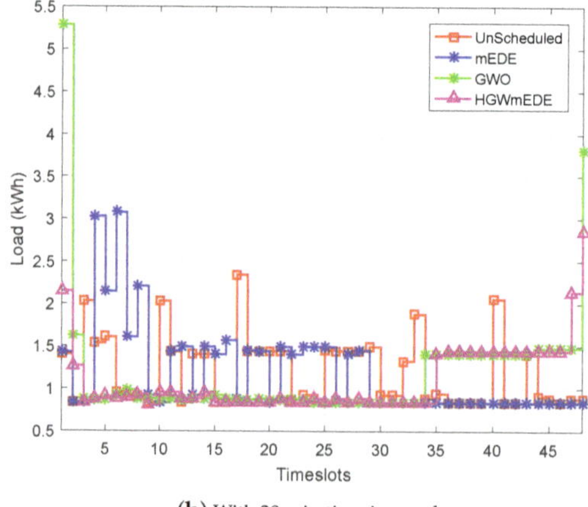

(**b**) With 30 min time interval

Figure 7. *Cont.*

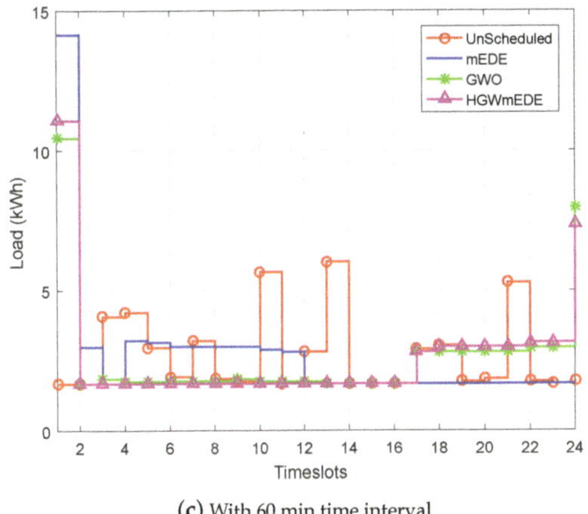

(**c**) With 60 min time interval

Figure 7. Smart home energy consumption per timeslot evaluation using CPPS.

The energy consumption for 60 min OTI is illustrated in Figure 7c. The energy consumption for all cases remain same but the status of appliances vary according to the schedule of HEMC based on our proposed HGWmEDE algorithm and existing algorithm.

The overall energy consumption with and without scheduling must remain same subjected to a fair comparison. The overall energy consumption for 15 min OTI is 56.3107 kWh for the proposed (HGWmEDE) and existing (GWO and mEDE) schemes. Similarly, the energy consumption for 30, and 60 min OTI are 57.7656 kWh and 64.5661 kWh, respectively. Thus, both electricity cost and peaks in demand are reduced by scheduling the smart home appliance while keeping the energy consumption constant.

6.4. Peaks in Demand

Peaks in demand are defined as the value of peak load emerged during a time horizon of 24 h or the maximum load switched on by user during 24 h time horizon. Our objective is to reduce the peaks in demand and to ensure smooth demand curve for 24 h time horizon. Various DSM programs can be applied to alleviate peaks in demand such as peak clipping, load shifting, and price-based DR to eliminate peaks in demand and smooth the demand curve. Peaks elimination in the demand reduces the electricity cost and burden on the power company. The peaks reduction in demand is evaluated for both RTPS and CPPS in the succeeding section.

6.4.1. Peaks Reduction in Demand Evaluation under RTPS

The peaks reduction in demand evaluation under RTPS for different OTI is depicted in Figure 8a. In case, when the load is not scheduled the peak emerged in demand is 10.9697. In case, when the load is scheduled based on mEDE and GWO, the peaks emerged in demand are 8.1722 and 5.6750, respectively. The peak emerged for proposed HGWmEDE scheme is 5.1530, which is low as compared to the both unscheduled and scheduled (GWO and mEDE) cases. The percentage reduction in peaks of the proposed HGWmEDE scheme is 53.02%, while the percentage reduction in peaks of the GWO and mEDE is 25.50%, 48.26%, respectively. Thus, the proposed HGWmEDE scheme outperforms the other schemes in terms of peaks reduction in demand.

In case of unscheduled load for 30 min OTI, the peak emerged in demand is recorded as 6.0257; however, the peak has reduced to 5.8424 and 5.9335 when the load is scheduled using mEDE and GWO algorithms. The proposed HGWmEDE scheme outperforms both mEDE and GWO schemes by reducing peak in demand to 3.6209. Similarly, for 60 min OTI, mEDE, GWO, and HGWmEDE reduced the peaks in demand by a value of 3.6559, 4.3510 and 2.5370 as compared to without scheduling case, which is 5.0245. Thus, it is obvious from the aforementioned statistical analysis that the proposed HGWmEDE scheme outperforms both mEDE and GWO schemes.

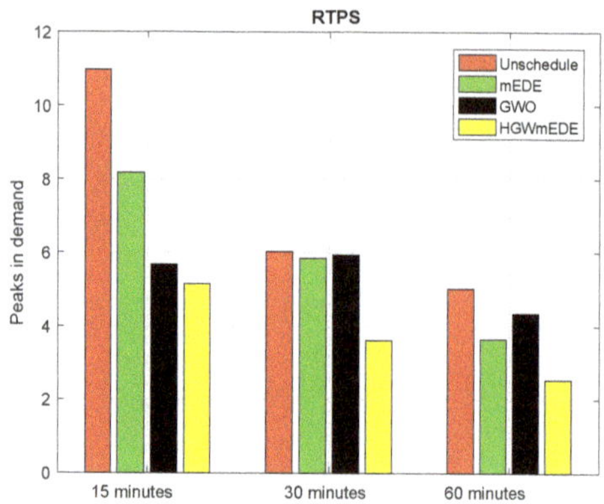

(**a**) Peaks in demand with RTPS

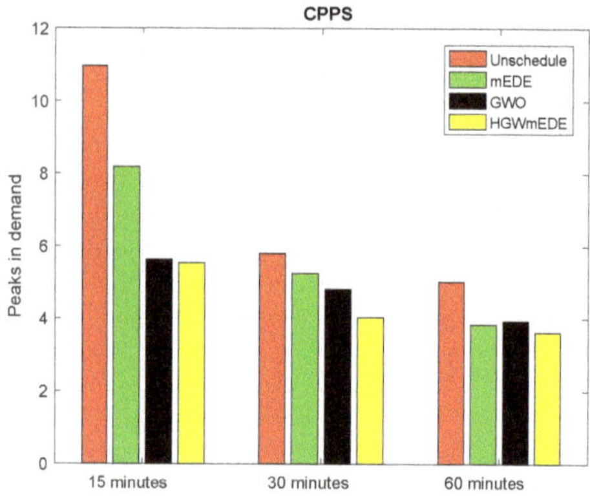

(**b**) Peaks in demand with CPPS

Figure 8. Peaks in demand evaluation under forecasted RTPS and CPPS for different OTI.

6.4.2. Peaks in Demand Evaluation under CPPS for Different OTI

Evaluation of peaks in demand under CPPS is illustrated in Figure 8b. In case of unscheduled load for 15 min OTI, peak in demand is recorded as 10.9697. However, after performing load scheduling, peak in demand is alleviated to 5.5416 with HGWmEDE, 5.627 with GWO and 8.1723 with mEDE. Peaks reduction in terms of percentage for the proposed HGWmEDE, mEDE, GWO, are 49.4836%, 48.7120%, and 25.5012%, respectively.

In case of unscheduled load for 30 min OTI, the peak emerged is 5.86. After load scheduling with mEDE, GWO, and HGWmEDE, peaks in demand are recorded as 5.2536, 4.8165, and 4.0215, respectively. Thus, the proposed HGWmEDE scheme outperforms the existing (mEDE and GWO) schemes in terms of peaks reduction in demand. The comparative evaluation of the proposed HGWmEDE and existing (mEDE and GWO) schemes in terms of peaks reduction in demand under RTPS and CPPS for different OTI is listed in Table 4.

Table 4. Peaks in demand evaluation of the proposed and existing schemes for 24 h.

Scenarios	Peaks in Demand under RTPS with Different OTI			Peaks in Demand under CPPS with Different OTI		
	15 min	30 min	60 min	15 min	30 min	60 min
Without scheduling	10.9698	6.0258	5.0258	10.9698	5.8035	5.0258
mEDE	8.1723	5.8425	3.6558	8.1723	5.2537	3.8425
GWO	5.676	5.9336	4.3509	5.6265	4.8166	3.9336
HGWmEDE	5.1531	3.6210	2.5369	5.5416	4.0264	3.6210

6.5. User-Comfort Evaluation in Terms of Waiting Time

In this subsection, the user-comfort in terms of waiting time is evaluated. In nature, always there is a trade-off between different conflicting parameters. In this paper, a trade-off between electricity cost and waiting time (user-comfort) exist. To reduce electricity cost, the residents must wait for timeslots where the electricity price is low, i.e., OFF-peak hours to switch on their load. Thus, user-comfort and electricity cost are directly related to [47]. In without scheduling scenario, waiting time is almost zero because the smart home appliances are operated according to the resident choice and priority. However, the case when the load is scheduled based on mEDE, GWO, and the proposed HGWmEDE, the user switched on their appliances according to the schedule provided by HEMC to reduce electricity cost. Thus, user-comfort is compromised while reducing the electricity cost due to the trade-off. Evaluation of appliances waiting time under RTPS and CPPS are as follows:

6.5.1. Smart Home Appliances Waiting Time Evaluation under RTPS

The appliance waiting time for 15, 30, and 60 min OTI is illustrated in Figure 9a. Waiting time of the proposed HGWmEDE, mEDE, and GWO for 15 min OTI are calculated as 10.4 h, 4.3 h, and 9.7 h, respectively. Waiting time for 30 min OTI of the proposed HGWmEDE, mEDE, and GWO are calculated as 12.7007 h, 4.5394 h, and 10.0262 h. Similarly, for 60 min OTI of the proposed HGWmEDE, mEDE, and GWO, waiting time is 3.8 h, 2.6560 h, and 2.2397 h, respectively.

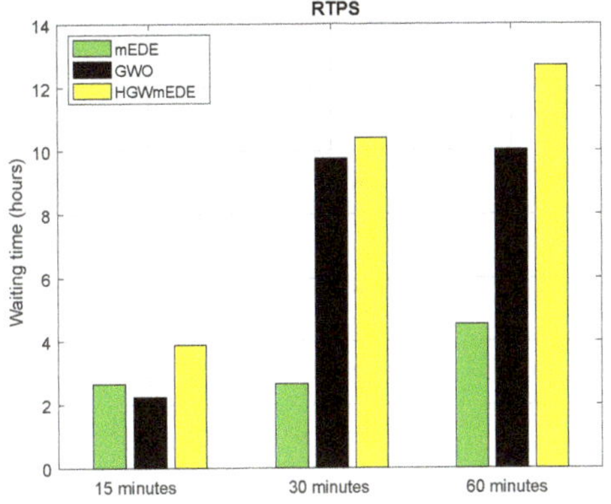

(**a**) User-comfort (waiting time) using RTPS

Figure 9. *Cont.*

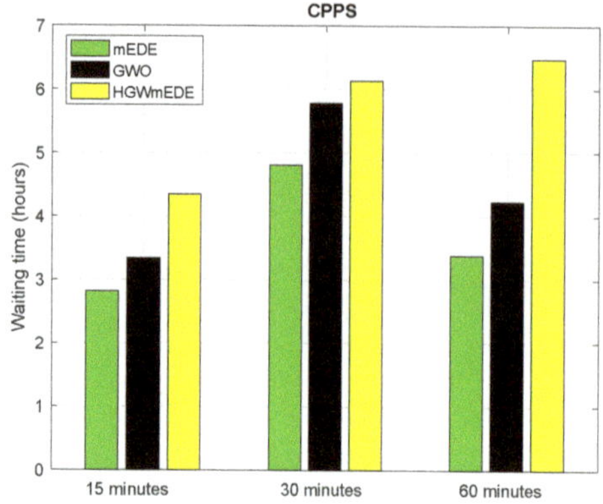

(**b**) User-comfort (waiting time) using CPPS

Figure 9. User-comfort (waiting time) of the scheduled load based on the proposed HGWmEDE, mEDE, and GWO using RTPS and CPPS.

It is concluded from the above results and discussion that user frustration in terms of waiting time always reduces by keeping the OTI smaller. For example, for 15 min OTI, if the operational time of a toaster is 12 min, then HEMC will allocate 15 min timeslot to toaster; in this way, only 3 min of a timeslot is wasted because this slot is not allocated to other appliances. In contrast, in case of 60 min OTI, the HEMC will allocate 60 min timeslot to an electric kettle, which has the operational time of only 5 min, then rest of 55 min timeslots will be wasted. Thus, in larger OTI, the user frustration increase in terms of waiting time.

6.5.2. Smart Home Appliances Waiting Time Evaluation under CPPS

Waiting time of the proposed HGWmEDE and existing (mEDE and GWO) under CPPS is illustrated in Figure 9b. The recorded value of waiting time for mEDE, GWO, and the proposed HGWmEDE is as 3.39 h, 4.23 h, and 6.49 h, respectively. It is obvious that the load schedule return by HMEC based on HGWmEDE algorithm has more waiting time, which indicates that user-comfort is compromised for the purpose to reduce electricity cost. The statistical analysis of the proposed and existing algorithms in terms of waiting time for different OTI under RTPS and CPPS is listed in Table 5.

Table 5. Comparative evaluation of the proposed HGWmEDE and existing (mEDE and GWO) algorithm in terms of waiting time under RTPS and CPPS for different OTI.

Scenarios	Evaluation of Waiting under RTPS for Different OTI			Evaluation of Waiting Time under CPPS for Different OTI		
	15 min	30 min	60 min	15 min	30 min	60 min
mEDE	4.3781 h	4.5394 h	2.6560 h	3.3826 h	4.8012 h	2.8158 h
GWO	9.7494 h	10.0262 h	2.2397 h	4.2293 h	5.7853 h	3.3346 h
HGWmEDE	10.4249 h	12.7007 h	3.8793 h	6.4814 h	6.1335 h	4.3408 h

6.6. Convergence Evaluation of the Proposed HGWmEDE Algorithm-Based Fitness Function

The convergence evaluation of the fitness function of the proposed HGWDE algorithm is depicted in Figure 10. X-label of the plot is the number of iterations and Y-label is the value of the fitness function. Figure 10 illustrates that the solution converges after 100 iterations, which indicates that the global maximum is achieved. The proposed HGWmEDE algorithm converging behavior for each iteration is plotted. The value of cost is constantly decreasing from 0 to 10 iterations and after these iterations, the graph slight varies. The behavior of convergence of the proposed HGWDE algorithm iterations is observed for 100 iterations. Finally, a straight line is achieved, which means that the solution is converged, and this is the most optimal point.

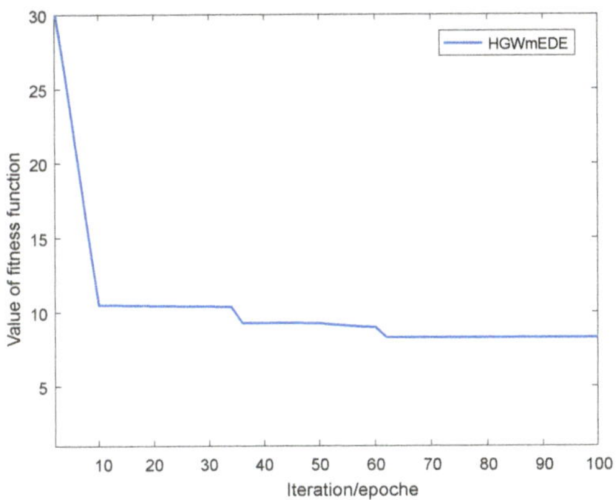

Figure 10. Convergence evaluation of the fitness function.

6.7. Performance Trade-Off

The performance trade-off exists in various conflicting parameters of the system. In the proposed modular framework, a trade-off between electricity cost and user-comfort exist. When the residents wants to reduce their electricity cost, he must face frustration in the form of user discomfort. Residents frustration increasing when the difference between the resident preferred time and the HEMC scheduled time is more. It is observed in Figure 4a,b that the electricity cost is reduced by applying HGWmEDE for scheduling at the expense of maximum average waiting time.

7. Conclusions and Future Research Directions

In this paper, first a modular framework is introduced, and then an algorithm HGWmEDE is proposed, which with the help of forecasted price-incentive DR scheme schedule the household load to maximize the aggregate utility of both resident and power company. The proposed approach is beneficial for residents because it reduces the electricity bill with an affordable waiting time of smart home appliances. In addition, it is beneficial for the power company because it reduces peaks in demand, which smooth out the demand curve and increase the stability of the power system. The integration of the forecaster module to the home energy management framework provides optimal load schedule, which not only facilitates residents but also power companies. Simulation analysis validated the proposed approach by comparing with two other approaches, a system without HEMC and a system with HEMC based on GWO and mEDE algorithms. Surely, the proposed framework can be applied for DSM and reliable operation of the SG. The idea of this paper in the future can be extended to various directions:

1. A system with renewable and non-renewable energy provider can be considered.
2. A system with a deep neural network can be used to optimize home energy management.
3. A system with multiple power companies and multiple homes and effect malicious residents can be considered.
4. To implement fog and cloud concept for household load scheduling instead of using a HEMC.
5. The same framework can be extended for other heuristics, deterministic, and stochastic techniques under RES.

Author Contributions: Conceptualization, G.H.; Formal analysis, S.A. and M.U.; Investigation, G.H.; Methodology, G.H. and K.S.A.; Software, G.H.; Supervision, K.S.A.; Validation, K.S.A., N.I. and A.A.; Writing—original draft, G.H.; Writing—review and editing, K.S.A., S.A., and M.U.

Funding: This research received no external funding.

Conflicts of Interest: The authors declare no conflicts of interest.

Nomenclature

Acronym	Meaning
RTPS	Real-Time Pricing Scheme
TOUPS	Time of Use Pricing Scheme
CPPS	Critical Peak Pricing Scheme
FRPS	Flat-Rate Pricing Scheme
DAPS	Day-Ahead Pricing Scheme
HEMC	HEM controller
DSM	Demand-Side Management
HGWmEDE	Hybrid Gray Wolf-Modified Differential Evolution
RBM	Restricted Boltzmann Machine
LOT	Length of Operational Time
OTI	Operation Time Interval
GA	Genetic Algorithm
BPSOA	Binary-Particle Swarm Optimization Algorithm
DEA	Differential Evolution Algorithm
IDSS	Intelligent Decision Support System
WDO	Wind-Driven Optimization
K-WDO	Algorithm with Knapsack
ACOA	Ant Colony Optimization Algorithm
MOEA	Multi-Objective Evolutionary Algorithm
PS	Pareto Sets
GMI	Generalized Mutual Information
TLBOA	Teaching and Learning-Based Optimization Algorithm
SFL	Shuffled-Frog Leaping
ANN-GA	Artificial Neural Network
LP	Linear Programming
NLP	Non-LP
BFA	Bacterial Foraging Algorithm
DR	Demand Response
AMI	Advanced Metering Infrastructure
ICT	Information and Communication Technology
RERs	Renewable Energy Resources
SG	Smart Grid

References

1. Mhanna, S.; Chapman, A.C.; Verbi, G. A fast distributed algorithm for large-scale demand response aggregation. *IEEE Trans. Smart Grid* **2016**, *7*, 2094–2107. [CrossRef]
2. Energy Reports. Available online: http://www.enerdata.net/enerdatauk/press-and-publication/energy-features/enerfuture-2007.php (accessed on 10 January 2019).
3. Logenthiran, T.; Srinivasan, D.; Shun, T.Z. Demand side management in smart grid using heuristic optimization. *IEEE Trans. Smart Grid* **2012**, *3*, 1244–1252 [CrossRef]
4. Shirazi, E.; Jadid, S. Optimal residential appliance scheduling under dynamic pricing scheme via HEMDAS. *Energy Build.* **2015**, *93*, 40–49. [CrossRef]
5. Bradac, Z.; Kaczmarczyk, V.; Fiedler, P. Optimal scheduling of domestic appliances via MILP. *Energies* **2014**, *8*, 217–232. [CrossRef]

6. Katz, J.; Andersen, F.M.; Morthorst, P.E. Load-shift incentives for household demand response: Evaluation of hourly dynamic pricing and rebate schemes in a wind-based electricity system. *Energy* **2016**, *115*, 1602–1616. [CrossRef]

7. Rabiya, K.; Javaid, N.; Rahim, M.H.; Aslam, S.; Sher, A. Fuzzy energy management controller and scheduler for smart homes. *Sustain. Compu. Inf. Syst.* **2019**, *21*, 103–118.

8. Adika, O.C.; Wang, L. Autonomous appliance scheduling for household energy management. *IEEE Trans. Smart Grid* **2014**, *5*, 673–682. [CrossRef]

9. Li, C.; Yu, X.; Yu, W.; Chen, G.; Wang, J. Efficient computation for sparse load shifting in demand side management. *IEEE Trans. Smart Grid* **2017**, *8*, 250–261. [CrossRef]

10. Javaid, N.; Naseem, M.; Rasheed, M.B.; Mahmood, D.; Khan, S.A.; Alrajeh, N.; Iqbal, Z. A new heuristically optimized Home Energy Management controller for smart grid. *Sustain. Cities Soc.* **2017**, *34*, 211–227. [CrossRef]

11. Bilal, H.; Javaid, N.; Hasan, Q.; Javaid, S.; Khan, A.; Malik, S. An inventive method for eco-efficient operation of home energy management systems. *Energies* **2018**, *11*, 3091.

12. Huang, Y.; Wang, L.; Guo, W.; Kang, Q.; Wu, Q. Chance constrained optimization in a home energy management system. *IEEE Trans. Smart Grid* **2018**, *9*, 252–260. [CrossRef]

13. Ogwumike, C.; Short, M.; Abugchem, F. Heuristic Optimization of Consumer Electricity Costs using a Generic Cost Model. *Energies* **2016**, *9*, 6. [CrossRef]

14. Rasheed, M.B.; Javaid, N.; Ahmad, A.; Khan, Z.A.; Qasim, U.; Alrajeh, N. An efficient power scheduling scheme for residential load management in smart homes. *Appl. Sci.* **2015**, *5*, 1134–1163. [CrossRef]

15. Tuaimah, F.M.; Abd, Y.N.; Hameed, F.A. Ant Colony Optimization based Optimal Power Flow Analysis for the Iraqi Super High Voltage Grid. *Int. J. Comput. Appl.* **2013**, *67*, 13–18.

16. Logenthiran, T.; Srinivasan, D.; Vanessa, K.W.M. Demand side management of smart grid: Load shifting and incentives. *J. Renew. Sustain. Energy* **2014**, *6*, 033136. [CrossRef]

17. Islam, S.M.; Das, S.; Ghosh, S.; Roy, S.; Suganthan, P.N. An adaptive differential evolution algorithm with novel mutation and crossover strategies for global numerical optimization. *IEEE Trans. Syst. Man Cybern. Part B* **2012**, *42*, 482–500. [CrossRef] [PubMed]

18. Li, H.; Zhang, Q. Multiobjective optimization problems with complicated Pareto sets, MOEA/D and NSGA-II. *IEEE Trans. Evol. Comput.* **2009**, *13*, 284–302. [CrossRef]

19. Setlhaolo, D.; Xia, X. Optimal scheduling of household appliances incorporating appliance coordination. *Energy Proc.* **2014**, *61*, 198–202. [CrossRef]

20. Osório, G.J.; Matias, J.C.O.; Catalão, J.P.S. Electricity prices forecasting by a hybrid evolutionary-adaptive methodology. *Energy Convers. Manag.* **2014**, *80*, 363–373. [CrossRef]

21. Shayeghi, H.; Ghasemi, A.; Moradzadeh, M.; Nooshyar, M. Simultaneous day-ahead forecasting of electricity price and load in smart grids. *Energy Convers. Manag.* **2015**, *95*, 371–384. [CrossRef]

22. Derakhshan, G.; Shayanfar, H.A.; Kazemi, A. The optimization of demand response programs in smart grids. *Energy Policy* **2016**, *94*, 295–306. [CrossRef]

23. Soares, J.; Ghazvini, M.A.F.; Vale, Z.; de Moura Oliveira, P.B. A multi-objective model for the day-ahead energy resource scheduling of a smart grid with high penetration of sensitive loads. *Appl. Energy* **2016**, *162*, 1074–1088. [CrossRef]

24. Ghasemi, A.; Shayeghi, H.; Moradzadeh, M.; Nooshyar, M. A Novel hybrid algorithm for electricity price and load forecasting in smart grids with demand-side management. *Appl. Energy* **2016**, *177*, 40–59. [CrossRef]

25. Jayabarathi, T.; Raghunathan, T.; Adarsh, B.R.; Suganthan, P.N. Economic dispatch using hybrid grey wolf optimizer. *Energy* **2016**, *111*, 630–641. [CrossRef]

26. Asif, K.; Javaid, N.; Khan, M.I. Time and device based priority induced comfort management in smart home within the consumer budget limitation. *Sustain. Cities Soc.* **2018**, *41*, 538-555.

27. Geem, Z.W.; Kim, J.H.; Loganathan, G.V. A new heuristic optimization algorithm: Harmony 578 search. *Simulation* **2001**, *76*, 60–68. [CrossRef]

28. Storn, R.; Price, K. *Differential Evolution—A Simple and Efficient Adaptive Scheme for Global 580 Optimization over Continuous Spaces*; International Computer Science Institute: Berkeley, CA, USA, 1995.

29. Vardakas, J.S.; Zorba, N.; Verikoukis, C.V. Performance evaluation of power demand scheduling scenarios in a smart grid environment. *Appl. Energy* **2015**, *142*, 164–178. [CrossRef]

30. Awais, M.; Javaid, N.; Ullah, I.; Abdul, W.; Almogren, A.; Alamri, A. An intelligent hybrid heuristic scheme for smart metering based demand side management in smart homes. *Energies* **2017**, *10*, 1258.
31. Yuce, B.; Rezgui, Y.; Mourshed, M. ANN–GA smart appliance scheduling for optimised energy management in the domestic sector. *Energy Build.* **2016**, *111*, 311–325. [CrossRef]
32. Reka, S.S.; Ramesh, V. A demand response modeling for residential consumers in smart grid environment using game theory based energy scheduling algorithm. *Ain Shams Eng. J.* **2016**, *7*, 835–845. [CrossRef]
33. Erdinc, O. Economic impacts of small-scale own generating and storage units, and electric vehicles under different demand response strategies for smart households. *Appl. Energy* **2014**, *126*, 142–150. [CrossRef]
34. Agnetis, A.; de Pascale, G.; Detti, P.; Vicino, A. Load scheduling for household energy consumption optimization. *IEEE Trans. Smart Grid* **2013**, *4*, 2364–2373. [CrossRef]
35. Setlhaolo, D.; Xia, X.; Zhang, J. Optimal scheduling of household appliances for demand response. *Electr. Power Syst. Res.* **2014**, *116*, 24–28. [CrossRef]
36. Pradhan, M.; Roy, P.K.; Pal, T. Grey wolf optimization applied to economic load dispatch problems. *Int. J. Electr. Power Energy Syst.* **2016**, *83*, 325–334. [CrossRef]
37. Vardakas, J.S.; Zorba, N.; Verikoukis, C.V. Power demand control scenarios for smart grid applications with finite number of appliances. *Appl. Energy* **2016**, *162*, 83–98. [CrossRef]
38. Ogunjuyigbe, A.S.O.; Ayodele, T.R.; Akinola, O.A. User satisfaction-induced demand side load management in residential buildings with user budget constraint. *Appl. Energy* **2017**, *187*, 352–366. [CrossRef]
39. Javaid, N.; Ullah, I.; Akbar, M.; Iqbal, Z.; Khan, F.A.; Alrajeh, N.; Alabed, M.S. An intelligent load management system with renewable energy integration for smart homes. *IEEE Access* **2017**, *5*, 13587–13600. [CrossRef]
40. Yao, E.; Samadi, P.; Wong, V.W.S.; Schober, R. Residential demand side management under high penetration of rooftop photovoltaic units. *IEEE Trans. Smart Grid* **2016**, *7*, 1597–1608. [CrossRef]
41. Nadeem, J.; Hafeez, G.; Iqbal, S.; Alrajeh, N.; Alabed, M.S.; Guizani, M. Energy efficient integration of renewable energy sources in the smart grid for demand side management. *IEEE Access* **2018**, *6*, 77077–77096.
42. Ghulam, H.; Javaid, N.; Iqbal, S.; Khan, F. Optimal residential load scheduling under utility and rooftop photovoltaic units. *Energies* **2018**, *11*, 611.
43. Ashfaq, A.; Javaid, N.; Guizani, M.; Alrajeh, N.; Khan, Z.A. An accurate and fast converging short-term load forecasting model for industrial applications in a smart grid. *IEEE Trans. Ind. Inform.* **2017**, *13*, 2587–2596.
44. Muqaddas, N.; Iqbal, Z.; Javaid, N.; Khan, Z.; Abdul, W.; Almogren, A.; Alamri, A. Efficient power scheduling in smart homes using hybrid grey wolf differential evolution optimization technique with real time and critical peak pricing schemes. *Energies* **2018**, *11*, 384.
45. Singh, N.; Singh, S.B. Hybrid Algorithm of Particle Swarm Optimization and Grey Wolf Optimizer for Improving Convergence Performance. *J. Appl. Math.* **2017**, *2017*, 2030489. [CrossRef]
46. Waterloo North Hydro. Available online: https://www.wnhydro.com/en/your-home/time-of-use-rates. asp (accessed on 9 January 2019).
47. Muralitharan, K.; Sakthivel, R.; Shi, Y. Multiobjective optimization technique for demand side management with load balancing approach in smart grid. *Neurocomputing* **2016**, *177*, 110–119. [CrossRef]

Article

Investigating the Dynamic Impact of CO_2 Emissions and Economic Growth on Renewable Energy Production: Evidence from FMOLS and DOLS Tests

Muhammad Waris Ali Khan [1,*], Shrikant Krupasindhu Panigrahi [2], Khamis Said Nasser Almuniri [2], Mujeeb Iqbal Soomro [3], Nayyar Hussain Mirjat [4] and Eisa Salim Alqaydi [5]

[1] Faculty of Industrial Management, Universiti Malaysia Pahang, Gambang 26300, Malaysia
[2] College of Business, University of Buraimi, Alburaimi, P.O. box 890, P.C.512, Oman
[3] Department of Mechanical Engineering, Mehran University of Engineering & Technology, SZAB Campus, Khairpur 66020, Pakistan
[4] Department of Electrical Engineering, Mehran University of Engineering and Technology, Jamshoro 76062, Pakistan
[5] Faculty of Health Sciences, DeMontfort University, Leicester LE8 4AA, UK
* Correspondence: waris@ump.edu.my; Tel.: +60-136-091-898

Received: 26 April 2019; Accepted: 28 July 2019; Published: 1 August 2019

Abstract: Understanding the dynamic nexus between CO_2 emissions and economic growth in the sustainable environment helps the economies in developing resources and formulating apposite energy policies. In the recent past, various studies have explored the nexus between CO_2 emissions and economic growth. This study, however, investigates the nexus between renewable energy production, CO_2 emissions, and economic growth over the period from 1995 to 2016 for seven Association of Southeast Asian Nations (ASEAN) countries. Fully Modified Ordinary Least Square (FMOLS) and Dynamic Ordinary Least Square (DOLS) methodologies were used to estimate the long- and short-run relationships. The panel results revealed that renewable energy production has a significant long term effect on CO_2 emissions for Vietnam (t = −2.990), Thailand (t = −2.505), and Indonesia (t = −2.515), and economic growth impact for Malaysia (t = 2.050), Thailand (t = −2.001), and the Philippines (t = −2.710). It is, therefore, vital that the ASEAN countries implement policies and strategies that ensure energy saving and continuous economic growth without forsaking the environment. This study, as such, recommends that ASEAN countries should take measures to decrease the reliance on fossil fuels for achieving these objectives. Future research should consider the principles of circular economy and clean energy development mechanisms integrated with renewable energy technologies.

Keywords: carbon emissions; economic growth; energy; renewable energy; Fully Modified Ordinary Least Square (FMOLS); dynamic panel cointegration model

JEL Classification: C22; C33; Q20; Q43

---

1. Introduction

The promotion of sustainable development and combating climate change is altogether an integral and challenging aspect of energy planning and policy development. Over the past decades, global warming has been rising, which has emerged as one of the key challenges that humanity is facing. The increased emissions of greenhouse gases (GHGs), such as carbon dioxide (CO_2), methane (CH_4), ozone, and nitrous oxide, are causing severe damage to the global environment. Among other greenhouse gases, CO_2 emissions are considered as the principal cause of global warming, thereby toppling the climate [1–3]. The CO_2 emissions due to excessive burning of fossil fuels, such as coal, oil, and gas, along with increased deforestation, have considerably contributed to climate change. There is a great abundance of CO_2 in the atmosphere compared to other greenhouse gases and it is estimated rise in the next 100 years [4]. This situation has prompted the consideration and research on the measures to reduce CO_2 emissions. One of the key measures to reduce CO_2 emissions is to reduce fossil fuel consumption and maintain economic growth by meeting part of the energy needs by harnessing renewable energy sources [3]. In order to substantiate this measure, the literature has sufficiently maintained the establishment of relationships among renewable energy production, CO_2 emissions, and economic growth [5–7]. However, the higher production cost of renewable energy compared to fossil fuel energy hinders its full-scale commercialization and, hence, has not achieved the much-needed success so far. The researchers are, nevertheless, attempting to establish the much required positive relationship between environmental quality and economic growth [8]. As such, it is essentially required that energy policymakers should come up with robust, sustainable policy endeavors to address the environmental and economic challenges with the effective utilization of renewable energy resources [8]. According to the Environmental Kuznets Curve (EKC) hypothesis [9], income and emissions are directly proportional to the threshold level. In this context, various empirical studies have so far only focused on CO_2 emissions as a pollutant in the industrialized world. A review of previous studies in a similar context reveals that there is lack of studies performed on developing countries especially for the Association of Southeast Asian Nations (ASEAN) countries, including the emerging economies of the region, such as Myanmar and Vietnam, see for example [1,4,10,11]. These studies, however, have not produced clarity in their work for neglecting Myanmar and Vietnam as they have only considered five leading ASEAN countries (Malaysia, Thailand, Indonesia, the Philippines, and Singapore) for their past economic growth [10]. As such, the impact of the emerging economies of Myanmar and Vietnam for their renewable energy production has consistently gone unnoticed. As such, in this study, a summary of percentages of renewable energy production, CO_2 emissions per capita, and Gross Domestic Product (GDP) per capita for seven selected ASEAN countries for the study period of 1995–2016 is thoroughly analyzed. The panel data plots of the CO_2 emissions per capita for ASEAN countries is shown in Figure 1a while the real GDP for the ASEAN countries as shown in Figure 1b indicate that Singapore and Malaysia have been growing consistently from 1995 until 2016. These details signify the importance of the relationships between renewable energy production, CO_2 emissions, and the economies for the selected ASEAN countries, which require essential investigations. The dynamic relationship between energy consumption and economic growth has been thoroughly investigated in the past few decades [12]. However, the impact of harnessing renewable energy on the reduction of CO_2 emissions and thereby bringing about sustainable economic growth has been overlooked by previous studies. It is pertinent to mention that renewable energy resources are now emerging as the mainstream sources of energy in various developed economies and co-operative ownership of these resources has also increased and expanded rapidly in the past few years. The contribution of low growing economies like Myanmar and Vietnam towards renewable energy production, as shown in Figure 1c, has been consistently unnoticed [12].

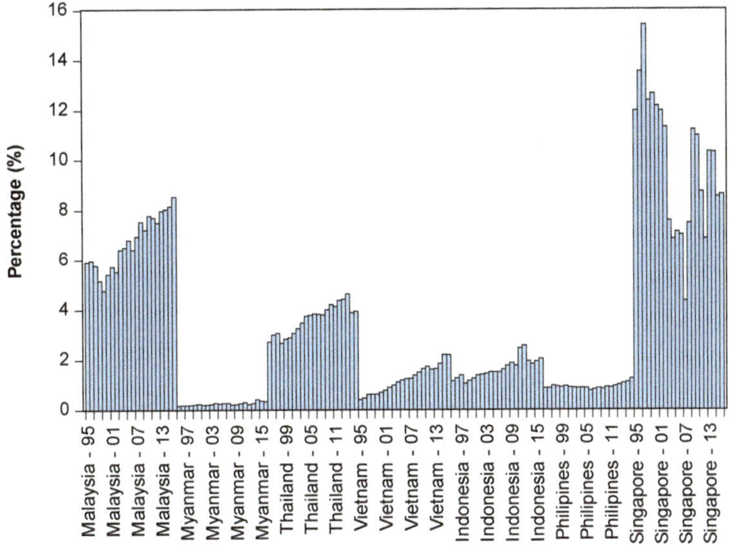

(a)

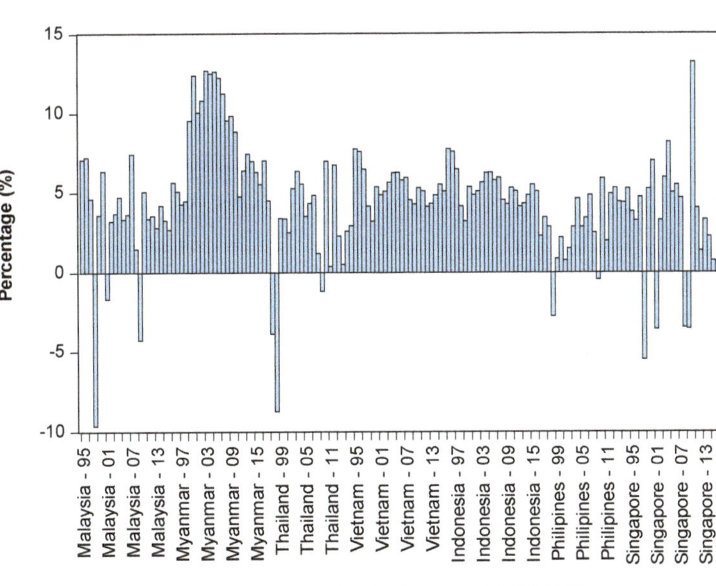

(b)

Figure 1. *Cont.*

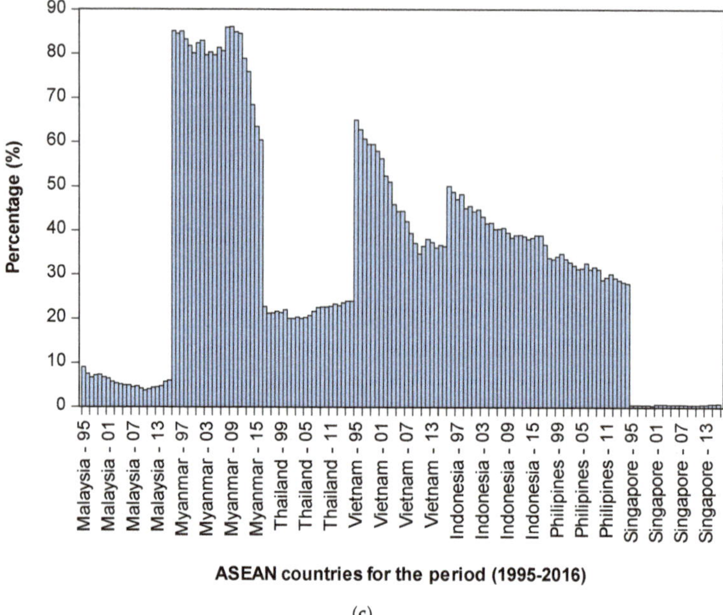

(c)

Figure 1. (a) Panel data plots of the CO_2 emissions per capita for Association of Southeast Asian Nations (ASEAN) countries; (b) Panel data plots of the real GDP (Gross Domestic Product) per capita for ASEAN countries; (c) Panel data plots of renewable energy production in % for ASEAN countries.

Given the above facts, the main objective of this study is to investigate the impact of renewable energy production on CO_2 emissions and economic growth for seven ASEAN countries from 1995 to 2016. This objective is achieved by employing a panel co-integration technique that followed the growth model framework. Furthermore, an attempt has been made to validate the EKC hypothesis for the selected seven ASEAN countries using panel data analysis with an approach of co-integration and Granger causality tests. The consistency of the results was assured by the inclusion of each country, thereby analyzing the environmental complexities in the respective countries. The country-specific investigation helped to merge the complexities of renewable energy production, the CO_2 emission relationship, and economic growth nexus, which has been ignored in previous studies. As such, contrary to previous studies, this study explored the causality relationships among renewable energy production, CO_2 emissions, and economic growth.

ASEAN Countries as the Context

Historically, the consumption and reliance on fossil fuels in ASEAN countries have been generally on the rise [13]. ASEAN countries are experiencing the world's most significant jump and are considered as regions vulnerable to causing climate change [14]. Every decade, since 1960, the average temperature in the Southeast Asian region has risen, which has adversely added to climate change. The economic progress of five of the ASEAN countries: Malaysia, Indonesia, the Philippines, Singapore, and Thailand, is commendable compared to the other two [4]. Nevertheless, Vietnam and Myanmar have shown the highest renewable energy production among the other regional counties. As such, these two developing economies of ASEAN, for renewable energy initiatives, were also included in this study. It is pertinent to mention that 57% of the total electricity production in Myanmar has been based on its indigenous hydropower resources, thereby tackling, effectively, the emission of greenhouse gases [15]. Moreover, Myanmar, together with Vietnam, has also focused on implementing the climate-resilient de-carbonization pathways, thus assuring its significance in the renewable energy market. It is estimated that Myanmar can further install electricity plants of 104,000 MW capacity based on the wind and hydropower plants and that it can produce another 40 TWh/year (TWh/year stands for terawatt-hour per year) from solar power plants [16]. As per a World Bank Database report (2016), Singapore (98%), the Philippines (46%), Thailand (42%), and Cambodia (33%) were the four major energy-importing ASEAN countries [17]. Meanwhile, other countries like Brunei, Indonesia, Myanmar, and Vietnam are well-known energy resource exporters. In most of the ASEAN countries, renewable energy resources are abundant and are sufficiently consumed, such as in Indonesia, Malaysia, Thailand, and Vietnam [18]. As such, these countries with the further harnessing of renewable energy resources can significantly reduce greenhouse gas emissions. On the other hand, Singapore being a highly urbanized and energy-intensive nation (Figure 1a), aims at reducing its emissions by 7%–11% by 2020 using urban renewable energy applications [19]. The environmental quality in many large ASEAN cities has also considerably declined, which is alarming, and thus requires the development of appropriate energy strategies.

The leading contributors of greenhouse gas emissions in the ASEAN counties are the cement industry, palm oil, and chemical plants, petroleum refineries, power generating plants, and wood-based industries [20]. In general, almost 90% of the energy requirements of ASEAN economies are fulfilled by fossil fuels [21]. The electricity generation using these fossil fuels is one of the key processes which results in GHG emissions. Apart from urbanization, rapid industrialization over the past decades has also caused enormous demands and consumption of electricity which, being produced from fossil fuels, has caused an environmental imbalance. Energy consumption in the selected ASEAN countries of this study grew by 261 Million Tons of Oil Equivalent (MTO) throughout 1990–2013, and the majority of this consumption was attributed to Indonesia, Malaysia, Thailand, and Vietnam [22]. The leading contributors for CO_2 emissions among ASEAN countries include Malaysia and Singapore for their rapid economic growth. It has been documented that the air quality in the capital cities of ASEAN countries, such as Bangkok, Jakarta, Kuala Lumpur, and Manila has severally deteriorated and is very poor, which is unhealthy for human beings [23]. This poor air quality in the environment can be attributed to the booming industrialization in the ASEAN countries, causing irreversible environmental damage. It is quite evident that the major industrial processes utilize non-renewable fossil fuels as their primary energy source, releasing a high amount of greenhouse gases (carbon dioxide, methane, nitrous oxide, etc.) in the surrounding environment. Moreover, some industries, such as palm oil production plants, release highly hazardous and toxic effluents in the nearby water bodies, considerably affecting the water quality and its inhabitant's survival. It is, therefore, essential to investigate and develop innovative techniques which suggest harnessing renewable and environmentally friendly energy fuels on a large scale. This study employed panel cointegration and causal relationships among renewable energy production, CO_2 emissions, and economic growth for seven key ASEAN countries for the period of 1995–2016. The recently developed Fully Modified Ordinary Least Square (FMOLS) testing approach and Granger causality tests were used to achieve the objective of this study.

The following section of the paper further provides an insight into renewable energy production, CO_2 emissions, and economic relationship analysis under this study for the selected ASEAN countries of this study.

2. Review of the Literature

The fourth industrial revolution has transformed various aspects of the industry, such as super-computing, cellular phones, robot intelligence, self-driving cars, genetic editing, and much more happening around us at an exponential speed. It is apparent that all fields of knowledge will be further revolutionized by future technologies in the next 50 years. The energy sector, therefore, has also somewhat adjusted to these changes and requires essential efforts to transform and diversify energy sources to increase energy security and reliability while reducing GHG emissions. Conservation of energy is also one of the essential aspects that the economies need to undertake to cope with climate change. In particular, with respect to mitigation efforts to contain climate change, there has been increased attention on improving energy efficiency worldwide, which can effectively reduce the energy demand [24,25]. As such, with the conservation measures realized, the consumption of energy in the future is likely to lower at 1.3% compared to 2.2% per annum during 1995–2015 [26]. However, China and India are the leading, fast-growing, and emerging economies have now almost accounted for over half of the increased global energy demand [27]. More energy will also be required for the economic growth of developing countries. It is anticipated that electricity generation capacity will grow to 90% by 2040, and around 80% of this will be from developing countries [28]. Over the next 20 years, it is also expected that energy supply growth will be sufficiently contributed by the renewable source together with nuclear and hydroelectric power. This anticipated situation is likely to help in offsetting GHG emissions. In 2016, the top CO_2 emitters were China, India, Japan, Russia, and the United States, whereas the European nations were the top contributors to renewable power generation. However, the pace of the transition to a lower-carbon economy is insignificant and uncertain without key initiatives by leading GHG emitting countries in the future. It is anticipated that the speeding up of the said transition will also significantly impact future economic growth. It is, therefore, expected that by 2035, non-fossil fuels and renewables will form a greater share of the overall rising energy mix. It is also anticipated that natural gas will grow twice the rate of oil and coal in the future, but the demand could be slower if coal consumption is not prioritized appropriately [29].

Various studies [30,31] have been alarmed that GHGs and, in particular, CO_2 adversely impact human activities and thus, hamper the natural ecosystem globally. Climate change and energy problems have deeply threatened mankind and their sustainable existence. Globally, the intensity of primary energy declined continuously by 30% between 1990 and 2014. Nevertheless, the global economic growth was better, resulting in a steady net growth in energy demand of 56% between 1990 and 2014 with an annual growth of 1.9% [32]. In some recent studies, the relationship between renewable energy consumption and economic growth has been investigated. In studies which examined the energy consumption of EU countries throughout 1992–2010, [33,34], using the residual cointegration test, it was established that, on average, a 1% increase in energy consumption per capita increased per capita emissions by 0.56%. A group of studies in the literature has also prioritized the economic growth and environmental pollution nexus under the EKC framework. The use of the EKC hypothesis explains the relationships between natural resources, economic growth, and pollutant emissions. Kuznets [35] hypothesized that income inequality increases to a certain level and then falls with an increased income per capita. Various studies have focused on validating the EKC hypothesis (see for example [36–38]). The main focuses of these studies have been to validate the EKC hypothesis. However, there are other groups of studies in the literature which contradict the EKC hypothesis (see for example [39,40]). There are some other studies which validate the EKC hypothesis, too [33,41,42]. Based on the contradiction of the confirmation of the EKC hypothesis, this study contributes to validating the EKC hypothesis by examining broader literature in detail. For a review of the extensive literature on the causal relationship between energy consumption and economic growth see [20,43,44]. A summary of studies which employed the empirical linkage of CO_2 emissions, economic growth, and the energy consumption is provided in (Table 1) with the econometric method utilized by each of these studies. A review of these studies reveals that there has been a variance in the usage of the specific econometric method. Regarding the findings, some studies suggested that CO_2 emissions had a negative impact on the GDP, whereas energy consumption increased the CO_2 emissions [45].

Table 1. Summary of empirical relationships among CO_2 emissions, energy consumption, and economic growth from 2009 to 2015.

Authors	Variables	Methodology	Countries	Findings
Saidi and Hammami [46]	CO_2 emissions, energy consumption, and economic growth	Panel data using Generalized Method of moments	58 countries	Energy consumption increased economic growth; CO_2 emissions had a negative impact on economic growth
Chandran and Tang [10]	Transport energy consumption, foreign direct investment, and CO_2 emissions	Cointegration and Granger causality	Five ASEAN countries	CO_2 emissions were co-integrated only in Malaysia, Indonesia, and Thailand; economic growth contributed to CO_2 emissions
Jalil and Feridun [32]	Growth, energy and financial development, and CO_2 emissions	Autoregressive Distributed Lag (ARDL)	China	Income, trade openness, and energy consumption were able to determine CO_2 emissions
Narayan and Narayan [47]	CO_2 emissions and economic growth	Panel data	Developing countries	Individual countries showed that CO_2 emissions had fallen over the long run
Sharma [48]	Trade openness, per capita GDP, and energy consumption	Dynamic panel modeling	69 countries	Trade openness, per capita GDP, and energy consumption had positive effects on CO_2 emissions
Jaunky [49]	CO_2 emissions and income	VECM	Rich countries	Unidirectional causality from the real per capita GDP to per capita CO_2 emissions
Salahuddin, Gow [50]	Economic growth, electricity consumption, CO_2 emissions, and financial development	Dynamic OLS, fully modified OLS and dynamic fixed effect model	Gulf countries	Electricity consumption and economic growth stimulated CO_2 emissions; there was no causal link between financial development and CO_2 emissions
Ozturk and Acaravci [51]	CO_2 emissions, energy consumption, and economic growth	ARDL Cointegration test	Europe	There was a positive long-run elasticity between CO_2 emissions and economic growth
Leit [52]	Economic growth, CO_2 emissions, renewable energy, and globalization	GMM, Granger causality and ECM	Portugal	CO_2 emissions and renewable energy were positively related to economic growth
Saboori and Sulaiman [53]	CO_2 emissions, energy consumption, and economic growth	ARDL and VECM	ASEAN	There was a nonlinear relationship between CO_2 emissions and economic growth; bi-directional Granger causality between energy consumption and CO_2 emissions
Sadorsky [54]	Renewable energy consumption, CO_2 emissions, and oil prices	Cointegration	G7 countries	GDP and CO_2 emissions were major drivers of renewable energy consumption

It was apparent from the review of the literature, wherein the complex relationships among CO_2 emissions, economic growth, energy consumption, and other factors had been studied, that the evidence is still inconclusive. Furthermore, the evidence of the pollution heaven hypothesis was also limited. However, limited research efforts were the only evidence about the promotion of renewable energy sources as alternatives to fossil fuels to reduce CO_2 emissions and achieve economic growth. Finally, most of these studies utilized the panel data analysis and lack usage of extensive evidence on long-run relationships among the investigated variables.

3. Methodology

In the context of exploring the relationships among renewable energy production, CO_2 emissions, and the associated economic growth, this study employed the FMOLS model, [55], DOLS regression analysis [56], and Granger causality analysis, among others, using renewable energy production, CO_2 emissions, and economic growth as variables in the empirical model.

3.1. Empirical Model

The empirical model developed in this study examined the relationships among renewable energy production (RW_t), CO_2 emissions (CO_{2t}), and economic growth (G_t) for seven ASEAN countries: Indonesia, Malaysia, Myanmar, the Philippines, Singapore, Thailand, and Vietnam. The methodology employed for the model development was similar to earlier studies [10,11,33]. However, the parameters of this study differed from those studies.

It is an essential and well-realized fact that an effective economic policy formulation requires a better understanding of various interconnections in the economy, including the nexus among renewable energy production (dependent variable), CO_2 emissions, and economic growth. The empirical model representing the long-run relationships among these interconnections as variables is given in Equation (1) as follows:

$$RW_{it} = \pi_{0t} + \pi_1 CO_{2it} + \pi_2 GDP_{it} + \varepsilon_{it}. \tag{1}$$

Here, RW_t was the renewable energy production at time t, and for the specific country it was represented by i; CO_{2it} was the carbon dioxide emission in kilotons at time t, and for the specific country it was represented by i; GDP_{it} was the real GDP per capita at time t and for the specific country it was represented as i; and ε_{it} was the residual at time t and for the specific country it was represented as i. It was assumed that these variables were normally distributed. The long-term beta coefficient for CO_2 emissions and economic growth were π_1 and π_2 respectively. Since the study was based on panel data, as such, the long-run panel cointegration specifications applied in previous studies [57,58] were employed. The error term ε_t was assumed to be identically distributed. Finally, the signs for π_1 and π_2 were expected to be negative.

3.2. Econometric Methodology

The determination of the long-run relationships among renewable energy production, CO_2 emissions, and economic growth was undertaken in this study with the essential steps described as follows. Firstly, the stationarity properties of the panel data set variables were examined using panel unit root tests. When the data was non-stationary, the panel cointegration technique was generally used to test the co-integrating relationships in the variable series. Once the cointegration of the variables was confirmed, the long-run elasticities were estimated using the Fully Modified OLS (also known as FMOLS) test. In the third and final steps, the short- and long-run dynamics of the series were examined using panel error correction models.

3.2.1. Panel Unit Root Tests

The panel unit root tests were performed to analyze the stationarity variable with the null hypothesis of a series having a unit root. It was imperative to detect the issue of spurious correlations.

The series was assumed using the intercept, constant, and trend. The equations for the series are provided in the following relationships in Equations (2)–(9) as follows:

Without constant and trend

$$\Delta Y_{it} = \delta Y_{it-1} + \mu_{it}. \tag{2}$$

With Constant

$$\Delta Y_{it} = \alpha + \delta Y_{it-1} + \mu_{it}. \tag{3}$$

With constant and trend

$$\Delta Y_{it} = \alpha + \beta T + \delta Y_{it-1} + \mu_{it}. \tag{4}$$

$H_0: = \delta = 0$ (Unit Root)

$H_1: = \delta \neq 0$

After the first differencing (Δ) of the series, the stationarity had to be achieved [59]. The most commonly used unit root tests, which are currently in practice, are Levin and Lin (LL), Im–Pesaran–Shin (IPS) and Maddala Wu (MW) [60]. Of the three popular unit root tests, LL has not been used widely in practice due to the unrealistic nature of the hypothesis. The model developed by Im, Pesaran [61] was used in this study as below:

$$y_{i,t} = \beta_i + \gamma_i y_{i,t-1} + \varepsilon_{i,t}. \tag{5}$$

For all $i = 1, \ldots , N$ and $t = 1, 2, 3, \ldots , T$

The study performed four different unit root tests; namely, the Phillips–Perron (PP), Augmented Dicker Fuller (ADF), Levin Liu Chu (LLC) and Im–Pesaran–Shin (IPS). The IPS and LLC were used to complement the widely used ADF and PP tests to arrive at the robust results. Further, in order to choose the optimal lag, the study used the Barlett Kernel Method following [62] for the estimation.

3.2.2. FMOLS Estimator

The Fully Modified Least Square (FMOLS) was developed by Phillips and Hansen [63] in order to administer an optimal co-integrating regression estimation. However, the study used the Pedroni [64] heterogeneous FMOLS estimator for the panel cointegration regression as it has the advantage of correcting endogeneity bias and serial correlation [44]. According to Hamit-Haggar [65], FMOLS is the most suitable technique for the panel, which includes heterogeneous cointegration.

Considering that a panel FMOLS estimator for the coefficient β of model 1 was:

$$\beta_{NT}^* - \beta = \left(\sum_{i=1}^{N} L_{22i}^{-2} \sum_{i=1}^{T} (\chi_{it} - \overline{\chi}_{it})^2 \right) \sum_{i=1}^{N} L_{11i}^{-1} L_{22i}^{-1} \left(\sum_{i=1}^{T} (\chi_{it} - \overline{\chi_i}) \mu_{it}^* - T\hat{\gamma}_i \right), \tag{6}$$

where,

$$\mu_{it}^* = \mu_{it} - \frac{\hat{L}_{21i}}{\hat{L}_{22i}} \Delta \chi_{it}, \ \hat{\gamma}_i = \hat{\Gamma}_{21i} \hat{\Omega}_{21i}^0 - \frac{\hat{L}_{21i}}{\hat{L}_{22i}} (\Gamma_{22i} + \hat{\Omega}_{22i}^0),$$

and $\hat{L}_i$ was the lower triangulation of $\hat{\Omega}_i$.

The Dynamic OLS estimator had the same asymptotic distribution as that of the panel FMOLS estimation derived by Pedroni [66]. Both the DOLS and FMOLS estimations were performed as shown to confirm the consistency of the outcome.

3.2.3. Granger Causality Test

According to the cointegration theory, if the variables are co-integrated, then the short- and long-run equilibrium can be described using an error correction model (ECM). The panel residual cointegration test confirmed that renewable energy production, CO_2 emissions, and economic growth had co-integrating relationships. However, the co-integrating relationships were unable to provide information on the direction. Therefore, a panel-based error correction model with error correction representation was used to investigate the short- and long-run causal relationships. The Granger causality test was performed within the Vector Error Correction Model (VECM) framework [67]. The Granger causality test, together with the error correction term (ECT), is stated as follows:

$$\Delta RW_{it} = \varnothing_0 + \sum_{i=1}^{p} \varnothing_{1i}\Delta CO_{2t-1} + \sum_{i=1}^{p} \varnothing_{2i}\Delta GDP_{t-1} + \sum_{i=1}^{p} \varnothing_{3i}\Delta RW_{t-1} + \rho_1 \epsilon_{t-1} + \mu_{1t} \tag{7}$$

$$\Delta CO_{2it} = \varnothing_0 + \sum_{i=1}^{p} \varnothing_{1i}\Delta CO_{2t-1} + \sum_{i=1}^{p} \varnothing_{2i}\Delta GDP_{t-1} + \sum_{i=1}^{p} \varnothing_{3i}\Delta RW_{t-1} + \rho_2 \epsilon_{t-1} + \mu_{2t}, \tag{8}$$

$$\Delta GDP_{it} = \varnothing_0 + \sum_{i=1}^{p} \varnothing_{1i}\Delta CO_{2t-1} + \sum_{i=1}^{p} \varnothing_{2i}\Delta GDP_{t-1} + \sum_{i=1}^{p} \varnothing_{3i}\Delta RW_{t-1} + \rho_3 \epsilon_{t-1} + \mu_{3t}, \tag{9}$$

where, Δ and ρ denote the first difference operator and lag structure. The residuals ($\mu_1, \mu_2, \mu_3, \mu_4$ and μ_5) were assumed to be serially independent with a zero mean and ϵ_{t-1} was the one period lagged error correction term.

4. Statistical Results

The study used the dataset over the period from 1995 to 2016 extracted from the World Development Indicators (WDI). A total of seven ASEAN countries were selected for the study.

Table 2 provides a summary of the statistics (mean, standard deviation, minimum, and maximum) associated with the CO_2 emissions per capita for an individual country with the panel set throughout 1995–2016. The mean value for the CO_2 emissions per capita was in the range of between 0.2362 in Myanmar and 9.8729 in Singapore. As regarding the GDP per capita, Myanmar showed the least GDP mean value of 710.8, whereas Singapore achieved the highest GDP per capita of 40,232.7. However, concerning renewable energy production, Myanmar exhibited surplus production with a mean value of 79.87%, whereas Singapore was the least renewable energy producer with only 0.51%.

Table 2. Descriptive statistics for CO_2 emissions, real GDP, and renewable energy production.

Countries	Mean	S.D.	Minimum	Maximum
Panel A: CO_2 emissions per capita				
Malaysia	6.713767	1.08802	4.763997	8.530658
Myanmar	0.236277	0.063911	0.16097	0.4166
Thailand	3.610162	0.590426	2.668098	4.62186
Vietnam	1.208286	0.539194	0.404057	2.205355
Indonesia	1.624109	0.401152	1.041246	2.55975
Philippines	0.922408	0.103728	0.770906	1.235657
Singapore	9.872934	2.743237	4.342606	15.39236
Panel A: Real GDP per capita				
Malaysia	2.989936	3.881041	−9.65575	7.445581
Myanmar	8.605125	2.989286	4.274814	12.68542
Thailand	2.748366	3.755799	−8.734046	7.047772
Vietnam	5.348793	1.105453	3.213679	7.758815
Indonesia	5.348793	1.105453	3.213679	7.758815
Philippines	2.904749	2.096575	−2.74496	5.903121
Singapore	3.004872	4.330412	−5.491059	13.21649
Panel A: Renewable energy production				
Malaysia	5.638253	1.340974	3.819042	9.029613
Myanmar	79.87185	6.997041	60.4532	86.11957
Thailand	21.88286	1.305616	20.02467	24.0143
Vietnam	47.06246	10.44879	34.7959	65.12578
Indonesia	42.38311	3.78953	38.06614	50.09815
Philippines	31.93819	2.784079	28.0034	38.942
Singapore	0.517375	0.088952	0.325119	0.715843

Note: S.D., CO_2, GDP stand for standard deviation, per capita carbon dioxide emissions, and per capita real GDP. Data period was 1995–2016 for the ASEAN countries.

It was further determined that the mean value of renewable energy production was high in low-income countries, such as Myanmar, followed by middle- and high-income countries. For both the energy variables (CO_2 emissions and renewable energy production), it can be conferred that the low-income countries were more volatile with the highest coefficients of variations.

This study investigated the dynamic relationships among renewable energy production, CO_2 emissions, and economic growth over the period of 1995–2016 from a panel of seven ASEAN countries. In this context, the EKC hypothesis was employed using panel cointegration estimation methods. In the first step, to tackle homogeneity issues, the study used the unit root test as suggested by [61] for a cointegration modeling process. Once it was confirmed that there was no homogeneity issue and the variables were in the order of interest, a test of the cointegration was examined to know whether the variables were co-integrated or not [58].

The panel unit root tests undertaken in this study are provided in Table 3. The validity of the EKC hypothesis for the ASEAN countries was examined using a bivariate framework. In [68] it was argued that testing for cointegration is crucial to determine the appropriateness of the model, together with verifying the causal relationships. The study employed the Augmented Dickey Fuller (ADF) unit root test to check the integration of each series. Table 3 reports the results of the ADF test at the level and first difference. At the 1% significance level, the study found that there was no stationarity issue for any of the variables. Thus, the study proceeded to examine the presence of cointegration among renewable energy production, CO_2 emissions, and economic growth. The results also indicated that the series for each country was not spurious and had a unit root. Thus, it can be concluded that the panel data variables were characterized as I (1) process.

Table 3. Panel unit root test results for ASEAN countries.

| | | Unit Root Methods | | | |
| | | PP | | ADF | |
Country	Variables	Level	First Difference	Level	First Difference
	Renewable	−3.004	−3.885 ***	−1.404	−3.815 ***
Malaysia	CO_2 emissions	0.151	−5.004 ***	0.159	−3.644 ***
	GDP	−8.489 ***	−12.199 ***	−5.327 ***	−6.260 ***
	Renewable	0.846	−2.801 **	1.500	−2.778 ***
Myanmar	CO_2 emissions	−1.599	−8.669 ***	−1.802	−6.060 ***
	GDP	−1.476	−3.675 ***	−1.433	−3.719 ***
	Renewable	−0.731	−5.439 ***	−0.733	−5.405 ***
Thailand	CO_2 emissions	−1.509	−4.928 ***	−1.509	−4.925 ***
	GDP	−3.727 ***	−9.767 ***	−3.721 ***	−5.707 ***
	Renewable	−1.829	−3.219 **	−1.932	−3.219 **
Vietnam	CO_2 emissions	1.308	−4.078 ***	0.575	−6.373 ***
	GDP	−3.039 **	−3.863 ***	−2.961 *	−3.857 ***
	Renewable	−3.245 **	−6.827 ***	−2.704 **	−6.827 ***
Indonesia	CO_2 emissions	−1.222	−7.289 ***	−1.427	−4.845 ***
	GDP	−3.039 **	−3.863 ***	−2.961 *	−3.857 ***
	Renewable	−2.577	−6.491 ***	−2.500	−5.772 ***
Philippines	CO_2 emissions	1.461	−3.720 ***	1.143	−3.857 ***
	GDP	−3.399 **	−14.868 ***	−3.399 **	−5.788 ***
	Renewable	−1.206	−5.487 ***	−1.308	−5.487 ***
Singapore	CO_2 emissions	−1.717	−5.414 ***	−1.792	−4.871 ***
	GDP	−7.688 ***	−13.506 ***	−4.692 ***	−6.483 ***

Note: *** denotes the significance level at 1%. Δ stands for first difference, ADF—Augmented Dicker Fuller.

The residual cointegration results are provided in Table 4 as suggested by [64,69]. The statistical results found that the majority of the tests were significant and, therefore, the null hypothesis of having no cointegration was not established. Thus, the variables were co-integrated at significant levels.

Table 4. Residual cointegration test.

Statistics	Within Dimensions		Statistics	Between Dimensions	
	Value	*p*-Value		Value	*p*-Value
Panel v-Statistic	1.264 *	0.006			
Panel rho-Statistic	−1.421	0.922	Group rho-Statistic	2.534	0.994
Panel PP-Statistic	−4.961 *	0.000	Group PP-Statistic	−1.985 **	0.023
Panel ADF-Statistic	−5.060 *	0.000	Group ADF-Statistic	−2.323 **	0.010

Note: the selection of Lag was based on the Akaike Information Criterion (AIC). The null hypothesis was that the variables were not co-integrated. * and ** denote the significance levels at 1% and 5%, respectively.

Finally, the Pedroni [64] FMOLS estimator was selected to determine the long-run relationships among the constructs. The long-run results of the FMOLS estimation for the model are provided in (Table 5) is considered in this study.

The long-run elasticities were interpreted for all the variables which were expressed in natural logarithms. The study used the FMOLS as a robustness test. The first and second differences of the variables were all stationary as provided in Table 3. The quadratic term of per capita renewable energy production was used to record a possible country-specific non-linear relationship between per capita CO_2 emissions and per-capita GDP. The main result confirmed the significant inverse relationship between CO_2 emissions and real GDP for countries like Malaysia and Singapore; whereas, there was a positive relationship between CO_2 emissions and real GDP for Myanmar. This result indicated that in the long run, the EKC hypothesis was justified. An increase of 1% in CO_2 emissions would lead to a decrease in real GDP per capita by 0.08% for Malaysia and 0.11% for Singapore. The results of this study were duly in line with the other literature findings (see for example [1,70]). The results further indicated that the coefficients of renewable energy production were statistically significant and negative, indicating a nonlinear relationship with per capita CO_2 emissions and per capita GDP in the sampled ASEAN countries considered in this study.

Table 5. Long-run elasticity results (dependent variable: renewable energy production).

FMOLS Estimation			DOLS Estimation		
Country	GDP	CO$_2$ Emissions	Country	GDP	CO$_2$ Emissions
Malaysia	1.653 [b] (2.050)	−0.593 [b] (−0.912)	Malaysia	2.006 [b] (2.684)	−1.359 [b] (−2.232)
Myanmar	0.092 [b] (2.845)	−0.126 [a] (−1.845)	Myanmar	0.064 [b] (4.678)	−0.027 [a] (−0.723)
Thailand	0.549 (−2.001)	−0.393 (−2.505)	Thailand	0.337 (−1.183)	−0.699 (−2.504)
Vietnam	0.193 [a] (−0.37)	−0.579 (−2.990)	Vietnam	1.049 [a] (−1.639)	−1.164 (−5.833)
Indonesia	−0.251 [a] (−1.254)	−0.217 (−2.515)	Indonesia	−0.485 [a] (−2.290)	−0.092 (−0.394)
Philippines	−0.392 (−2.710)	−0.042 (−0.703)	Philippines	0.592 (−2.031)	−0.288 (−3.332)
Singapore	0.18 (0.233)	−0.163 [b] (−0.904)	Singapore	−2.856 (−2.164)	−0.619 [b] (−2.638)

Notes: [a], [b], and [c] indicate the significance levels at 1%, 5%, and 10%, respectively. Intercepts and linear trends are included in the regressions. Barlett Kernel with a fixed bandwidth of 6 was used following [71]. Figures in the parentheses are the t-statistics. FMOLS—fully modified ordinary least square; DOLS—dynamic ordinary least square; GDP—gross domestic product.

The inverse U-shaped relationships between per capita renewable energy production and per capita GDP were detected from the sign of the parameters. However, the long-run estimations provided in Table 5 did not provide the direction of causality among the variables. The results also indicated that there were negative relationships between per capita renewable energy production and per capita CO_2 emissions for Indonesia, Malaysia, Myanmar, and Vietnam. These results were also supported by the previous literature [46,72]. In the globalization era, with the rapid increase in demand for energy and cleanup of the environment, there is a need for renewable energy sources as an alternative [47].

Next, by employing the FMOLS estimator, proposed by Pedroni [64], and the single equation DOLS estimator, proposed by Phillips and Loretan [73], the U-shaped relationship among renewable energy production, CO_2 emissions, and economic growth was supported and realized by the EKC hypothesis. The FMOLS and DOLS results provided in Table 5 show the parameter estimation of the model for interpreting long-run elasticities. Hence, it can be stated that renewable energy production decreased with economic growth, stabilized, and then, gradually increased. The country-specific income-energy elasticities using FMOLS and DOLS confirmed the statistical significance at 5% for the long-run relationships among the variables. Furthermore, the country-specific long-run analysis demonstrated U-shaped curves for three countries, namely, Malaysia, Thailand, and the Philippines. On the other hand, there was no significant relationship found between renewable energy production and economic growth for the four remaining ASEAN countries of this study.

This research also analyzed the results using the Granger causality panel error correction model to identify the direction of the long-run and short-run causalities in the renewable energy production, CO_2 emissions, and economic growth nexus and the interactions among them. The error correction model estimation result indicated the presence of unidirectional causality between renewable energy production and CO_2 emissions. However, in the applied research, it is obviously of interest to know the response of one variable to an impulse in another variable [74]. From the Granger causality, it was noticed that there were high but negative correlations among renewable energy production, CO_2 emissions, and economic growth. This indicated that innovation in renewable energy production would possess high effects on CO_2 emissions and economic growth. As such, increasing renewable energy production would gradually decrease the CO_2 emissions together with slowly contributing to economic growth in the long run. In the initial stages of renewable energy production, the minor pressure on the GDP may be explained by the additional production cost imposed by the renewable energy sources and their plant setups.

The study explored the causal relationship between the constructs using error correction Granger Causality test for short run Granger causality. The results for Granger causality model (see Table 6) can be summarized as:

(i) Real GDP per capita does not cause carbon emissions per capita for Myanmar, Thailand, Vietnam, Indonesia, the Philippines, and Singapore except for Malaysia.

(ii) It was found that renewable energy production Granger causes real GDP per capita for Myanmar, Indonesia, and the Philippines only; whereas it was also found that renewable energy production Granger causes carbon emissions per capita for Myanmar and Vietnam only.

(iii) There is no causal evidence between carbon emissions per capita and real GDP per capita for all the ASEAN countries; whereas only in Thailand, there was a causal relationship between carbon emissions and renewable energy production.

Table 6. Results of panel Granger causality test identifying long-run relationships between CO_2 emissions, real GDP, and renewable energy production for ASEAN countries.

				Short-Run Granger Causality					
Country					**Country**				
Malaysia	Variable	ΔGDP	ΔRNW	ΔCO_2	Indonesia	Variable	ΔGDP	ΔRNW	ΔCO_2
	ΔGDP	1	0.455	4.229 **		ΔGDP	1	10.30 **	1.177
	ΔRNW	0.252	1	0.937		ΔRNW	4.367 *	1	0.902
	ΔCO_2	0.358	1.083	1		ΔCO_2	0.731	0.849	1
Myanmar	Variable	ΔGDP	ΔRNW	ΔCO_2	Philippines	Variable	ΔGDP	ΔRNW	ΔCO_2
	ΔGDP	1	0.635	0.417		ΔGDP	1	1.220	2.115
	ΔRNW	22.83 **	1	3.055 *		ΔRNW	4.590 *	1	2.627
	ΔCO_2	0.788	2.092	1		ΔCO_2	1.213	1.064	1
Thailand	Variable	ΔGDP	ΔRNW	ΔCO_2	Singapore	Variable	ΔGDP	ΔRNW	ΔCO_2
	ΔGDP	1	2.537	0.583		ΔGDP	1	1.663	1.745
	ΔRNW	1.102	1	0.691		ΔRNW	0.124	1	0.285
	ΔCO_2	1.921	21.59 **	1		ΔCO_2	2.988	0.453	1
Vietnam	Variable	ΔGDP	ΔRNW	ΔCO_2					
	ΔGDP	1	20.61 **	1.068					
	ΔRNW	1.359	1	3.319 *					
	ΔCO_2	1.190	0.121	1					

Note: The t-statistics are provided in the parentheses, respectively. *, **, and *** denote the significance levels at 1%, 5%, and 10%, respectively. Δ denotes change; GDP stands for the gross domestic product; RNW stands for renewable; CO_2 stands for carbon emissions.

Therefore, from the overall results, it can be concluded that (1) energy conservation policies have no adverse effect on climate change and GDP growth of the ASEAN countries and (2) controlling carbon emissions has no adverse effect on real GDP per capita of the ASEAN countries. Figure 2 summarizes the interactions among renewable energy production, CO_2 emissions, and economic growth developed from the findings provided in Table 6.

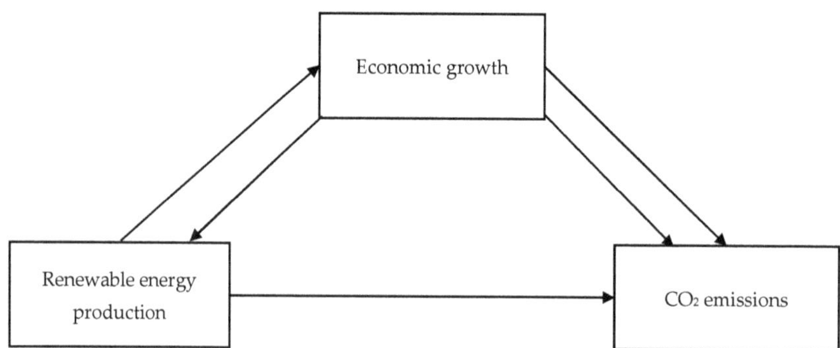

Figure 2. The long-run relationship between renewable energy production, CO_2, and economic growth for ASEAN countries.

The results corroborated the three-way link among renewable energy production, CO_2 emissions, and economic growth. Our results suggest that an increase in renewable energy usage and production will diminish CO_2 emissions, leading to clean economic growth as per the Granger causality test. Our result rejects the neoclassical assumption that energy is neutral to economic growth. Thus, this study considers renewable energy to be the most critical determinant in the long run to economic growth and vice versa. It is very crucial to take into account the adverse effect of CO_2 energy consumption on economic growth in establishing energy conservation policies.

5. Conclusions and Policy Implications

This study examined the cointegration and causal relationships among renewable energy production, CO_2 emissions, and economic growth for the ASEAN countries, namely, Malaysia, Myanmar, Thailand, Vietnam, Indonesia, the Philippines, and Singapore for the period 1995–2016. The FMOLS and DOLS analyses were undertaken to examine whether, in the long run, renewable energy production, CO_2 emissions, and economic growth were co-integrated or not. The FMOLS results suggest that renewable energy production was positive and significant in relation to real GDP at the 0.001 level for Malaysia (t value of 2.050) and for Myanmar (t value of 2.845), whereas there was a negative and significant relationship between renewable energy production and real GDP for Thailand (t value of −2.505) and the Philippines (t value of −2.710). There was no relationship between renewable energy production and real GDP for Singapore, Indonesia, and Vietnam due to limited renewable energy options. The results of the panel FMOLS and DOLS also provided significant shreds of evidence to support U-Shaped relationships among the long-run renewable energy production, CO_2 emissions, and economic growth contrary to the EKC hypothesis. The empirical results from the granger causality test (Table 6) also revealed the substantial negative nexus between renewable energy production and CO_2 emissions.

Estimation of a panel error correction model (ECM) was performed in order to define the long-run and short-run causalities. The results from the ECM indicated the presence of short-run unidirectional causality running from renewable energy production to economic growth. Whereas, in the long-run causalities, renewable energy production and CO_2 emissions were inversely related, thus confirming that an increase in the usage of renewable energy sources as an alternative to fossil fuels will offset CO_2 emissions. Besides, carbon footprints can be reduced, leading to clean energy development and economic growth. The results, therefore, implied that the efficient use of renewable energy would reduce the rate of GHG emissions in the long run.

The precedent literature, as presented in the earlier sections of the study, have generally found a positive relationship between energy consumption and CO_2 emissions; however, there is a lack of studies which performed the relationship analysis of renewable energy production and CO_2 emissions. As a novelty, this research gap was addressed and found the overall relationship between renewable energy production and CO_2 emissions, which came to be negative. The relationship between renewable energy production and CO_2 emissions for Malaysia, Thailand, Vietnam, Indonesia, and Singapore were found to be inverse in the short run. This showed that an increase in renewable energy production would lead to a reduction in CO_2 emissions. Even though, renewable energy is expensive as compared to fossil fuels or other non-renewable sources, the initiatives to produce and consume more renewable energy should be focused on in developed and developing countries, like in Singapore and Malaysia, to contain climate change by reducing environmental pollution.

The rapid growth in the renewable energy production can be driven by many factors, including cost reduction due of renewable energy technologies, dedicated policy initiatives, better access to energy resources, environmental concerns, and growing energy demand of emerging economies. Effective energy policies, therefore, can play a significant role in providing energy consumption direction and in increasing public awareness. The pedagogical innovation, therefore, suggests accelerating the renewable energy usage campaigns in the high carbon emission countries like Malaysia and Singapore for a more environmentally-friendly and sustainable future. The renewable energy sources like hydropower, solar power, wind energy, biomass, and geothermal energy would significantly condense the pollution and reduce the reliance on fossil fuels, which are the leading causes of the CO_2 emissions.

As a practical implication, energy saving and conservation policies can be adopted by the ASEAN countries quickly in order to limit environmental pollution. It is crucial that the ASEAN countries, especially Malaysia and Singapore, implement policies and strategies that ensure continuous economic growth without forsaking the environment. This study recommends that in order to decrease the reliance on fossil fuels, that have impacted the environment adversely, future research should consider and focus on the principles of the circular economy and clean energy development mechanisms integrated with renewable energy technologies.

Author Contributions: All authors contributed to this study. M.W.A.K. and S.K.P. formulated the study design. M.W.A.K., S.K.P., M.I.S. and N.H.M. conceived and designed the research methodology. M.W.A.K., S.K.P., K.S.N.A., M.I.S. and E.S.A. collected and analyzed the data. M.W.A.K., S.K.P. and N.H.M. finalized the paper.

Funding: The authors would like to acknowledge the financial support by Universiti Malaysia Pahang under the FRGS Grant, i.e., RDU170134, for this research work.

Acknowledgments: The authors highly acknowledge the Ministry of Higher Education Malaysia for the Fundamental Research Grant Scheme (FRGS) under grant No. RDU170134 and Universiti Malaysia Pahang for their financial and technical support for completing this research work.

Conflicts of Interest: The authors declare no conflict of interest.

References

1. Heidari, H.; Katircioglu, S.T.; Saeidpour, L. Economic growth, CO_2 emissions, and energy consumption in the five ASEAN countries. *Int. J. Electr. Power Energy Syst.* **2015**, *64*, 785–791. [CrossRef]

2. Zhang, X.-P.; Cheng, X.-M. Energy consumption, carbon emissions, and economic growth in China. *Ecol. Econ.* **2009**, *68*, 2706–2712. [CrossRef]

3. Mengal, A.; Mirjat, N.H.; Das Walasai, G.; Khatri, S.A.; Harijan, K.; Uqaili, M.A. Modeling of Future Electricity Generation and Emissions Assessment for Pakistan. *Processes* **2019**, *7*, 212. [CrossRef]

4. Lean, H.H.; Smyth, R. CO_2 emissions, electricity consumption and output in ASEAN. *Appl. Energy* **2010**, *87*, 1858–1864. [CrossRef]

5. Chang, C.-C. A multivariate causality test of carbon dioxide emissions, energy consumption and economic growth in China. *Appl. Energy* **2010**, *87*, 3533–3537. [CrossRef]

6. Dhakal, S. Urban energy use and carbon emissions from cities in China and policy implications. *Energy Policy* **2009**, *37*, 4208–4219. [CrossRef]

7. Fei, L.; Dong, S.; Xue, L.; Liang, Q.; Yang, W. Energy consumption-economic growth relationship and carbon dioxide emissions in China. *Energy Policy* **2011**, *39*, 568–574. [CrossRef]

8. Hussain, N. *Development of Energy Modeling and Decision Support Framework for Sustainable Electricity System of Pakistan*; Mehran University of Eng. & Technology: Jamshoro, Pakistan, 2019.

9. Panayotou, T. Demystifying the environmental Kuznets curve: Turning a black box into a policy tool. *Environ. Dev. Econ.* **1997**, *2*, 465–484. [CrossRef]

10. Chandran, V.; Tang, C.F. The impacts of transport energy consumption, foreign direct investment and income on CO_2 emissions in ASEAN-5 economies. *Renew. Sustain. Energy Rev.* **2013**, *24*, 445–453. [CrossRef]

11. Saboori, B.; Sulaiman, J. CO_2 emissions, energy consumption and economic growth in Association of Southeast Asian Nations (ASEAN) countries: A cointegration approach. *Energy* **2013**, *55*, 813–822. [CrossRef]

12. Omri, A.; Daly, S.; Rault, C.; Chaibi, A. Financial development, environmental quality, trade and economic growth: What causes what in MENA countries. *Energy Econ.* **2015**, *48*, 242–252. [CrossRef]

13. Vilaysouk, X.; Schandl, H.; Murakami, S. Improving the knowledge base on material flow analysis for Asian developing countries: A case study of Lao PDR. *Resour. Conserv. Recycl.* **2017**, *127*, 179–189. [CrossRef]

14. IMF. Boiling Point. Finance & Development 2018. Available online: https://www.imf.org/external/pubs/ft/fandd/2018/09/southeast-asia-climate-change-and-greenhouse-gas-emissions-prakash.htm (accessed on 15 July 2019).

15. Chan, S.K.-L. Dams on a Myanmar—Thai transboundary river: Unequal hydropower exchange model in critical hydropolitical perspective. *Int. J. Dev. Issues* **2017**, *16*, 147–160. [CrossRef]

16. Amara, T. Where Will Myanmar's Energy Come From? Available online: https://thediplomat.com/2017/08/where-will-myanmars-energy-come-from(accessed on 19 January 2019).

17. Worldbank. World Development Indicators 2016. 2016. Available online: http://documents.worldbank.org/curated/en/805371467990952829/World-development-indicators-2016 (accessed on 19 January 2019).

18. Chang, T.; Fang, W.; Wen, L.-F. Energy consumption, employment, output, and temporal causality: Evidence from Taiwan based on cointegration and error-correction modelling techniques. *Appl. Econ.* **2001**, *33*, 1045–1056. [CrossRef]

19. Karthikeya, B.; Negi, P.S.; Srikanth, N. Wind resource assessment for urban renewable energy application in Singapore. *Renew. Energy* **2016**, *87*, 403–414. [CrossRef]
20. Daniel, G.R.; Wang, C.; Berthelsen, D. Early school-based parent involvement, children's self-regulated learning and academic achievement: An Australian longitudinal study. *Early Child. Res. Q.* **2016**, *36*, 168–177. [CrossRef]
21. Karki, S.K.; Mann, M.D.; Salehfar, H. Energy and environment in the ASEAN: Challenges and opportunities. *Energy Policy* **2005**, *33*, 499–509. [CrossRef]
22. Schernikau, L. *Economics of the International Coal Trade: Why Coal Continues to Power the World*; Springer: Berlin/Heidelberg, Germany, 2017.
23. Lee, H.-H.; Bar-Or, R.Z.; Wang, C. Biomass burning aerosols and the low-visibility events in Southeast Asia. *Atmos. Chem. Phys. Discuss.* **2017**, *17*, 965–980. [CrossRef]
24. Riahi, K.; Van Vuuren, D.P.; Kriegler, E.; Edmonds, J.; O'Neill, B.C.; Fujimori, S.; Bauer, N.; Calvin, K.; Dellink, R.; Fricko, O.; et al. The Shared Socioeconomic Pathways and their energy, land use, and greenhouse gas emissions implications: An overview. *Glob. Environ. Chang.* **2017**, *42*, 153–168. [CrossRef]
25. Mirjat, N.H.; Uqaili, M.A.; Harijan, K.; Das Walasai, G.; Mondal, M.A.H.; Sahin, H. Long-term electricity demand forecast and supply side scenarios for Pakistan (2015–2050): A LEAP model application for policy analysis. *Energy* **2018**, *165*, 512–526. [CrossRef]
26. Berardi, U. A cross-country comparison of the building energy consumptions and their trends. *Resour. Conserv. Recycl.* **2017**, *123*, 230–241. [CrossRef]
27. Nejat, P.; Jomehzadeh, F.; Taheri, M.M.; Gohari, M.; Majid, M.Z.A. A global review of energy consumption, CO_2 emissions and policy in the residential sector (with an overview of the top ten CO_2 emitting countries). *Renew. Sustain. Energy Rev.* **2015**, *43*, 843–862. [CrossRef]
28. Soytas, U.; Sari, R.; Ewing, B.T. Energy consumption, income, and carbon emissions in the United States. *Ecol. Econ.* **2007**, *62*, 482–489. [CrossRef]
29. Ang, J.B. Economic development, pollutant emissions and energy consumption in Malaysia. *J. Policy Model.* **2008**, *30*, 271–278. [CrossRef]
30. Ozcan, B.; Tzeremes, P.G.; Tzeremes, N.G. Energy consumption, economic growth and environmental degradation in OECD countries. *Econ. Model.* **2019**. [CrossRef]
31. Acheampong, A.O. Economic growth, CO_2 emissions and energy consumption: What causes what and where? *Energy Econ.* **2018**, *74*, 677–692. [CrossRef]
32. Jalil, A.; Feridun, M. The impact of growth, energy and financial development on the environment in China: A cointegration analysis. *Energy Econ.* **2011**, *33*, 284–291. [CrossRef]
33. Kasman, A.; Duman, Y.S. CO_2 emissions, economic growth, energy consumption, trade and urbanization in new EU member and candidate countries: A panel data analysis. *Econ. Model.* **2015**, *44*, 97–103. [CrossRef]
34. Dogan, E.; Seker, F. Determinants of CO_2 emissions in the European Union: The role of renewable and non-renewable energy. *Renew. Energy* **2016**, *94*, 429–439. [CrossRef]
35. Balsalobre-Lorente, D.; Shahbaz, M.; Roubaud, D.; Farhani, S. How economic growth, renewable electricity and natural resources contribute to CO_2 emissions? *Energy Policy* **2018**, *113*, 356–367. [CrossRef]
36. Begum, R.A.; Sohag, K.; Abdullah, S.M.S.; Jaafar, M. CO_2 emissions, energy consumption, economic and population growth in Malaysia. *Renew. Sustain. Energy Rev.* **2015**, *41*, 594–601. [CrossRef]
37. Adu, D.T.; Denkyirah, E.K. Economic growth and environmental pollution in West Africa: Testing the Environmental Kuznets Curve hypothesis. *Kasetsart J. Soc. Sci.* **2018**. [CrossRef]
38. Rasli, A.M.; Qureshi, M.I.; Isah-Chikaji, A.; Zaman, K.; Ahmad, M. New toxics, race to the bottom and revised environmental Kuznets curve: The case of local and global pollutants. *Renew. Sustain. Energy Rev.* **2018**, *81*, 3120–3130. [CrossRef]
39. Ozatac, N.; Gokmenoglu, K.K.; Taspinar, N. Testing the EKC hypothesis by considering trade openness, urbanization, and financial development: The case of Turkey. *Environ. Sci. Pollut. Res.* **2017**, *24*, 16690–16701. [CrossRef] [PubMed]
40. Twerefou, D.K.; Adusah-Poku, F.; Bekoe, W. An empirical examination of the Environmental Kuznets Curve hypothesis for carbon dioxide emissions in Ghana: An ARDL approach. *Environ. Socio-Econ. Stud.* **2016**, *4*, 1–12. [CrossRef]

41. Alam, M.M.; Murad, M.W.; Noman, A.H.M.; Ozturk, I. Relationships among carbon emissions, economic growth, energy consumption and population growth: Testing Environmental Kuznets Curve hypothesis for Brazil, China, India and Indonesia. *Ecol. Indic.* **2016**, *70*, 466–479. [CrossRef]
42. Al-Mulali, U. The impact of biofuel energy consumption on GDP growth, CO_2 emission, agricultural crop prices, and agricultural production. *Int. J. Green Energy* **2015**, *12*, 1100–1106. [CrossRef]
43. Omri, A. CO_2 emissions, energy consumption and economic growth nexus in MENA countries: Evidence from simultaneous equations models. *Energy Econ.* **2013**, *40*, 657–664. [CrossRef]
44. Özcan, B. The nexus between carbon emissions, energy consumption and economic growth in Middle East countries: A panel data analysis. *Energy Policy* **2013**, *62*, 1138–1147. [CrossRef]
45. Mi, Z.-F.; Pan, S.-Y.; Yu, H.; Wei, Y.-M. Potential impacts of industrial structure on energy consumption and CO_2 emission: A case study of Beijing. *J. Clean. Prod.* **2015**, *103*, 455–462. [CrossRef]
46. Saidi, K.; Hammami, S. The impact of CO_2 emissions and economic growth on energy consumption in 58 countries. *Energy Rep.* **2015**, *1*, 62–70. [CrossRef]
47. Narayan, P.K.; Narayan, S. Carbon dioxide emissions and economic growth: Panel data evidence from developing countries. *Energy Policy* **2010**, *38*, 661–666. [CrossRef]
48. Sharma, S.S. Determinants of carbon dioxide emissions: Empirical evidence from 69 countries. *Appl. Energy* **2011**, *88*, 376–382. [CrossRef]
49. Jaunky, V.C. The CO_2 emissions-income nexus: Evidence from rich countries. *Energy Policy* **2011**, *39*, 1228–1240. [CrossRef]
50. Salahuddin, M.; Gow, J.; Ozturk, I. Is the long-run relationship between economic growth, electricity consumption, carbon dioxide emissions and financial development in Gulf Cooperation Council Countries robust? *Renew. Sustain. Energy Rev.* **2015**, *51*, 317–326. [CrossRef]
51. Ozturk, I.; Acaravci, A. CO_2 emissions, energy consumption and economic growth in Turkey. *Renew. Sustain. Energy Rev.* **2010**, *14*, 3220–3225. [CrossRef]
52. Leit, N.C. Economic growth, carbon dioxide emissions, renewable energy and globalization. *Int. J. Energy Econ. Policy* **2014**, *4*, 391.
53. Saboori, B.; Sulaiman, J. Environmental degradation, economic growth and energy consumption: Evidence of the environmental Kuznets curve in Malaysia. *Energy Policy* **2013**, *60*, 892–905. [CrossRef]
54. Sadorsky, P. Renewable energy consumption, CO_2 emissions and oil prices in the G7 countries. *Energy Econ.* **2009**, *31*, 456–462. [CrossRef]
55. Long, X.; Naminse, E.Y.; Du, J.; Zhuang, J. Nonrenewable energy, renewable energy, carbon dioxide emissions and economic growth in China from 1952 to 2012. *Renew. Sustain. Energy Rev.* **2015**, *52*, 680–688. [CrossRef]
56. Sebri, M.; Ben-Salha, O. On the causal dynamics between economic growth, renewable energy consumption, CO_2 emissions and trade openness: Fresh evidence from BRICS countries. *Renew. Sustain. Energy Rev.* **2014**, *39*, 14–23. [CrossRef]
57. Arellano, M.; Bond, S. Some Tests of Specification for Panel Data: Monte Carlo Evidence and an Application to Employment Equations. *Rev. Econ. Stud.* **1991**, *58*, 277. [CrossRef]
58. Panigrahi, S.K. Economic Value Added and Traditional Accounting Measures for Shareholder's Wealth Creation. *Asian J. Account. Gov.* **2017**, *8*, 125–136. [CrossRef]
59. Perron, P. The Great Crash, the Oil Price Shock, and the Unit Root Hypothesis. *Econometrica* **1989**, *57*, 1361. [CrossRef]
60. Hoang, N.T.; McNown, R.F. Panel data unit roots tests using various estimation methods. *Univ. Colo. Bull.* **2006**, *6*, 33–66.
61. Im, K.S.; Pesaran, M.; Shin, Y. Testing for unit roots in heterogeneous panels. *J. Econ.* **2003**, *115*, 53–74. [CrossRef]
62. Newey, W.K.; West, K.D. *A Simple, Positive Semi-Definite, Heteroskedasticity and Autocorrelationconsistent Covariance Matrix*; National Bureau of Economic Research: Cambridge, MA, USA, 1986.
63. Phillips, P.C.B.; Hansen, B.E. Statistical Inference in Instrumental Variables Regression with I(1) Processes. *Rev. Econ. Stud.* **1990**, *57*, 99. [CrossRef]
64. Pedroni, P. Fully modified OLS for heterogeneous cointegrated panels. In *Nonstationary Panels, Panel Cointegration, and Dynamic Panels*; Emerald Group Publishing Limited: Bingley, UK, 2001; pp. 93–130.

65. Hamit-Haggar, M. Greenhouse gas emissions, energy consumption and economic growth: A panel cointegration analysis from Canadian industrial sector perspective. *Energy Econ.* **2012**, *34*, 358–364. [CrossRef]

66. Pedroni, P. *Fully Modified OLS for Heterogeneous Cointegrated Panels and the Case of Purchasing Power Parity*; Department of Economics, Indiana University: Bloomington, IN, USA, 1996.

67. Granger, C. Some recent development in a concept of causality. *J. Econ.* **1988**, *39*, 199–211. [CrossRef]

68. Perman, R. Cointegration: An Introduction to the Literature. *J. Econ. Stud.* **1991**, *18*, 3–30. [CrossRef]

69. Pedroni, P. Panel cointegration: asymptotic and finite sample properties of pooled time series tests with an application to the ppp hypothesis. *Econ. Theory* **2004**, *20*, 597–625. [CrossRef]

70. Arouri, M.E.H.; Youssef, A.B.; M'henni, H.; Rault, C. Energy consumption, economic growth and CO_2 emissions in Middle East and North African countries. *Energy Policy* **2012**, *45*, 342–349. [CrossRef]

71. Kao, C.; Chiang, M.-H.; Chen, B. International R&D Spillovers: An Application of Estimation and Inference in Panel Cointegration. *Oxf. Bull. Econ. Stat.* **1999**, *61*, 691–709.

72. Zhang, B.; Wang, Z.; Wang, B. Energy production, economic growth and CO_2 emission: Evidence from Pakistan. *Nat. Hazards* **2018**, *90*, 1–24.

73. Phillips, P.C.B.; Loretan, M. Estimating Long-Run Economic Equilibria. *Rev. Econ. Stud.* **1991**, *58*, 407. [CrossRef]

74. Rossi, E. *Impulse Response Functions*; Notes for a Lecture on Econometrics; University of Pavia: Pavia, Italy, 2004.

 *processes*

Article

Temporal Feature Selection for Multi-Step Ahead Reheater Temperature Prediction

Ning Gui [1], Jieli Lou [2], Zhifeng Qiu [3,*] and Weihua Gui [3]

[1]   School of Comuputer Science and Engineering, Central South University, Changsha 410000, China
[2]   School of Mechanical Engineering and Automation, Zhejiang Sci-Tech. University, Hangzhou 310000, China
[3]   School of Automation, Central South University, Changsha 410000, China
*   Correspondence: zhifeng.qiu@csu.edu.cn

Received: 27 March 2019; Accepted: 5 June 2019; Published: 22 July 2019

Abstract: Accurately predicting the reheater steam temperature over both short and medium time periods is crucial for the efficiency and safety of operations. With regard to the diverse temporal effects of influential factors, the accurate identification of delay orders allows effective temperature predictions for the reheater system. In this paper, a deep neural network (DNN) and a genetic algorithm (GA)-based optimal multi-step temporal feature selection model for reheater temperature is proposed. In the proposed model, DNN is used to establish a steam temperature predictor for future time steps, and GA is used to find the optimal delay orders, while fully considering the balance between modeling accuracy and computational complexity. The experimental results for two ultra-super-critical 1000 MW power plants show that the optimal delay orders calculated using this method achieve high forecasting accuracy and low computational overhead. Moreover, it is argued that the similarities of the two reheater experiments reflect the common physical properties of different reheaters, so the proposed algorithms could be generalized to guide temporal feature selection for other reheaters.

Keywords: reheat steam temperature; temporal feature selection; delay order prediction; deep neural network; genetic algorithm

---

1. Introduction

Steam reheating plays an important role in power plants. It can increase thermal efficiency by 2% and it can also reduce steam humidity and improve the safety of the final stage's blade [1,2]. However, due to the complexity of the many influential factors, it is difficult to maintain the reheat steam temperature within a certain range [3]. For instance, the reheater steam temperature of two ultra-super-critical 1000 MW units investigated in this paper may fluctuate between 565 °C and 610 °C, while the normal reheater outlet steam temperature is 603 °C with tolerable fluctuation within the range of 503 to 608 °C [4] (the specific threshold may vary with the type of reheater). A temperature that is too high will cause damage to the metal material, while a temperature that is too low will reduce the thermal cycle efficiency [5]. Therefore, finding features that affect the modeling target and analyzing the extent of these features are crucial for the system's safety and efficiency.

A reheater system is a typical nonlinear hysteresis thermal system, which is highly coupled, complex, and impacted by many factors [6,7]. The selection of the most related features from a large variety of sensors is important for the realization of effective control [8]. Traditional feature selections are normally developed on the basis of mass balance, energy balance, and dynamic principles, which rely greatly on human expertise and normally require a long modeling time [9–11]. Recently, researchers have increasingly adopted the data-driven methodology that extracts features directly from huge amounts of accumulated process data [12–14]. Li et al. [15] analyzed operation parameters in power plants by correlation analysis to improve boiler efficiency. Wei et al. [16] used principle component analysis to transform higher-dimensional original data to lower-dimensional principle components, which were employed as the inputs to the NO_x emission model to reduce memory storage requirements and computational costs for data analytics. Buczyński et al. [17] judged whether features could exert substantial effects on a CFD (phase fluidised bed)-based model using sensitivity analysis to predict the performance of a domestic central-heating boiler fired with solid fuels. Pisica et al. [18] chose mutual information to assess the relevance of feature subsets in order to determine the operating states of power systems. Wang et al. [19] utilized the outputs of an improved random forest algorithm as inputs of a back propagation neural network to weight the importance of features and to improve the prediction accuracy of NO_x.

The above research works mainly focused on finding the most related features with respect to the modeling target, which only explores one dimension from all possible relationships. In practice, for the complex process, each feature may have a temporal effect on the modeling target [20]. For instance, some features might have a rapid impact on the target, while some other features might only display certain time-delay effects, i.e., effects after a certain period of time. In order to cover the temporal effects, multi-step features are often accumulated for data-driven modeling in the feature engineering process. Normally, the larger the delay order (number of steps selected) of a feature is, the more information it contains [21]. However, overly large delay orders of features may lead to overfitting, which may cause poor performance on unseen instances [22] and significantly increase memory storage and computational complexity for data analysis [23]. Therefore, it is necessary to find an optimal delay order set for each feature while maintaining a good balance between modeling accuracy and computational economy.

A few researchers have investigated the temporal feature selection problem. Lv et al. [21] used particle swarm optimization to determine delay orders and used a least square support vector machine (SVM) to predict the bed temperature of circulating fluidized bed boilers. However, this method suffered from computational complexity when modeling large-scale data sets. Shakil et al. [24] applied genetic algorithms to estimate the time delay of soft sensors for NO_x and O_2. Although these studies achieved good results on the delay order selection, their modeling targets were only for one particular future time instance, which has the potential not to include the features that impose too rapid or too slow impacts on the target. These approaches also provided little discussion on whether the generated delay order could be used to guide future modeling processes for similar equipment.

To address the optimal feature selection of delay orders for multi-step prediction, a method that combines a deep neural network (DNN) and a genetic algorithm (GA) is proposed. A prediction target with multiple future time steps is introduced to explore features that have rapid or slow effects. A DNN model is used to establish a steam temperature predictor [25] for the next 20 steps. A GA is proposed to find the optimal delay orders with the objective function of balancing modeling accuracy and computational complexity. The proposed method is tested in two 1000 MW coal-fired power plants, namely unit 3 and unit 4, which use more than two million records. The results of the two units display similar sets of delay orders for each feature, reflecting that the physical properties of reheater steam systems are similar to some extent.

The rest of this paper is organized as follows: Section 2 briefly describes the reheater system and proposes the problem statement. Section 3 establishes an objective function for model evaluation. The detailed introduction of the delay order selection mechanism is provided in Section 4. Section 5 presents experiments and discussions. Discussions and possible directions for future work are provided in the final section.

2. System Description and Problem Statement

2.1. Description of Reheater System

A reheater is a set of tubes located in a boiler, the main purpose of which is to avoid excess moisture in steam at the end of expansion to protect the turbine. The exhaust steam from the high-pressure turbines passes through these heated tubes to collect more energy before driving the intermediate- and then low-pressure turbines. The conceptual structure of the reheater unit is shown in Figure 1.

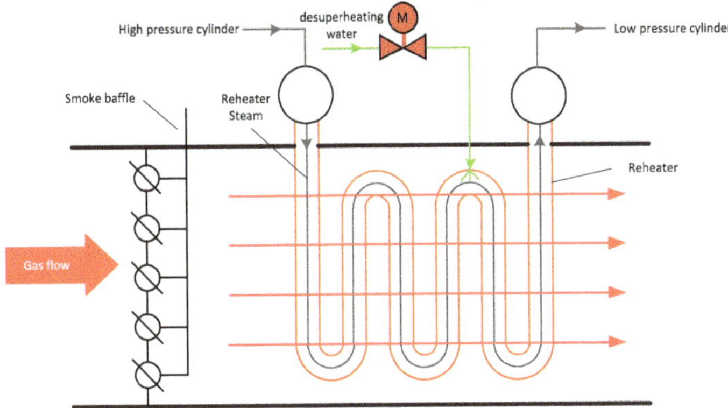

Figure 1. Reheater structure.

After the high-pressured turbine, the exhaust pressure and temperature at the inlet of the reheater are about 35–37 kg/cm^2 and 345–355 °C, respectively. A reheater is designed in the shape of a serpentine tube in order to increase the heated area. The hot smoke generated by the combustion of coal transfers heat to the reheater, meaning that the temperature of steam in the reheater rises. The steam temperature at the outlet of the reheater is kept around 603 °C. Reheater steam with high-temperature and high-pressure characteristics is collected into the high-temperature reheat steam container. A similar process is performed again in the low-pressure cylinder.

Table 1 denotes the influential features of our modeling target, which is the outlet steam temperature of the reheater. Many features affect the reheat steam temperature, such as the inlet steam temperature, inlet gas temperature, smoke baffle opening, etc. Also, these variables have different inertias toward the reheater outlet steam temperature. Therefore, these variables and their hysteresis times should be considered in the prediction model. Here, the previous values of the steam outlet temperature are also used in the modeling process and the multi-step steam temperatures are used as the outputs of the model. In order to simplify our discussion, the major factors are referred to by the notations shown in Table 1.

Table 1. Influential parameters for the temperature of the outlet steam.

Feature	Unit	Inertia	Not.
Inlet steam temperature	°C	small	$Steam^t$
Inlet steam pressure	Mpa	small	$Steam^P$
Inlet smoke temperature	°C	large	$Smoke^t$
Inlet smoke pressure	Kpa	small	$Smoke^P$
Smoke baffle opening	%	small	$Baffle^o$
Desuperheated water flow	t/h	large	D^{water}
Reheater steam temperature	°C	-	$Steam^o$

2.2. Problem Statement

One of the major control concerns of a reheater is the stability of steam^o. In respect to the reheater, some features are reheater-uncontrollable, e.g., smoke temperature and pressure. These features might influence the reheater wall temperature and then change the outlet steam temperature. Steam^o has the characteristics of being non-linear and having a large inertia. Due to the change in operation conditions, it may deviate from the expected range. The normal operation changes the smoke flow toward the reheater by adjusting the smoke baffle opening degree. This operation exhibits a long delay before it imposes impacts on temperature. Another method is to spray the desuperheated water to the reheater steam. This method promptly lowers steam temperature, but also reduces the boiler's efficiency. Considering the economic benefits, the first method is always used. The second method is employed only in an emergency, such as when the steam temperature is too high or the working condition is changing.

Similar to the control variables mentioned above, other features also have impacts characterized by different inertias toward the steam temperature. One major concern is the complexity of accurately determining the impact inertia of different features, which highly depends on the physical nature laws of the reheater as well as the operational conditions of the reheater, e.g., combustion stability. One natural choice is to use long delay orders to compose the model inputs. However, the indiscriminate delay order settings make the feature dimension very high and introduce considerable overheads for both storage and computation. Thus, it is important to select the most cost-effective delay order for features while keeping the system model accurate enough.

3. Objective Function for Model Evaluation

3.1. Multi-Step Prediction

In order to predict the temperature trend of steam°, the nonlinear autoregressive exogenous model is presented. Differing from other approaches, the proposed model predicts values not for any given time, but for a set of future moments.

Since the reheater system displays different hysteresis characteristics toward different features, modeling the steam° with both short and long hysteresis parameters is important. A multi-step steam° prediction model, which generates a serial of predictions for the next $n + 1$ time steps, is given in Equation (1).

$$\begin{bmatrix} \hat{y}(t) \\ \hat{y}(t+1) \\ ... \\ \hat{y}(t+n) \end{bmatrix} = f(x_1(t-1), \cdots, x_1(t-\tau_1), \cdots, x_k(t-1), \cdots, x_k(t-\tau_k), y(t-1), \cdots, y(t-\tau_y)) \ , \quad (1)$$

where t is the current time, $t + n$ is the n-th future moment, x_k is the k-th independent variable, y is a dependent variable, τ_k represents the time delay order corresponding to x_k, and τ_y is the time delay order of dependent variable y.

3.2. Optimization Function

The prediction target increases the forecast performance for the next $n + 1$ time steps by selecting the most appropriate delay order. However, the total number of delay orders is proportional to the computational complexity and opposite to the model accuracy. Thus, the optimization goal defined is to strike a balance between the computational complexity and modeling accuracy. Accordingly, the objective function is used to minimize the total number of delay orders to minimize the computational complexity. Furthermore, the total number of delay orders is kept as high as possible but within a certain range in order to keep the prediction error low enough. Let ε be the maximum acceptable prediction error for the modeling target; thus, another optimization goal is transferred as one constraint, i.e., the prediction error is smaller than or equivalent to ε. Thus, a constrained optimization problem is formulated as Equation (2).

$$
\begin{aligned}
\min \quad J \;&=\; \tau_y + \sum_{k=1}^{K} \tau_k \\
\text{s.t.} \quad e \;&=\; \frac{1}{m\cdot(n+1)}\|\hat{Y} - Y\|_1 \\
e \;&\leq\; \varepsilon \\
e_{l+1} \;&\leq\; e_l,\ \forall l = 1,2,\cdots,L \\
0 \;&\leq\; \tau_y \\
\tau_k \;&\leq\; C
\end{aligned}
\tag{2}
$$

where K is the total delay orders of inputs, m is the total of test data, n is the n-th future moment, τ_k is the delay order of x_k, and τ_y is the delay order of the dependent variable. J is the total of delay orders. e is the error in total m samples and $n + 1$ prediction numbers in the form of mean absolute error (MAE). e_l is the error generated by the l-th iteration. C is the max delay order. ε is the upper limit of MAE. $\hat{Y}$ is the prediction value vector and $\hat{Y} = [\hat{y}(t),\ \hat{y}(t+1),\ldots,\hat{y}(t+n)]^T$, Y is the actual value vector, and $Y = [y(t),\ y(t+1),\ldots,y(t+n)]^T$; $\hat{Y}$ and Y have m samples.

4. Delay Order Selection

In order to accurately select the temporal features, two parts—i.e., the DNN-based prediction model and the GA-based optimal feature selection algorithm—are designed. First of all, the GA generates the individuals of different delay order combinations, which are used as the inputs to the DNN. Then, the DNN outputs the multi-step predictions, which are evaluated by the test sets. The evaluated values are employed as fitness values, which are used in the GA.

4.1. Delay Order Optimization

Delay order optimization is performed by the GA algorithm. The schema of GA is shown in Figure 2. The algorithm starts from an initial population with 20 individuals and each individual has 28 genes. These randomly generated genes are divided into seven sections. Each section represents an input parameter and has 4 binary numbers which can delay the order range from 0 to 15. Then, the individuals are evaluated by the fitness function, which returns two fitness values (MAE and the total of orders). The different fitness values are assigned different fitness scores. The smaller the MAE value, the higher the fitness scores. In a case in which the MAE values are very close (the difference is below a certain threshold), the smaller the total number of delay orders, the higher the fitness scores. The fitness score determines the probability of being selected as a parent. The probability of being selected is according to the roulette wheel selection, shown in Equation (3).

$$
p_i = \frac{f_i}{\sum\limits_{j=1}^{N} f_j},
\tag{3}
$$

where N is the number of individuals in the population, f_i is the fitness of individual i in the population, and p_i is the probability of individual i being selected in the population.

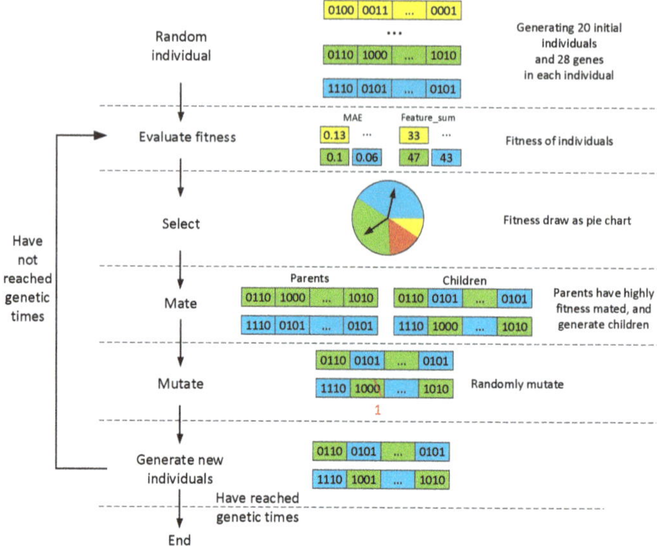

Figure 2. The schematic schema of the genetic algorithm (GA)-based optimized feature selection algorithm. MAE(mean absolute error).

Once the parents are selected, they have a certain probability (p_c) of being mated randomly and generating new individuals. If the parents are not mated, they become new individuals in the new population. Then, the new population has a certain probability (p_m) of deciding whether the individual is mutated. Mutating changes (0 changes to 1, or 1 to 0) randomly. The new individuals are evaluated, selected, mated, and mutated until the number of cycles is reached. At the end of the cycle, the GA obtains the best individuals [26,27].

4.2. Prediction Model

DNN is used to fit the correlation between the future steam° and the historical reheater inlet variables with the accumulated data sets. Figure 3 is the structure of the steam° trend prediction model. Let $m = \tau_1 + \tau_2 + \ldots + \tau_k$ be the total of input dimensions to DNN. The outputs of DNN are $n + 1$ values of steam°. DNN has one input layer, two hidden layers, one output layer, and a large number of neurons. The hypothesis function is shown in Equation (4).

$$h(X) = g(\Theta^3 \cdot g(\Theta^2 \cdot g(\Theta^1 \cdot X))), \tag{4}$$

where X is a vector with m dimensions and Θ^1, Θ^2, and Θ^3 are the weight matrixes between four layers, respectively. $g(\bullet)$ is the activation function.

The cost function of DNN is shown in Equation (5).

$$J(\theta) = \frac{1}{2m \cdot n} \sum_{i=1}^{m} \sum_{j=1}^{n} \left[h(X_j^i) - Y_j^i \right]^2 + \lambda \cdot L2, \tag{5}$$

where m is the total number of samples, n is the total number of output variables, l_k is the number of neurons in the k-th layer, and $h(X_j^i)$ is the prediction value in the i-th sample and the j-th predict value. Y_j^i is the prediction value in the i-th sample and the j-th actual value, λ is the regularization parameter, and $L2$ is the regularization term to limit over-fitting. The goal of the DNN is to minimize Equation (5) with the given sets of features and training samples.

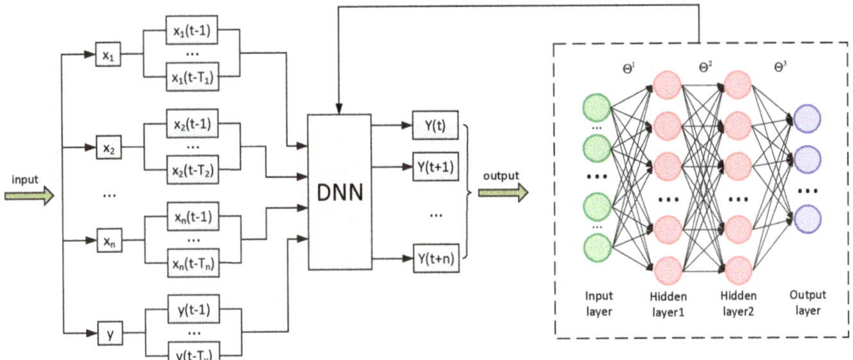

Figure 3. Multi-step prediction model for steam°. DNN—deep neural network.

5. Experiments and Discussion

The data for modeling are collected every 3 s from unit 3 and unit 4 by the distributed control system (DCS). Unit 3 and unit 4 are two ultra-super-critical 1000 MW power plants with the same structure. In our experiment, in total, 7,084,800 records are used for evaluation, in which unit 3 and unit 4, respectively, have 3,542,400 records from 1 May 2016 to 31 August 2016.

5.1. Data Preprocessing

In the data preprocessing process, two steps are taken: Outlier removal and standardization.

Outlier removal: The outliers that violate the physical or technical limitations might affect the model's performance and should be removed before modeling. (1) The points out of the normal range of physical or technical are replaced with the average of adjacent points. For instance, for a certain period, the temperature of steam° should be around 600 °C; thus, the points below 594 °C that violate the steady change characteristics of temperature should be replaced. (2) The errors of D^{water} control time should be modified. Under normal circumstances, the D^{water} control time (more than 0) takes a few minutes. For instance, if the collected data shows that the control time lasts for several hours, the abnormal control time will be modified to a maximum of 3 min.

Standardization: The different features might have different range of values. If these variables are used directly, the feature data with small values may be ignored, while the ones with large dimensions will be selected. Therefore, the Z-score standardization technique [28] is used to scale the data to the ones with a mean value of 0 and a standard deviation of 1, which will speed up the iteration rate of the optimization and convergence.

5.2. Experiment Settings

The parameters of DNN and GA are shown in Table 2. The DNN is a 2-hidden-layer neural network, and the learning rate is set to 0.001. MAE, which is the average absolute differences between predictions and actual observations, is used to evaluate the modeling error. Tanh is chosen as the activation function since it achieves the smallest average MAE compared to other activation functions (e.g., identity, logistic, relu) for the chosen data set.

The 4-month data for unit 3 and unit 4 are divided into 20 different sets. Each set consists of training data from 7 days (about 201,600 records) and test data from 1 day (about 28,800 records).

Table 2. The parameters of DNN and GA.

Neutral Network	Value	GA	Value
Number of hidden layers	2	Number of initial individuals	20
Number of first/second layer neurons	42/23	Mate rate	0.5
Number of outputs	20	Mutate rate	0.2
Activation function	tanh	Number of genes	0–15
Solver	sgd	Iterations	100
Learning_rate	0.001	E	0.14
Λ	0.0001	-	-

5.3. Results and Discussion

This proposed method is evaluated from three different perspectives: Firstly, a one-round simulation is performed with a set of data to demonstrate its capability for finding the optimal delay order for different features; secondly, the experiment is implemented on unit 3 and unit 4 at different times to demonstrate the adaptability of the presented method; finally, the delay order identified with data from the unit 3 is directly used in the modeling process for unit 4 to check its capability for generalization.

(1)   Results of the one-round simulation

As for getting the preliminary delay order in unit 3, the data from ~23 July 2016–30 July 2016 is selected as the experiment data. The changes of MAE and the total number of selected orders during the iteration process are shown in Figure 4a. The accuracy level of MAE is set as 0.001. In the early iterations, MAE begins to decrease while the total delay order increases. Then, until MAE stabilizes at 0.13—i.e., the lower limit of MAE—the total delay order decreases. In the later iterations, these criteria remain constant, which indicates that the algorithm is converged. Figure 4b shows each feature's delay order. It can be seen that some features have a larger delay order, e.g., smoke^p, which indicates large hysteresis, while in contrast, the order of D^{water} shows timely but transient impacts.

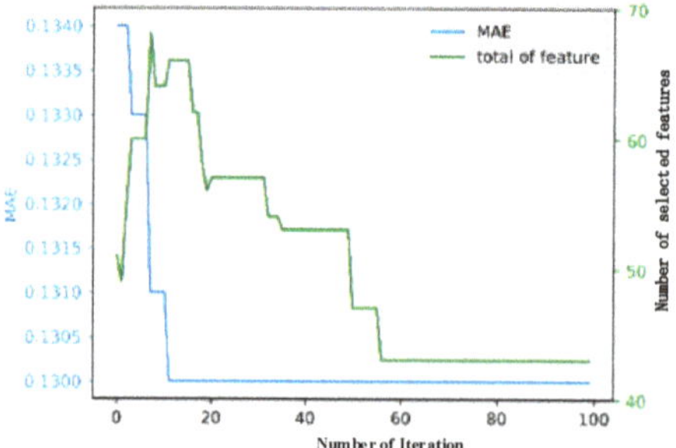

steam^t	steam^p	smoke^t
1	6	8
smoke^p	baffle^o	D^{water}
10	4	1
steam^o		
13		

(**a**) Fitness curve during the experiment   (**b**) Optimal delay order

Figure 4. Results for the one-round simulation for unit 3 (~23 July 2016–30 July 2016).

In Figure 5, the forecasting errors in one-minute periods with 20 points in 30 July 2016 are plotted in a box plot which displays the distribution of five different metrics, i.e., minimum, first quartile, median, third quartile, and maximum. Figure 5 shows that MAE increases with the increase in the predicting time step. This is normal, as timely response factors, such as steam^p, smoke^t, and baffle^o, cannot be captured by predictor. However, the median MAE in one minute is less than 0.3 °C, and the average is near 0.1 °C. According to Figure 4b, the maximum delay order of the reheater steam temperature steam^o is 13. This means that the historical data of steam^o have major impacts on the accuracy of the model. It also shows that, in the current system, steam^o is not well controlled, as it should kept steady around 600 °C.

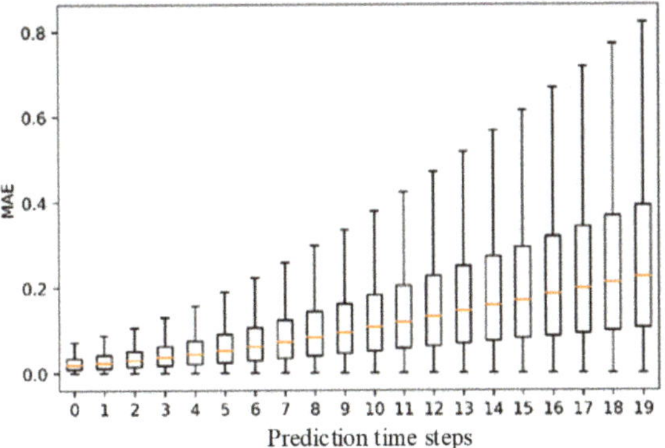

Figure 5. The box error curve.

(2) Comparisons of unit 3 and unit 4 from different perspectives

The feature selection method is tested for both unit 3 and unit 4 based on the operational data from 1 May 2016 to 31 August 2016. Since the records from some days contain too many abnormal data, the data from those days are not used for the model training. As shown in Table 3, the data periods are closed from the intra-comparisons within unit 3 or unit 4 or the inter-comparison between those two units.

Table 3 shows that the general range of the seven studied features has the corresponding length of delay orders with respect to their inertia toward steam^o. For all 20 tests, there is no significant deviation regarding MAE. This means that the designed DNN with the selected features as the inputs achieves good convergence. It also shows that the delay orders of smoke^t and smoke^P are larger than those of steam^t and steam^P, as the smoke has indirect impacts toward the steam^o. Thus, their delay orders are much larger than those of the feature of the inlet steam. D^{water} has a very small delay order due to the fast temporal response toward steam^o. For certain periods, the delay orders of D^{water} are zero, e.g., in tests 9, 10, 16, and 18. The zero value is due to the lack of training data for D^{water}. In those periods, the action of spraying de-superheated water is seldom performed. This is due to the insufficient training samples. At these stages, the numbers of sprays are, respectively, 31, 22, 26, and 18, while other tests have about 60 actions, owing to the comparable steam^o which is more stable. A similar phenomenon can also be observed for the optimal delay order for baffle^o. These results show the importance of the data coverage for the accuracy of feature selection.

Table 3. Results for both unit 3 and 4 (value before "/" is for unit 3 and after is for unit 4). MAE—Mean absolute error.

Test	Sample Date	Steam^t	Steam^P	Smoke^t	Smoke^P	Baffle^o	D^{water}	Steam^o	MAE
1/11	8 May–15 May/1 May–8 May	1/1	6/6	9/8	10/10	4/4	1/1	15/13	0.095/0.116
2/12	16 May–23 May/17 May–24 May	1/2	6/2	9/8	10/11	4/0	1/1	15/15	0.088/0.094
3/13	20 May–27 May/24 May–31 May	1/1	6/6	8/8	10/10	4/4	1/1	13/13	0.129/0.123
4/14	9 June–16 June/5 June –12 June	3/1	6/6	12/8	10/13	0/4	1/1	13/13	0.118/0.111
5/15	17 June–24 June/8 June–15 June	1/1	6/4	9/14	13/10	4/4	1/1	15/15	0.086/0.101
6/16	1 July–8 July/16 June–23 June	3/1	6/6	9/8	15/10	4/0	1/0	15/13	0.100/0.101
7/17	22 July–29 July/17 July–24 July	1/1	6/2	12/11	10/10	0/0	1/2	13/15	0.128/0.117
8/18	6 August–13 August/24 July–31 July	1/1	6/2	9/12	10/13	4/0	1/0	15/13	0.103/0.095
9/19	10 August–17 August/5 August –12 August	1/2	6/6	8/10	13/10	4/2	0/1	13/15	0.132/0.115
10/20	13 August–20 August/17 August–24 August	1/2	6/6	8/9	10/10	4/2	0/1	15/14	0.115/0.115

(3) Determination of delay order

For the purpose of controlling steam° changes within the ideal range, properly finding a delay order is crucial to accurately describing the hysteresis of features for a prediction model. The variations of delay orders for each feature are shown in Figure 6; the shadow ranges from the maximum to minimum delay order. There is a large overlap between two units, which indicates the existence of common delay orders. The medians of overlap (2, 6, 10, 10, 2, 1, and 14) represent the general level of intervals and may serve as the references for delay orders regarding the steam° system of ultra-super-critical 1000 MW power plants.

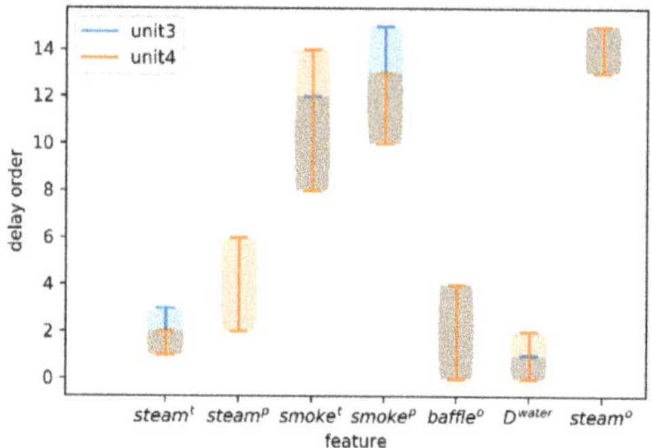

Figure 6. Delay order distribution of the seven features.

The features with delay orders of 2, 6, 10, 10, 2, 1, and 14 generated from the data from unit 3 are used as selected features for the reheater steam temperature prediction. We also adopt the same methods to find the optimal feature distributed for the unit 4. Then, those results are compared with the dataset of test 1 to test 20, which are from unit 4. The orange bars indicate the MAE with the identified delay order. The directly calculated optimal solution is shown by the blue bars. Figure 7 shows the comparisons, which obviously indicate that the MAEs of two cases are approximately equal. The maximum error is only 0.9% (on the 16th day), which means that it is almost the same as the results from the optimal solutions. This shows that the selected delay orders (2, 6, 10, 10, 2, 1, and 14) have good generalization capability, and can, it is argued, represent the physical characteristics of two reheaters.

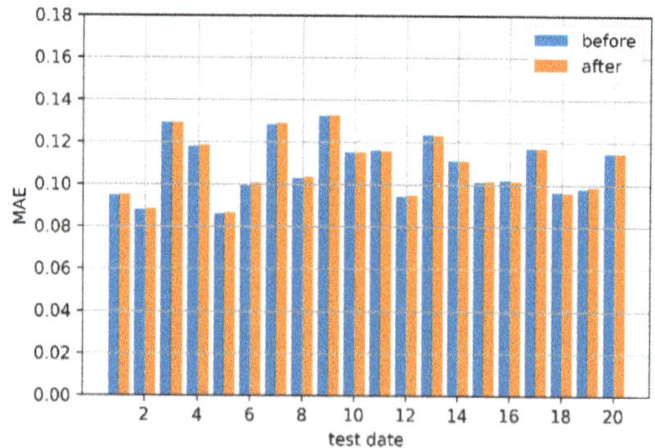

Figure 7. Comparisons between optimal and proposed methods. Blue bars are the MAE of the optimal solution and the orange bars indicate the MAE of the proposed method.

6. Conclusions

For many industrial processes, it is important to find the best feature delay orders as well as features that are most correlated with the prediction targets. In this paper, a delay order identification method based on GA and DNN is proposed. This method adopts the GA to generate candidate feature sets which try to find minimal numbers of features while keeping the MAE of the prediction model low enough. The DNN model is used for modeling processes that generate the multi-step predictions typically demanded in many industrial processes. This method is evaluated with experiments from different perspectives; data from two similar units are used to check whether the found time delays indeed demonstrate the physical characteristics of the underlying systems. The experimental results indicate that two units have similar delay orders and the delay order can be directly used for modeling similar devices with little loss of accuracy.

Of course, many interesting issues still need to be investigated. For instance, our solution limits the temporal feature selection. It is important for the delay order selection method to support both spatial and temporal feature selection. We are investigating the use of an attention mechanism to find the optimal solution for both dimensions. In addition, the GA demands considerable resources and computational costs. We are working to design more computationally efficient methods, e.g., filter-based feature selection for industrial feature processing.

Author Contributions: N.G. proposed the main idea of the method; J.L. and Z.Q. implemented the model and validated the field test. W.G. provided the funding.

Funding: This work is funded by the Nature Science Foundation of China 61403429, 61621062 and 61772473.

Acknowledgments: This work is funded by the Nature Science Foundation of China 61403429, 61621062 and 61772473.

Conflicts of Interest: The authors declare no conflict of interest.

References

1. Zhang, L. *Principle of Boiler*; China Machine Press: Beijing, China, 2011.
2. Ge, Z.; Zhang, F.; Sun, S.; He, J.; Du, X. Energy Analysis of Cascade Heating with High Back-Pressure Large-Scale Steam Turbine. *Energies* **2018**, *11*, 119. [CrossRef]
3. Lee, K.Y.; Ma, L.; Boo, C.J.; Jung, W.H.; Kim, S.H. Intelligent modified predictive optimal control of reheater steam temperature in a large-scale boiler unit. In Proceedings of the 2009 IEEE Power & Energy Society General Meeting, Calgary, AB, Canada, 26–30 July 2009.
4. Fan, Q.G. *Principle of Boiler*; China Electric Power Press: Beijing, China, 2014.
5. Khartchenko, N.V.; Kharchenko, V.M. *Advanced Energy Systems*; CRC Press: Cleveland, OH, USA, 2013.
6. Hogg, B.W.; El-Rabaie, N.M. Multivariable generalized predictive control of a boiler system. *IEEE Trans. Energy Convers.* **1991**, *6*, 82–288. [CrossRef]
7. Liu, X.J.; Kong, X.B.; Hou, G.L.; Wang, J.H. Modeling of a 1000 MW power plant ultra super-critical boiler system using fuzzy-neural network methods. *Energy Convers. Manag.* **2013**, *65*, 518–527. [CrossRef]
8. Suryanarayana, G.; Lago, J.; Geysen, D.; Aleksiejuk, P.; Johansson, C. Thermal load forecasting in district heating networks using deep learning and advanced feature selection methods. *Energy* **2018**, *157*, 141–149. [CrossRef]
9. Staehelin, C.; Schultze, M.; Kondorosi, É.; Mellor, R.B.; Boiler, T.; Kondorosi, A. Structural modifications in Rhizobium meliloti Nod factors influence their stability against hydrolysis by root chitinases. *Plant J.* **1994**, *5*, 319–330. [CrossRef]
10. Gnanapragasam, N.V.; Reddy, B.V. Numerical modeling of bed-to-wall heat transfer in a circulating fluidized bed combustor based on cluster energy balance. *Int. J. Heat Mass Transf.* **2008**, *51*, 5260–5268. [CrossRef]
11. Black, S.; Szuhánszki, J.; Pranzitelli, A.; Ma, L.; Stanger, P.J.; Ingham, D.B.; Pourkashanian, M. Effects of firing coal and biomass under oxy-fuel conditions in a power plant boiler using CFD modelling. *Fuel* **2013**, *113*, 780–786. [CrossRef]
12. Guyon, I.; Elisseeff, A. An introduction to variable and feature selection. *J. Mach. Learn. Res.* **2003**, *3*, 1157–1182.
13. Saeys, Y.; Inza, I.; Larrañaga, P. A review of feature selection techniques in bioinformatics. *Bioinformatics* **2007**, *23*, 2507–2517. [CrossRef]

14. Chandrashekar, G.; Sahin, F. A survey on feature selection methods. *Comput. Electr. Eng.* **2014**, *40*, 16–28. [CrossRef]
15. Li, J.Q.; Gu, J.J.; Niu, C.L. The Operation Optimization based on Correlation Analysis of Operation Parameters in Power Plant. In Proceedings of the 2008 International Symposium on Computational Intelligence and Design, Wuhan, China, 17–18 October 2008.
16. Wei, Z.; Li, X.; Xu, L.; Cheng, Y. Comparative study of computational intelligence approaches for NOx reduction of coal-fired boiler. *Energy* **2013**, *55*, 683–692. [CrossRef]
17. Buczyński, R.; Weber, R.; Szlęk, A. Innovative design solutions for small-scale domestic boilers: Combustion improvements using a CFD-based mathematical model. *J. Energy Inst.* **2015**, *88*, 53–63. [CrossRef]
18. Pisica, I.; Taylor, G.; Lipan, L. Feature selection filter for classification of power system operating states. *Comput. Math. Appl.* **2013**, *66*, 1795–1807. [CrossRef]
19. Wang, F.; Ma, S.; Wang, H.; Li, Y.; Qin, Z.; Zhang, J. A hybrid model integrating improved flower pollination algorithm-based feature selection and improved random forest for NO_X emission estimation of coal-fired power plants. *Measurement* **2018**, *125*, 303–312. [CrossRef]
20. Sun, L.; Li, D.; Lee, K.Y. Enhanced decentralized PI control for fluidized bed combustor via advanced disturbance observer. *Control Eng. Pract.* **2015**, *42*, 128–139. [CrossRef]
21. Lv, Y.; Hong, F.; Yang, T.; Fang, F.; Liu, J. A dynamic model for the bed temperature prediction of circulating fluidized bed boilers based on least squares support vector machine with real operational data. *Energy* **2017**, *124* (Suppl. C), 284–294. [CrossRef]
22. Galicia, H.J.; He, Q.P.; Wang, J. A reduced order soft sensor approach and its application to a continuous digester. *J. Process Control* **2011**, *21*, 489–500. [CrossRef]
23. Souza, F.; Santos, P.; Araújo, R. Variable and delay selection using neural networks and mutual information for data-driven soft sensors. In Proceedings of the 2010 IEEE 15th Conference on Emerging Technologies & Factory Automation (ETFA 2010), Bilbao, Spain, 13–16 September 2010.
24. Shakil, M.; Elshafei, M.; Habib, M.A.; Maleki, F.A. Soft sensor for NOx and O_2 using dynamic neural networks. *Comput. Electr. Eng.* **2009**, *35*, 578–586. [CrossRef]
25. Xia, C.; Wang, J.; McMenemy, K. Short, medium and long term load forecasting model and virtual load forecaster based on radial basis function neural networks. *Int. J. Electr. Power Energy Syst.* **2010**, *32*, 743–750. [CrossRef]
26. Gosselin, L.; Tye-Gingras, M.; Mathieu-Potvin, F. Review of utilization of genetic algorithms in heat transfer problems. *Int. J. Heat Mass Transf.* **2009**, *52*, 2169–2188. [CrossRef]
27. Woodward, R.I.; Kelleher, E.J.R. Towards 'smart lasers': Self-optimisation of an ultrafast pulse source using a genetic algorithm. *Sci. Rep.* **2016**, *6*, 37616. [CrossRef] [PubMed]
28. Kreszig, E. *Advanced Engineering Mathematics*, 4th ed.; Wiley: Weinheim, Germany, 1979; p. 880.

Article

Optimized Energy Management Strategies for Campus Hybrid PV–Diesel Systems during Utility Load Shedding Events

Jacques Maritz

Department of Engineering Sciences, University of the Free State, P.O. Box 339, Bloemfontein 9300, South Africa; maritzjm@ufs.ac.za; Tel.: +27-51-401-2076

Received: 12 June 2019; Accepted: 1 July 2019; Published: 8 July 2019

Abstract: The unique situation of utility power curtailment unveils opportunities in the fields of energy management and digital resource management. During utility load shedding events, campuses are typically driven as Photo Voltaic (PV)–diesel generator hybrid systems, of which the main fossil resource driver is diesel. With the appropriate Supervisory Control and Data Acquisition (SCADA) systems, discrete departmental energy policies along with control, forecasting and Internet of Things (IoT) infrastructure, the campus hybrid system could be optimized on a short timescale during the shedding event. In this paper the optimization methodology, required technology infrastructure, possible forecasting algorithms and potential implementation will be discussed.

Keywords: hybrid PV–diesel generator systems; digital resource management; energy management

---

1. Introduction

The scenario of power loss from the grid, whether for short or long periods, places a unique operational constraint on the electric network of a campus [1]. Typically, this mode of operation is governed by a hybrid PV–diesel generator system, that allows ample room for optimization of its main resource metric, namely fossil fuel [2]. Campus curtailment and demand side management programs could benefit from this scheme, especially when energy managers engage in campus disconnect from utility. The choice of hybrid Photo Voltaic (PV)–diesel generator system for optimization in this paper is based on local utilization trends and that diesel generators are considered an underutilized asset. The optimization framework of modern hybrid systems is fed by several elements that culminate in the field of digital resource management, such as energy Internet (EI) [3], energy Internet of Things (EIoT), energy forecasting techniques (EFT) [4], intelligent electrical control systems [5], the ability to digitally manage the consumer's energy purchasing strategy [5] and coherent energy data formats generated by both loads and generators. However, few high-resolution digital energy management strategies exist that make use of energy forecasting (with forecasting uncertainties) to optimize fuel usage, while simultaneously increasing PV penetration of the electrified campus grid on a small time-step basis.

Hybrid generation systems on campuses mimic the setup of micro-grids, in islanded operational mode (loss of utility power), where loads and generators interact in an intelligent mode to generate savings and possibly other avenues of revenue for campuses [6], without feeding back into the powerless grid (for situations where feedback is not allowed). As part of such a campus micro grid of the future, it is assumed that the hypothetical campus presented in this paper, is interwoven with IoT sensors (multi-faceted) and equally capable communication networks [7]. The EIoT infrastructure on campus injects unique data streams into the energy management (EM) local/cloud server that could be leveraged for the purposes of enhancement of decision-making structures regarding energy usage, dispatch and forecasting on campus. For example, the EIoT network could feed departmental energy managers with information about campus dynamics that would, in turn, influence their scheduling of lecture halls (or partial use) and the utilization of research lab equipment. The same argument holds for Heating, Ventilation and Air-conditioning (HVAC) scheduling of lecture halls based on the local weather. The campus EIoT network therefore would be able to supply rich layers of information that could feed decision support systems used by departmental energy managers (DEMs) and lead campus energy managers (LCEMs). An EIoT network on campus also feeds the concept of the energy Internet, whereby all consumers and generators are connected by some efficient communication network (such as Zigbee) and allows for full control, high resolution data acquisition, energy data analysis, energy management and departmental energy policy monitoring (all while communicating in an energy efficient manner) [7].

Real-time and optimized energy management of smart campuses relies on the full control of the campus energy system, real-time energy data acquisition and demand/generation horizon forecasting. This type of energy generation/demand forecasting is driven by several state-of-the-art machine learning techniques that leverage the EI [8]. The forecasting techniques rely on historical data with different training set lengths (depending on the forecasting method) in order to forecast energy usage/generation with uncertainty based on a plethora of different training metrics and unique data streams. A typical example of such a forecasting strategy is that of solar irradiance based on the underlying metrics of humidity, temperature and cloud cover as can be found in Reference [9]. Complex connections exist between different metrics that can only be recovered using machine learning techniques that learn from experience, given a typical large energy dataset. As a state-of-the-art example, consider the scenario where LCEMs strategize a particular renewable dispatch strategy based on historic data and half-hourly newly acquired data. The LCEMs train models on historic data and uses the newly acquired data to firm up on the energy dispatch system. It remains inefficient to retrain the entire model on the historic data and newly added data, thus the LCEMs use online forecasting methods (such as online Gaussian Process (GP) forecasting methods that are based on multi in–multi out systems) to retrain only on the new acquired data. In this manner, LCEMs can firm up their energy dispatch system, whilst remaining computationally efficient with quick response strategies [10].

The nervous system of a smart campus will typically by governed by an electrical control system (Building Automated/Management System), to ensure that a high resolution of control is achieved when imposing energy management strategies and critical control. This nervous system interconnects the largest consumers such as HVAC, hot water supply (part of HVAC) and research labs [11]. High-resolution low-level departmental control becomes more expensive due to either non-existing control circuits or mandatory control retrofits, thus in many cases a control retrofit is needed [12]. However, building automated systems (BASs) will be able to perform macro control of the largest energy consumers.

A critical concept in digital resource management of smart campuses is that of coherent energy data structures and exchange formats. As an example, the campus building control systems typically generate data that aids decision support systems in shifting or eliminating certain electrical loads during critical periods. Separate from the building control systems, the PV generation data is shared on a separate platform and so is the distributed diesel generators (both control and electrical output). However, when dealing with such large energy ecosystems such as campuses, the energy management strategies become largely data driven and depend on coherent multi-faceted data streams culminated on a single platform. The energy data analysis backend is hosted on the same platform, allowing for easy debugging and amendment [13].

Smart trading of resources feature in several markets and will be equally beneficial as part of the smart campus scenario (one smart campus selling to a neighboring smart campus). A smart campus with the necessary digital infrastructure (EIoT, EI, EM etc.) will easily overcome the requirements to manage its own resources and interact with other receiving clients [5]. In order to trade resources in real-time, a campus needs to be able to control and forecast resource usage and generation with certainty. Added to this requirement, the LCEMs need to utilize the EIoT data influx in a real-time manner that adapts to market responses and predictions [14].

This paper proposes an umbrella scheme/concept that utilizes smart campus elements in order to optimize fossil fuel usage associated with PV–diesel generator hybrid systems, from the perspective of energy forecasting techniques, energy campus IoT and the necessary control systems. Special focus was given towards the critical role-players and data flow within the proposed scheme. The emphasis of this paper is therefore on the formulation and discussion of the umbrella optimization scheme (not to determine the equation of energy utilization) and attempt to include all the possible elements that are considered role-players within the final optimization. Future work will include the process of finding the function of energy utilization based on real campus data that will be conditioned to feed the hypothetical scheme. The mathematical framework (or data driven methodology) will follow future testing of the scheme on real campus data.

2. Holistic Optimization Framework

The opportunity generated by emergency utility curtailment for a scheduled period realizes in the interplay of the two main non-utility contributing sources of energy on campus, namely that of PV generation and fossil fuel generators, or PV–diesel hybrid systems. Battery storage is omitted from this proposed framework, due to non-favorable economic feasibility when considering large PV plants. Assuming that the diesel generators are of high enough quality to allow synchronicity with that of the renewable generator, and that the fossil fuel generators are distributed in such a manner to allow connection to any load collection on campus; typically, a characteristic of future smart campuses from the perspective of resources. Campus electrical networks can be classified as centralized PV with decentralized diesel generators islands or centralized PV with centralized diesel generators, however, the technical influence of the configuration on the proposed optimization framework is not part of the scope of the paper. The following scenario is proposed for the optimization of fuel usage, the reader is referred to Figures 1–3 for a visual summary of each step of the global optimization scheme.

The entire network (generation and load) is optimized in discrete time-steps (that could become real-time in future) based on an influx of data from the PV plant, diesel generators, load banks and the campus EIoT in which the elements are embedded. The main aims of the optimization scheme will be to determine the fixed critical load bank and fixed serving diesel generator bank, after which the remaining variable banks (both load and generator) could be optimized based on a list of critical metrics harvested from the PV plant and campus IoT network. The main energy savings metric in this proposed scenario is driven by the savings of fuel (diesel) and possibly the energy saved by comparing to "normal operations" if the utility curtailment event did not occur. The optimization scheme of the load/generation banks is driven from the viewpoint of the diesel generators with the added benefit of critical heat/electric generation from hydrogen plants on campus.

The optimization of the load banks is done on an energy manager server that utilizes building automated systems (BAS), building control systems (BCS), EIoT data influx (departmental and campus wide), renewable power generation and departmental energy policies, all of which are embedded on a local or cloud framework that holds enough resources and memory capacity to perform the optimization process (typically a server located on campus or cloud driven services). The local EM server (if this is the case) will also be restricted to a campus EM policy to perform computations during the appropriate tariff times (as not to be power intrusive itself). It is assumed that the PV plant is connected to the direct current (DC) bus and both the loads and diesel generators are connected to the alternating current (AC) bus. Converters reside between the AC and DC bus. Furthermore, it is assumed that generator-load pairs are not islanded and that every component in the network is linked.

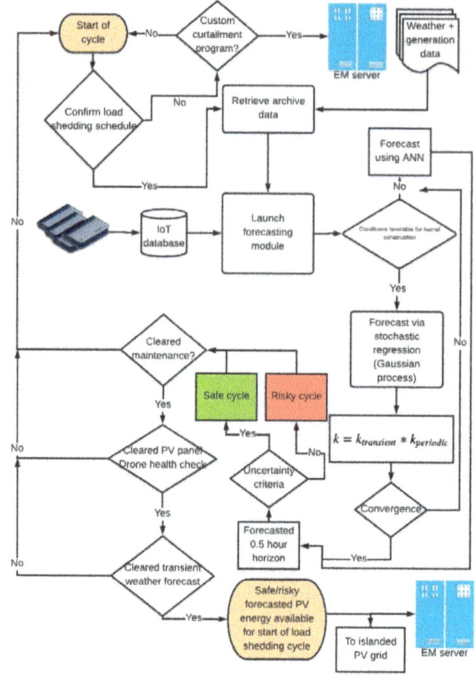

Figure 1. Optimization scheme for the determination of the available PV power output for a 0.5 h horizon. The PV scheme feeds both load and diesel generator schemes.

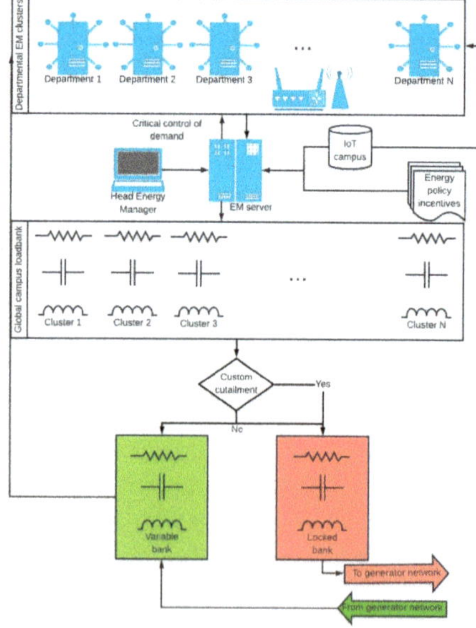

Figure 2. Optimization scheme to determine the variable/locked load banks to be fed to the diesel generator optimization scheme. This scheme depends on low-level departmental control and energy management policies.

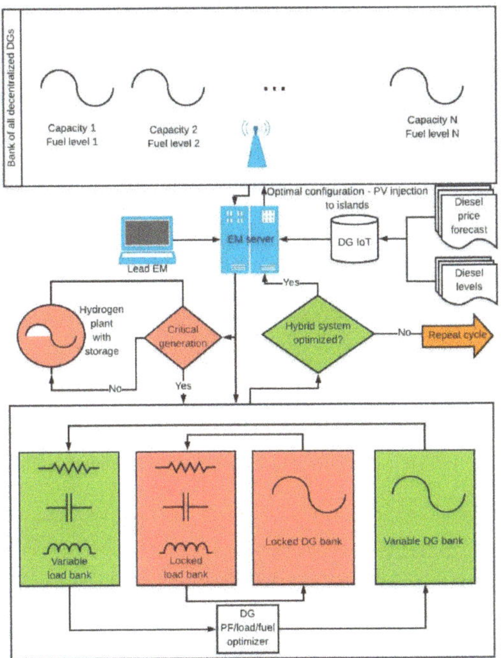

Figure 3. Optimization scheme for diesel generator—load interactions based two classes, namely variable and locked banks.

The PV optimization scheme (see Figure 1) uses actual environmental data together with generation forecasting to produce an estimated available PV energy over a 0.5 h horizon. The estimated available energy is complimented with a known uncertainty and will classify the possible PV output as either a risky cycle or a safe cycle (depending on the chosen critical uncertainty). The PV optimization scheme uses several streams of unique data, including historic weather and generation data, IoT sensor network data associated with the PV plant, facilities management software (maintenance programs) and weather forecasts spanning the prediction horizon. The PV optimization scheme must be the first step to be completed as part of the holistic optimization scheme, since the chosen dispatch strategy primarily manages the remaining demand.

Prior to the utility curtailment event, all curtailment schedules and priorities are checked and the necessary data is gathered in order to initialize the forecasting module. The forecasting module is based on either a hybrid scheme that initializes neural nets (if no kernel can be constructed) or a Gaussian process when data conditions are favorable for kernel construction. The forecasting module is trained to achieve a 0.5 h horizon with forecasting uncertainty. Depending on the length of the training data set and regression resolution, forecasting algorithms can become computationally expensive and should be included as part of a dedicated EM server with enough computational capacity for quick response times on forecasting requirements. The output of the forecasted module will be the total renewable energy that can be dispatched to the campus electrical network with associated uncertainty (based on the classification of safe or risky cycle). The PV module (see Figure 1) acts as the first step to initialize the energy dispatch strategy within the horizon of 0.5 h. The optimization scheme will critically act on maintenance conditions that are flagged and if the utility curtailment process seized. Transient weather events will increase the uncertainty of the cycle and could lead to the determination of the optimization cycle.

The load bank optimizer (see Figure 2) initiates at departmental level by checking the current state of each department's digital EM strategy. Departmental EM strategies are typically schedule driven (lecture halls, research labs and offices), thus each departmental EM strategy could deliver a load bank comprising of variable and locked banks as preconditioned by the departmental EM champion or framework. A typical response from the departmental EM framework can be seen in Tables 1 and 2 (color coding corresponds to safe or risky cycle in optimization schemes).

Table 1. Simplistic response of a departmental energy policy framework over a 0.5 h time-step (without campus IoT intervention).

Load Type	Locked/Variable	Energy Estimation of Load (kWh)
Lecture Hall A	Locked	1.8 kWh
Offices 1–5	Locked	1.25 kWh
Offices 5–10	Variable (HVAC)	1.25 kWh
Research Lab A	Locked	10 kWh
Research Lab B	Variable (fume hoods)	10 kWh

Table 2. Simplistic response of a departmental energy policy framework over a 0.5 h time-step (with campus IoT intervention).

	Locked/Variable	Energy Estimation of Load (kWh)
Lecture Hall A	Locked	1.8 kWh
Offices 1–5	Variable (Lecturers not yet entered main gate of campus)	1.25 kWh
Offices 5–10	Variable (HVAC)	1.25 kWh
Research Lab A	Variable (Flagged due to power quality issue)	10 kWh
Research Lab B	Variable (fume hoods)	10 kWh

The lead EM framework accounts for all departmental locked/variable load banks, as well as the relative sizes of the loads. It is assumed that the departmental EM framework can account for the size, type of load, power factor and associated schedules of each load in the department (elements that future smart campuses will have). The final load bank that will partake in the 0.5 h cycle of energy optimization is finalized into a compatible data structure that awaits input from the fossil fuel generator optimization scheme (see Figure 3).

The fossil fuel optimization scheme is based on the assumption that all fossil fuel generators are equipped with two-way communication and control. Furthermore, any load could be connected to any generator on campus (dismissing any practical considerations). The fossil fuel IoT network of sensors captures data associated with the generators such as fuel levels, operating conditions and location. The lead EM server supplies additional information such as maintenance and refueling schedules to the global optimizer based on fuel price forecasting and diesel levels.

At this point global energy accounting can be performed (matching supply with load) and critical energy generation via hydrogen plants could be initiated to supply critical loads. A bank of variable/locked fossil fuel generators can now be generated and passed on to the final optimization phase. The optimizer then proceeds to match the locked/variable loads with the fossil fuel generators based on power factor and load size. Once the locked loads are matched with the locked fossil fuel generators, the algorithm proceeds to match the variable banks over the horizon of 0.5 h. This process repeats every 0.5 h with the aim of reducing fuel consumption for the span of the utility power loss.

3. Extremities of Optimization Framework

The following two scenarios are presented: (1) the remaining load (after PV quenching) is distributed among generators irrespectively of fuel levels, load type or fuel forecasts, (2) optimizer algorithm (see Figures 2 and 3) suggests two types of generators linked with two types of load collections (from either the locked or variable banks), calculated every 0.5 h over a 3 h horizon.

Two extreme poles/classes of option (2) are a generator with a low fuel level that is scheduled for refueling and a generator with high fuel levels. If the optimizer algorithm predicts favorable fuel, it will suggest that the low fuel level generator to be utilized at 50% and the high fuel level generator at the difference between maximum capacity and the spinning reserve. The aim here is to maximize fuel savings at the end of the three-hour utility curtailment cycle. Due to the efficiency curves of both typical generators described as extremities, fixed parameter loads are required to be served by both generators (having different fuel levels). The 50% utilized generator can serve loads from the variable bank (non-stringent constraints on power factor and load size requirements) and the fully utilized generator can serve loads from the locked bank (due to stringent power factor and load type requirements). The locked load bank can be assigned with certainty and all load sizes, types and Power Factor (PFs) are known.

Total savings generated by the optimization scheme is decomposed into the savings generated by refueling at a better tariff rate and actual fuel savings due to more efficient generator utilizations. Over the three-hour utility power loss event, the optimizer performs six optimization steps (0.5 h apart). The PV plant energy output could change over this horizon, thus changing the dynamics of the optimization. The spinning reserve of each generator is dependent on the uncertainty of the PV power output (risky or safe cycles, see Figure 1) divided among all the available generators. As an example, if the available PV power output contains large uncertainty, the spinning reserve for each generator will increase and vice versa. The spinning reserve of each generator is re-calculated after every cycle and determines the maximum utilization of the generators.

It is further assumed that the collection of generators are of variable type (constant frequency, but variable kW input). The typical variable generator is specified by a rating (e.g., 250 kVA, 0.8 PF, 200 kW and 150 kVAR) and will be connected to a collection of loads specified by type, size and PF. Specified by the generator's efficiency/PF curve, as the utilization of the generator increases, both the efficiency and PF requirements increases. Generators with a high forced utilization factor (determined by the optimization algorithm, see Figures 1–3), must be connected to cleaner collection of loads (from the perspective of PF) to safeguard possible breaching of the generator's kVAR rating. Thus, low fuel level generators could be utilized at e.g., 50% and will serve load collections with low power factors safely (vice versa for high fuel generators).

4. Conclusions

The holistic optimization framework exhibits potential of typical campus hybrid PV–fossil fuel generator systems generating fuel savings under powerless utility scenarios when optimized in discrete time steps using campus IoT, EM and generation/consumption forecasting techniques, e.g., forecasting of renewable power output (PV, wind and hydrogen) and campus load characteristics. This is an improved savings scheme as opposed to running the generators at fixed speed without any type of data/optimization feedback. The novelty of the scheme lies in the PV power output estimation with uncertainty. The known uncertainty associated with the available PV energy could classify the output as risky or safe. This will allow improved load distribution to generators that are sensitive to fuel price forecasts and refueling maintenance, as well as improved predictive maintenance schedules associated with the generators. The main aim of the optimization scheme is to maximize fuel savings during the utility power loss. Technical recommendations for a full-scale scenario, consisting of an EM server platform hosting all the forecasting algorithms, campus IoT and policy data will be discussed and commissioned in future.

The full-scale scenario will initially be commissioned using a campus department that utilizes decentralized renewable resources, together with fossil fuel generation and energy policies. The EM server that hosts all the tiers of the optimization scheme will be located in the department itself. Following the initial commissioning test, the platform will be rolled out to the entire campus network.

This full-scale scenario will require full integration of controls, metering installation/retrofit, IoT campus integration, SCADA system implementation, energy policy reviews, energy policy integration and decentralization/centralization of required generators. All of the last-mentioned elements will have to be interwoven on an EM server that contains the optimization scheme in algorithmic form, as well as the necessary maintenance and energy policy framework. Since all the discrete optimization steps will have associated uncertainty, the final savings will be able to be verified with ease. The optimization scheme presented in this paper is an example of generating savings in the absence of grid power.

Future work will be based on data produced by actual smart campus elements, sensors, generators and loads. From there a study and framework will be formulated regarding the system time dependencies. However, emphasis must be placed on the practicality in the hypothetical scheme, especially since the sampling time, signal changes and signal delays found in the forecasting techniques, campus IoT network and internal data transport play a critical role in the success of this proposed global scheme. The process of illustrating the improvement of the campus power supply will be based on a PV penetration metric, that needs to be evaluated in conjunction with the scheme.

Funding: We acknowledge the financial support of the Manufacturing, Engineering and Related Services Sector Education and Training Authority (merSETA).

Conflicts of Interest: The author declares no conflict of interest.

References

1. Van Merode, D.; Tabunshchyk, G.; Patrakhalko, K.; Yuriy, G. Flexible technologies for smart campus. In Proceedings of the 2016 13th International Conference on Remote Engineering and Virtual Instrumentation (REV 2016), Madrid, Spain, 24–26 February 2016.
2. Bajpai, P.; Dash, V. Hybrid renewable energy systems for power generation in stand-alone applications: A review. *Renew. Sustain. Energy Rev.* **2012**, *16*, 2926–2939. [CrossRef]
3. Appelrath, H.-J.; Terzidis, O.; Weinhardt, C. Internet of Energy. *Bus. Inf. Syst. Eng.* **2012**, *4*, 1–2. [CrossRef]
4. Suganthi, L.; Samuel, A.A. Energy models for demand forecasting-A review. *Renew. Sustain. Energy Rev.* **2012**, *16*, 1223–1240. [CrossRef]
5. Zhang, C.; Wu, J.; Long, C.; Cheng, M. Review of Existing Peer-to-Peer Energy Trading Projects. *Energy Procedia* **2017**, *105*, 2563–2568. [CrossRef]
6. Sikorski, J.J.; Haughton, J.; Kraft, M. Blockchain technology in the chemical industry: Machine-to-machine electricity market. *Appl. Energy* **2017**, *195*, 234–246. [CrossRef]
7. Xu, L.D.; He, W.; Li, S. Internet of things in industries: A survey. *IEEE Trans. Ind. Inform.* **2014**, *10*, 2233–2243. [CrossRef]
8. Deb, C.; Zhang, F.; Yang, J.; Lee, S.E.; Shah, K.W. A review on time series forecasting techniques for building energy consumption. *Renew. Sustain. Energy Rev.* **2017**, *74*, 902–924. [CrossRef]
9. Voyant, C.; Notton, G.; Kalogirou, S.; Nivet, M.L.; Paoli, C.; Motte, F.; Fouilloy, A. Machine learning methods for solar radiation forecasting: A review. *Renew. Energy* **2017**, *105*, 569–582. [CrossRef]
10. Csató, L.; Opper, M. Sparse on-line gaussian processes. *Neural Comput.* **2002**, *14*, 641–668. [CrossRef] [PubMed]
11. Laughman, C.; Lee, K.; Cox, R.; Shaw, S.; Leeb, S.; Norford, L.; Armstrong, P. Power signature analysis. *IEEE Power Energy Mag.* **2003**, *1*, 56–63. [CrossRef]
12. Agdas, D.; Srinivasan, R.S.; Frost, K.; Masters, F.J. Energy use assessment of educational buildings: Toward a campus-wide sustainable energy policy. *Sustain. Cities Soc.* **2015**, *17*, 15–21. [CrossRef]
13. Diamantoulakis, P.D.; Kapinas, V.M.; Karagiannidis, G.K. Big Data Analytics for Dynamic Energy Management in Smart Grids. *Big Data Res.* **2015**, *2*, 94–101. [CrossRef]
14. Siano, P. Demand response and smart grids-A survey. *Renew. Sustain. Energy Rev.* **2014**, *30*, 461–478. [CrossRef]

 *processes*

Article

Wind Energy Generation Assessment at Specific Sites in a Peninsula in Malaysia Based on Reliability Indices

Athraa Ali Kadhem [1,*], Noor Izzri Abdul Wahab [2] and Ahmed N. Abdalla [3]

[1] Center for Advanced Power and Energy Research, Faculty of Engineering, University Putra Malaysia, Selangor 43400, Malaysia
[2] Advanced Lightning, Power and Energy Research, Faculty of Engineering, University Putra Malaysia, Selangor 43400, Malaysia
[3] Faculty of Electronics Information Engineering, Huaiyin Institute of Technology, Huai'an 223003, China
* Correspondence: athraaonoz2007@yahoo.com

Received: 12 April 2019; Accepted: 22 May 2019; Published: 27 June 2019

Abstract: This paper presents a statistical analysis of wind speed data that can be extremely useful for installing a wind generation as a stand-alone system. The main objective is to define the wind power capacity's contribution to the adequacy of generation systems for the purpose of selecting wind farm locations at specific sites in Malaysia. The combined Sequential Monte Carlo simulation (SMCS) technique and the Weibull distribution models are employed to demonstrate the impact of wind power in power system reliability. To study this, the Roy Billinton Test System (RBTS) is considered and tested using wind data from two sites in Peninsular Malaysia, Mersing and Kuala Terengganu, and one site, Kudat, in Sabah. The results showed that Mersing and Kudat were best suitable for wind sites. In addition, the reliability indices are compared prior to the addition of the two wind farms to the considered RBTS system. The results reveal that the reliability indices are slightly improved for the RBTS system with wind power generation from both the potential sites.

Keywords: reliability indices; wind farms; Sequential Monte Carlo Simulation; Malaysia

---

1. Introduction

Recent environmental impacts and the depletion of fossil fuel reserves are the main concerns that have stimulated the integration of renewable energy power plants using solar power, wind power, biomass, biogas, etc. as alternative sources of electrical generation. This has inspired global concerns in energy balance, sustainability, security, and environmental preservation [1].

Wind energy is non-depletable, free, environmentally friendly, and almost available globally [2]. It is intermittent, though very reliable from a long-term energy policy viewpoint [3]. In the measure of adequacy, wind energy is regarded as a better choice compared with other energies.

Electric power systems continue to witness the penetration of high-level wind power into the system as a global phenomenon [4], due to the problems associated with power system planning and operation. This makes the assessment of wind power generation system capacities, and their impacts on reliability in the system by appropriate planning, in line with their power utilization and environmental benefits. Thus, high penetration of intermittent wind energy resources into the electric power system requires the need to investigate the system reliability while adding a large amount of varying wind power generation to the system [5].

Owing to the industrial development and growth in the economy, an increase in the demand for electricity is one of the major challenges faced by both developed and developing countries like Malaysia. This has precipitated the Malaysian electrical utility to integrate wind generation based renewable energy into the grid. Many studies have been carried out by researchers to identify the potential location of wind energy systems in Malaysia. This process has been encouraged by both public and private institutions, with the aim of producing green energy [6,7]. In addition, the extraction of power from wind energy is optimized, even in location with average wind speed, by the proper design of wind turbine models that can effectively trap power due to the advancement in technologies [8].

In general, Malaysia experiences low wind speeds, but some particular regions experience strong winds in specific periods of the year [9]. Locations like Mersing experience higher wind speed variations throughout the year, with average wind speeds ranging from 2 m/s to 5 m/s [10]. According to the literature, the wind in Malaysia could be able to generate a great quantity of electric energy despite its lower average wind speeds, especially at the eastern coastal areas or its remote islands [11]. Researchers in [12] applied the Weibull function to investigate the characteristics of wind speed and subsequently evaluated the wind energy generation potential at Chuping and Kangar in Perlis, Malaysia. Furthermore, small capacity wind turbine plants (5–100 kW) have been installed by the Ministry of Rural and Regional development in Sabah and Sarawak [11]. Researchers in [13] stated that ten units of wind turbines with three different rated powers (6, 10, and 15 kW) were used in energy calculation for the area in the north part of Kudat. As the wind turbine units are principally dependent on wind velocity and location, wind speed forecast is essential for siting a new wind generating turbine in a prospective location [14], as the study in Kudat location reveals. Moreover, another study was performed by a research group of the University of Malaysia (UM) using the Weibull distribution function for the analyses of wind energy potential at the sites in Kudat and Labuan in the Sabah region in Malaysia [6]. The outcome of this research demonstrated that Kudat and Labuan are suitable for sitting small-scale wind generating units [15].

The question to ask here is whether it is probable to harness small-scale wind generating units at selected locations in Malaysia for the purpose of electricity generation. So far, studies on wind power characteristics in Malaysia are limited and wind speed depends on geographical and meteorological factors. This study discusses the effect of potential wind power from various locations in Malaysia for adequately reliable power systems. Analysis of the wind speed data characteristics and wind power potential assessment at three given locations in Malaysia was done. The main objective of the paper is to examine the capacity contribution of wind power in generating system adequacy and its impact on generation system reliability. The Sequential Monte Carlo simulation (SMCS) technique and Weibull models are employed to demonstrate the impact of wind power in power system reliability. Also, the results presented in the paper could serve as preliminary data for the establishment of a wind energy map for Malaysia.

This paper is structured in six sections. The introduction includes a brief introduction of the concept for the wind energy potential in Malaysia. The next section describes related work adapted to enable estimation of the wind power potential of the region. Section 3 shows the fundamental reliability indices evaluated in this work, which are used by assessment policy makers to exploit the wind power potential of the region. Section 4 describes the wind speed data analysis at specific sites in Malaysia. Section 5 shows the obtained results of the simulation in the case study, which are also discussed. Finally, Section 6 summarizes the main conclusions of the study.

2. Related Work

2.1. Weibull Distribution for the Estimation of Wind Power and Energy Density

The Weibull distribution is the most well recognized mathematical description of wind speed frequency distribution. The value of the scale parameter c of the Weibull distribution is close to the mean wind speed in actual wind speed data, and because of that, the Weibull distribution is a reasonable fit for the data. Consequently, using the two parameters (shape parameter k and scale parameter c), the Weibull distribution can be used with acceptable accuracy to present the wind speed frequency distribution and to predict wind power output from wind energy conversion system (WECS) [16].

Many numerical methods are employed to estimate the values of the shape parameter k and scale c. The Empirical Method (EM) is used in this paper for calculating the Weibull parameters. The EM can be calculated by employing mean wind speed and the standard deviation, where the Weibull parameters c and k are given by the following equations [17].

$$k = \left(\frac{\sigma}{\bar{v}}\right)^{-1.089} \tag{1}$$

$$c = \frac{\bar{v}}{\Gamma\left(1 + \frac{1}{k}\right)} \tag{2}$$

where σ is standard deviation, $\bar{v}$ is the mean wind speed, Γ is the gamma function, and k can be determined easily from the values of σ and $\bar{v}$, which are computed from the wind speed data set provided with the following formulation [18].

$$\bar{v} = \frac{1}{n} \sum_{i=1}^{n} v_i \tag{3}$$

$$\sigma = \left[\frac{1}{n-1} \sum_{i=1}^{n} \left(v_i - \bar{v}\right)^2\right]^{\frac{1}{2}} \tag{4}$$

where $\bar{v}$ is the mean wind speed (m/s), n is the number of measured data, v_i is wind speed of the observed data in the form of time series of wind speed (m/s), and σ is standard deviation. Once k is obtained from the solution of the above numerical expression (1), the scale factor c can be calculated by the above Equation (2).

Wind power density is a beneficial way of evaluating wind source availability at a potential height. It indicates the quantity of energy that can be used for conversion by a wind turbine [19]. The power that is available in the wind flowing at mean speed, $\bar{v}$, through a wind rotor blade with sweep area, An (m^2), at any particular site can be projected as

$$P(v) = \frac{1}{2} \rho A \left(\bar{v}\right)^3.$$

The monthly or annual mean wind power density per unit area of any site on the basis of a Weibull probability density function can be displayed in [20] as follows:

$$P_D(w) = \frac{p(v)}{A} = \frac{1}{2} \rho c^3 \left(1 + \frac{3}{k}\right) \tag{5}$$

where $p(v)$ is the wind power (Watts), $P_D(w)$ is the mean wind power density (Watts/m^2), ρ is the air density at the site (1.225 kg/m^3), A is the sweep area of the rotor blades (m^2), and Γ (x) is the gamma function.

The extractible mean energy density over a time period (T) is calculated as

$$E_D = \frac{1}{2} \rho c^3 \Gamma \left(1 + \frac{3}{k}\right) T \tag{6}$$

where the time period (T) is expressed a daily, monthly or annual.

2.2. Estimation of Wind Turbine Output Power and Capacity Factor

The performance of how a wind machine located in a site performs can be assessed as mean power output $P_{e,ave}$ over a specific time frame and capacity factor, C_f, of the wind machine. $P_{e,ave}$ determines the total energy production and total income, whereas C_f is a ratio of the mean power output to the rated electrical power P_{rated} of the chosen wind turbine model [21]. Depending on the Weibull distribution parameters, the $P_{e,ave}$ and capacity factor C_f of a wind machine are computed according to the following equations.

$$P_{e,ave} = P_{rated} \left[\frac{e^{-\left(\frac{V_c}{c}\right)^k} - e^{-\left(\frac{V_r}{c}\right)^k}}{\left(\frac{V_r}{c}\right)^k - \left(\frac{V_c}{c}\right)^k} \right] - e^{-\left(\frac{V_c}{c}\right)^k} \tag{7}$$

$$C_f = \frac{P_{e,ave}}{P_{rated}} \tag{8}$$

where V_c is cut-in wind speed and V_r is the rated wind speed of the wind turbine generator (WTG). For an economical and viable investment in wind power, it is advisable that the capacity factor should exceed 25% and be maintained in the range of 25–45% [22].

2.3. Extrapolation of Wind Speed at Different Heights

Indirect wind speed estimation methods consist of measuring wind speed at a lower height and applying an extrapolation model to estimate the wind speed characterization at different elevations. The most commonly used model is the power law [23].

Wind speed increases significantly with the height above ground level, depending on the roughness of the terrain. Therefore, correct wind speed measurements must consider the hub height (H) for the WTG and the roughness of the terrain of the wind site. If measurements are difficult at high elevations, the standard wind speed height extrapolation formula, as in the power law Equation (9) [24], can be used to estimate wind speed at high elevations by using wind speed measured at a lower reference elevation, typically 10 m [25].

$$v = v_0 \left[\frac{H}{H_O} \right]^n \tag{9}$$

where v is the wind speed estimated at desired height, H; v_0 is the wind speed reference hub height H_o, and (n) is the ground surface friction coefficient. The exponent (n) is dependent on factors such as surface roughness and atmospheric stability. Numerically, it ranges from (0.05–0.5) [26]. The normal value of ground surface for every station is approximated 1/7 or 0.143, as suggested by [27] for neutral stability conditions.

3. Reliability Assessment for Generation Systems

3.1. Fundamental Reliability Indices

The Load and generation models are conjoined to produce the risk model of the system. Indices that evaluate system reliability and adequacy can be used to forecast the reliability of the power generating system.

The fundamental reliability indices evaluated in this work are adapted to enable the estimation of the reliability level of the power generating systems, comprised of Loss of Load Frequency (LOLF), Loss of Energy Expectation (LOEE), Loss of Load Duration (LOLD), and Loss of Load Expectation (LOLE).

At present, LOLE represents the reliability index of the electrical power systems used in many countries [3]. The standard level of LOLE is one-day-in ten years or less. This does not mean a full day of shortages once every ten years; rather, it refers to the total accumulated time of shortages, which should not exceed one day in ten years. Therefore, the level of LOLE in this study is used as a reliability index of the generation systems.

The combined Sequential Monte Carlo simulation (SMCS) method (or the Monte Carlo simulation method cooperate with Frequency and Duration method) in [28] enables accurate evaluation of reliability indices. To accurately evaluate the reliability assessment for the overall reliability of generating systems adequacy containing wind energy, an SMCS method was used alongside the Weibull distribution model to generate and repeat the wind speed. The Roy Billinton Test System (RBTS) is an essential reliability test system produced by the University of Saskatchewan (Canada) for educational and research purposes. The RBTS has 11 conventional generating units, each having a power capacity ranging of around 5–40 MW, with an installed capacity of 240 MW and a peak load of 185 MW. Figure 1 shows the single line diagram for the RBTS, and the detailed reliability data for the generating units in the test system are shown in Appendix A. The load model is generally represented as chronological Load Duration Curve (LDC), which is used along with different search techniques. The LDC will generate values for each hour, so there will be 8736 individual values recorded for each year. The chronological LDC hourly load model shown in Figure 2 was utilized, and the system peak load is 185 MW. Besides the traditional generators, the wind farm was comprised of 53 identical WTG units with a rated power of 35 kW, each of which was considered in the current study. A peak load of 1% penetrated wind energy in the RBTS system, which has a peak load of 185 MW.

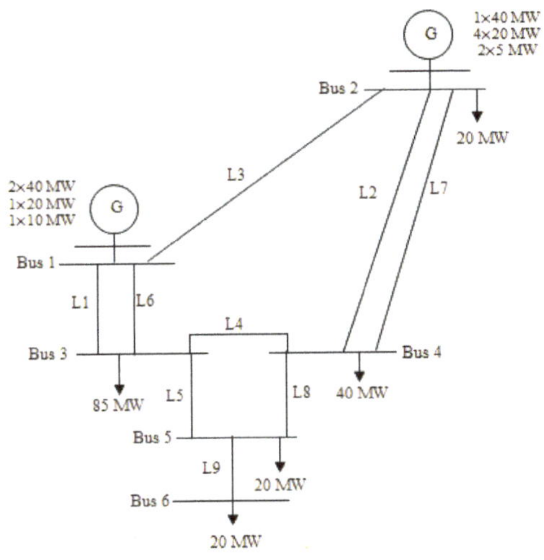

Figure 1. Single line diagrams of the Roy Billinton Test System (RBTS).

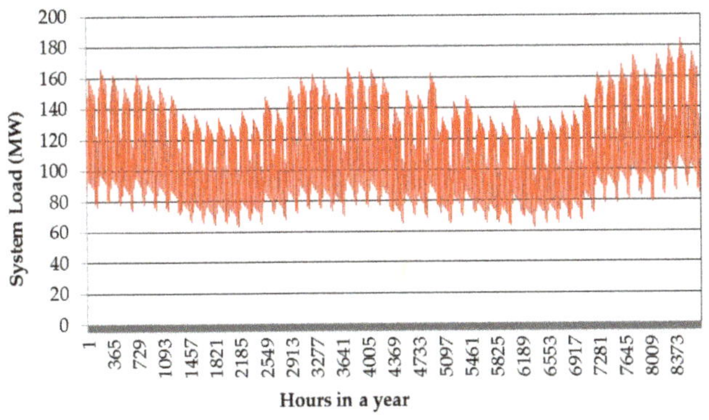

Figure 2. Chronological hourly load (Load Duration Curve (LDC)) model for the RBTS.

3.2. Proposed Methodology

The basic simulation procedures for applying the SMCS with the Weibull model in calculating reliability indices for the electrical power generating systems with wind energy penetration are based on the following steps.

Step 1: The generation of the yearly synthetic wind power time series employs a Weibull model, as follows:

- Set the Weibull distribution parameters k and c.
- Generate a uniformly distributed random number U between (0,1).
- Determined the artificial wind speed v with Equation (10).

$$v = c\left[-ln(U)^{\frac{1}{k}}\right] \tag{10}$$

- Set the WTG's Vci, Vr, and Vco wind speeds.
- Determine the constants A, Bx, and Cx with the equations below.

$$A = \frac{1}{(V_{ci} - V_r)^2}\left\{V_{ci}\left(V_{ci} + V_r\right) - 4V_{ci}V_r\left[\frac{V_{ci} + V_r}{2V_r}\right]^3\right\}$$

$$Bx = \frac{1}{(V_{ci} - V_r)^2}\left\{4\left(V_{ci} + V_r\right)\left[\frac{V_{ci} + V_r}{2V_r}\right]^3 - (3V_{ci} + V_r)\right\}$$

$$Cx = \frac{1}{(V_{ci} - V_r)^2}\left\{2 - 4\left[\frac{V_{ci} + V_r}{2V_r}\right]^3\right\}.$$

- Calculate the WTG output power using Equation (11),

$$P_{WTG} = \begin{cases} 0 & ws < V_{ci} \\ (A + Bx + Cx^2) \times P_r & V_{ci} \leq ws < V_r \\ P_r & V_r \leq ws < V_{co} \\ 0 & ws > V_{co} \end{cases} \tag{11}$$

where ws = wind speed (m/s), V_{ci} = WTG cut-in speed (m/s), V_{co} = WTG cut-out speed (m/s), V_r = WTG rated speed (m/s), and P_r = WTG rated power output (MW). The constants A, Bx, and Cx have previously been calculated by [3].

Step 2: Create the total available capacity generation by a combination of the synthetic generated wind power time series with a conventional chronological generating system model by employing SMCS, as follows:

- Define the maximum number of years (N) to be simulated and set the simulation time (h), (usually one year) to run with SMCS.
- Generate uniform random numbers for the operation cycle (up-down-up) for each of the conventional units in the system by using the unit's annual MTTR (mean time to repair) and λ (failure rate) values.
- The component's sequential state transition processes within the time of all components are then added to create the sequential system state.
- Define the system capacity by aggregating the available capacities of all system components by combining the operating cycles of generating units and the operating cycles with the WTG available hourly wind at a given load level.

- Superimpose the available system capacity curve on the sequential hourly load curve to obtain the available system margin. A positive margin denotes sufficient system generation to meet the system load whereas a negative margin suggests system load shedding.
- The reliability indices for a number of sample years (N) can be obtained using Equations (12)–(18).

$$\Phi_{LOLE}(s) = \begin{cases} 0 & if \ \ sj \ \in \ \ s_{success} \\ 1 & if \ \ sj \ \in \ \ s_{failure} \end{cases} \tag{12}$$

$$\tilde{E}(\Phi_{LOLE}(s)) = \frac{\sum_{i=1}^{N}\left\{\sum_{j=1}^{nj(s)} \Phi_{LOLE}(sji)\right\}}{N} \tag{13}$$

where $i = 1, 2 \dots N$, N = number of years simulated, $\phi(sji)$ = index function analogous to jth occurrence within the year i, $j = 1, 2 \dots, nj(s)$, $nj(s)$ is the number of system state occurrences of (sj) in the year i, $sj = s_{success} \cup s_{failure}$ is the set of all possible states (sj) (i.e., the state-space), and the content of two subspaces s_{sucess} of the success state and $s_{failure}$ of the failure states.

$$\Phi_{LOEE}(s) = \begin{cases} 0 & if \ \ sj \ \in \ \ s_{success} \\ \Delta Pj \times T & if \ \ sj \ \in \ \ s_{failure} \end{cases} \tag{14}$$

$$\tilde{E}(\Phi_{LOEE}(s)) = \frac{\sum_{i=1}^{N}\left\{\sum_{j=1}^{nj(s)} \Phi_{LOEE}(sji)\right\}}{N} \tag{15}$$

where $\Delta Pj \times T$ is the amount of curtailing energy in the failed state (sji).

$$\Phi_{LOLF}(s) = \begin{cases} 0 & if \ \ sj \ \in \ \ s_{success} \\ \Delta \lambda j & if \ \ sj \ \in \ \ s_{failure} \end{cases} \tag{16}$$

$$\tilde{E}(\Phi_{LOLF}(s)) = \frac{\sum_{i=1}^{N}\left\{\sum_{j=1}^{nj(s)} \Phi_{LOLF}(sji)\right\}}{N}. \tag{17}$$

$\Delta \lambda j$, is the sum of the transition rates between sj and all the $s_{success}$ states attained from sj in one transition.

$$LOLD = \frac{LOLE}{LOLF}. \tag{18}$$

- If (N) is equal to the maximum number of years, stop the simulation; otherwise, set (N = N + 1), (h = 0), then return to move 2 and repeat the attempt.

Step 3: Evaluate and update the outcome of the test function for the reliability indices evaluation. The above procedure is detailed in the form of flowchart, as represented in Figure 3.

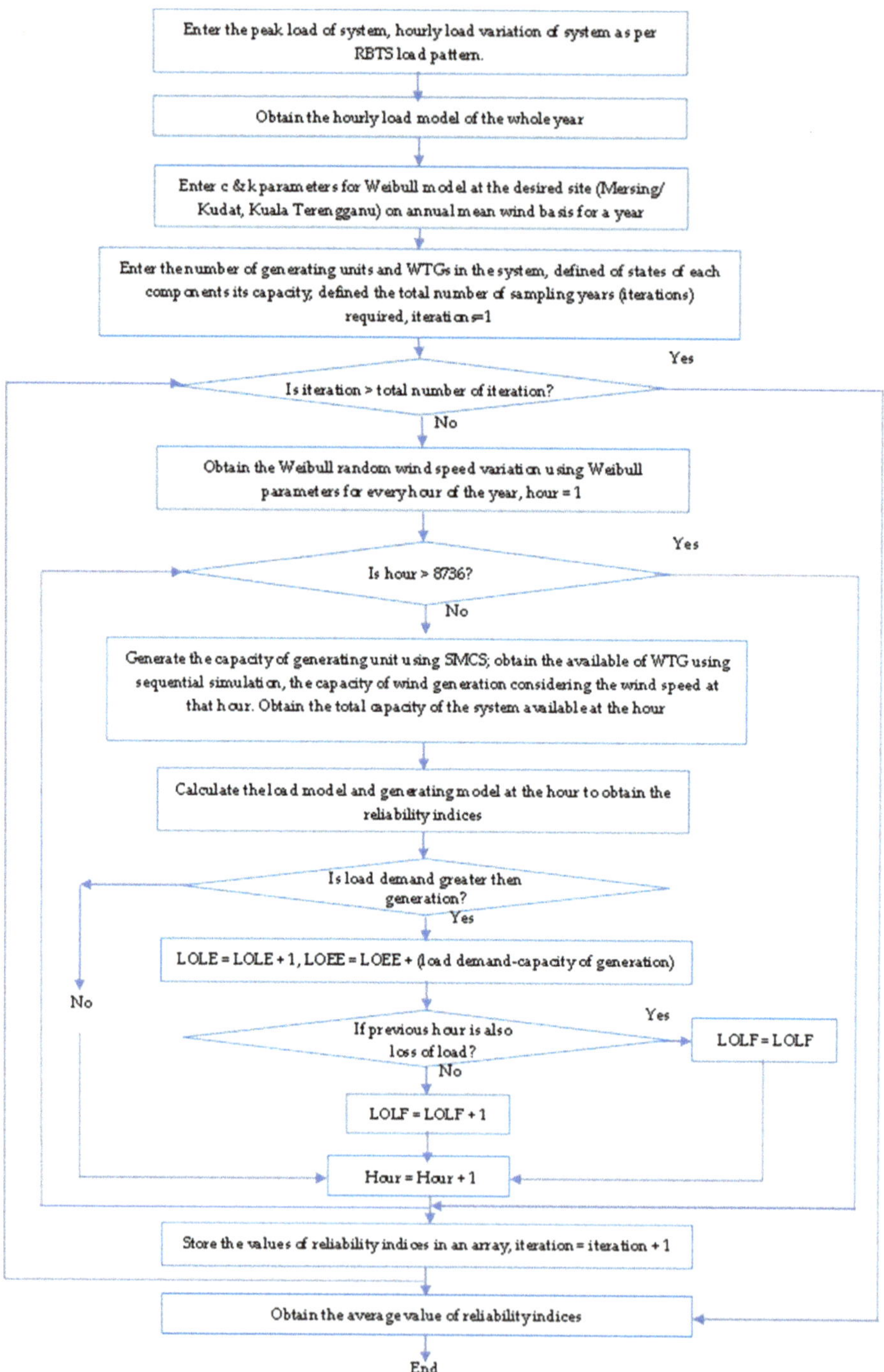

Figure 3. Flow chart showing reliability assessment for a generation system including wind generating sources.

4. Wind Speed Data Analysis at Specific Sites in Malaysia

To estimate the possible potential wind energy site in Malaysia, the analysis, correlation, and prediction of wind data from the location need to be done. It has been often recommended in the literature that making use of the wind data available from meteorological stations increases the vicinity of the proposed candidate site by preliminary estimates of the wind resource potential of the site. Meteorological data that are recorded for long periods need to be extrapolated to obtain an estimation of the wind profile of the site. In this study, wind speed data from Mersing, Kudat, and Kuala Terengganu have been statistically analyzed to propose the wind energy characteristics for these sites. The data for this study were gathered from the Malaysia Meteorological Department (MMD). The data recorded comprise three years of hourly mean surface wind speeds from 2013 to 2015 at three locations in Malaysia. The mean of the wind speed form the simulated process for each hour is calculated based on Weibull parameters. The hourly mean wind speed is then used in the sequential simulation process. Figure 4 shows the locations of MMD stations in Peninsular Malaysia. This map was drawn by using the Arc Graphical Information System (AGIS) software and depicts the strength of the wind speed distribution in Mersing, Kudat, and Kuala Terengganu. The area that showed the highest wind speed value is in red and orange, while other areas show moderate wind speeds. Table 1 presents a description of the selected regions in Malaysia, which consist of the latitude, longitude, and elevation of the anemometer.

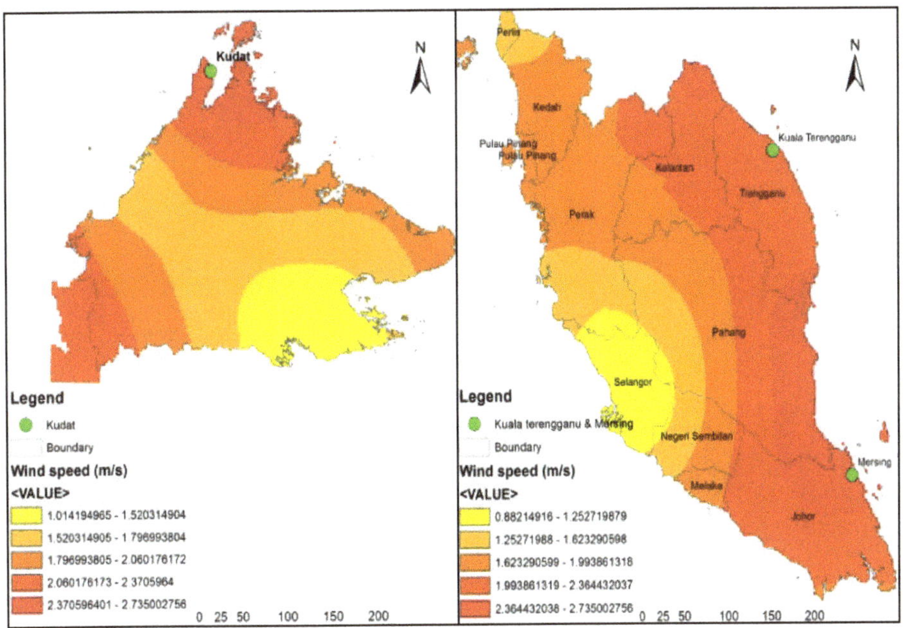

Figure 4. Wind speed distribution maps for station sites used in this study in Malaysia.

Table 1. Description of the wind speed stations at selected regions in Malaysia.

Station	Latitude	Longitude	Altitude (m)
Mersing	2°27′ N	103°50′ E	43.6
Kuala Terengganu	5°23′ N	103°06′ E	5.2
Kudat	6°55′ N	116°50′ E	3.5

4.1. Estimation of Average Wind Speed with Different Height

In this study, the scale parameter (c) of Weibull distribution was applied to test the generated wind speed data from the Weibull model at the Mersing, Kudat, and Kuala Terengganu sites. The monthly average wind speed data for the three sites within the three year period, with a scale parameter, are presented in Figure 5. It is clear from the figures that for the entire three years of the wind speed data and scale parameter c of the Weibull distribution they showed similar variations during one year.

Wind speed is typically measured at standard heights, such as 10 m, but there is the need to obtain wind speed values at a high level in cases where the electricity is generated by the mean power of a wind turbine [29]. The wind power law has been acknowledged to be a beneficial tool and is frequently employed in assessing the wind power where wind speed data at different elevations must be adjusted to a standard height before use. In this study, the exponent n of the power law is set to 0.143 for the Kudat and Kuala Terengganu sites, as suggested by [13]. Meanwhile, for the Mersing site, the exponent n of the power law is set to 0.5, according to the nature of the ground. Wind data taken from the MMD station were measured at a level height of 43.6 m for Mersing. However, wind data of the Kudat and Kuala Terengganu sites require the data to be converted at a height of 10 m above hub height, because the wind turbine always runs at elevations above 10 m height.

Figure 6 shows both monthly and annual mean wind speeds at the sites (Mersing, Kudat, and Kuala Terengganu) for the average years of 2013–2015; these results were extrapolated to a different height. In this study, the wind speed was extrapolated for various heights (mean wind observation station, 60 m, and 100 m) in the wind observation stations at Kudat, Kuala Terengganu, and Mersing (extrapolated). The obtained results are presented in Table 2.

All the tabulated values reveal that the wind speed increases with an increase in elevation. For Mersing, the annual average wind speed was 2.82 m/s, 3.31 m/s, and 4.27 m/s at wind observation station elevations of 60 m, and 100 m, respectively. Furthermore, the annual average wind speeds in Kudat and Kuala Terengganu were 2.45 m/s, 3.68 m/s, and 3.95 m/s; and 2.03 m/s, 2.89 m/s, and 3.10 m/s at wind observation station elevations of 60 m, and 100 m, respectively.

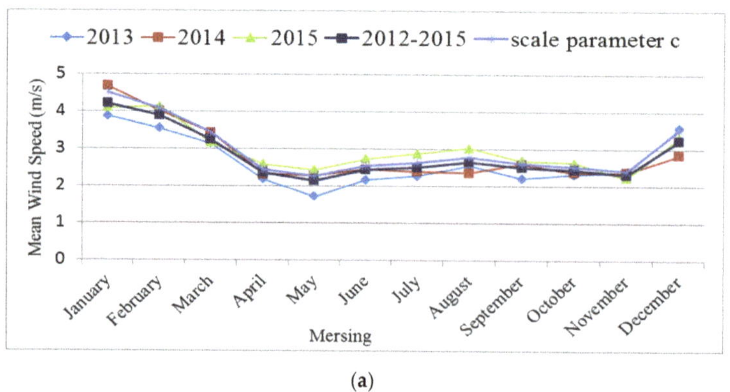

(a)

Figure 5. *Cont.*

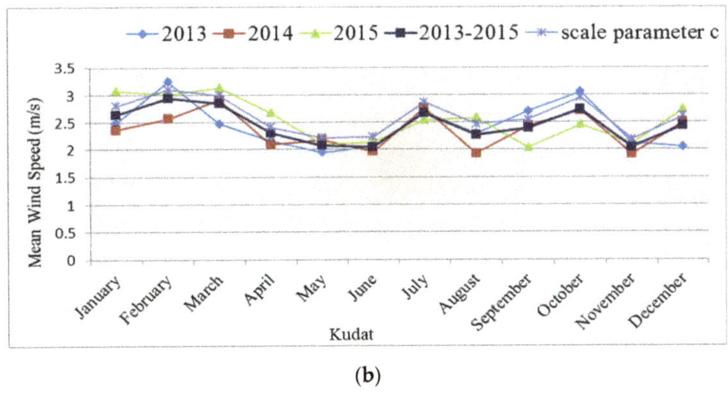

(**b**)

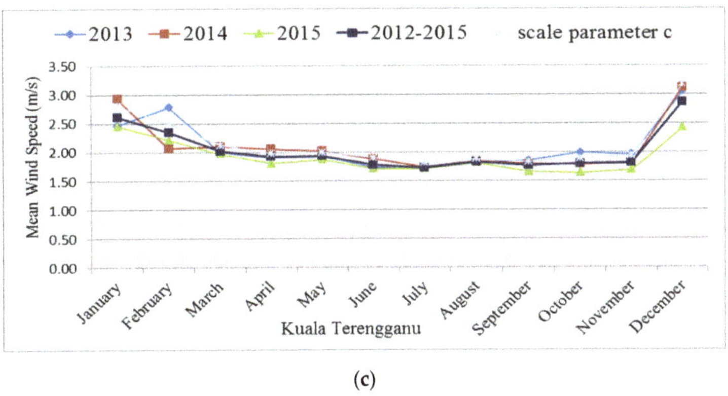

(**c**)

Figure 5. Comparison between monthly mean wind speed data and the scale parameter (c) over 3 years period at sites (**a**) Mersing, (**b**) Kudat, and (**c**) Kuala Terengganu.

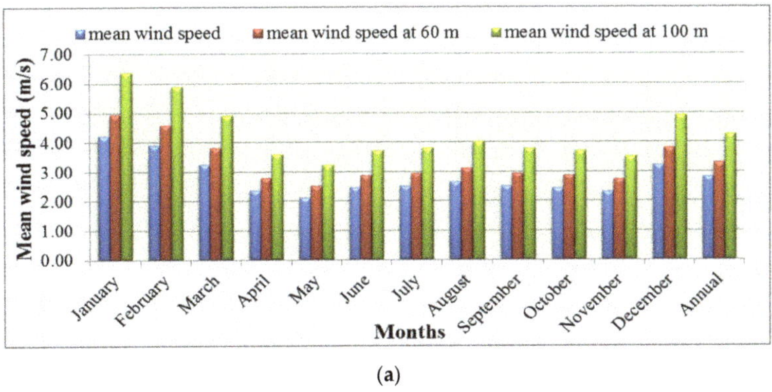

(**a**)

Figure 6. *Cont.*

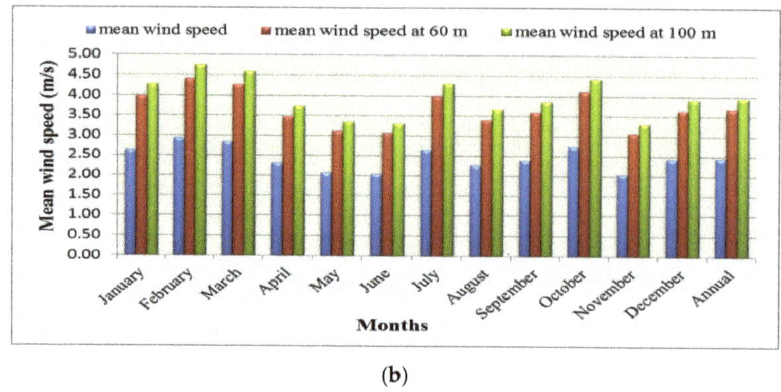

(b)

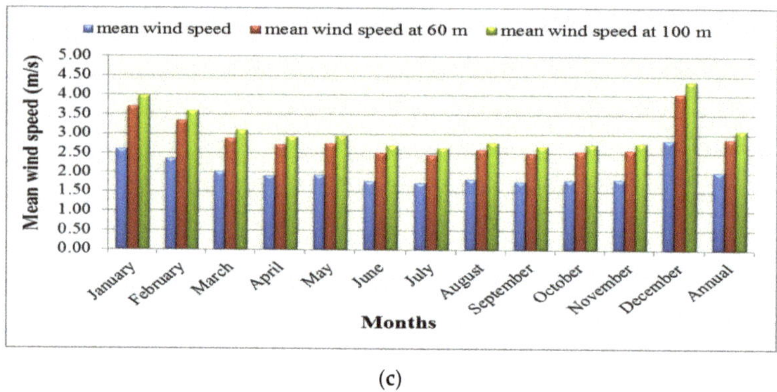

(c)

Figure 6. Monthly and annual mean wind speeds at sites (**a**) Mersing, (**b**) Kudat, and (**c**) Kuala Terengganu.

4.2. Estimation of Wind Power Density and Energy Density at Different Heights

The observed wind speed data at the station were converted to 100 m height wind speed data using Equation (9), and then the converted data were used to determine the wind potential. The scale c and shape k Weibull parameters were estimated using the EM. The wind power and energy density were measured, respectively, by Equations (5) and (6) at heights of (43.6 m) and (100 m) in Mersing, as shown in Table 3. The rest of the calculations for wind power and energy density at heights of 3.5 and 100 m and 5.2 and 100 m in Kudat and Kuala Terengganu, respectively, can be found in Appendix B.

From Table 3, it is observed that the maximum power density from the actual wind speed of Mersing, Kudat, and Kuala Terengganu was found to be 52 W/m², 19 W/m², and 20 W/m², respectively. However, the maximum power density of Mersing, Kudat, and Kuala Terengganu, when the actual wind speed data was converted to 100 m height, was calculated to be 180 W/m², 80 W/m², and 72 W/m², respectively. Here, it is evident that the Mersing site has a higher mean monthly power density compared to Kudat and Kala Terengganu under various heights.

Table 2. Monthly and annual mean wind speed (m/s) in Mersing, Kudat, and Kuala Terengganu at different heights above ground level.

Wind Observation Station	Months/Year												Annual Mean
	Jan.	Feb.	Mar.	Apr.	May	Jun.	Jul.	Aug.	Sep.	Oct.	Nov.	Dec.	
Mersing													
mean wind speed	4.22	3.89	3.24	2.36	2.13	2.45	2.51	2.65	2.50	2.44	2.32	3.24	2.82
60 m	4.94	4.57	3.81	2.77	2.50	2.87	2.95	3.11	2.94	2.87	2.73	3.80	3.31
100 m	6.38	5.89	4.91	3.58	3.23	3.71	3.80	4.01	3.79	3.70	3.52	4.91	4.27
Kudat													
mean wind speed	2.65	2.94	2.84	2.31	2.07	2.05	2.66	2.27	2.39	2.74	2.05	2.43	2.45
60 m	3.97	4.41	4.27	3.47	3.11	3.08	4.00	3.41	3.59	4.12	3.08	3.64	3.68
100 m	4.28	4.75	4.59	3.73	3.35	3.31	4.30	3.66	3.86	4.43	3.31	3.92	3.95
Kuala Terengganu													
mean wind speed	2.61	2.35	2.03	1.92	1.94	1.78	1.73	1.82	1.76	1.80	1.82	2.86	2.03
60 m	3.70	3.34	2.88	2.72	2.75	2.52	2.45	2.59	2.50	2.55	2.58	4.05	2.89
100 m	3.98	3.59	3.10	2.93	2.96	2.71	2.64	2.78	2.69	2.75	2.77	4.36	3.10

Table 3. Wind power and energy density characteristics at heights of 43.6 m and 100 m in Mersing.

Months/Year	A Height of 43.6 m						A Height of 100 m					
	$\bar{v}$	k	c	P_D (w/m²)	Hours	E_D (kWh/m²)	$\bar{v}$	k	c	P_D (w/m²)	Hours	E_D (kWh/m²)
January	4.22	5.37	4.572	52.070	747	38.896	6.38	5.37	6.922	180.702	747	134.9841
February	3.89	5.68	4.209	40.532	675	27.359	5.89	5.68	6.373	140.698	675	94.97137
March	3.24	3.84	3.588	26.214	747	19.582	4.91	3.84	5.432	90.960	747	67.94703
April	2.36	3.89	2.612	10.086	723	7.292	3.58	3.89	3.955	35.014	723	25.31527
May	2.13	3.26	2.381	8.010	747	5.984	3.23	3.26	3.605	27.803	747	20.76847
June	2.45	3.67	2.716	11.488	723	8.306	3.71	3.67	4.112	39.866	723	28.823
July	2.51	3.37	2.798	12.859	747	9.606	3.80	3.37	4.237	44.653	747	33.35577
August	2.65	3.00	2.968	16.014	747	11.962	4.01	3.00	4.494	55.591	747	41.52655
September	2.50	3.35	2.788	12.746	723	9.215	3.79	3.35	4.221	44.232	723	31.97939
October	2.44	3.67	2.710	11.412	747	8.525	3.70	3.67	4.103	39.605	747	29.58467
November	2.32	3.88	2.568	9.590	723	6.934	3.52	3.88	3.887	33.257	723	24.04445
December	3.24	3.45	3.604	27.287	747	20.384	4.91	3.45	5.456	94.674	747	70.72135
Annual	2.82	2.25	3.221	24.370	-	17.790	4.27	2.25	4.877	84.595	-	61.754

The annual mean power density of Mersing, Kudat, and Kuala Terengganu varies between 84.59 W/m^2, 79.28 W/m^2, and 33.36 W/m^2 at a height of 100 m. The annual power density is also less than 100 W/m^2 for all the locations, and, therefore, these locations can be categorized as a class 1 wind energy resource. This wind energy resource class, in general, is inappropriate for large-scale wind turbine applications. Nevertheless, the generation of small-scale wind energy at a turbine height of 100 m [6] is viable. However, for small-scale applications, and in the long-term with the development of wind turbine technology, the use of wind energy continues to hold great promise.

4.3. Estimation of the Suitable Wind Turbine Units at Malaysia Sites

The selection of the wind turbine should be made with a rated wind speed that corresponds to the maximum energy wind speed in order to maximize energy output. For the annual energy output, the selected wind turbine will have the maximum capacity factor, defined by the ratio of the actual power generated to the rated power output [30]. The average power output values, $P_{e,ave}$, and C_f, are crucial performance factors of the wind energy conversion system (WECS).

The technical data of six differently sized wind turbines are summarized in Table 4. The summarized information in Table 4 is obtained from [13,19]. The cut-in wind speed, or the speed at which the turbine commences power production, is 2.7 m/s for four of the six turbines, while for the other two turbines, the cut-in wind speed values are 2 and 3.5 m/s, respectively. The cut-out wind speed of 25 m/s applies to all the turbines. Table 4 represents the information pertaining to the rated speed, rated output power, hub height, and rotor diameter of the wind turbines analyzed.

Table 4. The technical data of wind turbines.

Characteristics	P10-20	591672E	P12-25	G-3120	P15-50	P25-100
Rated power (kw)	20	22	25	35	50	100
Hub height (m)	-	30	-	42.7	-	-
Rotor diameter (m)	10	15	12	19.2	15.2	25
Cut-in wind speed (m/s)	2.7	2	2.7	3.5	2.7	2.7
Rated wind speed (m/s)	10	10	10	8	12	10
Cut-off wind speed (m/s)	25	25	25	25	25	25

Depending on the turbine's characteristics in Table 4, and the Weibull parameters derived from applying EM using the Matlab toolbox, the electrical output of the wind turbines can be made available by using the formulation earlier defined in Equation (7).

Knowing the output power of the wind turbines, it is then possible to obtain a computation of the average output power value of each wind turbine. As the capacity factor of a wind turbine is the ratio of its average output power to its rated power, the energy output data are employed in calculating the capacity factor of the wind turbines, which are of sizes 20, 22, 25, 35, 50, and 100 kW. A comparison of the capacity factors computed for various wind turbines at different heights is presented in Figure 7.

From Figure 7, it can be seen that the capacity factor goes up as the hub height increases. Moreover, the capacity factor increases for wind turbines of a size of 35 kW. In Mersing, the maximum capacity factor is achieved as about 23.66% for the Endurance America model of the G-3120 kW wind turbine, whereas in Kuala Terengganu, the lowest capacity factor is achieved as approximately 7.82% for the Endurance America model of the G-3120 kW wind turbine. Kudat, with about 19.21%, ranks second in terms of capacity factors compared to the regions.

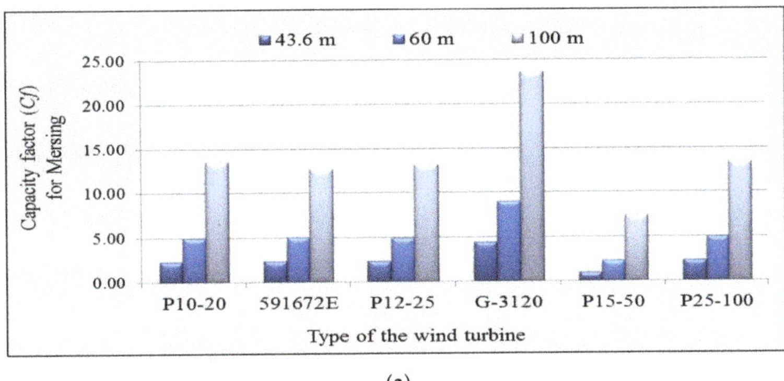

(**a**)

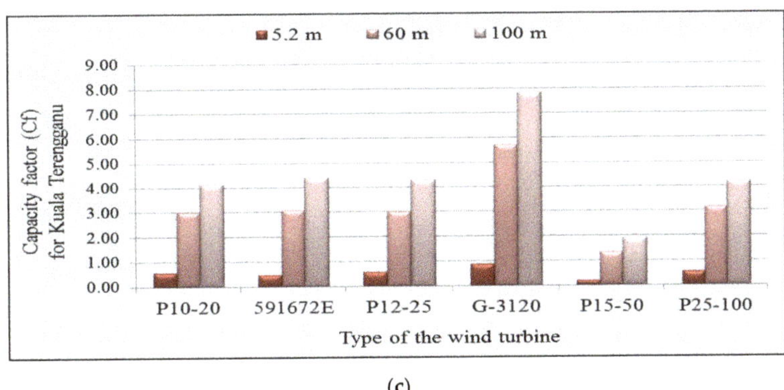

(**b**)

(**c**)

Figure 7. Comparison of the capacity factors obtained for different wind turbines at various heights for sites (**a**) Mersing, (**b**) Kudat, and (**c**) Kuala Terengganu.

The G-3120 (35 kW) wind turbine has the highest capacity factors of 23.66%, 19.21%, and 7.82%, at suggested heights of 100 m for Mersing, Kudat, and Kuala Terengganu, respectively, among the models considered. Therefore, the reliability analysis was carried out only for Mersing and Kudat, which have high capacity factors, whereas Kuala Terengganu was not considered as it has low capacity factors.

5. Results and Discussion

In this section, the reliability indices evaluation of generating systems for wind power generation using a sequential Monte Carlo simulation (SMCS) is presented. In addition, the strategies for wind farm operation at Malaysian sites (Mersing and Kudat) are presented and compared by assessing the reliability of wind energy generation when adding to the RBTS test system [31].

5.1. Case Studies

As reported in the literature, two wind generating stations suggested at the specific sites in Malaysia, Mersing and Kudatas, have low wind speed and thus require small-scale rated power wind turbines of around 35 kW for installation in two selected locations for reliability analysis.

Reliability analysis using the simulation technique suggested in this paper is applied to the RBTS, which contains the WECS. The hourly wind data obtained from the two locations—Mersing and Kudat—atre used for studying the hourly wind speed of the Weibull model considered for the simulation. Then, the Weibull parameters c and the k are obtained by the empirical method. The obtained values were used to generate hourly wind speed data for deducing the available wind power from the wind turbine generators (WTG) chosen for both of the sites for reliability assessment.

The values of c are around 4.88 and 4.46 m/s, and the values of k for wind speed distribution are 2.25 and 1.84 for Mersing and Kudat at the proposed height of 100 m, respectively; these values were obtained by simulation. The WTG unit that is selected for installation in the farm has the following specifications: $Vci = 3.5$ m/s, $Vr = 8$ m/s, and $Vco = 25$ m/s, and the rated power output of every WTG unit is $Pr = 35$ kW [19]. Figures 8 and 9 show the simulated output power with 35 kW for each WTG in the sampling year, and the simulation of the farm with output power is 1.85 MW for 53 WTG units in Mersing.

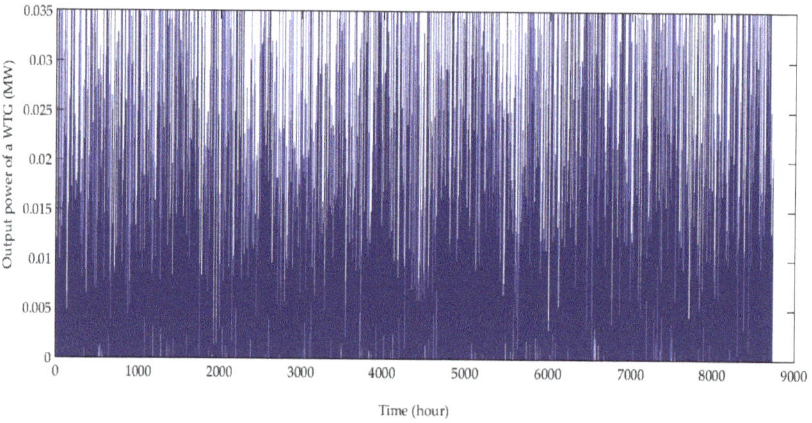

Figure 8. Simulation of the output power from WTG for the sampling year.

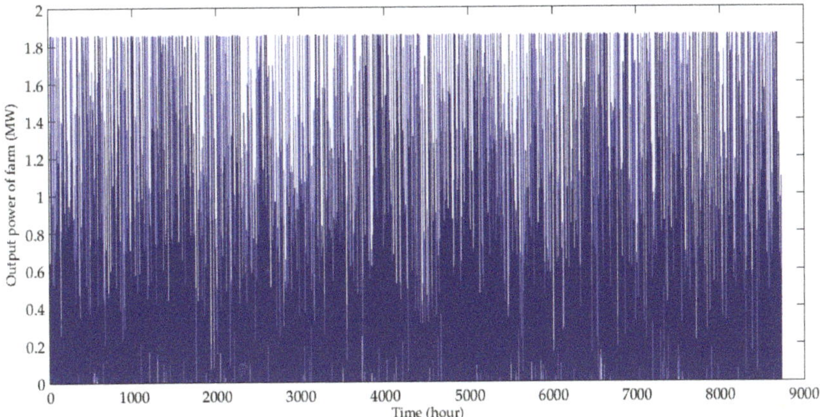

Figure 9. Simulation of the output power from wind farm for the sampling year.

The RBTS was simulated for 600 trials using the SMCS method. The simulation proceeded in chronological order from one hour to the next, repeatedly, using yearly samples until the specified convergence criteria were met. Figure 10 shows the available capacity for the power system containing wind power generation from the wind farm in Mersing during the simulated process for yearly samples and the superimposition of the available capacity with the chronological load model. It can be seen from this state of the system that the available capacity of the power generating system is not sufficient to meet the load demands. Thus, there are some intersections that are seen in the diagram. Figure 11 represents the reliability indices for simulation with (600) sampling years. The values of LOLE, the amount of the LOEE, and the frequency of losing a load during the simulation process are depicted in Figure 11 for wind power in the Mersing site.

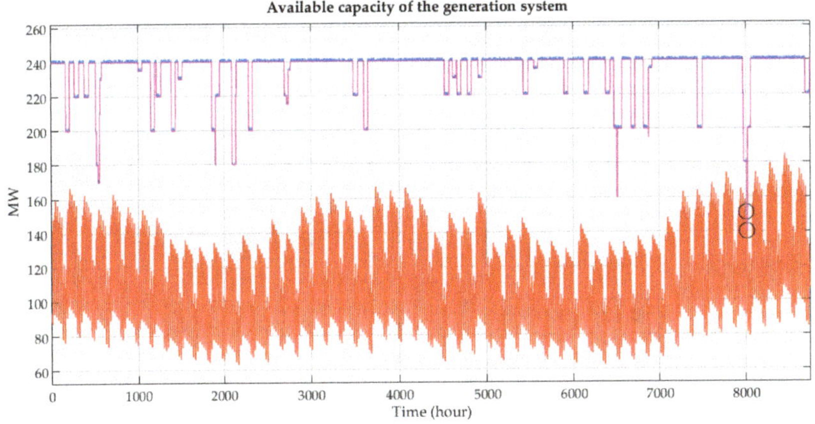

Figure 10. The available capacity of the generation system which is superimposed with the chronologically available load model.

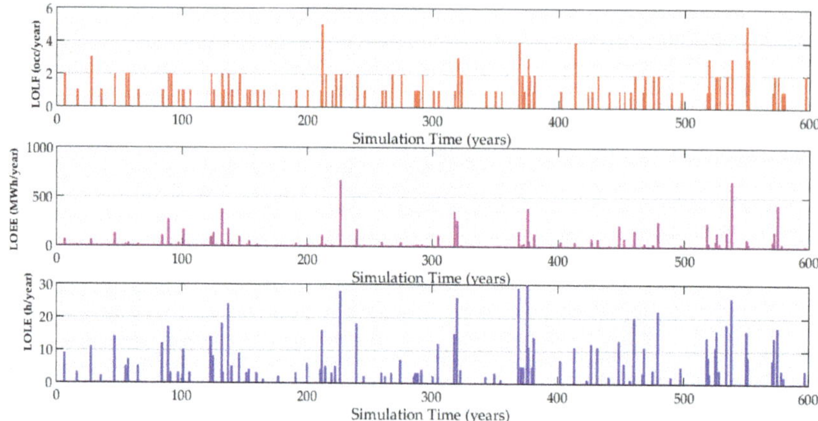

Figure 11. Simulation of reliability indices with (600) sampling years.

5.2. Calculated Reliability Indices for RBTS Including Wind Power Generation

To evaluate the contribution of wind energy to the overall reliability of the generating systems, Table 5 compares the reliability indices before and after adding the 53 WTGs and 106 WTGs to the conventional units of RBTS. The results obtained were compared with the results obtained from the SMCS method reported in [32]. The simulation process was terminated after a set number of samples (600 times) had been achieved. The results show that the reliability indices demonstrate a distinguished and slightly improved reliability of RBTS, including wind power from both locations (Mersing and Kudat) by the addition of 1.85 MW and 3.71 MW from the proposed wind farms. The LOLE and LOEE indices are typically employed to gauge the extent of benefits in assessing the wind energy of generating systems. Therefore, after adding the wind generating 1.85 MW to the system, the LOLE index was reduced to 1.115 and 1.131 h/year for Mersing and Kudat, respectively, when compared with the results from the base case, which shows the reliability assessment of the power generation system. Additionally, after adding the wind generating 3.71 MW to the system, the LOLE index wass reduced to 0.987 and 1.128 h/year for Mersing and Kudat, respectively, when compared with the results from the base case, which shows the reliability assessment of the power generation system.

Table 5. Reliability indices at various sites in Malaysia.

Name of Site	Reliability Indices			
	LOLE (hrs/year)	LOEE (MWh/year)	LOLF (occ/year)	LOLD (hrs/occ)
Basic RBTS system without wind generation (published)	1.152	11.78	0.229	4.856
Basic RBTS system without wind generation (computed)	1.161	10.191	0.230	5.05
Basic RBTS system and (53 × 0.035 = 1.85 MW) wind generators at Mersing site	1.115	9.744	0.225	4.944
Basic RBTS system and (106 × 0.035 = 3.71 MW) wind generators at Mersing site	0.987	7.357	0.220	4.486
Basic RBTS system and (53 × 0.035 MW) wind generators at Kudat site	1.131	10.948	0.225	5.012
Basic RBTS system and (106 × 0.035 = 3.71 MW) wind generators at Kudat site	1.128	10.018	0.236	4.779

Loss of Load Expectation (LOLE)/ LOLE (hour/year); Loss of Energy Expectation (LOEE)/ LOEE (MWh/year); Loss of Load Frequency (LOLF)/ LOLF (occurrence/year); Loss of Load Duration (LOLD)/ LOLD (hour/occurrence).

Further, it can be observed that this study was done with a small percentage of peak load reduction at around 1% and a small number of wind turbines, to demonstrate the primary effect of wind energy penetration from selected locations in Malaysia in the reliability of the generation system for the RBTS.

6. Conclusions

In this paper, analyses of the wind speed data characteristics and wind power potential assessment at three given locations in Malaysia were done. In addition, this study tests the effects of the potential wind power from different locations. An SMCS technique is used to show the effects of wind energy for the RBTS test system by a set of reliability indices. The results reveal that the wind power connected to the RBTS test system is only from two locations in Malaysia. Further, the reliability indices are compared prior to and after the addition of the two farms to the considered system. The results show that the reliability indices are slightly improved for RBTS, including wind power from both locations, as suggested. Moreover, the wind resources at specific sites in Malaysia are more suitable for small-scale standalone energy conversion systems and could also be hybrid energy systems.

Recommendations for future studies include extending the statistical analysis model used for different sites in Malaysia to include more relevant factors for wind farms and evaluating their impact on wind power potential for these sites, such as wind speeds at the installation site, types of wind turbine offshore and onshore, and numbers of wind turbines installed, according to the size of the farm.

Author Contributions: This work was part of the Ph.D. research carried out by A.A.K. The research is supervised by N.I.A.W. and A.N.A. The project administration and funding acquisition by University Putra Malaysia.

Funding: The authors are grateful for financial support from the University Putra Malaysia (UPM), Malaysia, under Grant No. GP-IPB-9630000.

Conflicts of Interest: The authors declare no conflict of interest.

Appendix A

Table A1. The RBTS generating unit ratings and reliability data.

Units No.	Unit Size (MW)	FOR	MTTF (Hours)	MTTR (Hours)
1	5	0.01	4380	45
2	5	0.01	4380	45
3	10	0.02	2190	45
4	20	0.02	3650	55
5	20	0.02	3650	55
6	20	0.02	3650	55
7	20	0.02	3650	55
8	20	0.03	1752	45
9	40	0.02	2920	60
10	40	0.03	1460	45
11	40	0.03	1460	45

Forced outage rate (FOR); Mean time to failure (MTTF); Mean time to repair (MTTR).

Appendix B

Table A2. Wind power and energy density characteristics at heights of 3.5 m and 100 m in Kudat.

	A Height of 3.5 m						A Height of 100 m					
Months/Year	$\bar{v}$	k	c	P_D (w/m²)	Hours	E_D (kWh/m²)	$\bar{v}$	k	c	P_D (w/m²)	Hours	E_D (kWh/m²)
January	2.65	3.77	2.931	14.347	747	10.717	4.28	3.77	4.734	60.450	747	45.156
February	2.94	4.00	3.244	19.217	675	12.972	4.75	4.00	5.239	80.946	675	54.639
March	2.84	3.55	3.158	18.213	747	13.605	4.59	3.55	5.100	76.709	747	57.301
April	2.31	2.87	2.594	10.905	723	7.884	3.73	2.87	4.190	45.957	723	33.227
May	2.07	2.48	2.339	8.682	747	6.485	3.35	2.48	3.777	36.555	747	27.307
June	2.05	1.99	2.314	10.142	723	7.333	3.31	1.99	3.738	42.753	723	30.911
July	2.66	2.43	3.004	18.648	747	13.930	4.30	2.43	4.851	78.528	747	58.660
August	2.27	2.12	2.562	12.922	747	9.653	3.66	2.12	4.137	54.406	747	40.641
September	2.39	2.16	2.698	14.835	723	10.725	3.86	2.16	4.356	62.433	723	45.139
October	2.74	2.86	3.077	17.015	747	12.710	4.43	2.86	4.969	76.777	747	57.353
November	2.05	2.80	2.303	7.723	723	5.584	3.31	2.80	3.719	32.524	723	23.515
December	2.43	3.27	2.708	11.772	747	8.794	3.92	3.27	4.373	49.574	747	37.032
Annual	2.45	1.84	2.759	18.819	-	13.738	3.95	1.84	4.456	79.284	-	57.877

Table A3. Wind power and energy density characteristics at height of 5.2 m and 100 m in Kuala Terengganu.

	A Height of 5.2 m						A Height of 100 m					
Months/Year	$\bar{v}$	k	c	P_D (w/m²)	Hours	E_D (kWh/m²)	$\bar{v}$	k	c	P_D (w/m²)	Hours	E_D (kWh/m²)
January	2.6097	3.24	2.912	14.685	747	10.969	3.98	3.24	4.443	52.157	747	38.961
February	2.3543	3.10	2.632	11.020	675	7.439	3.59	3.10	4.017	39.177	675	26.445
March	2.0300	2.54	2.287	7.991	747	5.969	3.10	2.54	3.490	28.396	747	21.212
April	1.9203	2.63	2.161	6.601	723	4.772	2.93	2.63	3.298	23.463	723	16.964
May	1.9381	3.16	2.165	6.089	747	4.548	2.96	3.16	3.303	21.622	747	16.152
June	1.7760	2.68	1.997	5.154	723	3.726	2.71	2.68	3.048	18.324	723	13.248
July	1.7277	3.14	1.930	4.324	747	3.230	2.64	3.14	2.946	15.378	747	11.487
August	1.8242	3.61	2.024	4.774	747	3.566	2.78	3.61	3.088	16.954	747	12.664
September	1.7642	3.16	1.971	4.594	723	3.322	2.69	3.16	3.007	16.314	723	11.795
October	1.7993	2.81	2.020	5.202	747	3.886	2.75	2.81	3.083	18.496	747	13.816
November	1.8165	2.87	2.038	5.288	723	3.823	2.77	2.87	3.110	18.793	723	13.587
December	2.8554	2.88	3.203	20.496	747	15.310	4.36	2.88	4.887	72.799	747	54.381
Annual	2.0338	2.09	2.293	9.390	-	6.854	3.10	2.09	3.499	33.366	-	24.356

References

1. Ouammi, A.; Dagdougui, H.; Sacile, R.; Mimet, A. Monthly and seasonal of wind energy characteristics at four monitored locationals in Liguria region (Italy). *Renew. Sustain. Energy Rev.* **2010**, *14*, 1959–1968. [CrossRef]
2. Chaiamarit, K.; Nuchprayoon, S. Modeling of renewable energy resources for generation reliability evaluation. *Renew. Sustain. Energy Rev.* **2013**, *26*, 34–41. [CrossRef]
3. Shi, S.; Lo, K.L. An Overview of Wind Energy Development and Associated Power System Reliability Evaluation Methods. In Proceedings of the 2013 48th International Universities Power Engineering Conference (UPEC), Dublin, Ireland, 2–5 September 2013; pp. 1–6.
4. Benidris, M.; Mitra, J. Composite Power System Reliability Assessment Using Maximum Capacity Flow and Directed Binary Particle Swarm Optimization. In Proceedings of the IEEE Conference, Manhattan, KS, USA, 22–24 September 2013; pp. 1–6.
5. Padma, L.M.; Harshavardham, R.P.; Janardhana, N.P. Generation Reliability Evaluation of Wind Energy Penetrated Power System. In Proceedings of the International Conference on High Performance Computing and Applications (ICHPCA), Odisha, India, 22–24 December 2014; pp. 1–4.
6. Islam, M.R.; Saidur, R.; Rahim, N.A. Assessment of wind energy potentiality at Kudat and Labuan, Malaysia using Weibull distribution function. *Energy* **2011**, *36*, 985–992. [CrossRef]
7. Taylor, P.; Khatib, T.; Sopian, K.; Ibrahim, M.Z. Assessment of electricity generation by wind power in nine costal sites in Malaysia. *Int. J. Ambient Energy* **2013**, 37–41. [CrossRef]
8. Gebrelibanos, K.G. *Feasibility Study of Small Scale Standalone Wind Turbine for Urban Area: Case Study: KTH Main Campus*; KTH School of Industrial Engineering and Management, Energy Technology EGI: Stockholm, Sweden, 2013.
9. Siti, M.R.S.; Norizah, M.; Syafrudin, M. The Evaluation of Wind Energy Potential in Peninsular Malaysia. *Int. J. Chem. Environ. Eng.* **2011**, *2*, 284–291.
10. Kadhem, A.A.; Abdul, W.N.I.; Aris, I.; Jasni, J.; Abdalla, A.N. Advanced Wind Speed Prediction Model Based on Combination of Weibull Distribution and Artificial Neural Network. *Energies* **2017**, *10*, 1744. [CrossRef]
11. Borhanazad, H.; Mekhilef, S.; Saidur, R.; Boroumandjazi, G. Potential application of renewable energy for rural electrification in Malaysia. *Renew. Energy* **2013**, *59*, 210–219. [CrossRef]
12. Irwanto, M.; Gomesh, N.; Mamat, M.R.; Yusoff, Y.M. Assessment of wind power generation potential in Perlis, Malaysia. *Renew. Sustain. Energy Rev.* **2014**, *38*, 296–308. [CrossRef]
13. Albani, A.; Ibrahim, M.Z.; Yong, K.H. Wind Energy Investigation in Northern Part of Kudat, Malaysia. *Int. J. Eng. Appl. Sci.* **2013**, *2*, 14–22.
14. Goh, H.H.; Lee, S.W.; Chua, Q.S.; Teo, K.T.K. Wind energy assessment considering wind speed correlation in Malaysia. *Renew. Sustain. Energy Rev.* **2016**, *54*, 1389–1400. [CrossRef]
15. Hwang, G.H.; Lin, N.S.; Ching, K.B.; Wei, L.S. Wind Farm Allocation In Malaysia Based On Multi-Criteria Decision Making Method. In Proceedings of the National Postgraduate Conference (NPC), Seri Iskandar, Malaysia, 19–20 September 2011; pp. 1–6.
16. Chang, T.P. Performance comparison of six numerical methods in estimating Weibull parameters for wind energy application. *Appl. Energy* **2011**, *88*, 272–282. [CrossRef]
17. Kaoga, D.K.; Serge, D.Y.; Raidandi, D.; Djongyang, N. Performance Assessment of Two-parameter Weibull Distribution Methods for Wind Energy Applications in the District of Maroua in Cameroon. *Int. J. Sci. Basic Appl. Res.* **2014**, *17*, 39–59.
18. Kidmo, D.K.; Danwe, R.; Doka, S.Y.; Djongyang, N. Statistical analysis of wind speed distribution based on six Weibull Methods for wind power evaluation in Garoua, Cameroon. *Rev. Des. Energ. Renouv.* **2015**, *18*, 105–125.
19. Adaramola, M.S.; Oyewola, O.M.; Ohunakin, O.S.; Akinnawonu, O.O. Performance evaluation of wind turbines for energy generation in Niger. *Sustain. Energy Technol. Assess.* **2014**, *6*, 75–85. [CrossRef]
20. Ohunakin, O.S.; Adaramola, M.S.; Oyewola, O.M. Wind energy evaluation for electricity generation using WECS in seven selected locations in Nigeria. *Appl. Energy* **2011**, *88*, 3197–3206. [CrossRef]
21. Oyedepo, S.O.; Adaramola, M.S.; Paul, S.S. Analysis of wind speed data and wind energy potential in three selected locations in south-east Nigeria. *Int. J. Energy Environ. Eng.* **2012**, *3*, 1–11. [CrossRef]

22. Anurag, C.; Saini, R.P. Statistical Analysis of Wind Speed Data Using Weibull Distribution Parameters. In Proceedings of the 1st International Conference on Non-Conventional Energy (ICONCE 2014), Kalyani, India, 16–17 January 2014; pp. 160–163.

23. Molina-García, A.; Fernández-Guillamón, A.; Gómez-Lázaro, E.; Honrubia-Escribano, A.; Bueso, M.C. *Vertical Wind Profile Characterization and Identification of Patterns Based on a Shape Clustering Algorithm*; IEEE: New York, NY, USA, 2019; Volume 7, pp. 30890–30904.

24. Ahmed, A.S. Wind energy as a potential generation source at Ras Benas, Egypt. *Renew. Sustain. Energy Rev.* **2010**, *14*, 2167–2173. [CrossRef]

25. Hussain, M.Z.; Roy, D.K.; Khan, S.; Sharma, P.K.; Talukdar, R. Wind Energy Potential at Different Cities of Assam Using Statistical Models. *Int. J. Adv. Res. Innov.* **2018**, *6*, 38–43.

26. Ayodele, T.R.; Ogunjuyigbe, A.S.O.; Amusan, T.O. Wind power utilization assessment and economic analysis of wind turbines across fifteen locations in the six geographical zones of Nigeria. *J. Clean. Prod.* **2016**, *129*, 341–349. [CrossRef]

27. Chandel, S.S.; Ramasamy, P.; Murthy, K.S.R. Wind power potential assessment of 12 locations in western Himalayan region of India. *Renew. Sustain. Energy Rev.* **2014**, *39*, 530–545. [CrossRef]

28. Shi, S. Operation and Assessment of Wind Energy on Power System Reliability Evaluation. Ph.D. Thesis, Department of Electronic and Electrical Engineering, University of Strathclyde, Glasgow, Scotland, 2014.

29. Dursun, B.; Alboyaci, B. An Evaluation of Wind Energy Characteristics for Four Different Locations in Balikesir. *Energy Sour. Part A: Recovery Util. Environ. Eff.* **2011**, *33*, 1086–1103. [CrossRef]

30. Ayodele, T.R.; Jimoh, A.A.; Munda, J.L.; AgeeWind, J.T. distribution and capacity factor estimation for wind turbines in the coastal region of South Africa. *Energy Convers Manag.* **2012**, *64*, 614–625. [CrossRef]

31. Heshmati, A.; Najafi, H.R.; Aghaebrahimi, M.R.; Mehdizadeh, M. Wind Farm Modeling For Reliability Assessment from the Viewpoint of Interconnected Systems. *Electr. Power Compon. Syst.* **2012**, *40*, 257–272. [CrossRef]

32. Kadhem, A.A.; Abdul Wahab, N.I.; Aris, I.; Jasni, J.; Abdalla, A.N. Computational techniques for assessing the reliability and sustainability of electrical power systems: A review. *Renew. Sustain. Energy Rev.* **2017**, *80*, 1175–1186. [CrossRef]

 *processes*

Article

A Model for Optimizing Location Selection for Biomass Energy Power Plants

Chia-Nan Wang, Tsang-Ta Tsai * and Ying-Fang Huang

National Kaohsiung University of Science and Technology, Kaohsiung 80778, Taiwan;
cn.wang@newfancy.com (C.-N.W.); winner@nkust.edu.tw (Y.-F.H.)
* Correspondence: tsangtatsai2019@gmail.com

Received: 9 May 2019; Accepted: 4 June 2019; Published: 8 June 2019

Abstract: In addition to its potential for wave power, wind power, hydropower, and solar power, it can be said that Vietnam is a country with great potential for biomass energy derived from agricultural waste, garbage, and urban wastewater, which are resources widely available across the country. This huge amount of biomass, however, if left untreated, could become a major source of pollution and cause serious impacts on ecosystems (soil, water, and air), as well as on human health. In this research, the authors present a fuzzy multicriteria decision-making model (FMCDM) for optimizing the site selection process for biomass power plants. All of the criteria affecting location selection are identified by experts and literature reviews; in addition, the fuzzy analytic hierarchy process (FAHP) method was utilized so as to identify the weight of all of the criteria in the second stage. Furthermore, the Technique for Order Preference by Similarity to an Ideal Solution (TOPSIS) is applied for ranking potential locations in the final stage of this research. As a result, Long An (DMU/005) was found to be the best location for building biomass energy in Vietnam. The main contributions of this work include modeling the site selection decision process under fuzzy environment conditions. The proposed approaches also can address the complex problems in site selection; it is also a flexible design model for considering the evaluation criteria, and is applicable to location selection for other industries.

Keywords: biomass energy; site selection; optimization; MCDM; FMCDM; FAHP; TOPSIS

---

1. Introduction

In the context of increasingly depleting domestic fossil fuels, rising world oil prices, and increasing reliance on world energy prices, the ability to meet energy requirements for domestic demand is increasingly becoming difficult, as well as a major challenge. Thus, considering that the exploitation of clean renewable energy (RE) has important meaning in terms of economy, society, food security, and sustainable development, the energy demand in Vietnam has increased at twice the rate of the gross domestic product (GDP) growth, while this rate is only approximately 1% in developed countries. Vietnam's energy consumption has increased four times since 2005. From 1998 to 2008, the total electricity consumption demand in Vietnam increased by about 400% compared with the prior period. If this trend continues, Vietnam will become an energy importer in the future [1]. Vietnam's power transmission system includes voltage levels of 500, 220, and 110 kV. A power transmission system of 500 KV with a total length of 4670 km from north to south facilitates the transmission of electricity exchange among the north, central, and south of Vietnam. Circuit 1 of the 500-kV line was put into operation in September 1994, while Circuit 2 was put into operation in late 2005. Although the power distribution system is in relatively good condition, it still has high power loss. Overloaded lines, transformers operating with low efficiency, and poor-quality cables are the main causes of high losses [2].

In addition to the potential for wave power, wind power, hydropower, and solar power, it can be said that Vietnam is a country with great potential for creating biomass energy from agricultural waste, garbage, and urban wastewater, which are widely distributed across the country. This huge amount of biomass, however, if left untreated, will become a major source of pollution and continue to cause serious impacts on ecosystems (soil, water, and air), as well as on human health [1].

In recent years, interest in developing renewable energy technologies to replace fossil energy sources has increased globally, as fossil energy sources are in danger of depletion and fuel costs are increasing. Vietnam is a long-term agricultural country; furthermore, raw materials needed for producing biomass energy are about 118 million tons/year [3]. Vietnam has favorable natural conditions, such as a hot and humid climate, heavy rain, and fertile land; as a result, biomass grows quickly. Byproducts from agriculture and forestry are abundant and constantly increasing. However, these byproducts are typically considered natural waste, thus becoming more dangerous and causing environmental pollution. Therefore, Vietnam has many favorable conditions for developing biomass energy. Taking advantage of biomass energy will simultaneously provide energy for economic development and ensure environmental protection [3].

Vietnam's impressive economic reform over the past two decades has been accompanied by a sharp increase in energy demand. The Vietnamese government's strategy is to ensure sustainable economic growth in the future. Therefore, the Vietnamese government has set renewable energy development targets, including those for solar energy, wind energy, and biomass energy. Biomass energy is targeted to achieve 2.1% of the total electricity output by 2030 [4].

As an agricultural country, Vietnam has great potential to develop biomass energy. Accordingly, the ability to sustainably exploit biomass for energy production in Vietnam is about 150 million tons/year [5]. The main types of energy biomass include wood, waste byproducts from crops, livestock waste, municipal waste, and other organic waste. Sources of renewable energy can be burned directly or be used to create biomass fuel pellets [5]. Currently, on the global scale, biomass is the fourth-largest energy resource, accounting for 14%–15% of the world's total energy consumption. In developing countries, biomass is often the largest energy resource, averaging about 35% of the total energy supply [6].

One of the most important activities of the biomass energy project is site selection for building plants. A good location of a biomass plant is of vital importance for the project's economic survival, and it includes solving complex location and transportation problems. In this research, we propose an MCDM model for biomass power plant location selection in Vietnam. All of the criteria affecting the location selection are identified by experts and literature reviews; furthermore, the fuzzy analytic hierarchy process (FAHP) method was utilized in order to identify the weight of all of the criteria in the second stage. TOPSIS was applied for ranking the potential locations in the final stage of this research. The general flow of the MCDM model is shown in Figure 1.

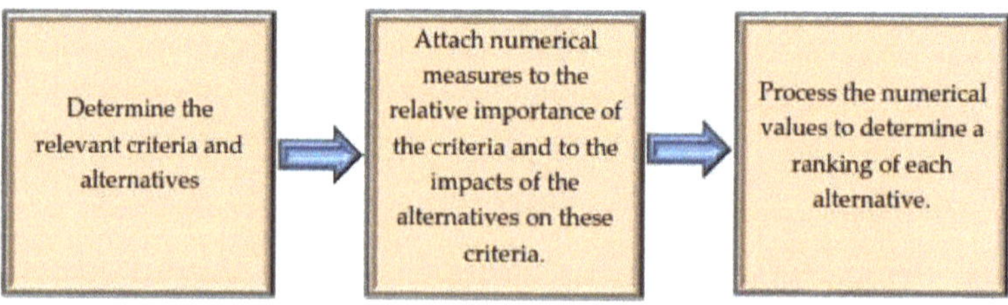

Figure 1. General flow of multicriteria decision-making model (MCDM) model [7,8].

The primary goal of this research is to propose a useful fuzzy MCDM model for biomass power plant location selection. The proposed approaches can also address different complex problems in site selection; in addition, it is also a flexible design model for considering the evaluation criteria, and it is applicable to location selection for other industries.

The remainder of the research is to provide background to assist the authors in building the fuzzy MCDM model. Then, an integrated model using the FAHP–TOPSIS approach is introduced so as to select the best site for biomass power plant from eight potential locations in Vietnam. The results and contributions will be presented at the end of this paper.

2. Literature Review

Nowadays, there are many studies that have applied the MCDM model to select locations in the energy sector and some other industries, such as Roghayeh Ghasempour et al. [9], who have reviewed an application of the MCDM model for solar plant location selection and technology. In this paper, a wide variety MCDM methods, investigated by various researchers, are presented so as to obtain effective criteria for selecting solar plant sites and solar plant technologies. C. Chen et al. [10] proposed an optimization model for biomass location selection. In this research, the number of biomass power plants, optimum site, and transportation schemes were obtained with minimum cost, minimum energy consumption, and the lowest effect on the environment. Jakkawan Patomtummakan et al. [11] presented a mixed-integer linear programming for biomass power plant location selection. The objective of the optimization model is to minimize the total cost at selected bio-power plants by considering four cost components, namely, the fixed opening of the proposed bio-power plants, the material cost from purchasing biomass, the transportation cost between suppliers and bio-power plants, and the inventory holding cost.

Massimiliano Cattafi et al. [12] proposed an integer linear programming approach for defining the energy and cost-efficient biomass plant location, along with the corresponding provisioning basin. This optimization tool is just a small part of a wider perspective that is aimed at defining the decision support tools for the improvement of regional planning and its precise strategic environmental assessment. Heesung Woo et al. [13] integrated a multi-criteria analysis (MCA) and geographical information systems (GIS) for optimizing the site of the biomass energy plants. The research results show that land use, resource availability, and supply chain cost data can be integrated and mapped using GIS, in order to facilitate the determination of different sustainable factors weightings, and to ultimately generate optimal candidate sites for biomass energy plants.

Hao Lv et al. [14] proposed a multi-objective mixed-integer programming approach to solve the location selection issue for a straw-based power generation plant with CO_2 emissions. It is anticipated that this paper will make a contribution to the current scientific knowledge by presenting innovative approaches for the sustainable utilization of forest harvest residues as a resource for the generation of bioenergy in Tasmania. Wang et al. [15] applied an MCDM model for solid waste to energy plant location selection. Jin Su Jeong et al. [16] proposed a multicriteria GIS assessment with a weighted linear combination (WLC) to various disciplines, using suitable criteria to optimize a biomass facility location. This assessment could be used in studies to verify suitable biomass plant sites with corresponding geographical and spatial circumstances and available spatial data necessary in various governmental and industrial sectors. Ali Asghar Isalou et al. [17] proposed an integrated fuzzy logic and analytic network process (FANP) to locate a suitable location for the landfilling municipal solid waste generated in Iran. Their findings revealed that the integration of fuzzy logic and ANP can give a better idea compared with other models like AHP, fuzzy logic, and ANP (individually). Therefore, this model can be applied in site selection for landfill of other similar places.

Wang et al. [18] proposed an MCDM model for solar panel site selection in Vietnam. In this work, the authors applied a data envelopment analysis (DEA) model for the selection of potential locations, FAHP was used for determining the weight of sub-factors, and TOPSIS was applied for ranking potential options. Wang et al. [19] proposed an MCDM model for wind power plant location selection. This study resided in the evolution of a new model that is flexible and practical for a decision-maker for site selection in the energy sector. Abdolvahhab Fetanat et al. [20] proposed a hybrid MCDM model for offshore wind power plants location selection in Iran. The evaluation factors and this model could be used in other coastal cities for promoting the progress of integrated coastal management (ICM), towards the goal of sustainability. Mostafa Rezaei-Shouroki [21] proposed an MCDM model by using a hybrid data envelopment analysis (DEA), AHP, and fuzzy TOPSIS model for wind farm location selection. The purpose of this study is to prioritize and rank 13 cities of the Fars province in Iran in terms of their suitability for the construction of a wind farm.

Fahime Heidarzade et al. [22] applied a MCDM model for wind farm site selection in Iran. In this research, a step-wise weight assessment ratio analysis (SWARA) is employed to rank factor affects to wind power plant site, and the weighted aggregates sum product assessment (WASPAS) is utilized to evaluate the decision-making unit (DMU). Halil Ibrahim Cobuloglu et al. [23] proposed a MCDM model for biomass location selection. In this work, the authors used an FANP for identifying the weight of all of the criteria. Yasir Ahmed Solangi et al. [24] integrated Delphi and AHP and fuzzy TOPSIS for the ranking and selection of renewable energy resources. The study provides important insights related to the prioritizing of RE resources for electricity generation, and can be used to undertake policy decisions toward sustainable energy planning in Pakistan. Mohammad Alhuyi Nazari et al. [25] used a TOPSIS model for analyzing solar farm location selection. The primary aim of this paper is to select suitable sites for photovoltaic installation in Iran. S. Saelee et al. [26] applied a TOPSIS multi-criteria approach for biomass type selection for boilers. Babak Daneshvar Rouyendegh et al. [27] applied intuitionistic fuzzy TOPSIS in the location selection of wind power plants in Turkey. The main purpose of the TOPSIS method is to rank the alternatives in the worst way. The intuitionistic fuzzy set (IFS) is used to reflect the approval, rejection, and hesitation of decision makers by dealing with real life uncertainty, imprecision, vagueness, and linguistic human decisions. Choudhary et al. [28] used a hybrid MCDM model including a social, technical, economic, environmental, and political (STEEP), fuzzy AHP and TOPSIS model for thermal power plant location selection in India. The paper presents a more accurate, effective, and systematic decision support tool for decision makers to conduct the evaluation process and to select optimal locations for TPPs.

Based on the literature review and experts' opinion, there are some factors that must be considered in the biomass power plant location selection process, such as economic, environmental, technical, and social-political factors, and there are many researchers who have applied the MCDM model to various fields of science and engineering—a trend that has been increasing for many years—but very few works have focused on this problem in a fuzzy environment.

3. Methodology

3.1. Research Development

Many researchers have applied the MCDM model to various fields of science and engineering, more so over the last few years. One field in the MCDM approach has identified location selection problems; thus, especially in the renewable energy sector, decision-makers have to evaluate both qualitative and quantitative criteria. Although some studies have reviewed applications of MCDM approaches in biomass power plant location selection, to the best of our knowledge, few works have focused on this problem in a fuzzy environment. This is why the authors proposed an MCDM model for optimizing location selection for biomass energy power plants in this study. There are three main steps in this research.

Step 1: All of the criteria and subcriteria affecting the site evaluation and selection processes are determined based on experts and literature reviews.

The processes of the selection criteria are shown in Figure 2.

Step 2: The FAHP model was utilized to identify the weight of all of the subcriteria in the second stage.

The FAHP method has many advantages compared with other multi-objective decision-making methods. First, many multicriteria decision-making methods face difficulties in determining the importance of each criterion, while FAHP is a well-known method of determining these weights. Therefore, FAHP can be combined with other easy methods to take advantage of each method in problem-solving. In addition, FAHP can check consistency in decision-makers' judgment. Moreover, the hierarchical analysis process is easy to understand, considering many small criteria and analyzing all of the qualitative and quantitative factors.

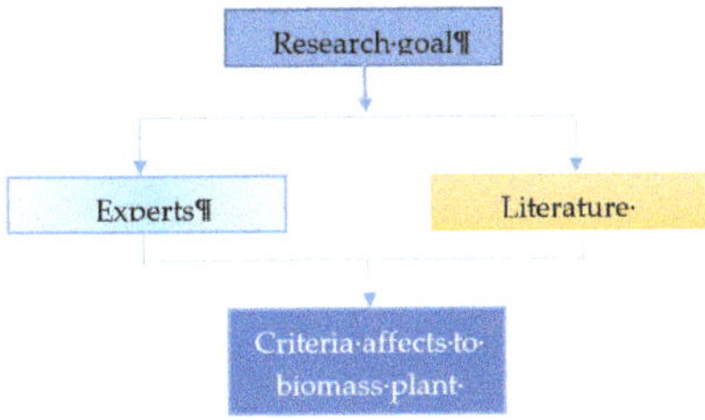

Figure 2. Processes of selection criteria.

Step 3: The TOPSIS model is used for ranking potential locations in the final stage.

The advantages of the TOPSIS methods are simplicity, rationality, comprehensibility, good computational efficiency, and the ability to measure the relative performance for each alternative in a simple mathematical form. TOPSIS is based on the concept that the chosen alternative should have the shortest geometric distance from the positive ideal solution (PIS) and the longest geometric distance from the negative ideal solution (NIS). Diagram of research as shown in Figure 3.

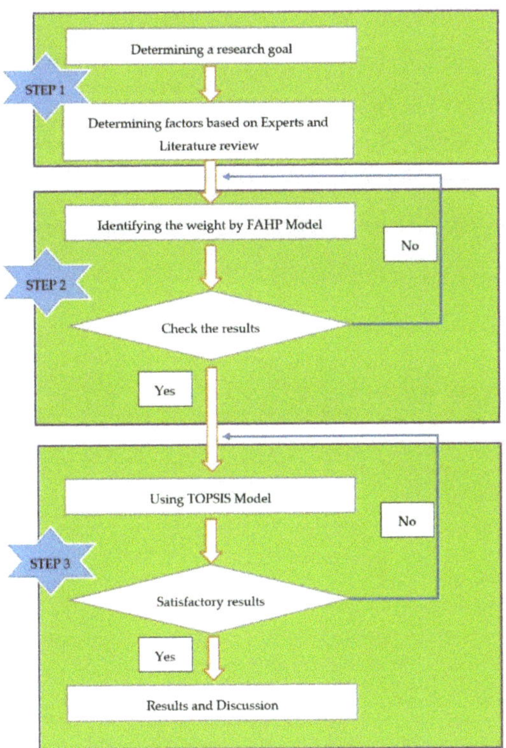

Figure 3. Diagram of research. FAHP—fuzzy analytic hierarchy process.

3.2. Fuzzy Sets, AHP, and TOPSIS Model

3.2.1. Fuzzy Sets and Fuzzy Number

In 1965, Lotfi A. Zadeh published the article with the title "Fuzzy Set", which describes the mathematics of "fuzzy set" and "fuzzy logic" theory. The triangular fuzzy number (TFN) can be defined as (g, p, s). TFNs are shown in Figure 4.

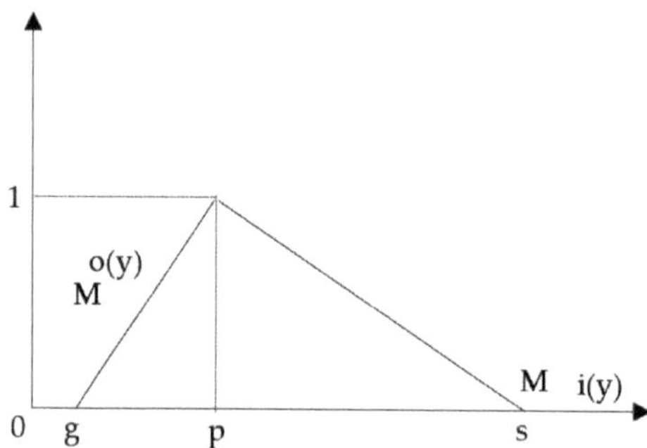

Figure 4. Traingular fuzzy number.

TFN also can be defined as follows:

$$\left(\frac{x}{M}\right) = \begin{cases} 0, & x < o, \\ \frac{x-g}{p-g} & g \leq x \leq p, \\ \frac{s-x}{s-p} & p \leq x \leq s, \\ g, & x > s, \end{cases} \tag{1}$$

$$\mu$$

The representatives of each level of membership give a fuzzy number, as follows:

$$\widetilde{M} = (M^{o(y)}, M^{i(y)}) = [g + (p-k)y, \ s + (p-s)y], y \in [0,1] \tag{2}$$

$o(y)$, $i(y)$ indicates both the left side and the right side of a fuzzy number, respectively, as follows:

$$\begin{aligned} (g_1, \ p_1, \ s_1) + (g_2, \ p_2, \ s_2) &= (g_1 + g_2, \ p_1 + p_2, s_1 + s_2) \\ (g_1, \ p_1, \ s_1) - (g_2, \ p_2, \ s_2) &= (g_1 - g_2, \ p_1 - p_2, s_1 - s_2) \\ (g_1, \ p_1, \ s_1) \times (g_2, \ p_2, \ s_2) &= (g_1 \times g_2, \ p_1 \times p_2, s_1 \times s_2) \\ \frac{(g_1, \ p_1, \ s_1)}{(g_2, \ p_2, \ s_2)} &= (k_1/k_2, \ p_1/p_2, s_1/s_2) \end{aligned} \tag{3}$$

3.2.2. Analytic Hierarchy Process (AHP) Model

AHP is presented by Saaty. AHP is an MCDM that simplifies complex problems by sorting criteria and options in a hierarchical structure. Let $F = \{F_a | a = 1, 2, \ldots, m\}$ is a factor set. The pairwise comparison metrics on m criteria will be presented in an m × m evaluation matric (D). Every element, h_{ab}, is the quotient of the weights of the factors, as follows:

$$D = (h_{ab}), a, b = 1, \ldots, m \tag{4}$$

The relative priorities are given by the Eigenvector (l) corresponding to the largest eigenvector (λ_{max}), as follows:

$$D_l = \lambda_{max} l \tag{5}$$

The consistency is determined by the relation between the entries of D and CI, as follows:

$$CI = \frac{(\lambda_{max} - m)}{(m - 1)} \tag{6}$$

CR is calculated as the ratio of the CI and the RI, as shown in Equation (7):

$$CR = \frac{CI}{RI} \tag{7}$$

CR value must less than (or equal) 0.1. If the CR >0.1, the evaluation needs to be repeated again for improving consistency.

3.2.3. Technique for Order Preference by Similarity to an Ideal Solution (TOPSIS)

TOPSIS assumes that we have m alternatives (options) and n attributes/criteria, and we have the score of each option with respect to each criterion.

- Construct the normalized decision matrix

$$e_{ij} = \frac{X_{ij}}{\sqrt{\sum_{i=1}^{m} X_{ij}^2}}. \tag{8}$$

with $i = 1, 2, \ldots, m$; and $j = 1, 2, \ldots, n$.
- Construct the weighted normalized decision matrix

$$S_{ij} = W_i e_{ij}. \tag{9}$$

with $i = 1, 2, \ldots, m$ and $j = 1, 2, \ldots, n$.
- Determine the ideal and negative ideal solutions

$$A^+ = s_1^+, s_2^+, \ldots, s_n^+; \\ A^- = s_1^-, s_2^-, \ldots, s_n^-; \tag{10}$$

- Calculate the separation measures for each alternative

$$D_i^+ = \sqrt{\sum_{j=1}^{m} \left(s_i^+ - s_{ij}\right)^2}; \ i = 1, 2, \ldots, m. \tag{11}$$

options to NIS.

$$D_i^- = \sqrt{\sum_{j=1}^{m} \left(s_{ij} - s_i^-\right)^2}; \ i = 1, 2, \ldots, m. \tag{12}$$

- Calculate the relative closeness to the ideal solution (G_i)

$$G_i = \frac{D_i^-}{D_i^- + D_i^+} \ i = 1, 2, \ldots, m. \tag{13}$$

4. Case Study

The total potential of Vietnam's biomass energy source is more than 99 million tons/year, corresponding to the electric energy source of more than 340,000 GWh, of which the largest is the

Mekong River Delta (Mekong Delta), accounting for 33.4%, followed by the North Central Coast and South Central Coast with 21.8% [29].

According to the Vietnam Institute of Energy, agricultural byproducts from the Mekong Delta amount for about 23 million tons/year, which includes about 3.8 million tons of rice husk, nearly 17 million tons of rice straw, over 372,000 tons of corn, and nearly 1.4 million tons of bagasse [30]. Thus, the authors proposed an MCDM model for optimizing location selection for biomass energy power plants in the Mekong Delta. In a renewable energy project, selecting a location to build a power plant includes complex decision-making, which includes economic, natural, and social factors. The material source of this project is rice husk. After preliminary studies, eight potential locations in the Mekong Delta (DMUs) were selected (see Table 1 and Figure 5).

Table 1. A potential locations list. DMU—decision making unit.

No.	Location's Name	Symbol
1	Kien Giang	DMU/001
2	An Giang	DMU/002
3	Dong Thap	DMU/003
4	Soc Trang	DMU/004
5	Long An	DMU/005
6	Tra Vinh	DMU/006
7	Tien Giang	DMU/007
8	Can Tho	DMU/008

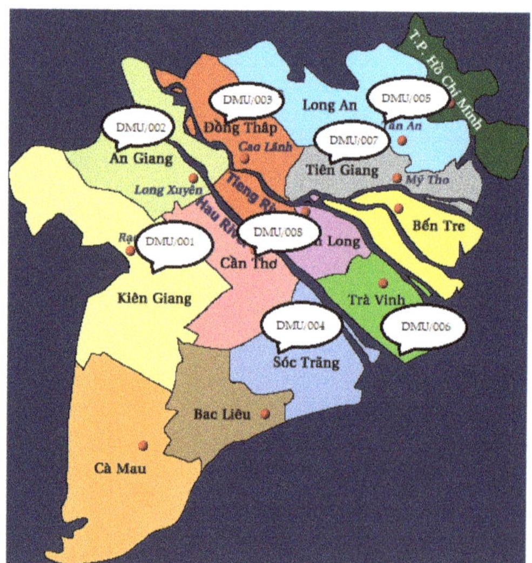

Figure 5. Mekong Delta map.

In the first stage of this work, the main factors and subcriteria affects on location selection are identified by experts and literature reviews. The general hierarchy structure of the proposed model is shown in Figure 6.

The fuzzy analytic hierarchy process (FAHP) method was utilized to identify the weight of all of the criteria in the second stage. The results are shown in Table 2.

The TOPSIS is applied for ranking potential locations in the final stage of this research. The normalized matrix and normalized weight matrix are shown in Tables 3 and 4, respectively.

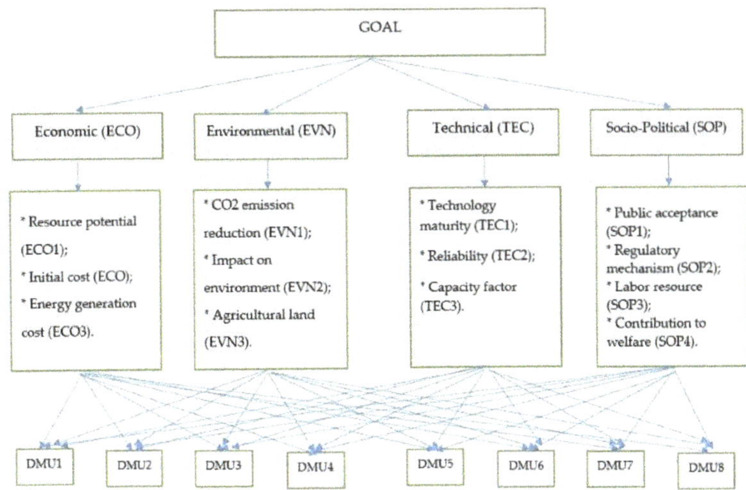

Figure 6. General hierarchy structure of proposed model.

Table 2. Weight of all of the subcriteria. ECO—economic; EVN—environmental; TEC—technical; SOP—socio-political.

No.	Criteria	Weight
1	ECO1	0.1650
2	ECO2	0.1042
3	ECO3	0.1968
4	EVN1	0.0589
5	EVN2	0.0572
6	EVN3	0.0404
7	TEC1	0.0547
8	TEC2	0.1283
9	TEC3	0.1163
10	SOP1	0.0236
11	SOP1	0.0210
12	SOP3	0.0144
13	SOP4	0.0193

Table 3. Normalized matrix.

	DMU/001	DMU/002	DMU/003	DMU/004	DMU/005	DMU/006	DMU/007	DMU/008
ECO1	0.2851	0.3326	0.3801	0.3801	0.4276	0.3326	0.2851	0.3801
ECO2	0.3567	0.3567	0.3121	0.4013	0.3567	0.2675	0.3567	0.4013
ECO3	0.3472	0.3038	0.3472	0.3906	0.3472	0.3472	0.3906	0.3472
EVN1	0.3845	0.3417	0.3417	0.2990	0.3845	0.3417	0.3845	0.3417
EVN2	0.3636	0.3636	0.3182	0.4091	0.3636	0.3182	0.3182	0.3636
EVN3	0.3219	0.3219	0.3678	0.4138	0.3219	0.2759	0.3678	0.4138
TEC1	0.3845	0.3417	0.3417	0.3417	0.2990	0.3417	0.3845	0.3845
TEC1	0.3786	0.3366	0.3786	0.2945	0.3786	0.3366	0.3786	0.3366
TEC3	0.3522	0.3522	0.3082	0.3962	0.3962	0.3522	0.3522	0.3082
SOP1	0.3581	0.3134	0.4029	0.3581	0.3581	0.3134	0.3581	0.3581
SOP2	0.3038	0.3472	0.3472	0.3906	0.3472	0.3906	0.3472	0.3472
SOP3	0.3581	0.3581	0.4029	0.3581	0.3581	0.3581	0.3134	0.3134
SOP4	0.3465	0.3898	0.3465	0.3898	0.3032	0.3465	0.3898	0.3032

Table 4. Normalized weight matrix.

	DMU/001	DMU/002	DMU/003	DMU/004	DMU/005	DMU/006	DMU/007	DMU/008
ECO1	0.0470	0.0549	0.0627	0.0627	0.0706	0.0549	0.0470	0.0627
ECO2	0.0372	0.0372	0.0325	0.0418	0.0372	0.0279	0.0372	0.0418
ECO3	0.0683	0.0598	0.0683	0.0769	0.0683	0.0683	0.0769	0.0683
EVN1	0.0226	0.0201	0.0201	0.0176	0.0226	0.0201	0.0226	0.0201
EVN2	0.0208	0.0208	0.0182	0.0234	0.0208	0.0182	0.0182	0.0208
EVN3	0.0130	0.0130	0.0149	0.0167	0.0130	0.0111	0.0149	0.0167
TEC1	0.0210	0.0187	0.0187	0.0187	0.0164	0.0187	0.0210	0.0210
TEC1	0.0486	0.0432	0.0486	0.0378	0.0486	0.0432	0.0486	0.0432
TEC3	0.0410	0.0410	0.0358	0.0461	0.0461	0.0410	0.0410	0.0358
SOP1	0.0085	0.0074	0.0095	0.0085	0.0085	0.0074	0.0085	0.0085
SOP2	0.0064	0.0073	0.0073	0.0082	0.0073	0.0082	0.0073	0.0073
SOP3	0.0052	0.0052	0.0058	0.0052	0.0052	0.0052	0.0045	0.0045
SOP4	0.0067	0.0075	0.0067	0.0075	0.0059	0.0067	0.0075	0.0059

Vietnam is a developing agricultural country and has great potential to create biomass sources for energy production. If effectively exploited, biomass energy will help reduce dependence on traditional power sources, reduce carbon emissions, reduce pollution, and bring direct profits to establishments; furthermore, farmers can participate in the biomass fuels supply chain via selling waste and agricultural byproducts. According to the calculation theory, the total potential of Vietnam's biomass energy source is more than 99 million tons/year, corresponding to the electric energy source of more than 340,000 GWh, of which the Mekong River Delta (Mekong Delta) accounts for 33.4%, followed by the North Central Coast and Central Coast with 21.8% [31]. The primary goal of this research is to propose a useful fuzzy MCDM model for biomass power plant location selection. In the first stage of this work, all of the criteria affecting the location selection are identified by experts and literature reviews; in addition, the FAHP method was utilized to identify the weight of all of the criteria in the second stage. The TOPSIS is applied for ranking the potential locations in the final stage of this research; furthermore, TOPSIS is based on the concept that the chosen alternative should have the shortest geometric distance from the PIS and the longest geometric distance from the NIS. As per the results shown in Figure 7 and Table 5, Long An (DMU/005) is found to be the best location for building biomass energy in Vietnam.

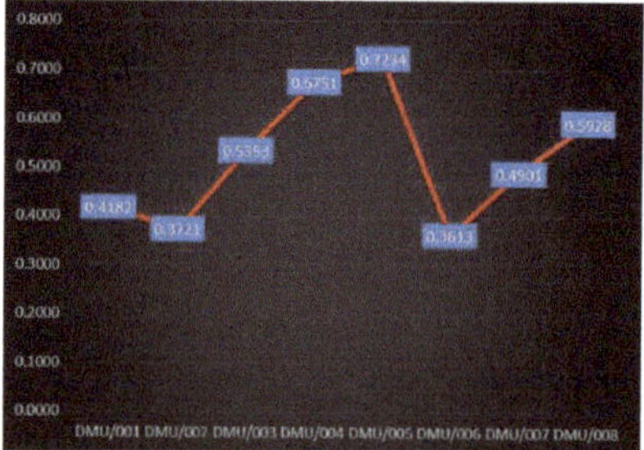

Figure 7. Final ranking list.

Table 5. Negative ideal solution (NIS) and positive ideal solution (PIS) values in the Technique for Order Preference by Similarity to an Ideal Solution (TOPSIS) model.

DMU	Di+	Di−
DMU/001	0.0265	0.0190
DMU/002	0.0255	0.0151
DMU/003	0.0192	0.0221
DMU/004	0.0145	0.0301
DMU/005	**0.0119**	**0.0312**
DMU/006	0.0254	0.0144
DMU/007	0.0252	0.0242
DMU/008	0.0170	0.0247

5. Conclusions

As an agricultural country, Vietnam is an ideal place for producing biomass energy. Agricultural waste is the most abundant in the Mekong Delta region, which makes up about 50% of the country. Major biomass resources include straw, rice husks from rice mills, bagasse from sugar mills, coffee husks from coffee-processing plants, and wood chips from wood-processing industries. Vietnam has set a target of a combined capacity of 500 MW of biomass energy by 2020, rising to 2000 MW by 2030 [32].

Therefore, Vietnam has many favorable conditions for developing biomass energy. Taking advantage of biomass energy will simultaneously provide energy for economic development, and ensure environmental protection. Although there are many studies in regard to developing methods and using different criteria for evaluating renewable energy location, research on biomass location evaluation and selection remains limited, especially in regard to the use of integrating economic and environmental criteria. In this research, the authors presented a fuzzy multicriteria decision-making model (FMCDM) for optimizing the site selection process for biomass power plants under fuzzy environment conditions. All of the criteria affecting location selection are identified by experts and literature reviews; in addition, the fuzzy analytic hierarchy process (FAHP) method was utilized to identify the weight of all of the criteria in the second stage. As a result, Long An (DMU/005) was found to be the best location for building biomass energy plants in Vietnam.

The main contribution of this work includes modeling the site selection decision process under fuzzy environment conditions. This paper also considers the evolution of a new model that is flexible and practicable to the decision-maker. This research also provides a useful guideline for biomass power plant location selection in many countries, as well as providing a guideline for location selection in related industries. Furthermore, for future research, the application of other MCDM approaches can be utilized so as to compare the results in the search of any changes.

Author Contributions: Conceptualization, C.-N.W., T.-T.T., and Y.-F.H.; formal analysis, C.-N.W. and Y.-F.H.; funding acquisition, C.-N.W.; methodology, T.-T.T.; project administration, C.-N.W. and Y.-F.H.; resources, T.-T.T.; software, T.-T.T.; supervision, C.-N.W.; writing (original draft), T.-T.T.; writing (review and editing), C.-N.W. and Y.-F.H.

Funding: This research received no external funding.

Conflicts of Interest: The authors declare no conflict of interest.

References

1. Điện sinh khối-nguồn năng lượng tái tạo hữu ích. Available online: http://chatdotxanh.com/chi-tiet-tin/dien-sinh-khoi-nguon-nang-luong-tai-tao-huu-ich.html (accessed on 12 February 2019).
2. Trung, N.Q. Tổng quan về lưới điện trong hệ thống điện Quốc gia. Trung tâm điều độ hệ thống điện QG. Available online: https://www.nldc.evn.vn/newsg/1/86/Tong-quan-ve-luoi-dien-trong-he-thong-dien-Quoc-gia/default.aspx (accessed on 20 February 2019).
3. NĂNG LƯỢNG SINH KHỐI Ở VIỆT NAM: VẪN CHỈ LÀ TIỀM NĂNG. Available online: https://www.pvpower.vn/nang-luong-sinh-khoi-o-viet-nam-van-chi-la-tiem-nang/ (accessed on 18 February 2019).
4. Ward, A.; Le, H.; Tran, T.; Pham, N. *Tạo sự hấp dẫn cho năng lượng sinh khối*; Viện Tăng trưởng xanh toàn cầu: Seoul, Korea, 2018.
5. Thanh, T. Nguồn năng lượng sinh khối ở Việt Nam khoảng 150 triệu tấn/năm. Available online: https://petrotimes.vn/nguon-nang-luong-sinh-khoi-o-viet-nam-khoang-150-trieu-tannam-516007.html (accessed on 12 February 2019).
6. Phát triển năng lượng sinh khối của Việt Nam: Tiềm năng và thách thức. Available online: http://chatdotxanh.com/chi-tiet-tin/phat-trien-nang-luong-sinh-khoi-cua-viet-nam-tiem-nang-va-thach-thuc.html (accessed on 15 February 2019).
7. Pamučar, D.; Gigović, L.; Bajić, Z.; Janošević, M. Location Selection for Wind Farms Using GIS Multi-Criteria Hybrid Model: An Approach Based on Fuzzy and Rough Numbers. *Sustainability* **2017**, *9*, 1315. [CrossRef]

8. Villacreses, G.; Gaona, G.; Martínez-Gómez, J.; Jijón, D.J. Wind farms suitability location using geographical information system (GIS), based on multi-criteria decision making (MCDM) methods: The case of continental Ecuador. *Renew. Energy* **2017**, *109*, 275–286. [CrossRef]

9. Ghasempour, R.; Nazari, M.A.; Ebrahimi, M.; Ahmadi, M.H.; Hadiyanto, H. Multi-Criteria Decision Making (MCDM) Approach for Selecting Solar Plants Site and Technology: A Review. *Int. J. Renew. Energy Dev.* **2019**, *8*, 15–25. [CrossRef]

10. Chen, C.; Li, W.; Li, Y.; Zhu, Y. Biomass power plant site selection modeling and decision optimization. *Trans. Chin. Soc. Agric. Eng.* **2011**, *27*, 255–260.

11. Patomtummakan, J.; Nananukul, N. Biomass Power Plant Location and Distribution Planning System. *GMSARN Int. J.* **2018**, *12*, 11–18.

12. Cattafi, M.; Gavanelli, M.; Cagnoli, P.; Massimiliano, C. Sustainable Biomass Power Plant Location in the Italian Emilia-Romagna Region. *ACM Trans. Intell. Syst. Technol.* **2011**, *2*, 23. [CrossRef]

13. Woo, H.; Acuna, M.; Moroni, M.; Taskhiri, M.S.; Turner, P. Optimizing the Location of Biomass Energy Facilities by Integrating Multi-Criteria Analysis (MCA) and Geographical Information Systems (GIS). *Forests* **2018**, *9*, 585. [CrossRef]

14. Lv, H.; Ding, H.; Zhou, D.; Zhou, P. A Site Selection Model for a Straw-Based Power Generation Plant with CO_2 Emissions. *Sustainability* **2014**, *6*, 7466–7481. [CrossRef]

15. Wang, C.-N.; Nguyen, V.T.; Duong, D.H.; Thai, H.T.N. A Hybrid Fuzzy Analysis Network Process (FANP) and the Technique for Order of Preference by Similarity to Ideal Solution (TOPSIS) Approaches for Solid Waste to Energy Plant Location Selection in Vietnam. *Appl. Sci.* **2018**, *8*, 1100. [CrossRef]

16. Jeong, J.S.; Ramírez-Gómez, Á. A Multicriteria GIS-Based Assessment to Optimize Biomass Facility Sites with Parallel Environment—A Case Study in Spain. *Energies* **2017**, *10*, 2095. [CrossRef]

17. Isalou, A.A.; Zamani, V.; Shahmoradil, B. Alizadeh Landfill site selection using integrated fuzzy logic and analytic network process (F-ANP). *Environ. Earth Sci.* **2012**, *68*, 1745–1755. [CrossRef]

18. Wang, C.-N.; Nguyen, V.T.; Thai, H.T.N.; Duong, D.H. Multi-Criteria Decision Making (MCDM) Approaches for Solar Power Plant Location Selection in Viet Nam. *Energies* **2018**, *11*, 1504. [CrossRef]

19. Wang, C.-N.; Huang, Y.-F.; Chai, Y.-C.; Nguyen, V.T. A Multi-Criteria Decision Making (MCDM) for Renewable Energy Plants Location Selection in Vietnam under a Fuzzy Environment. *Appl. Sci.* **2018**, *8*, 2069. [CrossRef]

20. Fetanat, A.; Khorasaninejad, E. A novel hybrid MCDM approach for offshore wind farm site selection: A case study of Iran. *Ocean Coast. Manag.* **2015**, *109*, 27–28. [CrossRef]

21. Rezaei-Shouroki, M. The location optimization of wind turbine sites with using the MCDM approach: A case study. *Energy Quipsys* **2017**, *5*, 165–187.

22. Heidarzade, F.; Varzandeh, M.H.M.; Rahbari, O.; Zavadskas, E.K.; Vafaeapour, M. Placement of Wind Farms Based on a Hybrid Multi Criteria. In Proceedings of the World Sustainability Forum, Oshawa, ON, Canada, 1–30 November 2014.

23. Cobuloglu, H.I.; Büyüktahtakin, I.E. A Multi-Criteria Approach for Biomass Crop Selection under Fuzzy Environment. In Proceedings of the 2014 Industrial and Systems Engineering Research Conference, Montreal, QC, Canada, 31 May–3 June 2014.

24. Solangi, Y.A.; Tan, Q.; Mirjat, N.H.; Valasai, G.D.; Khan, M.W.A.; Ikram, M. An Integrated Delphi-AHP and Fuzzy TOPSIS Approach toward Ranking and Selection of Renewable Energy Resources in Pakistan. *Processes* **2019**, *7*, 118. [CrossRef]

25. Nazari, M.A.; Aslani, A.; Ghasempour, R. Analysis of Solar Farm Site Selection Based on TOPSIS Approach. *Int. J. Soc. Ecol. Sustain. Dev.* **2018**, *9*, 14. [CrossRef]

26. Nazari, M.A.; Aslani, A.; Ghasempour, R. Biomass Type Selection for Boilers Using TOPSIS Multi-Criteria Model. *Int. J. Environ. Sci. Dev.* **2014**, *2*, 5.

27. Daneshvar Rouyendegh, B.; Yildizbasi, A.; Arikan, Ü.Z.B. Babak Daneshvar Rouyendegh. Using Intuitionistic Fuzzy TOPSIS in Site Selection of Wind Power Plants in Turkey. *Adv. Fuzzy Syst.* **2018**, *2018*. [CrossRef]

28. Choudhary, D.; Shankar, R. An STEEP-fuzzy AHP-TOPSIS framework for evaluation and selection of thermal power plant location: A case study from India. *Energy* **2012**, *42*, 510–521. [CrossRef]

29. Nhung, H. ĐBSCL có tiềm năng năng lượng sinh khối lớn nhất nước. Tuổi Trẻ. Available online: https://tuoitre. vn/dbscl-co-tiem-nang-nang-luong-sinh-khoi-lon-nhat-nuoc-1076932.htm (accessed on 27 May 2019).

30. Nguồn năng lượng nào cho Đồng bằng sông Cửu Long? Kinh Tế Sài Gòn. Available online: https://www.thesaigontimes.vn/121298/Nguon-nang-luong-nao-cho-Dong-bang-song-Cuu-Long (accessed on 27 May 2019).
31. ĐBSCL có tiềm năng năng lượng sinh khối lớn nhất nước. Available online: http://m.icon.com.vn/vi-VN/c620/127733/DBSCL-co-tiem-nang-nang-luong-sinh-khoi-lon-nhat-nuoc.aspx (accessed on 24 February 2019).
32. Biomass Energy in Vietnam. Available online: https://www.bioenergyconsult.com/biomass-energy-vietnam/ (accessed on 10 April 2019).

Article

The Influence and Optimization of Geometrical Parameters on Coast-Down Characteristics of Nuclear Reactor Coolant Pumps

Yuanyuan Zhao, Xiangyu Si, Xiuli Wang *, Rongsheng Zhu, Qiang Fu and Huazhou Zhong

National Research Center of Pumps, Jiangsu University, Zhenjiang 212013, China; zyy-michelle@163.com (Y.Z.); sixiangyu01@163.com (X.S.); ujs_zrs@163.com (R.Z.); ujsfq@sina.com (Q.F.); huazhou.zhong@turbotides.com.cn (H.Z.)

* Correspondence: ujswxl@ujs.edu.cn; Tel.: +86-1860-5110-959

Received: 22 March 2019; Accepted: 7 May 2019; Published: 1 June 2019

Abstract: Coast-down characteristics are the crucial safety evaluation factors of nuclear reactor coolant pumps. The energy stored at the highest moment of inertia of the reactor coolant pump unit is utilized to maintain a normal coolant supply to the core of the cooling loop system for a short period of time during the coast-down transition. As a result of the high inertia moment of the rotor system, the unit requires a high reliability of the nuclear reactor coolant pump and consumes considerable energy in the start-up and normal operation. This paper considers the operational characteristics of the coast-down transition process based on the existing hydraulic model of the nuclear reactor coolant pump. With the implementation of an orthogonal test, the hydraulic performance of the nuclear reactor coolant pump was optimized, and the optimal combination of impeller geometrical parameters was selected using multivariate linear regression to prolong the coast-down time of the reactor coolant pump and to avoid serious nuclear accidents.

Keywords: reactor coolant pump; coast-down characteristics; geometrical parameters; multiple linear regression; transition process

---

1. Introduction

The nuclear reactor coolant pump is the only rotating piece of equipment in the primary loop cooling system, and thus can be called the 'heart' of the nuclear power plant. In the unfortunate event of a power failure at the nuclear power plant, the reactor coolant pump loses its power source and it will enter a coast-down state. For a short period of time, the inertia of the inert wheel provides power for the reactor coolant pump and the coolant continues to cool the reactor core. The coast-down process of the nuclear reactor coolant pump can be seen as a typical transient process. The nuclear reactor has a short coast-down transient process and cooling circuit working time. As the heat of the reactor cannot be discharged in a short time the temperature tends to rise sharply, leading to the potential decomposition of the coolant and generating a large amount of hydrogen, which is not at all conducive to the safety of the system [1]. Hence, the study of the effect of dynamic characteristics on the nuclear reactor coolant pump during the coast-down transient process becomes imperative.

Very limited literature is available related to the transient process of domestic and foreign nuclear reactor coolant pumps. Nevertheless, the transient processes of the centrifugal pumps and the mixed flow pumps have been widely studied. The transient characteristics of the centrifugal pump in its acceleration and deceleration process were determined by Tsukamoto et al. [2] using a theoretical analysis. Wu and Li et al. [3,4] examined the starting and stopping transients of the centrifugal pumps and mixed flow pumps by combining experimental and numerical calculations. A systematic study on the start-up process of centrifugal pumps with different valve opening degrees, different impeller outside diameters, different blade widths, and different starting descending speeds was conducted by Elaoud et al. [5,6]. By establishing a mathematical start-up model for the primary circuit cooling system, Farhadi et al. [7–9] studied the effects of the ratio between the inertial energy of the unit and the fluid mass inertia energy of the pipe coolant during the start-up process of the nuclear reactor coolant pump. Whether the pumping capability meets the coast-down half time requirement as prescribed by safety analyses during the coast-down period was studied by Alatrash et al. [10] using experiments. Yonggang [11–13] studied the third and fourth generation nuclear main pumps, including gas–liquid two-phase flow and structural optimization. There have been numerous valuable studies on the coast-down characteristics of nuclear reactor coolant pumps, but comparatively little research details are available on the factors that affect the coast-down characteristics.

Geometric parameters of the impeller are one of the main factors affecting pump performance. Based on a numerical analysis of a 3D viscous flow, Hyuk et al. [14] designed a high-efficiency mixed-flow pump and the results suggested that the hydraulic efficiency of a mixed-flow pump at the design level can be improved by modifying its geometry. The impact of the geometry parameters of the impeller on the hydraulic performance of mixed flow pumps was studied by Varchola et al. [15], and several different designs were also compared. Long et al. [16] studied how the blade numbers of the impeller and the diffuser influence the reactor coolant pump performances using the numerical simulation method. The effect of the blade stacking lean angle on the hydraulic performance of a 1400 MW nuclear reactor coolant pump was studied by Zhou et al. [17], and it was determined that the geometric parameters such as the blade stacking lean angle highly influences the hydraulic efficiency of different flow intervals. Evidently, although the influence of the geometrical parameters on the pump has been studied, the research on the coupling effect of the nuclear reactor coolant pump has been very limited.

With regards to the factors influencing the coast-down characteristics of reactor coolant pumps, the indirect coupling effects between the different geometric parameters and combinations have been examined in depth by this paper. Changing the pump performance by changing the size of a certain geometric parameter virtually changes the direct impact of this parameter on performance, with an indirect impact on the performance of the other parameters simultaneously. This paper employs the multiple linear regression to analyze the optimal geometrical parameters of the impeller, based on the relationship between geometric parameters of the impeller and its efficiency, and the head.

2. Research Method

In the event of an unfortunate loss of power source to the nuclear reactor coolant pump, the unit utilizes its own moment of inertia to store energy to ultimately maintain the operation of the reactor coolant pump for a longer stretch of period, this phenomenon is known as the coast-down characteristic of the nuclear reactor coolant pumps [18]. In the coast-down transition process, the energy stored by the flywheel ensures that the time of the nuclear reactor coolant pump flow decreases by half within the specified safe time margin. The flywheel of an AP1000 nuclear reactor coolant pump is usually split into the upper flywheel and the lower flywheel while being fixed on the main shaft. To maximize improvement in the moments of inertia at a limited volume to maintain the coast-down characteristics of the reactor coolant pump, the flywheels were usually encompassed with high-density heavy metal tungsten alloy blocks and high-quality stainless-steel wheels. With respect to the issue of energy consumption, the main function of the flywheel was to provide energy to keep the nuclear reactor coolant pump running during the coast-down transition, and the flywheel should consume enough amounts of energy to maintain the start-up process and normal operation. It was suggested in the combination of Equations (1) and (2) that the time of coast-down was not only affected by the moment of inertia, but the efficiency of the rated operating point and the energy loss of the coast-down transition were also significant factors. As the rotational inertia of the impeller was obviously smaller than that of the flywheel, the moment of inertia of the rotor basically remains unchanged when the geometric parameters of the impeller were changed [18–20]. Accordingly, this study endeavors to increase the efficiency of the hydraulic model rated point by optimizing the impeller geometric parameters to reduce any extra energy loss in the coast-down transition, to extend the time of coast-down transition, to increase the system reliability, and to reduce the cost thereon.

$$E_T = 2\pi^2 \int (\oint \rho n^2(t) A dz) dt + E_f \tag{1}$$

$$t = \frac{P_0}{4\pi^2 I \eta_0 n_0^2} [(\frac{n_0}{n(t)} - 1)] \tag{2}$$

In Equations (1) and (2), $E_T = \frac{1}{2} I \rho \omega_0^2 + \frac{1}{2} \oint \rho \omega_0^2 A dz$ denotes the total energy stored by the moment of inertia of the unit and the inertia of the conveying liquid, E_f refers to the energy of various losses in the coast-down transition, ρ represents the density of the conveying liquid with units in kg/m^3, A denotes the average sectional area of the loop pipe with the unit of m^2, z refers to the effective pipeline length for the whole circuit with the unit of m, t represents the time since the outage began with the unit of s, and I refers to the total moment of inertia of the unit. With the unit of kg m^2, P_0, η_0, and n_0 are the effective power, efficiency, and rated speed of the nuclear reactor coolant pump under rated conditions, respectively, $n(t)$ refers to the rotational speed at different points in the coast-down transition with the unit of r/min.

Figure 1 shows the three-dimensional fluid calculation domain of the reactor coolant pump. In Figure 2, γ denotes the outlet inclination of the impeller, β_2 denotes the outlet angle of the impeller, φ denotes the wrap angle of the impeller, Z denotes the blade numbers of the impeller, D_2 denotes the outlet diameter of the impeller in mm, b_2 denotes the outlet width of the impeller in mm, and D_j denotes the inlet diameter in mm. Area ratio Y represents the ratio of the impeller outlet area to the volute throat area.

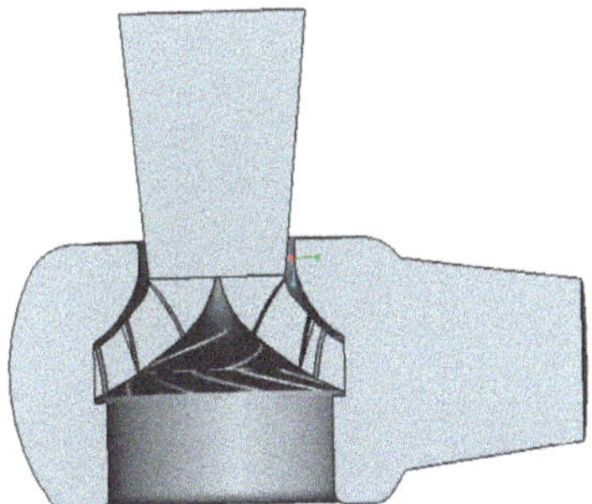

Figure 1. Three-dimensional fluid calculation domain of reactor coolant pump.

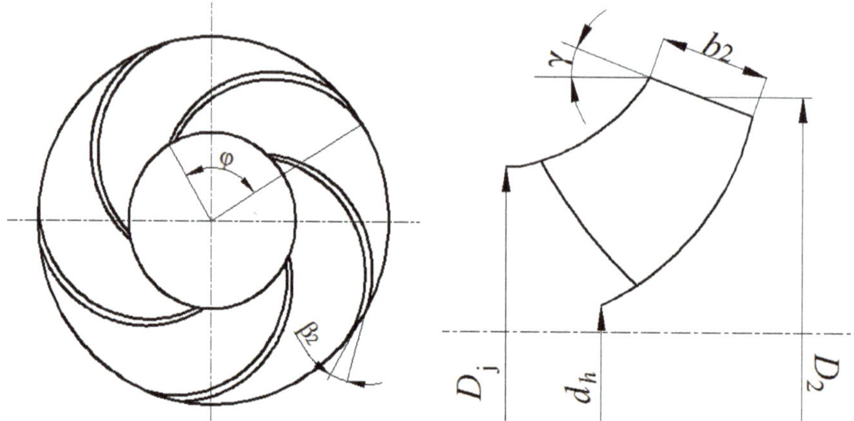

Figure 2. Schematic diagram of the main structural parameters of the impeller.

Table 1 presents that the efficiencies and heads obtained by the different combinations of eight different impeller geometric parameters, which suggest the influence of the different geometric parameters, parameter combination on efficiency, and the difference in the head [21]. Multiple linear regression was selected by this paper to analyze the relationship among each parameter, efficiency, and the head.

Table 1. Test scheme and performance calculation results.

	$\gamma/°$	$\beta_2/°$	$\varphi/°$	Z	D_2/mm	b_2/mm	D_j/mm	Area Ratio Y	Index η_1	H
1	20	20	115	4	760	190	555	0.927	79.45	94.89
2	20	25	120	5	765	190	560	0.916	81.09	109.48
3	20	30	125	6	770	195	550	0.916	82.76	134.33
4	23	30	125	5	765	195	550	0.932	84.38	111.69
5	23	25	120	6	760	200	560	1.002	84.70	138.02
6	23	20	115	4	760	200	555	0.952	84.04	100.39
7	26	20	115	4	770	190	555	0.914	82.24	101.80
8	26	25	120	5	765	190	560	0.925	82.16	112.89
9	26	30	120	6	765	195	550	0.935	82.32	136.20
10	20	20	125	6	770	195	550	0.916	80.67	101.59
11	23	30	125	4	770	200	560	0.931	83.10	95.47
12	26	25	115	5	760	200	555	0.956	82.90	119.49
13	20	30	115	6	770	190	555	0.905	84.22	137.83
14	23	30	120	5	765	190	560	0.921	83.69	124.90
15	26	25	120	6	760	195	550	0.946	82.27	120.81
16	20	20	115	4	770	195	550	0.916	83.98	101.89
17	23	20	125	5	765	200	560	0.941	83.11	100.60
18	26	25	125	4	760	200	555	0.956	81.31	103.49

2.1. Data Normalization

Data normalization was an important step in multiple linear regression. Given that each variable was different in the physical properties, they usually have different orders of magnitude and dimensions. When the regression equation was established directly, the regression coefficient was very often not directly comparable [21]. When variables vary largely in level, the role of the value of the higher numerical variable in the comprehensive analysis shall be highlighted, and the role of the comparatively lower numerical variable will be weakened if the raw data is directly used for analysis. Accordingly, to ensure the reliability of the results, it was imperative for standardizing the table and simulation data to eliminate the dimensional influence between variables so that the data could be comparable. Data normalization involved the centralization and compression processing of the data simultaneously, the specific principles were as follows

$$x_{ij}^* = \frac{x_{ij} - \bar{x}}{s_j}, \qquad \begin{array}{l} i = 1,2,3,\ldots,n \\ j = 1,2,3,\ldots,p \end{array} \tag{3}$$

where, x_{ij} represents the value of the i-th row and the j-th column, x_{ij} * represents the normalized data of x_{ij}, i refers to the i-th row, and j refers to the j-th column, s_j represents the normalized parameter of column j.

2.2. Path Analysis

Path analysis was conducted to analyze the direct relationship between the impeller geometrical parameters and the pump performance, as well as the indirect coupling relationship between the parameters. Assuming that the p independent variable can be established, $x_1, x_2, \ldots, x_p$, the simple correlation coefficients between each of the two variables and the dependent variable y were capable of forming the normalized normal equation to solve the path coefficient:

$$\begin{array}{l} r_{11}\rho_1 + r_{12}\rho_2 + \ldots + r_{1p}\rho_p = r_{1y} \\ r_{21}\rho_1 + r_{22}\rho_2 + \ldots + r_{2p}\rho_p = r_{2y} \\ \cdots \quad \cdots \quad \cdots \quad \cdots \\ r_{p1}\rho_1 + r_{p2}\rho_2 + \ldots + r_{pp}\rho_p = r_{py} \end{array} \tag{4}$$

where, $\rho_1, \rho_2, \ldots, \rho_p$ was the direct path coefficient. The direct path coefficient represented the direct effect size of the independent variable, while the indirect path coefficient suggested that the independent variable influences the dependent variable by impacting other independent variables. Besides, such coefficients could be calculated using the correlation coefficient r_{ij} and the direct path coefficient ρ_i. The direct path coefficient can be obtained by calculating the inverse matrix of the noted correlation matrix. With the assumption that B_{ij} was the inverse matrix of the correlation matrix r_{ij}, and then the direct path coefficient ρ_i ($i = 1, 2, \ldots, p$) was expressed as follows

$$\begin{bmatrix} \rho_1 \\ \rho_2 \\ \cdots \\ \rho_p \end{bmatrix} = \begin{bmatrix} B_{11} & B_{12} & B_{13} & \cdots & B_{1p} \\ B_{21} & B_{22} & B_{23} & \cdots & B_{2p} \\ \cdots & \cdots & \cdots & \cdots & \cdots \\ B_{p1} & B_{p2} & B_{p3} & \cdots & B_{pp} \end{bmatrix} \begin{bmatrix} r_{1y} \\ r_{2y} \\ \cdots \\ r_{py} \end{bmatrix} \tag{5}$$

The path coefficient ρ_{ye} of the remaining term is as expressed in Equation (6). In case the path coefficient ρ_{ye} of the remaining term was smaller, the impeller geometrical parameters and the performance would be well satisfied with the linear relation. Conversely, when the path coefficient ρ_{ye} of the remaining term was larger, it suggested that the test error was larger or other important factors were not introduced.

$$\rho_{ye} = \sqrt{1 - \left(\sum_{i=1}^{p} r_{iy}\rho_i \right)} \tag{6}$$

By path analyzing the impeller geometrical parameters and the pump head and the efficiency performance, the results were as listed in Table 2.

Table 2. Path analysis results between impeller geometric parameters and performance.

Factor	Direct Effect	Indirect Effect							
		$\gamma \to H$	$\beta_2 \to H$	$\varphi \to H$	$Z \to H$	$D_2 \to H$	$b_2 \to H$	$D_0 \to H$	$Y \to H$
γ	0.0076		0.0424	0.0000	−0.0502	−0.0748	−0.0365	−0.0031	0.1825
β_2	0.5091	0.0006		−0.1244	0.2510	0.0299	0.0122	0.0000	−0.0349
φ	−0.3731	0.0000	0.1697		0.1506	0.0299	−0.0487	0.0000	0.0295
Z	0.6025	−0.0006	0.2121	−0.1041		0.015	0.0122	0.0092	0.0322
D_2	0.1795	−0.0032	0.0849	−0.0622	0.0502		0.0487	0.0061	−0.3235
b_2	−0.1461	0.0019	−0.0424	−0.1244	−0.0502	−0.0598		0.0000	0.3087
D_0	−0.0368	0.0006	0.0000	0.0000	−0.1506	−0.0299	0.0000		0.1007
Y	0.4401	0.0032	−0.0404	−0.0250	0.0441	−0.1320	−0.1025	−0.0084	

Factor	Direct Effect	Indirect Effect							
		$\gamma \to \eta_1$	$\beta_2 \to \eta_1$	$\varphi \to \eta_1$	$Z \to \eta_1$	$D_2 \to \eta_1$	$b_2 \to \eta_1$	$D_0 \to \eta_1$	$Y \to \eta_1$
γ	−0.0528		0.0400	0.0000	0.0001	−0.2726	0.1070	0.0040	0.2509
β_2	0.4803	−0.0044		−0.1636	−0.0005	0.1090	−0.0269	0.0000	−0.048
φ	−0.4908	0.0000	0.1601		−0.0003	0.1090	0.1076	0.0000	0.0406
Z	−0.0011	0.0044	0.2001	−0.1227		0.0545	−0.0269	−0.0120	0.0443
D_2	0.6542	0.022	0.0800	−0.0818	−0.0001		−0.1076	−0.0080	−0.4445
b_2	0.3228	−0.0132	−0.0400	−0.1636	0.0001	−0.2181		0.0000	0.4242
D_0	0.0480	−0.0044	0.000	0.0000	0.0003	−0.1090	0.0000		0.1383
Y	0.6047	−0.0219	−0.0381	−0.0329	−0.0001	−0.4809	0.2265	0.0110	

3. Results

3.1. The Direct Impact Analysis of the Main Geometric Parameters of Nuclear Reactor Coolant Pump on Its Performance

As is evident from Table 2 the blade number, blade outlet angle, blade wrap angle, area ratio, impeller outlet diameter, and the blade outlet width in the eight impeller geometrical parameters had a large direct impact on the design point head of the nuclear reactor coolant pump, and the other two parameters had very little impact on the head. Among the six parameters with larger direct impact, the blade numbers were the largest; the blade outlet angle, blade wrap angle, area ratio, and impeller outlet diameter were ranked second; and the blade outlet width was the smallest. The direct path coefficient of the impeller blade number was 0.6025, suggesting that the blade number was the most critical in the geometrical parameters of the reactor coolant pump. When blade numbers were changed, the work efficiency of the impeller changed greatly, and its head also changed significantly. In a particular range, the head of the pump would definitely rise with the increase in the blade numbers. The direct path coefficients of the impeller blade outlet angle and area ratio were 0.5091 and 0.4401, respectively, which confirmed that the blade outlet angle and the area ratio on the head of the nuclear reactor coolant pump also had a major role. When the blade outlet angle was changed, the circumferential component of the absolute velocity at the impeller outlet was also changed. Thereafter, the circumferential component of the absolute velocity at the impeller outlet increased, while the head of the pump increased with the increase of blade outlet angle. As for area ratio the reaction of impeller and guide vane matching relationship, one of the important physical quantities, the performance of the pump was not unilaterally decided by the impeller but by the impeller, guide blade, and volute both (in this study, the volute remains the same, so it need not be considered). In the design process, the inlet area of the guide vane had hardly changed; hence the increase of area ratio in a certain scope was equivalent to the reduction in the impact loss and made the head increase. For the change of the blade wrap angle, the binding force of the fluid in the impeller channel was changed, and the relative velocity liquid angle of the impeller outlet was also changed. In a specific range, the restraint of the fluid in the impeller channel increased, while the blade wrap angle increased, however, the relative flow angle of the impeller outlet decreased, with the head. The direct path coefficient of the impeller outlet diameter was 0.1795, which indicated that in a certain range, the energy of the fluid increases, and the head also rises. The direct path coefficient of impeller outlet width minimum was −0.1461, which suggested that within a certain range, an increase of blade outlet width can make the impeller outlet edge overtilted, and a larger secondary flow would appear, resulting in the head decrease.

From Table 1, it can be observed that amongst the eight impeller geometric parameters, the blade outlet angle, blade wrap angle, blade outlet width, impeller outlet diameter, and area ratio had a greater direct impact on the efficiency performance of nuclear reactor coolant pump design point, while the other three parameters had a relatively smaller impact on its efficiency. Among the five parameters with larger direct impact, the impeller outlet diameter was the largest; the blade outlet angle, blade wrap angle, and area ratio ranked second; and blade outlet width was the smallest. The direct path coefficient of impeller outlet diameter reached 0.6542, which confirmed that impeller outlet diameter was the most critical of all the geometrical parameters of the nuclear reactor coolant pump.

The efficiency of nuclear reactor coolant pump varies with the impeller outlet diameter and the flow condition of the impeller outlet. This indicated that, within a certain range, with the increase of the impeller outlet diameter the impeller outlet speed is reduced, the impact loss between the blade wheel and guide vane decreases, and the hydraulic efficiency of the pump increases. The direct path coefficient of the area ratio between the impeller and the guide vane was 0.6047, which substantiated that increasing the impeller outlet area in a certain range makes the impeller and guide vane match better, reducing the impact loss and increasing the efficiency. The direct path coefficient of the impeller blade outlet angle and the blade wrap angle were 0.4803 and −0.4908, respectively, which confirmed that the blade angle and the blade wrap angle exert the main impact on the design point efficiency of the nuclear reactor coolant pump. Additionally, within a certain range, with the increase in the blade outlet angle, the circumferential component of the absolute speed at the impeller outlet increases, and the efficiency also increases. With the increase of the blade wrap angle, the fluid in the flow channel [22] gets restrained by the stronger blades, while the excessive flow channel increases the friction loss and decreases the efficiency of the pump. The direct path coefficient of the blade outlet width reached 0.3228, which suggested that in a particular range, with the increase of the blade outlet width, the pump can increase the efficiency.

3.2. Indirect Effect Analysis of the Geometric Parameters of the Nuclear Reactor Coolant Pump on Its Performance

Besides the direct impact on pump performance, the impeller geometry parameters have different degrees of mutual influence. The principle of path analysis is Correlation coefficient = direct path coefficient + indirect path coefficient, i.e., when a parameter changes, it not only has a direct impact on the performance, but also exerts an indirect impact on the performance by changing the other geometric parameters.

From the indirect path coefficient listed in Table 2, it is evident that for the head index, the indirect path coefficient of the outlet lean angle and the impeller inlet diameter were 0.0603 and −0.0792, respectively. This indicates that the outlet lean angle and the impeller inlet diameter exerted little indirect impact on the head by changing the other geometrical parameters, primarily by changing the area ratio ($\gamma \rightarrow Y \rightarrow H = 0.1825$, $D_0 \rightarrow Y \rightarrow H = 0.1007$). The indirect path coefficient of the blade outlet angle was 0.1344, indicating that blade outlet angle indirectly strengthens the head by changing the other geometrical parameters, and the outlet lean angle had the effect of reducing the head by changing the blade wrap angle ($\beta_2 \rightarrow \varphi \rightarrow H = -0.1244$). However, the outlet lean angle reinforced the head by blade numbers ($\beta_2 \rightarrow Z \rightarrow H = 0.251$). The indirect path coefficient of the blade wrap angle was 0.311, indicating that the blade wrap angle had an indirect strengthening effect on the head by changing the other geometrical parameters, and the blade wrap angle strengthens the head by changing the blade outlet angle and the blade numbers ($\varphi \rightarrow \beta_2 \rightarrow H = 0.1697$, $\varphi \rightarrow Z \rightarrow H = 0.1506$). The indirect path coefficient of the impeller blade numbers was 0.1868, suggesting that the impeller blade numbers exerted little indirect impact on the head by changing the other geometrical parameters, and the impeller blade numbers strengthened the head by changing the blade outlet angle ($Z \rightarrow \beta_2 \rightarrow H = 0.2121$), and yet the impeller blade numbers reduced the head by the blade wrap angle ($Z \rightarrow \varphi \rightarrow H$

= −0.1041). The indirect path coefficient of the blade outlet width was 0.0338, indicating that the blade outlet width had slightly enhanced the indirect impact on the head by changing the other geometrical parameters, and the blade outlet width reduced the head by changing blade wrap angle ($b_2 \rightarrow \varphi \rightarrow H$ = −0.1244), and yet the blade outlet width had the effect of reinforcing the head by area ratio ($b_2 \rightarrow Y \rightarrow H$ = 0.3087). The indirect path coefficient of the impeller outlet diameter was −0.199, which implied that the impeller outlet diameter had an abridged indirect effect on the head by changing the other geometrical parameters, and the impeller outlet diameter reduced the head by changing area ratio ($D_2 \rightarrow Y \rightarrow H$ = −0.3235). The indirect path coefficient of the area ratio was −0.261, which suggests that the area ratio had an abridged effect on the head, and the area ratio reduced the head by changing the impeller outlet diameter and the blade outlet width ($Y \rightarrow D_2 \rightarrow H$ = −0.132, $Y \rightarrow b_2 \rightarrow H$ = −0.3235).

For the efficiency index, the indirect path coefficient of impeller outlet diameter was 0.0252, which proved that the impeller inlet diameter had a small indirect impact on the efficiency by changing the other geometrical parameters. The impeller inlet diameter impacted the efficiency by the impeller outlet diameter and the area ratio, while the impeller inlet diameter reduced the efficiency by changing the impeller outlet diameter ($D_0 \rightarrow D_2 \rightarrow \eta_1$ = −0.109), the impeller inlet diameter increased the efficiency by the area ratio ($D_0 \rightarrow Y \rightarrow \eta_1$ = 0.1383). The indirect path coefficient of the outlet lean angle was 0.1031, which indicated that the outlet lean angle had little indirect impact on the increase of the efficiency by changing the other geometrical parameters, the outlet lean angle increased the efficiency by the blade outlet width and the area ratio ($\gamma \rightarrow b_2 \rightarrow \eta_1$ = 0.107, $\gamma \rightarrow Y \rightarrow \eta_1$ = 0.2509), while it decreased the efficiency by area ratio ($\gamma \rightarrow D_2 \rightarrow \eta_1$ = −0.2726). The indirect path coefficient of the blade outlet angle was −0.1344, showing that blade outlet angle had an indirect impact on the decrease of the efficiency by changing the other geometrical parameters, and the blade outlet angle decreased the efficiency by the blade wrap angle ($\beta_2 \rightarrow \varphi \rightarrow \eta_1$ = −0.1636); the blade outlet angle decreased the efficiency by the impeller outlet diameter ($\beta_2 \rightarrow D_2 \rightarrow \eta_1$ = 0.109). The indirect path coefficient of the blade wrap angle was −0.417, hinting that the blade wrap angle had an indirect impact on the increase of the efficiency by changing the other geometrical parameters, and the blade wrap angle increased the efficiency by blade outlet angle, impeller outlet diameter, and the blade outlet width ($\varphi \rightarrow \beta_2 \rightarrow \eta_1$ = 0.1601, $\varphi \rightarrow b_2 \rightarrow \eta_1$ = 0.109, $\varphi \rightarrow b_2 \rightarrow \eta_1$ = 0.1076). The indirect path coefficient of the impeller blade numbers was 0.1417, which demonstrated that the impeller blade numbers had an indirect impact on the increase of the efficiency by changing the other geometrical parameters and the impeller blade numbers increased the efficiency by the blade outlet angle ($Z \rightarrow \beta_2 \rightarrow \eta_1$ = 0.2001), while the impeller blade numbers decreased the efficiency by the blade wrap angle ($Z \rightarrow \varphi \rightarrow \eta_1$ = −0.1227). The indirect path coefficient of the impeller outlet diameter was −0.54, denoting that the impeller outlet diameter had an indirect impact on the decrease of the efficiency by changing the other geometrical parameters, and then it decreased the efficiency by the blade outlet width and the area ratio ($D_2 \rightarrow b_2 \rightarrow \eta_1$ = −0.1076, $D_2 \rightarrow Y \rightarrow \eta_1$ = −0.4445). The indirect path coefficient of the blade outlet width was −0.0106, which implied that the blade outlet width had an indirect impact on the decrease of the efficiency by changing the other geometrical parameters, and the blade outlet width decreased the efficiency by the area ratio ($b_2 \rightarrow Y \rightarrow \eta_1$ = 0.4242), while it decreased the efficiency by the blade wrap angle and the impeller outlet diameter ($b_2 \rightarrow \varphi \rightarrow \eta_1$ = −0.1636, $b_2 \rightarrow D_2 \rightarrow \eta_1$ = −0.2181).

It was acquired by analyzing the indirect path coefficient between the different geometric parameters that the influence weight of each parameter was different when the different performances served as the index ($\beta_2 \rightarrow Z \rightarrow H$ = 0.251, $\beta_2 \rightarrow Z \rightarrow \eta_1$ = −0.0005). Under the index of the same performance, the influence weight of each parameter was directional ($\beta_2 \rightarrow \varphi \rightarrow H$ = −0.1244, $\varphi \rightarrow \beta_2 \rightarrow H$ = 0.1697). Under the small indirect path coefficient, it did not mean that there was little interaction between the factor and other factors (the indirect path coefficient of the impeller outlet diameter reaches −0.0252, $D_0 \rightarrow D_2 \rightarrow \eta_1$ = −0.109, $D_0 \rightarrow Y \rightarrow \eta_1$ = 0.1383), which neutralized the indirect effect on each parameter.

3.3. Analysis of Residual Path Coefficient

The residual path coefficient is a critical value that determines the satisfaction of the linear relation between parameters and performances. In this paper, the determined path coefficients and the residual path coefficients between the eight geometrical parameters of the impeller and the performance of the nuclear reactor coolant pump are listed in Table 3.

Table 3. Residual path coefficient between impeller geometric parameters and performance.

Performance	Determine Path Coefficient	Residual Path Coefficient
H	0.8421	0.2909
η_1	0.6678	0.5541

It can be seen from the table that the determined path coefficient between the head and the performance reached 0.8421, which indicated that the eight parameters selected could calculate it accurately through the linear relationship, while the path coefficient of efficiency was determined as 0.6678. This proved that selecting the eight parameters and efficiency cannot satisfy the linear relationship, there may be larger errors, or the main parameter was not selected. However, according to the actual situation, as the flow of the components of nuclear reactor coolant pump, the impeller was employed to convert mechanical energy into potential energy, so that the head primarily becomes dependent on the impeller geometric parameters. For efficiency, the impeller was only a part of the nuclear main pump flow components, and the influence of the impeller geometry on the efficiency was greater than the selected seven. Thus, the determined path coefficient was normally not large, and hence the calculation process was accurate. The eight parameters selected could have been used as a representation of efficiency by linear. To visually represent the relationship between the efficiency index, the lift index, and the parameters, please refer the path diagram (shown in Figure 3).

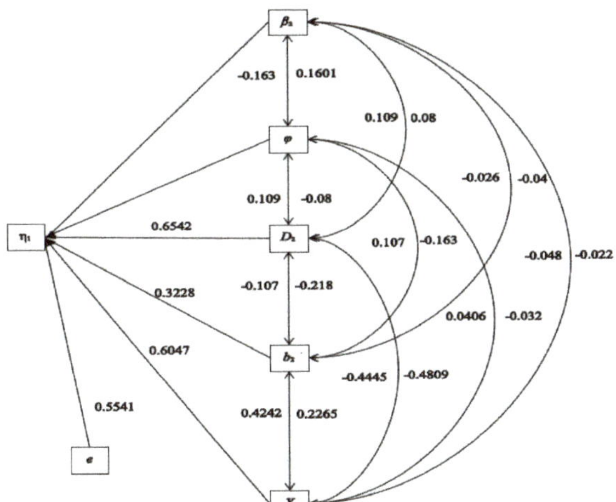

(**a**) The path diagram between the impeller geometrical parameters and the efficiency.

Figure 3. *Cont.*

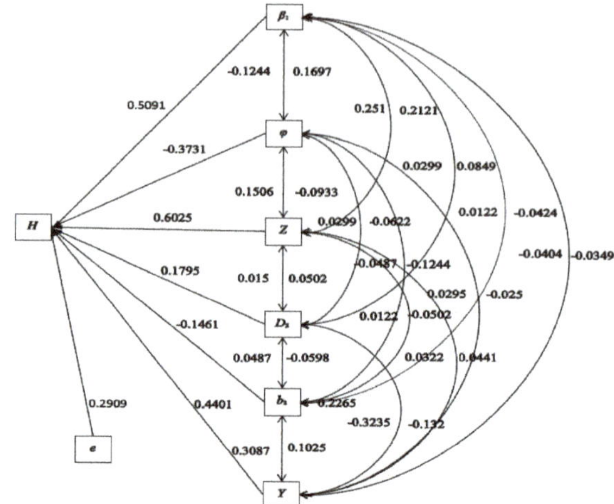

(**b**) The path diagram between the impeller geometrical parameters and the head

Figure 3. The path diagram between the impeller geometrical parameters and the performance.

3.4. Optimal Parameter Selection

By selecting the eight major parameters of the impeller, efficiency, and the head, respectively, as the performance evaluation indices, the geometrical parameters of the different indices have proved to have different effect sizes. When efficiency was the performance index, five geometrical parameters exerted the greatest influence on efficiency in line with the size of influence weight: blade outlet angle > impeller outlet diameter > blade wrap angle > area ratio > blade outlet width. When the head was the performance index, the six geometric parameters had the greatest impact on efficiency; according to influence weight: impeller blade numbers > blade outlet angle > blade wrap angle > area ratio > impeller outlet diameter > blade outlet width. This paper considers efficiency as the main performance index, the optimal parameters were selected in line with the results of partial correlation analysis and path analysis, and the results were as listed under Table 4.

Table 4. Optimal combination of impeller geometrical parameters.

Factor	γ	β_2	φ	Z	D_2	b_2	D_0	Y
Optimal results	23°	30°	115°	5	770 mm	200 mm	555 mm	1.002

4. Experiment Verification

To verify the effectiveness of this optimization method, the model pump, developed as per the specified parameters, was tested and verified. The model pump had the following specifications; design flow Q_M = 104 m³/h, head H_M = 3.6 m, rotating speed n = 1480 r/min, and specific speed n_s = 351. The model pump and the reactor coolant pump size had a ratio of 5.56. The transient performance testbed for reactor coolant pump was presented in Figure 4. To complete the collection, the flow measurement used the LWGY-type turbine flow sensor instrument supplied by the Nanjing Ditai Electromechanical Equipment Co. Ltd (Nanjing, China), turbine flow meter diameter of DN125, output current signal of 4–20 mA, and acquisition accuracy of 0.5 grade. The instantaneous flow signal was acquired from Beijing Altai Technology Development Co., Ltd. We used a production USB3200 type data acquisition card with rotating speed function and a moment sensor: ZJ-type rotating speed and the moment sensor supporting WJCG dynamometer acquisition pump shaft speed, moment, and other data; its working principle was magnetoelectric conversion and electric phase difference, the measurement range of the moment was 0–50 NM, the number of teeth was 180, the precision was 0.2%, and the speed range was 0–5000 r/min. In the test process, the model pump was first, and then the outlet valve was adjusted after its operations were stabilized so that the pump could be shut down under the rated working condition. The start-up curves of the three groups with starting time of approximately 2 s, 4.5 s, and 8.5 s, respectively, were obtained through the coupling to connect the flywheels with different moments of inertia to reduce the starting acceleration of the unit. Through the similar conversion of the model pump, the starting characteristic curves of the three groups of different starting accelerations of the nuclear reactor coolant pump were determined as in Figure 5.

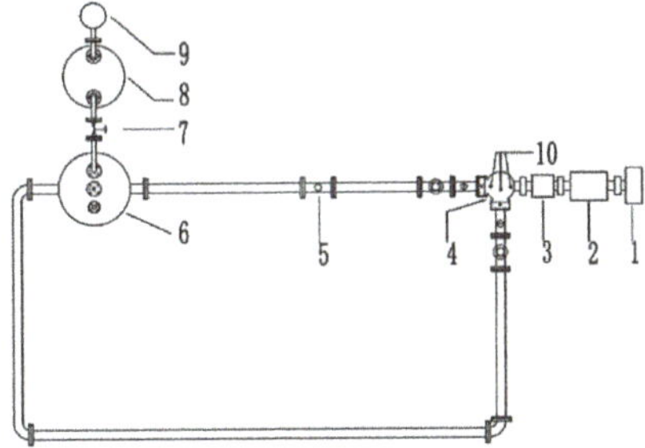

Figure 4. Test device for the coast-down transition process of the nuclear reactor coolant pump.

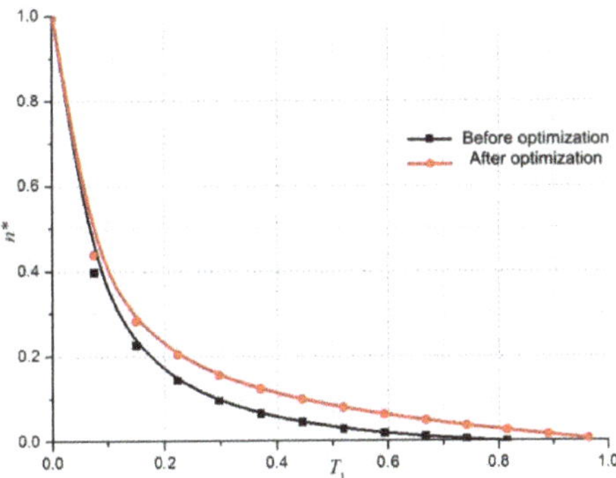

Figure 5. Change of rotational speed before and after the optimization of the nuclear reactor coolant pump during the coast-down transition process.

A dimensionless conversion was carried out with the optimized coast-down time of the pump as a cycle. From Figure 5, it is evident that the optimization had a great influence on the change of the rotation speed [23] in the coast-down transition of the reactor coolant pump. Because of the small moment of inertia of the impeller itself, the small changes in the impeller geometrical parameters and the amount of change in the moment of inertia can be neglected. Figure 3 illustrates that the curves of speed variations in the coast-down transition were different before and after the optimization. The preoptimization speed decreased to zero at nearly 0.8 T_1, and the optimized speed dropped to zero at nearly 1 T_1. Throughout the coast-down transition, the size of the speed before optimization was greater than the size of the speed of coast-down after optimization. This proved that almost the entire coast-down characteristics had been improved through the structural optimization of the nuclear reactor coolant pump.

5. Conclusions

The inertia moment energy stored in the rotor system was utilized to keep the nuclear reactor coolant pump running during coast-down transition. Considering the additional losses of the impeller were primarily caused by the nuclear reactor coolant pump under the off-design condition, the energy loss was reduced and the time of coast-down was delayed by optimizing the main structural parameters of the impeller.

(1) According to the energy conservation law, the calculation equation of coast-down time and the energy utilization of the inertia moment storage were listed, and the basis of coast-down optimization was given based on the reducing energy loss in the coast-down transition. Hydraulic optimization design of the reactor coolant pump impeller was carried out by combining orthogonal optimization test and CFD simulation software, while the hydraulic characteristics of the calculation model were also completed.

(2) The correlation between the impeller geometric parameters and efficiency, head, and different geometric parameters was computed, and the main parameters affecting efficiency and the pressure head were determined. By the path analysis on the results of hydraulic characteristics, the direct influence of geometric parameters on efficiency and head and the indirect influence of the geometric parameters on other parameters that changed the efficiency and head were ascertained.

(3) The efficiency of the pump was the target, the head was the constraint condition, and, combined with the calculation results of the partial correlation analysis and path analysis, the optimal parameters were selected as $\gamma = 23°$, $\beta_2 = 30°$, $\varphi = 115°$, $Z = 5$, $b_2 = 200$ mm, $D_2 = 770$ mm, $D_0 = 555$ mm, and $Y = 1.002$, respectively.

Author Contributions: Y.Z.: Experiments and simulations; X.S.: The writing and revision of the paper; X.W.: Ideas and fund support of the paper; R.Z., Q.F. and H.Z.: Experimental and simulated data processing.

Funding: The National Natural Science Foundation of China (51379091); a science and technology program funded by the Natural Science Foundation of Jiangsu Province (BK20130516); the National Youth Natural Science Foundation of China (51509112); key R & D programs of Jiangsu Province of China (BE2015129 & BE2016160); and a prospective joint research project of Jiangsu Province (BY2016072-02).

Conflicts of Interest: The authors declare no conflict of interest.

References

1. Gao, H.; Gao, F.; Zhao, X.; Chen, J.; Gao, X. Analysis of reactor coolant pump transient performance in primary coolant system during start-up period. *Ann. Nucl. Energy* **2013**, *54*, 202–208. [CrossRef]

2. Tsukamoto, H.; Matsunaga, S.; Yoneda, H.; Hata, S. Transient characteristics of a centrifugal pump during starting/stopping periods: Analysis based on rotating circular cascade model and criterion for quasi-steady change. *Trans. Jpn. Soc. Mech. Eng. B* **1986**, *52*, 1291–1299. [CrossRef]

3. Wu, D.; Wang, L.; Hao, Z.; Li, Z.; Bao, Z. Experimental study on hydrodynamic performance of a cavitating centrifugal pump during transient operation. *J. Mech. Sci. Technol.* **2010**, *24*, 575–582. [CrossRef]

4. Li, Z.; Wu, D.; Wang, L.; Huang, B. Numerical simulation of the transient flow in a centrifugal pump during starting period. *J. Fluids Eng.* **2010**, *132*, 081102. [CrossRef]

5. Chalghoum, I.; Elaoud, S.; Akrout, M.; Taieb, E.H. Transient behavior of a centrifugal pump during starting period. *Appl. Acoust.* **2016**, *109*, 82–89. [CrossRef]

6. Elaoud, S.; Hadj-Taïeb, E. Influence of pump starting times on transient flows in pipes. *Nucl. Eng. Des.* **2011**, *241*, 3624–3631. [CrossRef]

7. Farhadi, K.; Bousbia-Salah, A.; D'Auria, F. A model for the analysis of pump start-up transients in Tehran Research Reactor. *Prog. Nucl. Energy* **2007**, *49*, 499–510. [CrossRef]

8. Farhadi, K. Transient behaviour of a parallel pump in nuclear research reactors. *Prog. Nucl. Energy* **2011**, *53*, 195–199. [CrossRef]

9. Farhadi, K. The effect of retarding torque during a flow transient for Tehran research reactor. *Ann. Nucl. Energy* **2011**, *38*, 175–184. [CrossRef]

10. Alatrash, Y.; Kang, H.O.; Yoon, H.G.; Seo, K.; Chi, D.Y.; Yoon, J. Experimental and analytical investigations of primary coolant pump coastdown phenomena for the Jordan research and training reactor. *Nucl. Eng. Des.* **2015**, *286*, 60–66. [CrossRef]

11. Lu, Y.; Zhu, R.; Fu, Q.; Wang, X.; An, C.; Chen, J. Research on the structure design of the LBE reactor coolant pump in the lead base heap. *Nucl. Eng. Technol.* **2018**, *51*, 546–555. [CrossRef]

12. Lu, Y.; Zhu, R.; Wang, X.; Fu, Q.; Li, M.; Si, X. Study on gas-liquid two-phase all-characteristics of CAP1400 nuclear main pump. *Nucl. Eng. Des.* **2017**, *319*, 140–148. [CrossRef]

13. Lu, Y.; Zhu, R.; Wang, X.; Fu, Q.; Ye, D. Study on the complete rotational characteristic of coolant pump in the gas-liquid two-phase operating condition. *Ann. Nucl. Energy* **2019**, *123*, 180–189.

14. Kim, J.H.; Ahn, H.J.; Kim, K.Y. High-efficiency design of a mixed-flow pump. *Sci. China Technol. Sci.* **2010**, *53*, 24–27. [CrossRef]

15. Varchola, M.; Hlbocan, P. Geometry design of a mixed flow pump using experimental results of on internal impeller flow. *Procedia Eng.* **2012**, *39*, 168–174. [CrossRef]

16. Long, Y.; Yin, J.L.; Wang, D.Z.; Li, T.B. Study on the effect of the impeller and diffuser blade number on reactor coolant pump performances. In Proceedings of the 7th International Conference on Pumps and Fans (ICPF 2015), Hangzhou, China, 18–21 October 2015.

17. Zhou, F.M.; Wang, X.F. The effects of blade stacking lean angle to 1400 MW canned nuclear coolant pump hydraulic performance. *Nucl. Eng. Des.* **2017**, *325*, 232–244. [CrossRef]

18. Zhu, R.S.; Xing, S.B.; Fu, Q.; Li, T.B.; Wang, X.L. Study on transient flow characteristics of AP1000 nuclear reactor coolant pump under exhaust transit condition. *At. Energy Sci. Technol.* **2016**, *50*, 1040–1046.

19. Gao, H.; Gao, F.; Zhao, X.; Chen, J.; Cao, X. Transient flow analysis in reactor coolant pump systems during flow coastdown period. *Nucl. Eng. Des.* **2011**, *241*, 509–514. [CrossRef]

20. Long, Y.; Zhu, R.; Wang, D.; Yin, J.; Li, T. Numerical and experimental investigation on the diffuser optimization of a reactor coolant pump with orthogonal test approach. *J. Mech. Sci. Technol.* **2016**, *30*, 4941–4948.

21. Kang, C.; Mao, N.; Zhang, W.; Gu, Y. The influence of blade configuration on cavitation performance of a condensate pump. *Ann. Nucl. Energy* **2017**, *110*, 789–797. [CrossRef]

22. Zhang, K.; Yuan, J.; Sun, W.; Si, Q. Internal flow characteristics of residual heat removal pump during different starting periods. *J. Drain. Irrig. Mach. Eng.* **2017**, *35*, 192–199.

23. Fu, Q.; Zhang, B.; Zhu, R.; Cao, L. Effect of rotating speed on cavitation performance of nuclear reactor coolant pump. *J. Drain. Irrig. Mach. Eng.* **2016**, *34*, 651–656.

Article

Productivity Models of Infill Complex Structural Wells in Mixed Well Patterns

Liang Sun *, Baozhu Li and Yong Li

Research Institute of Petroleum Exploration & Development, PetroChina, Beijing 100083, China; lbz@petrochina.com.cn (B.L.); liyongph@petrochina.com.cn (Y.L.)
* Correspondence: sunliang328@petrochina.com.cn

Received: 14 May 2019; Accepted: 27 May 2019; Published: 31 May 2019

Abstract: The mathematical models of productivity calculation for complex structural wells mainly focus on the single well or the regular well pattern. Previous research on the seepage theory of complex structural wells and vertical wells in mixed well pattern is greatly insufficient. Accordingly, this article presents a methodology of evaluating the productivity of infill complex structural wells in mixed well patterns. On the basis of the mirror-image method and source–sink theory, two semi-analytical models are established. These models are applied to the productivity prediction of an infill horizontal well inhorizontal-vertical well pattern and an infill multilateral well inmultilateral-vertical well pattern, respectively, in which the interference of other wells, the randomicity of well patterns, and the pressure drawdown along the horizontal laterals are taken into account. The semi-analytical models' results are consistent with those calculated by the Eclipse reservoir simulator with the relative error of less than 15%. Results indicate that the bottom hole flowing pressure decreases logarithmically while the wellbore flow rate increases monotonically from the toe to the heel of the horizontal well. Due to the pseudo-hemispherical flow at each endpoint and the pseudo-linear flow at the center of the horizontal well, the drainage area at each endpoint is relatively larger than that at the center. The radial inflow at each endpoint of the horizontal segment is considerably greater than that at the center, which presents the U-shape distribution. The proposed methodology enhances and promotes the theory of productivity evaluation for complex structural wells in mixed well patterns.

Keywords: complex structural well; mixed well pattern; productivity evaluation; semi-analytical model; well location optimization

---

1. Introduction

Complex structural wells including horizontal wells and multilateral wells have become a popular alternative for the development of oil and gas fields around the world because of their high flow efficiency due to larger contact area made with the reservoir and lower pressure drawdown at the same liquid volume [1]. As the process of oilfield development enters into the intermediary and later phase, the mixed well patterns of complex structural wells and conventional vertical wells have been widely applied to the implementation of adjustment plans [2]. The infilling horizontal wells or multilateral wells in mixed well patterns are of great significance to optimization of development strategies. Due to the complexity of seepage mechanism near the horizontal wellbore, the coupling between reservoir flow and wellbore conduit flow, and the interference of other wells in mixed well patterns [3], the original mathematical models based on time invariant flow are no longer applicable as a result of the change of flow regimes. Therefore, it is necessary to establish new productivity models of infill complex structural wells in mixed well patterns.

The productivity evaluation of complex structural wells under different reservoir conditions is an important topic in the field of complex structural wells. At present, studies have been conducted on the methods of production calculation for horizontal wells [4–18] and multilateral wells [19–23], which are mainly divided into analytical methods and semi-analytical methods. The analytical model aims to directly build a calculation formula based on ideal assumptions. In the semi-analytical model, each branch of the complex structural well is divided into several infinitesimal sections, thus the productivity can be obtained by solving the system of linear equations combined with fluid flow rate and pressure drawdown for each infinitesimal in the wellbore. However, these methods mainly focus on building models of single well or regular well patterns, and little research has been conducted on the seepage theory of mixed well patterns of vertical wells and complex structural wells [24–30]. In view of the field problems involving the productivity prediction of infill wells in irregular well patterns, there are limitations in current mathematical models [31–35]. On the one hand, they are only suitable for the productivity calculation of the entire well pattern not for single infill wells; on the other hand, they are merely applied to the regular five-spot pattern, seven-spot pattern, and nine-spot pattern not for the irregularly mixed well pattern. More importantly, the pressure drop caused by wall friction and fluid acceleration along the horizontal lateral is not comprehensively taken into account during the coupling of reservoir seepage and wellbore conduit flow. In order to overcome the deficiencies of the existing models, Ye et al. [36] presented a productivity evaluation model for infill horizontal wells considering the interference of other wells and the wellbore friction, which is suitable for mixed horizontal injection and production patterns. Although this method did not take into consideration more complicated conditions, such as the mixed well pattern including multilateral-vertical wells, it provides us with an effective approach to solve these problems.

The objective of this article here is to present two semi-analytical models for the productivity evaluation of infill complex structural wells in mixed well patterns on the basis of the mirror-image method and source–sink theory. The first model is suitable for the infill horizontal well in horizontal-vertical well pattern, and the other for the infill multilateral well in multilateral-vertical well pattern. Then, the two models are applied to the study the seepage mechanism in terms of the bottom hole flowing pressure and the distribution of wellbore flow and radial flow along the horizontal segment. The main feature of this methodology is that the models take into account the interference of other wells, the randomicity of well patterns, and the pressure drawdown along the horizontal laterals. The application in the optimization of infill well location also indicates the significant practical value of the proposed models.

2. Productivity Model

2.1. Productivity Model of Mixed Horizontal-Vertical Well Pattern

2.1.1. Reservoir Flow Model

As is shown in Figure 1, in the deployment of mixed horizontal-vertical well pattern, we assume that the number of vertical producers including $P_1^z, P_2^z, \cdots, P_m^z$ is m^z, the number of horizontal producer including $P_1^s, P_2^s, \cdots, P_m^s$ is m^s, the number of vertical injectors including $I_1^z, I_2^z, \cdots, I_n^z$ is n^z, the number of horizontal injectors including $I_1^s, I_2^s, \cdots, I_n^s$ is n^s, and P_{new} is the infill horizontal producer. The top and bottom boundaries of the reservoir are both closed, and the surrounding area is infinite. In order to analyze the well performance, the vertical interval and horizontal interval are both divided into N segments (Figure 2) [2,36].

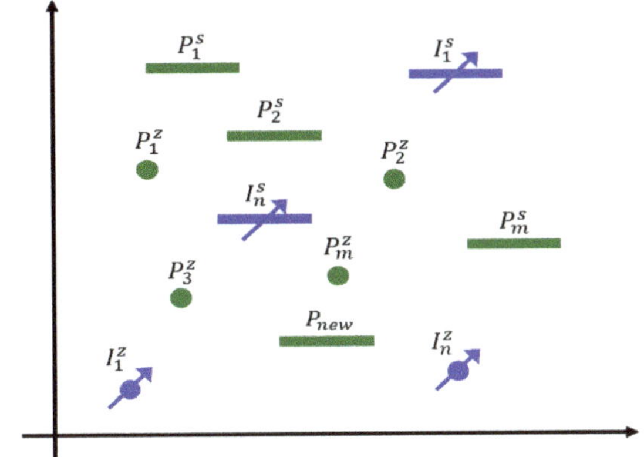

Figure 1. The deployment of mixed horizontal-vertical well pattern.

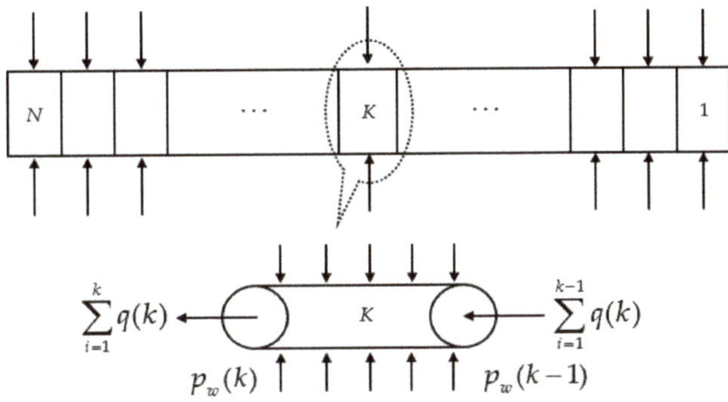

Figure 2. Schematic of the segmented horizontal well.

The coordinates of an arbitrary point M in the vertical segment k $(1 \leq k \leq N)$ of the uth vertical producer can be expressed as follows:

$$\begin{cases} x_z^p(u,k,t) = x_z^p, \\ y_z^p(u,k,t) = y_z^p, \\ \vdots \\ z_z^p(u,k,t) = z_z^p + (k+t-1)L/N, \quad 0 \leq t \leq 1 \end{cases} \tag{1}$$

The coordinates of an arbitrary point M in the horizontal segment k $(1 \leq k \leq N)$ of the uth horizontal producer can be expressed as follows:

$$\begin{cases} x_s^p(u,k,t) = x_s^p + (k+t-1)L/N, \\ y_s^p(u,k,t) = y_s^p, \\ \vdots \\ z_s^p(u,k,t) = z_s^p, \quad 0 \leq t \leq 1 \end{cases} \tag{2}$$

The coordinates of an arbitrary point M in the vertical segment k ($1 \leq k \leq N$) of the vth vertical injector can be expressed as follows:

$$
\begin{cases}
x_z^i(v,k,t) = x_z^i, \\
y_z^i(v,k,t) = y_z^i, \\
\vdots \\
z_z^i(v,k,t) = z_z^i + (k+t-1)L/N, \qquad 0 \leq t \leq 1
\end{cases}
\tag{3}
$$

The coordinates of an arbitrary point M in the horizontal segment k ($1 \leq k \leq N$) of the vth horizontal injector can be expressed as follows:

$$
\begin{cases}
x_s^i(v,k,t) = x_s^i + (k+t-1)L/N, \\
y_s^i(v,k,t) = y_s^i, \\
\vdots \\
z_s^i(v,k,t) = z_s^i, \qquad 0 \leq t \leq 1
\end{cases}
\tag{4}
$$

The coordinates of an arbitrary point M in the horizontal segment k ($1 \leq k \leq N$) of the infill horizontal producer can be expressed as follows [37]:

$$
\begin{cases}
x(new,k,t) = x_{new}^p + (k+t-1)L/N, \\
y(new,k,t) = y_{new}^p, \\
\vdots \\
z(new,k,t) = z_{new}^p, \qquad 0 \leq t \leq 1
\end{cases}
\tag{5}
$$

where $x_z^p(u,k,t), y_z^p(u,k,t), z_z^p(u,k,t)$ are the coordinates of an arbitrary point in the segment k of the uth vertical producer; x_z^p, y_z^p, z_z^p are the coordinates of the left end (well heel) of the uth vertical producer; $x_s^p(u,k,t), y_s^p(u,k,t), z_s^p(u,k,t)$ are the coordinates of an arbitrary point in the segment k of the uth horizontal producer; x_s^p, y_s^p, z_s^p are the coordinates of the left end (well heel) of the uth horizontal producer; $x_z^i(v,k,t), y_z^i(v,k,t), z_z^i(v,k,t)$ are the coordinates of an arbitrary point in the segment k of the vth vertical injector; x_z^i, y_z^i, z_z^i are the coordinates of the left end (well heel) of the vth vertical injector; $x_s^i(v,k,t), y_s^i(v,k,t), z_s^i(v,k,t)$ are the coordinates of an arbitrary point in the segment k of the vth horizontal injector; x_s^i, y_s^i, z_s^i are the coordinates of the left end (well heel) of the vth horizontal injector; $x(new,k,t), y(new,k,t), z(new,k,t)$ are the coordinates of an arbitrary point in the segment k of the infill horizontal producer; $x_{new}^p, y_{new}^p, z_{new}^p$ are the coordinates of the left end (well heel) of the infill horizontal producer. L is the length of the entire horizontal section, m; N is the total number of the segments, dimensionless.

The mirror-image method and source–sink theory have been widely applied to the boundary effect reduction. As is shown in Figure 3a, assuming the constant pressure boundary is a mirror, we can project the producer (we call sink in the method) into an image injector (we call source in the method) at the symmetric coordinate to counteract the constant pressure boundary effect. As is shown in Figure 3b, in the same way, we can project the producer into an image producer at the symmetric coordinate to counteract the closed boundary effect [2].

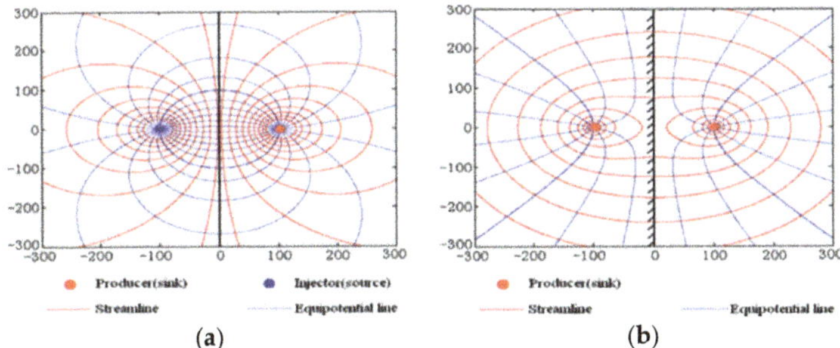

Figure 3. The mirror-image method and source–sink theory applied in different boundaries: (**a**) constant pressure boundary; (**b**) closed boundary.

On the basis of the method of images and the principle of superposition, we take the closed boundaries of the top and the bottom of the reservoir as mirrors. Thus, it is transformed into the problem of infinite well rows for horizontal producers, which is easy to solve. The potential of the producing segments of the infill horizontal well at any point $M\,(x, y, z)$ in the infinite formation is:

$$\Phi(x,y,z) = \sum_{k=1}^{N} \Phi_k(x,y,z) = \frac{1}{4\pi\Delta L_k} \sum_{k=1}^{N} [q_r(k)\varphi_k(x,y,z)] + C. \tag{6}$$

With:

$$\begin{aligned}
\varphi_i(x,y,z) = \sum_{n=-\infty}^{+\infty} \Big\{ &\xi_k[x(new,k,t), y(new,k,t), 2nh + z(new,k,t)] + \\
&\xi_k[x(new,k,t), y(new,k,t), 2nh - z(new,k,t)] + \\
&\sum_{u=1}^{m^z} \frac{q_{u,z}^p(k)}{q_r(k)} \xi_k[x_z^p(u,k,t), y_z^p(u,k,t), 2nh + z_z^p(u,k,t)] + \\
&\sum_{u=1}^{m^z} \frac{q_{u,z}^p(k)}{q_r(k)} \xi_k[x_z^p(u,k,t), y_z^p(u,k,t), 2nh - z_z^p(u,k,t)] + \\
&\sum_{u=1}^{m^s} \frac{q_{u,s}^p(k)}{q_r(k)} \xi_k[x_s^p(u,k,t), y_s^p(u,k,t), 2nh + z_s^p(u,k,t)] + \\
&\sum_{u=1}^{m^s} \frac{q_{u,s}^p(k)}{q_r(k)} \xi_k[x_s^p(u,k,t), y_s^p(u,k,t), 2nh - z_s^p(u,k,t)] - \\
&\sum_{v=1}^{n^z} \frac{q_{v,z}^i(k)}{q_r(k)} \xi_k[x_z^i(v,k,t), y_z^i(v,k,t), 2nh + z_z^i(v,k,t)] - \\
&\sum_{v=1}^{n^z} \frac{q_{v,z}^i(k)}{q_r(k)} \xi_k[x_z^i(v,k,t), y_z^i(v,k,t), 2nh - z_z^i(v,k,t)] - \\
&\sum_{v=1}^{n^s} \frac{q_{v,s}^i(k)}{q_r(k)} \xi_k[x_s^i(v,k,t), y_s^i(v,k,t), 2nh + z_s^i(v,k,t)] - \\
&\sum_{v=1}^{n^s} \frac{q_{v,s}^i(k)}{q_r(k)} \xi_k[x_s^i(v,k,t), y_s^i(v,k,t), 2nh - z_s^i(v,k,t)] \Big\}, \quad n = 0, \pm 1, \pm 2, \cdots
\end{aligned} \tag{7}$$

where $q_{u,z}^p(k) \approx \frac{Q_{u,z}^p}{N}$, $q_{v,z}^i(k) \approx \frac{Q_{v,z}^i}{N}$, $q_{u,s}^p(k) \approx \frac{Q_{u,s}^p}{N}$, $q_{v,s}^i(k) \approx \frac{Q_{v,s}^i}{N}$; $Q_{u,z}^p$ is the production of the uth vertical producer, m³/d; $Q_{v,z}^i$ is the injection rate of the vth vertical injector; $Q_{u,s}^p$ is the production of the uth horizontal producer, m³/d; $Q_{v,s}^i$ is the injection rate of the vth horizontal injector, m³/d.

For:

$$\xi_k[x(new,k,t), y(new,k,t), z(new,k,t)] = \ln \frac{r_1(new,k) + r_2(new,k) + \frac{L}{N}}{r_1(new,k) + r_2(new,k) - \frac{L}{N}},$$

$$r_1(new, k) = \sqrt{[x(new, k, t = 0) - x]^2 + [y(new, k, t = 0) - y]^2 + [z(new, k, t = 0) - z]^2},$$

$$r_2(new, k) = \sqrt{[x(new, k, t = 1) - x]^2 + [y(new, k, t = 1) - y]^2 + [z(new, k, t = 1) - z]^2}.$$

In the same way, we can obtain the expressions of $\xi_k[x(u, k, t), y(u, k, t), z(u, k, t)]$ and $\xi_k[x(v, k, t), y(v, k, t), z(v, k, t)]$.

Once the distributions of the flow rates of $q_r(k)$, $q_{u,z}^p(k)$, $q_{v,z}^i(k)$, $q_{u,s}^p(k)$, $q_{v,s}^i(k)$ are known, all terms of flow pressure $p_{wf}(k)$ of the infill horizontal producer can be calculated. Hence Equation (6) is simplified to a linear equation system, and can be arranged in the form as follows:

$$\begin{bmatrix} \varphi_{11} & \varphi_{12} & \varphi_{13} & \cdots & \varphi_{1N} \\ \varphi_{21} & \varphi_{22} & \varphi_{23} & \cdots & \varphi_{2N} \\ \varphi_{31} & \varphi_{32} & \varphi_{33} & \cdots & \varphi_{3N} \\ \vdots & \vdots & \vdots & \ddots & \vdots \\ \varphi_{N1} & \varphi_{N2} & \varphi_{N3} & \cdots & \varphi_{NN} \end{bmatrix} \begin{bmatrix} q_r(1) \\ q_r(2) \\ q_r(3) \\ \vdots \\ q_r(N) \end{bmatrix} = \frac{4\pi K \Delta L}{\mu_o} \begin{bmatrix} p_e - p_{wf}(1) \\ p_e - p_{wf}(2) \\ p_e - p_{wf}(3) \\ \vdots \\ p_e - p_{wf}(N) \end{bmatrix}, \tag{8}$$

where φ_{ij} is the value of φ_j at the midpoint of the segment i of the infill horizontal producer, dimensionless; $q_r(i)$ is the radial flow rate of the segment i of the infill horizontal producer, m^3/d; p_e is the reservoir pressure, MPa; $p_{wf}(i)$ is the flowing pressure of the segment i of the infill horizontal producer, MPa; K is the reservoir permeability, 10^{-3} μm^2; μ_o is the oil viscosity, mPa·s; h is the net pay thickness, m.

The wellbore flow rate along the horizontal well can be expressed as follows:

$$q_l(k) = \sum_{j=k}^{N} q_r(j), \tag{9}$$

where $q_l(k)$ is the wellbore flow rate of the segment k of the infill horizontal producer, m^3/d.

2.1.2. Wellbore Flow Model

The wellbore pressures of each segment along the horizontal segment are not independent of each other, instead, they are related to each other via wellbore hydraulics. More specifically, the pressure difference between two adjacent segment midpoints is dependent on the radial flow into the two segments, the local pressure, and the fluid property [2,36].

The pressure drop along the horizontal well is caused by the wall friction and the fluid acceleration [3,11,12,38], and can be calculated by the correlation of pressure drop in the horizontal interval under the condition of open-hole completion proposed by Liu et al. [12]:

$$\Delta p_{wf}(i) = \frac{L}{N} \left\{ \frac{2f\rho}{\pi^2 D^5} \left[2q_l(i) - q_r(i)\frac{L}{N} \right]^2 + \frac{16\rho q_r(i)}{\pi^2 D^4} \left[2q_l(i) - q_r(i)\frac{L}{N} \right] \right\},$$
$$p_{wf}(i) = p_{wf}(i - 1) + 0.5\big(\Delta p_{wf}(i - 1) + \Delta p_{wf}(i)\big), \quad (1 \leq i \leq N + 1) \tag{10}$$

where p_{wf} is the flowing pressure of the well heel, $p_{wf}(0) = p_{wf}$, $\Delta p_{wf}(0) = \Delta p_{wf}(N + 1) = 0$, MPa; ρ is the fluid density, g/cm^3; f is the wall friction factor, and can be calculated as follows:

$$f = \frac{64}{N_{Re}}, \qquad\qquad N_{Re} \leq 2100$$
$$\frac{1}{\sqrt{f}} = 1.14 - 2\lg\big(\frac{\tau}{D} + \frac{21.25}{N_{Re}^{0.9}}\big), \quad N_{Re} \geq 2100$$

By combing Equations (9) and (10), the following system of equations can be obtained:

$$f[q_r(i), p_{wf}(i)] = 0. \tag{11}$$

As expected, the total number of Equations (8) and (11) is equal to 2*N*, which is the same as that of unknowns. Thus, the model has a unique solution.

2.1.3. Solution Procedure

The iterative algorithm is an efficient way to solve this problem, and the solution procedure is shown in Figure 4.

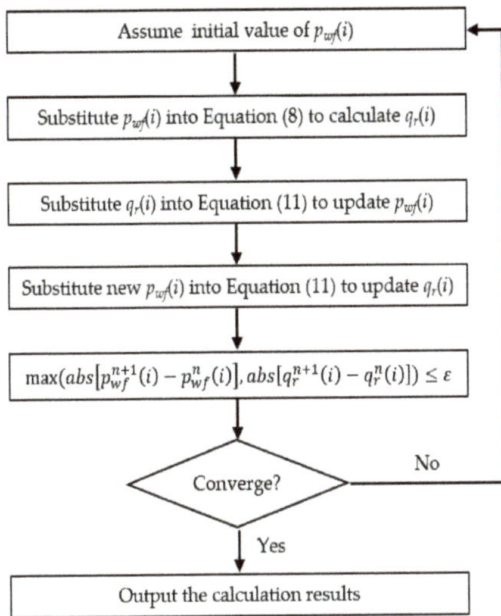

Figure 4. The flow chart of solution procedure.

2.2. Productivity Model of Mixed Multilateral-Vertical Well Pattern

2.2.1. Reservoir Flow Model

As is shown in Figure 5, in the deployment of mixed multilateral-vertical well pattern, we assume that the number of vertical producers including $P_1^z, P_2^z, \cdots, P_m^z$ is m^z, the number of vertical injectors including $I_1^z, I_2^z, \cdots, I_n^z$ is n^z, and P_{new} is the infill multilateral well. The top and bottom boundaries of the reservoir are both closed, and the surrounding area is infinite. In order to analyze the well performance, the vertical interval is divided into N segments.

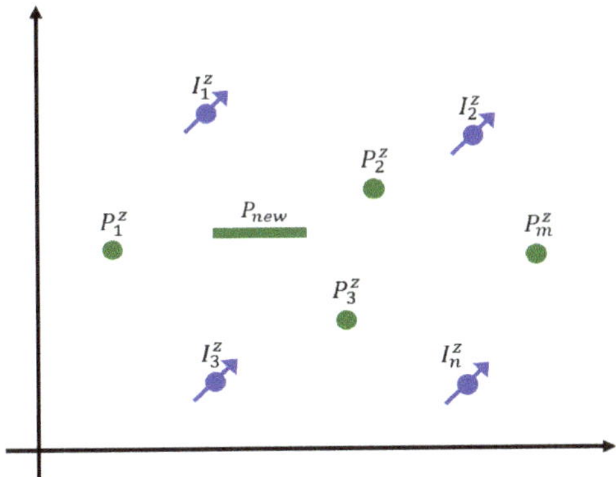

Figure 5. The mixed well pattern of multilateral-vertical wells.

The coordinates of an arbitrary point M in the vertical segment k ($1 \leq k \leq N$) of the uth vertical producer can be expressed as follows:

$$
\begin{cases}
x_z^p(u,k,t) = x_z^p, \\
y_z^p(u,k,t) = y_z^p, \\
\vdots \\
z_z^p(u,k,t) = z_z^p + (k+t-1)L/N, \quad 0 \leq t \leq 1
\end{cases}
\tag{12}
$$

The coordinates of an arbitrary point M in the vertical segment k ($1 \leq k \leq N$) of the vth vertical injector can be expressed as follows:

$$
\begin{cases}
x_z^i(v,k,t) = x_z^i, \\
y_z^i(v,k,t) = y_z^i, \\
\vdots \\
z_z^i(v,k,t) = z_z^i + (k+t-1)L/N, \quad 0 \leq t \leq 1
\end{cases}
\tag{13}
$$

where $x_z^p(u,k,t)$, $y_z^p(u,k,t)$, $z_z^p(u,k,t)$ are the coordinates of an arbitrary point in the segment k of the uth vertical producer; x_z^p, y_z^p, z_z^p are the coordinates of the left end (well heel) of the uth vertical producer; $x_z^i(v,k,t)$, $y_z^i(v,k,t)$, $z_z^i(v,k,t)$ are the coordinates of an arbitrary point in the segment k of the vth vertical injector; x_z^i, y_z^i, z_z^i are the coordinates of the left end (well heel) of the vth vertical injector.

Figure 6 is the three-dimensional (3D) schematic of the infill multilateral well, we assume that each branch is symmetrically distributed on the same horizontal plane, the length of which is identical. The xoy coordinate system is established by taking the subpoint of the main hole in the xoy plane as the origin of coordinates, and the direction of the main hole as the z axis. Each branch is distributed counterclockwise around the z axis with the x axis as the starting direction.

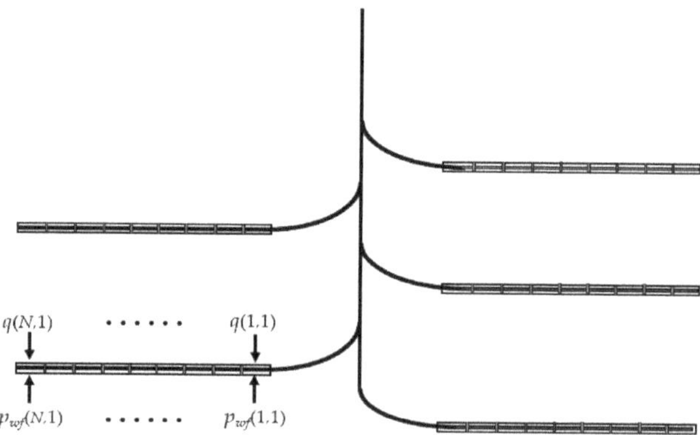

Figure 6. 3D schematic of infill multilateral well.

$M_0(x, y, z_0)$ is the bottom-hole coordinate of the main wellbore, and $M_i(x_{i,0}, y_{i,0}, z_{i,0})$ is the starting coordinate of the ith branch with the length of L_i ($1 \leq i \leq M$). M is the number of branches. The ith branch of the infill multilateral well is divided into N_i ($1 \leq i \leq M$) segments, thus the infinitesimal length of the ith branch is $\Delta L_i = L_i/N_i$.

Accordingly, the coordinates of an arbitrary point M in the horizontal segment i $(1 \leq i \leq M)$ of the infill multilateral well can be expressed as follows:

$$\begin{cases} x(i,j) = x_{i,0} + \Delta L_i\left[\sum_{k=1}^{j-1}(\sin\theta_{i,k}\cdot\cos\alpha_{i,k}) + t\sin\theta_{i,j}\cdot\cos\alpha_{i,k}\right], \\ y(i,j) = y_{i,0} + \Delta L_i\left[\sum_{k=1}^{j-1}(\sin\theta_{i,k}\cdot\sin\alpha_{i,k}) + t\sin\theta_{i,j}\cdot\sin\alpha_{i,k}\right], \\ z(i,j) = z_{i,0} + \Delta L_i\left[\sum_{k=1}^{j-1}\cos\theta_{i,k} + t\cos\theta_{i,j}\right], \quad (1 \leq i \leq M, 0 \leq j \leq N_i, 0 \leq t \leq 1) \end{cases} \quad (14)$$

where $x(i,j), y(i,j), z(i,j)$ are the coordinates of an arbitrary point in the ith branch of the infill multilateral producer; $x_{i,0}, y_{i,0}, z_{i,0}$ are the coordinates of the left end (well heel) in the ith branch of the infill multilateral producer; θ is the deviation angle of the ith segment of the infill multilateral producer; α is the azimuth angle of the ith segment of the infill multilateral producer.

As mentioned above, we can also transform this problem into the one of infinite well rows by using the method of images and the principle of superposition. Thus the potential of the producing segments of the infill multilateral well at any point $M(x, y, z)$ in the infinite formation is:

$$f(x,y,z) = \sum_{i=1}^{M} f_i(x,y,z) = \frac{1}{4\pi}\sum_{i=1}^{M}\sum_{j=1}^{N_j}\left[q_r(i,j)\varphi_{i,j}(x,y,z)\right] + C. \quad (15)$$

With:

$$\begin{aligned} \varphi_{i,j}(x,y,z) = \sum_{n=-\infty}^{+\infty} \Big[&\tfrac{1}{\Delta L_i}\xi_{i,j}\big(x_{i,j}, y_{i,j}, 2nh + z_{i,j}\big) + \tfrac{1}{\Delta L_i}\xi_{i,j}\big(x_{i,j}, y_{i,j}, 2nh - z_{i,j}\big) + \\ &\tfrac{1}{\Delta L}\sum_{u=1}^{m^z}\tfrac{q_{u,z}^p(k)}{q_r(i,j)}\xi_k[x_z^p(u,k,t), y_z^p(u,k,t), 2nh + z_z^p(u,k,t)] + \\ &\tfrac{1}{\Delta L}\sum_{u=1}^{m^z}\tfrac{q_{u,z}^p(k)}{q_r(i,j)}\xi_k[x_z^p(u,k,t), y_z^p(u,k,t), 2nh - z_z^p(u,k,t)] - \\ &\tfrac{1}{\Delta L}\sum_{v=1}^{n^z}\tfrac{q_{v,z}^i(k)}{q_r(i,j)}\xi_k[x_z^i(v,k,t), y_z^i(v,k,t), 2nh + z_z^i(v,k,t)] - \\ &\tfrac{1}{\Delta L}\sum_{v=1}^{n^z}\tfrac{q_{v,z}^i(k)}{q_r(i,j)}\xi_k[x_z^i(v,k,t), y_z^i(v,k,t), 2nh - z_z^i(v,k,t)] \Big] \end{aligned} \quad (16)$$

where $q_{u,z}^p(k) \approx \frac{Q_{u,z}^p}{N}, q_{v,z}^i(k) \approx \frac{Q_{v,z}^i}{N}$; $Q_{u,z}^p$ is the production of the uth vertical producer, m^3/d; $Q_{v,z}^i$ is the injection rate of the vth vertical injector, m^3/d.

Once the distributions of the flow rates of $q_r(i,j)$, $q_{u,z}^p(k)$, $q_{v,z}^i(k)$ are known, all terms of flow pressure $p_{wf}(i,j)$ of the infill multilateral producer can be calculated. Hence Equation (15) is simplified to a linear equation system, and can be arranged in the form as follows:

$$A\begin{bmatrix} q_r(1,1) \\ q_r(1,2) \\ \vdots \\ q_r(1,N_1) \\ \vdots \\ q_r(M,N_M) \end{bmatrix} = \frac{4\pi\sqrt{k_h k_v}}{\mu_o}\begin{bmatrix} P_e - P_{wf}(1,1) \\ P_e - P_{wf}(1,2) \\ \vdots \\ P_e - P_{wf}(1,N_1) \\ \vdots \\ P_e - P_{wf}(1,N_M) \end{bmatrix}. \quad (17)$$

With:

$$
A = \begin{bmatrix}
\phi_{(1,1)(1,1)} - \phi_{e(1,1)} & \cdots & \phi_{(1,1)(1,N_1)} - \phi_{e(1,N)} & \phi_{(1,1)(2,1)} - \phi_{e(2,1)} & \cdots & \phi_{(1,1)(M,N_M)} - \phi_{e(M,N_M)} \\
\vdots & \ddots & \vdots & \vdots & \ddots & \vdots \\
\phi_{(1,N_1)(1,1)} - \phi_{e(1,1)} & \cdots & \phi_{(1,N_1)(1,N_1)} - \phi_{e(1,N)} & \phi_{(1,N_1)(2,1)} - \phi_{e(2,1)} & \cdots & \phi_{(1,N_1)(M,N_M)} - \phi_{e(M,N_M)} \\
\phi_{(2,1)(1,1)} - \phi_{e(1,1)} & \cdots & \phi_{(2,1)(1,N_1)} - \phi_{e(1,N)} & \phi_{(2,1)(2,1)} - \phi_{e(2,1)} & \cdots & \phi_{(2,1)(M,N_M)} - \phi_{e(M,N_M)} \\
\vdots & \ddots & \vdots & \vdots & \ddots & \vdots \\
\phi_{(M,N_M)(1,1)} - \phi_{e(1,1)} & \cdots & \phi_{(M,N_M)(1,N_1)} - \phi_{e(1,N)} & \phi_{(M,N_M)(2,1)} - \phi_{e(2,1)} & \cdots & \phi_{(M,N_M)(M,N_M)} - \phi_{e(M,N_M)}
\end{bmatrix}
$$

where $q_r(i, j)$ is the radial flow rate of the ith branch of the infill multilateral producer, $1 \leq i \leq M$, $1 \leq j \leq N_i$, m^3/d; P_e is the reservoir pressure, MPa; $P_{wf}(i, j)$ is the flow pressure of the ith branch of the infill multilateral producer, $1 \leq i \leq M$, $1 \leq j \leq N_i$, MPa; k_h is the reservoir horizontal permeability, 10^{-3} μm^2; k_v is the reservoir vertical permeability, 10^{-3} μm^2; μ_o is the oil viscosity, mPa·s.

2.2.2. Wellbore Flow Model

1. Pressure drop model of the horizontal lateral

The pressure drop along the horizontal interval is caused by the wall friction and the fluid acceleration. According to Equation (11), the pressure drop of the ith branch of the multilateral well along the horizontal interval can be calculated by:

$$
F'_{i,j}\left[q_r(i, j), p_{wf}(i, j)\right] = 0. \tag{18}
$$

2. Pressure drop model of the deviated segment

The branch holes enter the reservoir through the deviated segment. Although each deviated segment does not contribute to the production, the pressure drop in the deviated segments cannot be ignored. According to fluid seepage theory in the elbow, the pressure drop can be calculated by the following formula:

$$
\Delta p_s = C_w \frac{4 \times 10^{-6} \rho \lambda_c R_c Q_W^2}{\pi d_w^5} + \frac{\rho g R_c}{10^6}, \tag{19}
$$

where ΔP_s the pressure drop along the deviated segment, MPa; R_c is curvature radius of the deviated segment, m; ρ is the fluid density, kg/m^3; d_w is the wellbore diameter of the deviated segment, m; C_w is the similarity factor of the wellbore and deviated segment, dimensionless; λ_c is the friction factor of the deviated segment, dimensionless.

3. Pressure drop model of the vertical segment

According to the law of conservation of energy, the equation of pressure gradient in the pipe with inclination can be expressed as follows:

$$
\frac{dP}{dz} = \rho g \sin \theta + \rho v \frac{dv}{dz} + f \frac{\rho v^2}{2d}, \tag{20}
$$

where dP/dz is the pressure loss per unit length, MPa/m; $\rho g \sin\theta$ is the pressure drop caused by the fluid gravity, MPa/m; $\rho v \times dv/dz$ is the pressure drop caused by the fluid acceleration, MPa/m; $f \times \rho v^2/2d$ is the pressure drop caused by the friction, MPa/m; ρ is the fluid density, kg/m^3; d is the pipe diameter, m. The friction factor f can be calculated by:

$$
f = \frac{64}{N_{Re}}, \quad N_{Re} \leq 2100
$$

$$
\frac{1}{\sqrt{f}} = 1.14 - 2\lg\left(\frac{\tau}{d} + \frac{21.25}{N_{Re}^{0.9}}\right), \quad N_{Re} \geq 2100
$$

where N_{Re} is the Reynolds number, dimensionless.

4. Internal pressure drop model of the branch hole

The flowing pressure at the heel of the *i*th branch can be calculated by simultaneous Equations (19) and (20):

$$P_{wf}(i) = P_{mb} - \Delta P_{vi} - \Delta P_{wi}(1 \le i \le M),\tag{21}$$

where P_{mb} is the flowing pressure of the main wellbore, MPa; ΔP_{vi} is the pressure drop of the vertical segment of the *i*th branch, MPa; ΔP_{wi} is the pressure drop of the deviated segment of the *i*th branch, MPa.

By combining Equations (20) and (24), the following system of equations can be obtained:

$$F_{i,j}\left[q_r(i,j), p_{wf}(i,j)\right] = 0, (1 \le i \le M, 1 \le j \le N_i).\tag{22}$$

As expected, the total number of equations of Equations (17) and (22) is equal to $2(M \times N_i)$, which is the same as that of unknowns. Therefore, the solution of the coupling model is unique.

2.2.3. Solution Procedure

The coupling model can also be solved by using the iterative algorithm, and the solution procedure is shown in Figure 7.

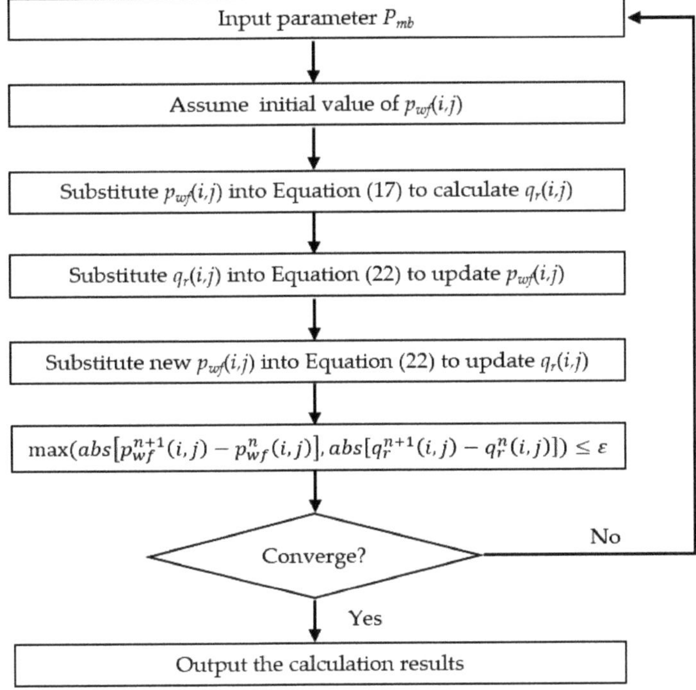

Figure 7. The flow chart of solution procedure.

3. Results and Discussion

3.1. Model Validation

The semi-analytical models can be applied to the productivity prediction of the infill horizontal well and the multilateral well in mixed well patterns. In order to verify the reliability of the models, two cases are considered, and the results calculated by the semi-analytical models are compared with those calculated by the Eclipse software (Schlumberger). In the first case, we selected a typical mixed well pattern of horizontal-vertical wells from a thin carbonate reservoir in the Middle East. There are three vertical producers (P_{11}, P_{12}, and P_{13}), one horizontal producer (P_{14}), one vertical injector (I_{11}), one horizontal injector (I_{12}), and P_{1new} is the infill horizontal producer. In the other case, the reservoir is a thick, multi-layer reservoir in the Middle East, multilateral wells are adopted to improve the degree of the reservoir development. The mixed well pattern of multilateral-vertical well includes two vertical producers (P_{21} and P_{22}), one vertical injector (I_{21}), and P_{2new} is the infill multilateral producer with two branch holes.

The productivity models can be solved by VB modular programming of the object-oriented technology. The reservoir properties and wellbore geometry are listed in Tables 1 and 2. By inputting the same model parameters, the productivity evaluation of the two wells were conducted respectively with the semi-analytical models and the Eclipse simulator. The calculation results are shown in Table 3.

Table 1. Productivity calculation parameters of P_{1new}.

Parameters (Unit)	Value
Net thickness (m)	6.0
Porosity (%)	15.3
Permeability (10^{-3} μm^2)	88.5
Reservoir pressure (MPa)	38.6
Oil viscosity (mPa·s)	8.1
Horizontal length of P_{1new} (m)	335
Wellbore radius of P_{1new} (m)	0.1
Bottom hole flowing pressure (MPa)	36.6
Production rate of P_{11} (m³/d)	18.5
Production rate of P_{12} (m³/d)	21.6
Production rate of P_{13} (m³/d)	24.3
Production rate of P_{14} (m³/d)	56.8
Injection rate of I_{11} (m³/d)	68.5
Injection rate of I_{12} (m³/d)	79.2

Table 2. Productivity calculation parameters of P_{2new}.

Parameters (Unit)	Value
Net thickness-the 1st branch of P_{2new} (m)	5.1
Net thickness-the 2nd branch of P_{2new} (m)	9.4
Porosity (%)	22.1
Ratio of vertical to horizontal permeability (-)	0.1
Reservoir pressure (MPa)	20.3
Oil viscosity (mPa·s)	6.2
Length of the 1st branch of P_{2new} (m)	200
Length of the 2nd branch of P_{2new} (m)	140
Bottom hole flowing pressure (MPa)	10.0
Production rate of P_{21} (m³/d)	52.9
Production rate of P_{22} (m³/d)	67.8
Injection rate of I_{21} (m³/d)	125.5

Table 3 indicates that the semi-analytical models have a high accuracy with the relative error of less than 15%. The results calculated by the models are consistent with those by the Eclipse numerical simulator, which verify the reliability of the productivity models. By taking into account the actual conditions, namely the interference of other wells in the well pattern, the coupling between the reservoir flow and wellbore flow, and the pressure drop along the horizontal lateral, the accuracy of the productivity prediction by the semi-analytical models is greatly improved.

Table 3. Results comparison of two methods.

Well Name		Actual Productivity (m³/d)	The Semi-Analytical Model		The Eclipse Simulator	
			Productivity (m³/d)	Relative Error (%)	Productivity (m³/d)	Relative Error (%)
P_{1new}		25.0	27.3	9.1	27.2	8.8
P_{2new}	1st branch	197.7	220.7	11.7	221.3	11.9
	2nd branch	147.1	166.5	13.2	168.0	14.2
	Total	344.7	387.2	12.3	389.3	12.9

3.2. Model Application

3.2.1. Study on Seepage Mechanism of Horizontal Well

In the first case of model validation, we can further analyze the bottom hole flowing pressure, the distribution of wellbore flow rate and radial flow rate along the horizontal segment. Figures 8 and 9 show that the bottom hole flowing pressure decreases logarithmically while the wellbore flow rate increases monotonically from the toe to the heel of the horizontal well. The reservoir flow coupled with variable mass wellbore flow results in this near wellbore dynamics. On one hand, the mass flow rate of fluid from the toe to the heel increases gradually. In this case, the fluid velocity along the main flow direction also increases with an accelerating pressure drop. The radial inflow of reservoir fluid along the horizontal wellbore disturbs the boundary layer of the main stream tube and affects its velocity profile, thus altering the wall friction resistance determined by the velocity distribution. These factors lead to the bottom hole flowing pressure decreasing in the form of logarithmic function. On the other hand, the radial inflow performance affects the distribution of pressure and pressure drop, and vice versa. Due to the pseudo-hemispherical flow at each endpoint (well heel and well toe) and the pseudo-linear flow at the center of the horizontal segment, the drainage area at each endpoint is relatively larger than that at the center. The radial flow rate at each endpoint of the horizontal segment is considerably greater than that at the center, and decreases quickly toward the intermediate section, which generally presents a U-shape distribution.

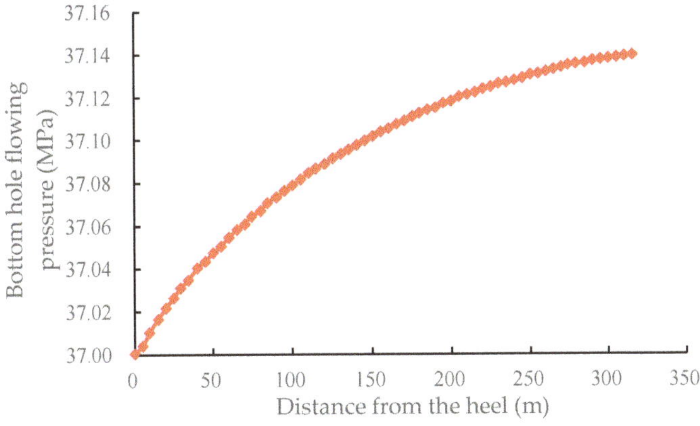

Figure 8. Distribution of bottom hole flowing pressure along the horizontal lateral.

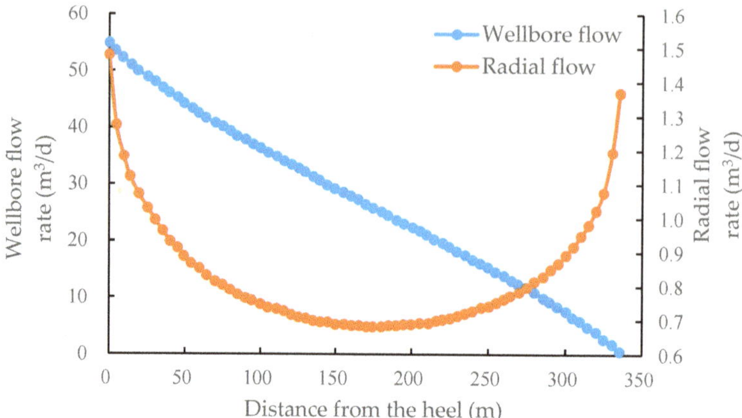

Figure 9. Distribution of wellbore flow and radial flow along the horizontal lateral.

3.2.2. Optimization of Infilling Well Location

In the intermediary and later phase of oilfield development, infill wells are usually needed to improve the well pattern and the effect of tapping potential of remaining oil. Therefore, the key is the optimization of infill well location. The productivity models provide a shortcut for this target. Based on the understanding of geological conditions and development status, the productivity of infill well can be accurately predicted, and in combination with the findings of remaining oil distribution, the infill well location could be optimized.

Taking the mixed five-spot well pattern of vertical injectors and horizontal producer as an example, we conducted a study on the productivity variation of horizontal well at different locations in the well pattern (Figure 10). The basic reservoir parameters refer to the first case of model validation. Figure 11 shows that the lateral shifting of horizontal well has some effects on the productivity, and the productivity at the center of the well pattern is relatively higher than that at other locations. However, this does not mean that the central location is the optimum selection for the horizontal well. In the adjustment of well pattern, geological condition and development factors, such as the distribution of flow field and remaining oil, should also be considered.

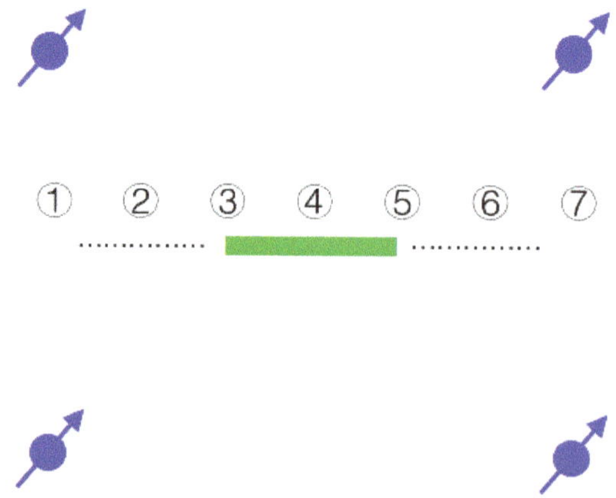

Figure 10. Mixed five-spot well pattern of vertical-injectors and horizontal producer (①–⑦ show the code of different horizontal locations).

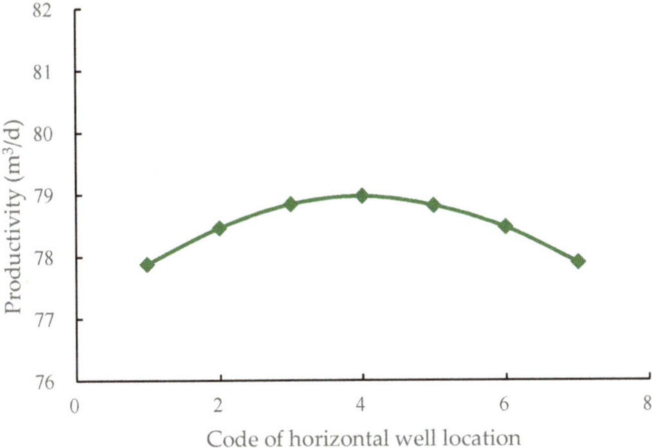

Figure 11. Productivity variation of horizontal well in mixed well pattern.

We made further study on the flow field distribution in the mixed well pattern so as to provide references for the optimization of infilling well location. Figures 12 and 13 indicate that the equipotential lines and stream lines are symmetrically distributed along the *x*-axis as the horizontal well moves laterally. When the horizontal well is located at the center of the well pattern, both the equipotential lines and stream lines present a uniform distribution; when the horizontal well moves toward the right side, both of the two types of the lines on the right side become dense, while those on the left side are sparsely distributed, and vice versa. The more the horizontal well location changes, the more obvious this phenomenon is. The flow pattern almost presents a linear flow as the horizontal well locates at the line of the two vertical wells on the same side, which indicates a relatively higher swept efficiency. Therefore, in combination with the findings of productivity variation and flow fluid distribution, the infilling well location could be optimized. This improves the effect of well pattern adjustment considerably.

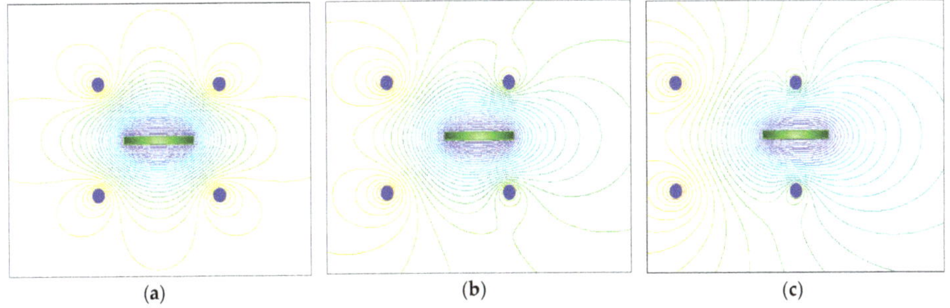

(a)   (b)   (c)

Figure 12. Equipotential line pattern for lateral shifting of the horizontal well. ((**a**–**c**) show the equipotential line pattern of different lateral location.)

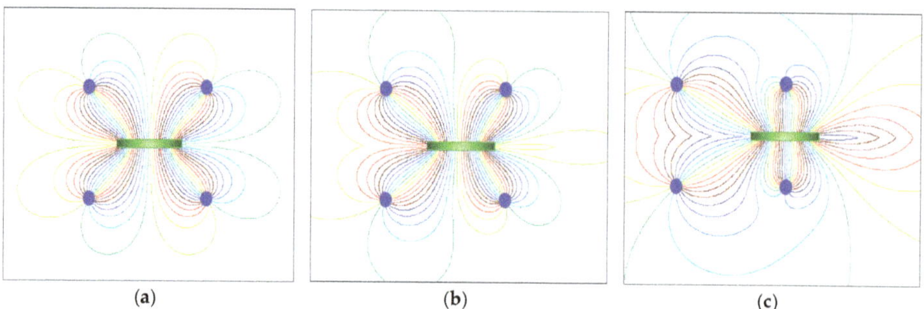

Figure 13. Stream line pattern for lateral shifting of the horizontal well. ((**a–c**) show the stream line pattern of different lateral location.)

3.3. Discussion

In this study, the semi-analytical models for productivity evaluation of infill complex structural wells are built with the method of images and source–sink theory. Compared with the existing models, the advantages of the semi-analytical model lie in the following three aspects. Firstly, the models are suitable for mixed well patterns of vertical wells and complex structure wells; Secondly, both the interference of other wells in the well pattern and the pressure drawdown along the horizontal segment are taken into account, which makes the models more reliable. In addition, the models provide some insight into the seepage mechanism of complex structural wells such as the distribution of wellbore flow and radial flow along the horizontal segment. However, the semi-analytical models also have some limitations, the basic assumptions of the models are steady-state flow and closed top and bottom boundaries, and the productivity evaluation of different reservoir boundaries under the condition of unsteady-state flow still needs to be further studied.

4. Conclusions

In this paper, on the basis of the mirror-image method and the source–sink theory, two semi-analytical models are established to predict the productivity of infill horizontal wells and multilateral wells in mixed well patterns respectively, in which the interference of other wells, the randomicity of well pattern, and the pressure drawdown along the horizontal lateral are taken into account.

The application of the productivity models verifies the reliability and practicability. The results calculated by the models are consistent with those by the Eclipse numerical simulator, and the semi-analytical models have a high accuracy with the relative error of less than 15%.

The results indicate that the bottom hole flowing pressure decreases logarithmically while the wellbore flow rate increases monotonically from the toe to the heel of the horizontal segment. Due to the pseudo-hemispherical flow at each endpoint and the pseudo-linear flow at the center of the horizontal segment, the drainage area at each endpoint is relatively larger than that at the center. The radial flow rate at each endpoint of the horizontal segment is considerably greater than that at the center, which generally presents a U-shape distribution.

This study also proposes a practical and efficient approach for the study on the horizontal seepage mechanism, and the optimization of infilling well locations.

Author Contributions: Conceptualization, L.S.; Data curation, L.S.; Formal analysis, L.S.; Funding acquisition, Y.L.; Methodology, L.S.; Project administration, B.L.; Supervision, Y.L.; Validation, L.S.; Writing – review & editing, L.S., Y.L.

Funding: This research was founded by the National Science and Technology Major Project of China, grant number 2016ZX05015-002, and the National Natural Science Foundation of China, grant number 2017ZX05030-001.

Acknowledgments: The authors are grateful for the technical and financial support from the Research Institute of Exploration and Development, PetroChina, Beijing 100083, China.

Conflicts of Interest: The authors declare no conflict of interest.

References

1. Adesina, F.; Paul, A.; Oyinkepreye, O.; Adebowale, O. An improved model for estimating productivity of horizontal drain hole. In Proceedings of the SPE Nigeria Annual International Conference and Exhibition, Lagos, Nigeria, 2 August 2016. [CrossRef]
2. Ye, S.J. A productivity evaluation model and its application for an infill horizontal well in different types of reservoirs. *Pet. Sci. Technol.* **2012**, *30*, 1677–1691.
3. Liangbiao, O.; Arbabi, S.; Aziz, K. General wellbore flow model for horizontal, vertical and slanted well completions. *SPE J.* **1998**, *3*, 124–133.
4. Merkulov, V.P. The flow of slanted and horizontal well. *Neft. Khoz* **1958**, *6*, 51–56.
5. Borisov, J.P. *Oil Production Using Horizontal and Multiple Deviation Wells*; Nedra: Moscow, Russia, 1964.
6. Giger, F.M. Horizontal wells production techniques in heterogeneous reservoirs. In Proceedings of the Middle East Oil Technical Conference and Exhibition, Manama, Bahrain, 14–17 March 1983. [CrossRef]
7. Joshi, S.D. Augmentation of well productivity using slant and horizontal wells. In Proceedings of the SPE Annual Technical Conference and Exhibition, New Orleans, Louisiana, 5 October 1986. [CrossRef]
8. Raghavan, R.; Joshi, S.D. Productivity of multiple drainholes or fractured horizontal wells. *SPE Format. Eval.* **1993**, *8*, 11–16. [CrossRef]
9. Thomas, L.K.; Todd, B.J.; Evans, C.E.; Pierson, R.G. Horizontal well IPR calculations. *SPE Reserv. Eval. Eng.* **1998**, *1*, 392–399. [CrossRef]
10. Zhang, W.; Han, D. 3D potential distribution and precise productivity equation of horizontal well. *Pet. Explor. Dev.* **1999**, *26*, 49–52.
11. Dikken, B.J. Pressure drop in horizontal wells and its effect on production performance. *J. Pet. Technol.* **1990**, *42*, 1426–1433. [CrossRef]
12. Liu, X.; Zhang, Z.S.; Liu, X.E.; Guo, S. A model to calculate pressure drops of horizontal wellbore variable mass flow coupled with flow in a reservoir. *J. Southwest Pet. Inst.* **2000**, *22*, 36–39.
13. Wang, R.H.; Zhang, Y.Z. A segmentally numerical calculation method for estimating the productivity of perforated horizontal wells. *Pet. Explor. Dev.* **2006**, *33*, 630.
14. Li, X.P.; Guo, C.Z.; Jiang, Z.X.; Liu, X.E.; Guo, S.P. The model coupling fluid flow in the reservoir with flow in the horizontal wellbore. *Acta Pet. Sin.* **1999**, *3*, 82–86.
15. Butler, R.M. Discussion of augmentation of well productivity with slant and horizontal wells. Author's reply. *J. Pet. Technol.* **1992**, *44*, 942–943.
16. Dang, L. Analysis of productivity formulae of horizontal well. *Pet. Explor. Dev.* **1997**, *5*, 21.
17. Jiang, H.; Ye, S.; Lei, Z.; Wang, X.; Zhu, G.; Chen, M. The productivity evaluation model and its application for finite conductivity horizontal wells in fault block reservoirs. *Pet. Sci.* **2010**, *7*, 530–535. [CrossRef]
18. Huang, S.J.; Cheng, L.S.; Zhao, F.L. The productivity evaluation model of the stepped horizontal well in thin interbeded reservoirs. *J. Southwest Pet. Univ.* **2007**, *3*, 16.
19. Борисов, Ю.П.; Табаков, В.П. *Расчет взаимодействия батарей наклонных и многозабойных скважин в слоистом пласте*; НТС по добыче нефти: Оренбургскаяобл, г. Оренбург, ул., Россия, 1961.
20. Wang, W.H.; Li, D. Productivity study on branch horizontal wells. *Oil Drill. Prod. Technol.* **1997**, *4*, 12.
21. Li, C.L. Derivation of productivity formulae of a fishbone well. *J. Southwest Pet. Inst.* **2005**, *27*, 36.
22. Cheng, L.S.; Li, C.L.; Lang, Z.X.; Zhang, L.H. The productivity study of branch a horizontal well with multiple branched wells. *Acta Pet. Sin.* **1995**, *16*, 49–55.
23. Zhao, L.X.; Jiang, M.H.; Zhao, X.F. Research on deliverability relationship of complicated horizontal well. *J. Univ. Pet. China (Ed. Nat. Sci.)* **2006**, *30*, 77–80.
24. Johansen, T.E.; James, L.; Cao, J. Analytical coupled axial and radial productivity model for steady-state flow in horizontal wells. *IJPE* **2015**, *1*, 290. [CrossRef]
25. Wang, H.; Guo, J.; Zhang, L. A semi-analytical model for multilateral horizontal wells in low-permeability naturally fractured reservoirs. *J. Pet. Sci. Eng.* **2017**, *149*, 564–578. [CrossRef]

26. Simonov, M.V.; Akhmetov, A.V.; Roshchektaev, A.P. Semi-analytical model of transient fluid flow to multilateral well. In Proceedings of the SPE Annual Caspian Technical Conference and Exhibition, Baku, Azerbaijan, 1–3 November 2017. [CrossRef]

27. Hassan, A.; Abdulraheem, A.; Elkatatny, S.; Ahmed, M. New approach to quantify productivity of fishbone multilateral well. In Proceedings of the SPE Annual Technical Conference and Exhibition, San Antonio, TX, USA, 9–11 October 2017. [CrossRef]

28. Liu, G.; Meng, Z.; Cui, Y.; Wang, L.; Liang, C.; Yang, S. A semi-analytical methodology for multiwell productivity index of well-industry-production-scheme in tight oil reservoirs. *Energies* **2018**, *11*, 1054. [CrossRef]

29. Vodorezov, D.D. Estimation of horizontal-well productivity loss caused by formation damage on the basis of numerical modeling and laboratory-testing data. *SPE J.* **2018**. [CrossRef]

30. Al-Rbeawi, S.; Artun, E. Fishbone type horizontal wellbore completion: A study for pressure behavior, flow regimes, and productivity index. *J. Pet. Sci. Eng.* **2019**. [CrossRef]

31. Suprunowicz, R.; Butler, R.M. The productivity and optimum pattern shape for horizontal wells arranged in staggered rectangular arrays. *J. Can. Pet. Technol.* **1992**, *31*. [CrossRef]

32. Chunlan, L.; Lingsong, C.; Lihua, Z.; Zhaoxin, L. The Study on Productivity of Horizontal Well 9-spot Patterns. *J. Southwest Pet. Inst.* **1998**, *20*, 56–58.

33. Cheng, L.S.; Zheng, J.Q.; Li, C.L. Productivity study of horizontal wells pattern. *Oil Drill. Prod. Technol.* **2002**, *24*, 39–41.

34. Liu, Y.T.; Zhang, J.C. Stable permeating flow and productivity analysis for anisotropic reservoir in horizontal well networks. *Pet. Explor. Dev.* **2004**, *31*, 94–96.

35. Yin, G.F.; Xu, H.M.; Ye, S.J.; Li, Y.R.; Qiu, J.P. Productivity evaluation model for infill horizontal well in horizontal injection and production pattern. *J. China Univ. Pet. (Ed. Nat. Sci.)* **2011**, *4*, 18.

36. Ye, S.J.; Jiang, H.Q.; Li, J.J. Productivity calculation of infill horizontal wells in mixed well pattern. *Chin. J. Comput. Phys.* **2011**, *28*, 693–697.

37. Huang, S.J.; Cheng, L.S.; Zhao, F.L.; Li, C.L. The flow model coupling reservoir percolation and variable mass pipe flow in production section of the stepped horizontal well. *J. Hydrodyn.* **2005**, *20*, 463–471.

38. Tabatabaei, M.; Ghalambor, A. A new method to predict performance of horizontal and multilateral wells. *SPE Prod. Oper.* **2011**, *26*, 75–87. [CrossRef]

MDPI
St. Alban-Anlage 66
4052 Basel
Switzerland
Tel. +41 61 683 77 34
Fax +41 61 302 89 18
www.mdpi.com

Processes Editorial Office
E-mail: processes@mdpi.com
www.mdpi.com/journal/processes